Frontiers in Protein and Peptide Sciences

(Volume 2)

Frontiers in Molecular Pharming

Edited by

Muhammad Sarwar Khan

Centre of Agricultural Biochemistry
and Biotechnology (CABB)
University of Agriculture
Faisalabad
Pakistan

Frontiers in Protein and Peptide Sciences

Volume # 2

Frontiers in Molecular Pharming

Editor: Muhammad Sarwar Khan

ISSN (Online): 2213-9877

ISSN (Print): 2589-2924

ISBN (Online): 978-981-5036-66-4

ISBN (Print): 978-981-5036-67-1

ISBN (Paperback): 978-981-5036-68-8

Published by Bentham Science Publishers Pte. Ltd. Singapore. All Rights Reserved.

need for a court order if at any point you breach any terms of this License Agreement. In no event will any delay or failure by Bentham Science Publishers in enforcing your compliance with this License Agreement constitute a waiver of any of its rights.

3. You acknowledge that you have read this License Agreement, and agree to be bound by its terms and conditions. To the extent that any other terms and conditions presented on any website of Bentham Science Publishers conflict with, or are inconsistent with, the terms and conditions set out in this License Agreement, you acknowledge that the terms and conditions set out in this License Agreement shall prevail.

Bentham Science Publishers Pte. Ltd.
80 Robinson Road #02-00
Singapore 068898
Singapore
Email: subscriptions@benthamscience.net

BENTHAM SCIENCE

CONTENTS

PREFACE

Manufacturing pharmaceuticals cost-effectively is one of the items on the wish list of biochemists and biotechnologists as drug regulatory authority in the USA has approved large-scale production and clinical trials of drugs developed through diverse production routes, including viruses, animals, and plants. Several factors are taken into account while selecting a production system of recombinant proteins since different expression systems have their own merits and demerits. The cost of expressed recombinant proteins includes production, processing, and purification costs. Normally, the production of expressed proteins costs around 70%, whereas purification costs around 30% of the total cost. Molecular pharming refers to the production of recombinant pharmaceutical proteins using plant biotechnology. This volume covers an array of topics relevant to structure, function, regulation, and mechanisms of action, biochemical significance, and usage of proteins and peptides as biomarkers, therapeutics, and vaccines for animals and human beings. Further, this book highlights the current progress from three directions, including system biology – *in silico* characterization of proteins and peptides, molecular pharming for animals, and molecular pharming for humans.

The book, Frontiers in Molecular Pharming, consists of 13 chapters subdivided into three sections. The chapters in the book are strategically organized to allow easy reading. Section I (System Biology – *in silico* Characterization of Proteins and Peptides) begins with Chapter 1 in which Dr. Rahman and his colleagues very comprehensively highlight various bioinformatics tools for predicting epitopic regions and a variety of immunological techniques to monitor the immune response generated against selected epitopic regions for the development of vaccines and diagnostics. Dr. Tahir ul Qamar and his colleagues in Chapter 2 have discussed the recent progress in the emerging field of immunoinformatics and its role in vaccine development. Dr. Ali explains the computational toolbox and its use in determining protein stability and analysis to improve thermostability in Chapter 3. In Chapter 4, Dr. Chen and her colleagues suggest how an evolving approach to Pan-proteomics is complementing our understanding of the functional complexity of emerging and highly virulent pathogens and their resistance development against drugs. Further, in Chapter 5, Drs. Haider and Niazi briefly overview the computational methods to predict the biological roles of peptides and proteins for medical or industrial applications.

Section II (Molecular Pharming for Human Beings) consists of six chapters, *i.e.*, Chapters 6 through 11. In Chapter 6, Dr. Khan and his team members explain comprehensively how diverse expression systems could be used to costeffectively develop recombinant pharmaceuticals and their application to control diseases in animals and human beings. Dr. Ahmad and his team provide a snapshot of different expression systems and argue that the plant-based expression system is highly commercially feasible not only for the production of high-value targets but also to address global challenges like COVID-19 in Chapter 7. In Chapter 8, Drs. Mangena and Mkhize explain the role of antibody cross-reactivity and specificity concerning basic principles, challenges, and detection for rapid and reliable assessment in *Fusarium* pathogens. Dr. Waheed and his team in Chapter 9 and Dr. Rashid and her team in Chapter10 have discussed how the requisition of plant-based medicine is increasing day-by-day with its perspective to human diseases, and several advantages owing to United Nations' sustainable development goals (SDGs). Dr. Qasim and his colleagues in Chapter 11 explain the importance of proteins and peptides as biomarkers for the diagnosis of cardiovascular diseases to improve the risk prediction at the population level. Further, the authors explore how new technologies and innovations can be applied to advance the science of vaccine-associated biomarkers.

Section III (Molecular Pharming for Animals) consists of two chapters. In Chapter 12, Dr. Aqib and his colleagues highlight the history and recent trends in veterinary pharmaceuticals and vaccines. They further discuss the nutraceutical potential of animal products as one of the fascinating areas of research with considerable anti-microbial, anti-cancer, anti-inflammatory, anti-diabetic, and neuroprotective functions. Dr. Khan and his team in Chapter 13 highlight the importance of plant-based gene expression systems that have been exploited as bioreactors for the cost-effective production of pharmaceuticals, predominantly for the expression and accumulation of antigenic proteins, to be used as vaccines for livestock and poultry. Further, they have discussed various types of vaccines keeping in view diseases like Infectious Bursal Disease (IBD), New Castle Disease (ND), and Foot and Mouth Disease (FMD).

Molecular farming is progressively reaching the stage of being considered as an economical alternative to established systems for the production of pharmaceuticals. Thus, this volume serves as a treasured resource for students and professionals of molecular biology, biotechnology, medicinal chemistry, and organic chemistry.

Muhammad Sarwar Khan
Centre of Agricultural Biochemistry and
Biotechnology (CABB)
University of Agriculture
Faisalabad
Pakistan

List of Contributors

Aamir Shehzad
Drug Discovery and Structural Biology Group, Health Biotechnology Division, National Institute for Biotechnology and Genetic Engineering (NIBGE), Faisalabad, Pakistan
Pakistan Institute of Engineering and Applied Sciences (PIEAS), National Institute for Biotechnology and Genetic Engineering (NIBGE), Faisalabad, Pakistan

Adnan K. Niazi
Centre of Agricultural Biochemistry and Biotechnology (CABB), University of Agriculture Faisalabad, Faisalabad, 38040, Pakistan

Aisha Mahmood
Department of Physiology, Faculty Of Veterinary And Animal Sciences, The Islamia University of Bahawalpur, Bahawalpur, Pakistan

Aisha Tarar
Centre of Excellence in Molecular Biology, University of the Punjab, Lahore, Pakistan

Amjad Ali
Atta-ur-Rahman School of Applied Biosciences (ASAB), National University of Sciences and Technology (NUST), Islamabad, Pakistan

Amjad Islam Aqib
Department of Medicine, Faculty of Veterinary Science, Cholistan University of Veterinary and Animal Sciences, Bahawalpur, 63100, Pakistan

Amna Bari
Hubei Key Laboratory of Agricultural Bioinformatics, College of Informatics, Huazhong Agricultural University, Wuhan, P. R. China

Amna Ramzan
Centre of Excellence in Molecular Biology, University of the Punjab, Lahore, Pakistan

Anam Naz
Institute of Molecular Biology and Biotechnology, The University of Lahore, Lahore, Pakistan

Ayesha Siddiqui
Agricultural Biotechnology Division, National Institute for Biotechnology & Genetic Engineering (NIBGE), Jhang Road, Faisalabad, 38000, Pakistan
Department of Biotechnology, Pakistan Institute of Engineering and Applied Sciences(PIEAS) , Nilore, Islamabad, Pakistan

Barira Zahid
Key Laboratory of Horticultural Plant Biology (Ministry of Education), College of Horticulture and Forestry, Huazhong Agricultural University, Wuhan, P. R. China

Bushra Rashid
Centre of Excellence in Molecular Biology, University of the Punjab, Lahore, Pakistan

Faisal Siddique
Department of Microbiology, Faculty of Veterinary Science, Cholistan University of Veterinary and Animal Sciences, Bahawalpur, 63100, Pakistan

Faiz Ahmad Joyia
Centre of Agricultural Biochemistry & Biotechnology (CABB), University of Agriculture, Faisalabad, Pakistan

Farah Shahid
Department of Bioinformatics and Biotechnology, Government College University Faisalabad, Faisalabad, Pakistan

Fatima Ijaz
Department of Biochemistry, Faculty of Biological Sciences, Quaid-i-Azam University, Islamabad, Pakistan

Fatima Khalid	Key Laboratory of Horticultural Plant Biology (Ministry of Education), College of Horticulture and Forestry, Huazhong Agricultural University, Wuhan, P. R. China
Feng Xing	College of Life Science, Xinyang Normal University, Xinyang 464000, P. R. China
Ghulam Mustafa	Centre of Agricultural Biochemistry & Biotechnology (CABB), University of Agriculture, Faisalabad, Pakistan
Huma Shakoor	Centre of Excellence in Molecular Biology, University of the Punjab, Lahore, Pakistan
Iqra Arshad	Centre of Excellence in Molecular Biology, University of the Punjab, Lahore, Pakistan
Iqra Mehmood	Department of Bioinformatics and Biotechnology, Government College University Faisalabad, Faisalabad, Pakistan
Iqra Muzammil	Department of Veterinary Medicine, Faculty of Veterinary Science, University of Veterinary and Animal Sciences, Lahore, 54000, Pakistan
Jia-Ming Song	State Key Laboratory for Conservation and Utilization of Subtropical Agro-bioresources, College of Life Science and Technology, Guangxi University, Nanning, P. R. China Hubei Key Laboratory of Agricultural Bioinformatics, College of Informatics, Huazhong Agricultural University, Wuhan, P. R. China
Kiran Saba	Department of Biochemistry, Faculty of Biological Sciences, Quaid-i-Azam University, Islamabad, Pakistan
Kishver Tusleem	Sir Ganga Ram Hospital, Fatima Jinnah Medical University, Lahore, Pakistan
Ling-Ling Chen	State Key Laboratory for Conservation and Utilization of Subtropical Agro-bioresources, College of Life Science and Technology, Guangxi University, Nanning, P. R. China Hubei Key Laboratory of Agricultural Bioinformatics, College of Informatics, Huazhong Agricultural University, Wuhan, P. R. China
Majid Ali Shah	Drug Discovery and Structural Biology Group, Health Biotechnology Division, National Institute for Biotechnology and Genetic Engineering (NIBGE), Faisalabad, Pakistan Pakistan Institute of Engineering and Applied Sciences (PIEAS), National Institute for Biotechnology and Genetic Engineering (NIBGE), Faisalabad, Pakistan
Maryam Zafar	Drug Discovery and Structural Biology Group, Health Biotechnology Division, National Institute for Biotechnology and Genetic Engineering (NIBGE), Faisalabad, Pakistan Pakistan Institute of Engineering and Applied Sciences (PIEAS), National Institute for Biotechnology and Genetic Engineering (NIBGE), Faisalabad, Pakistan

Mazhar Iqbal	Drug Discovery and Structural Biology Group, Health Biotechnology Division, National Institute for Biotechnology and Genetic Engineering (NIBGE), Faisalabad, Pakistan Pakistan Institute of Engineering and Applied Sciences (PIEAS), National Institute for Biotechnology and Genetic Engineering (NIBGE), Faisalabad, Pakistan
Moazur Rahman	Drug Discovery and Structural Biology Group, Health Biotechnology Division, National Institute for Biotechnology and Genetic Engineering (NIBGE), Faisalabad, Pakistan School of Biological Sciences, University of the Punjab, Lahore, Pakistan
Mohammad Tahir Waheed	Department of Biochemistry, Faculty of Biological Sciences, Quaid-i-Azam University, Islamabad, Pakistan
Mohsin Khurshid	Department of Microbiology, Government College University, Faisalabad, Pakistan
Mubashrah Mahmood	Department of Theriogenology, Faculty of Veterinary Science, University of Agriculture, Faisalabad, 38000, Pakistan
Muhammad Aamir Naseer	Department of Clinical Medicine and Surgery, Faculty of Veterinary Science, University of Agriculture, Faisalabad, 38000, Pakistan
Muhammad Omar Khan	Agricultural Biotechnology Division, National Institute for Biotechnology & Genetic Engineering (NIBGE), Jhang Road, Faisalabad, 38000, Pakistan Department of Biotechnology, Pakistan Institute of Engineering and Applied Sciences(PIEAS) , Nilore, Islamabad, Pakistan
Muhammad Sameeullah	Innovative Food Technologies Development Application and Research Center, Faculty of Engineering, Bolu Abant Izzet Baysal University, 14030, Bolu, Turkey
Muhammad Sarwar Khan	Center of Agricultural Biochemistry and Biotechnology, University of Agriculture, Faisalabad, Pakistan
Muhammad Shareef Masoud	Department of Bioinformatics and Biotechnology, Government College University, Faisalabad, Pakistan
Muhammad Shoaib	Institute of Microbiology, Faculty of Veterinary Science, University of Agriculture, Faisalabad, 38000, Pakistan
Muhammad Qasim	Department of Bioinformatics and Biotechnology, Government College University, Faisalabad, Pakistan
Muhammad Suleman Malik	Department of Biochemistry, Faculty of Biological Sciences, Quaid-i-Azam University, Islamabad, Pakistan
Muhammad Tahir ul Qamar	College of Life Science and Technology, Guangxi University, Nanning, P. R. China
Nazia Nahid	Department of Bioinformatics and Biotechnology, Government College University, Faisalabad, Pakistan
Niaz Ahmad	Agricultural Biotechnology Division, National Institute for Biotechnology & Genetic Engineering (NIBGE), Jhang Road, Faisalabad, 38000, Pakistan Department of Biotechnology, Pakistan Institute of Engineering and Applied Sciences(PIEAS) , Nilore, Islamabad, Pakistan

Phetole Mangena	Department of Biodiversity, School of Molecular and Life Sciences, Faculty of Science and Agriculture, University of Limpopo, Private Bag X1106, Sovenga 0727, Republic of South Africa
Phumzile Mkhize	Department of Microbiology, Biochemistry and Biotechnology, School of Molecular and Life Sciences, Faculty of Science and Agriculture, University of Limpopo, Private Bag X1106, Sovenga 0727, Republic of South Africa
Rabia Abbas	Centre of Excellence in Molecular Biology, University of the Punjab, Lahore, Pakistan
Rimsha Riaz	Center of Agricultural Biochemistry and Biotechnology (CABB), University of Agriculture, Faisalabad – 38040, Pakistan
Saad Ahmad	Lanzhou Institute of Husbandry and Pharmaceutical Sciences, Lanzhou, China
Saba Altaf	Centre of Excellence in Molecular Biology, University of the Punjab, Lahore, Pakistan
Saher Qadeer	Center of Agricultural Biochemistry and Biotechnology (CABB), University of Agriculture, Faisalabad – 38040, Pakistan
Sajjad Ahmad	Department of Health and Biological Sciences, Abasyn University, Peshawar, Pakistan
Samman Munir	Department of Bioinformatics and Biotechnology, Government College University, Faisalabad, Pakistan
Sara Latif	Department of Biochemistry, Faculty of Biological Sciences, Quaid-i-Azam University, Islamabad, Pakistan
Soban Tufail	Drug Discovery and Structural Biology Group, Health Biotechnology Division, National Institute for Biotechnology and Genetic Engineering (NIBGE), Faisalabad, Pakistan
	Pakistan Institute of Engineering and Applied Sciences (PIEAS), National Institute for Biotechnology and Genetic Engineering (NIBGE), Faisalabad, Pakistan
Sumera Rashid	Centre of Excellence in Molecular Biology, University of the Punjab, Lahore, Pakistan
Syed Farhat Ali	School of Life Sciences, Forman Christian College (A Chartered University), Lahore, Pakistan
Tean Zaheer	Department of Parasitology, Faculty of Veterinary Science, University of Agriculture, Faisalabad, 38000, Pakistan
Usman Ali Ashfaq	Department of Bioinformatics and Biotechnology, Government College University Faisalabad, Faisalabad, Pakistan
Xitong Zhu	Hubei Key Laboratory of Agricultural Bioinformatics, College of Informatics, Huazhong Agricultural University, Wuhan, P. R. China
Zainab Y. Sandhu	Montclair State University, New Jersey NJ 07043, USA
Zeshan Haider	Centre of Agricultural Biochemistry and Biotechnology (CABB), University of Agriculture Faisalabad, Faisalabad, 38040, Pakistan
	State Key Laboratory of Grassland Agro-Ecosystems, College of Pastoral Agriculture Science and Technology, Lanzhou University, Lanzhou, China

SECTION I: System Biology – *In silico* Characterization of Proteins and Peptides

Tools for Prediction and Validation of Epitopic Regions on Protein Targets for Vaccine Development and Diagnostics

Soban Tufail[1,2]**, Majid Ali Shah**[1,2]**, Maryam Zafar**[1,2]**, Mazhar Iqbal**[1,2]**, Amjad Ali**[3]**, Aamir Shehzad**[1,2,*] **and Moazur Rahman**[1,4,*]

[1] *Drug Discovery and Structural Biology Group, Health Biotechnology Division, National Institute for Biotechnology and Genetic Engineering (NIBGE), Faisalabad, Pakistan*

[2] *Pakistan Institute of Engineering and Applied Sciences (PIEAS), P.O. Nilore, Islamabad, Pakistan*

[3] *Atta-ur-Rahman School of Applied Biosciences (ASAB), National University of Sciences and Technology (NUST), Islamabad, Pakistan*

[4] *School of Biological Sciences, University of the Punjab, Lahore, Pakistan*

Abstract: Epitopes are parts of an antigen that are recognized by the immune system. Identification of epitopic regions on an immunogenic protein is important for several clinical and biotechnological applications. Various bioinformatics tools are currently available which can be used for the prediction of epitopic regions, and the immune response generated against selected epitopic regions can be monitored through a variety of immunological techniques. In this chapter, we provide an overview of widely used *in silico* tools for the prediction of epitopic regions, followed by biophysical methods used for their characterization. Furthermore, a brief description of important immunological approaches for measuring immune responses elicited by epitopes is also given. It is anticipated that the information provided in this chapter will help researchers in selecting appropriate tools for the prediction and validation of epitopes on a protein target for vaccine development and diagnostics.

Keywords: Diagnostics, Epitope prediction, Protein targets, Vaccine.

---

[*] **Corresponding authors Moazur Rahman & Aamir Shehzad:** School of Biological Sciences, University of the Punjab, Lahore, Pakistan; Tel: +92 42 99230960; Fax: +92 42 99230980; E-mails: moaz.sbs@pu.edu.pk, moazur.rahman@fulbrightmail.org, & Drug Discovery and Structural Biology group, Health Biotechnology Division, National Institute for Biotechnology and Genetic Engineering (NIBGE), Faisalabad, Pakistan; Tel: +92 41 9201316-9; Ext 242; Fax: +92 41 9201322; E-mail: aamiruopbbt@yahoo.com

1. IMMUNE SYSTEM-AN OVERVIEW

A properly functioning immune system plays a key role in neutralizing 'biological threats' posed by infectious diseases and cancer. Understanding the immune system is important for devising therapeutic interventions to cure various diseases. The immune system is categorized into innate and adaptive subsystems. The innate immune system, which is non-specific, is the first line of defense against infections. On the other hand, the adaptive immune system, which is highly specific, is only found in vertebrates. Adaptive immune responses are orchestrated by lymphocytes, namely B- and T-cells, which induce humoral and cell-mediated immunity.

Importantly, specific receptors present on the surface of B- and T-cells recognize molecular components, commonly known as antigens, of pathogens. B-cell receptors, which consist of membrane-bound immunoglobulins, usually recognize parts of antigens that are solvent-exposed. Activated B-cells produce soluble immunoglobulins, also known as antibodies, which are involved in humoral adaptive immunity. The humoral immune system not only enables the recognition of antigenic determinants in pathogenic proteins but also induces the formation of memory B-cells which generate a strong antibody-mediated immune response upon re-infection. On the other hand, T-cell receptors recognize antigens by binding to antigenic peptides attached to the groove of major histocompatibility complex (MHC) molecules, also known as human leukocyte antigen (HLA) in humans, on the surface of antigen-presenting cells (APCs). In humans, the HLA system is polygenic (encoded by 21 genes on chromosome 6) and highly polymorphic. There are two distinct subtypes of T-cells, phenotypically classified as CD8+ and CD4+ T-cells, which recognize linear antigenic peptides presented by MHC molecules. Activated CD8+ T-cells, also known as cytotoxic T lympho-cytes (CTLs), recognize peptides presented by MHC class I molecules (Fig. **1**). These peptides, which are typically 9 amino acids long, presented by MHC class I molecules originate from intracellular antigens degraded in the cytosol. Activated CD4+ T-cells, also known as T helper (Th) cells, recognize antigenic peptides presented by MHC class II molecules and are specific to extracellular antigens which have been endocytosed, degraded, and complexed to MHC class II molecules in endosomal compartments [1]. Typically, peptides attached to MHC class II molecules are 15 amino acids in length and protrude out of the peptide-binding groove of MHC class II molecules [2].

Portions of antigens that are recognized by B- and T-cells are known as epitopes. Discrete regions of antigens that are recognized by B-cell receptors or secreted antibodies and can evoke the humoral immune response are known as B-cell epitopes [3]. It has been found that B-cell epitopes predominantly consist of solvent-exposed (hydrophilic) regions which are located on the surface of antigens [4, 5]. B-cell epitopes can be linear (continuous) or conformational (discontinuous). Linear epitopes are comprised of a contiguous stretch of amino acids of an antigen [6, 7]. Conformational epitopes consist of amino acids that are not contiguous, and residues critical for recognition by antibodies are located nearby due to the folded three-dimensional structure of a given antigen. It has been observed that the majority of conformational epitopes (more than 70%) contain 1-5 short linear segments of amino acids [5]. Moreover, most B-cell epitopes (~90%) are conformational epitopes [8, 9]. T-cell epitopes are MHC binding peptides (ligands) that elicit a T-cell immune response. Upon recognition of a T-cell epitope, T-cells produce a long-lived memory population that confers on the host the ability to respond swiftly when the same epitope is encountered again [10, 11].

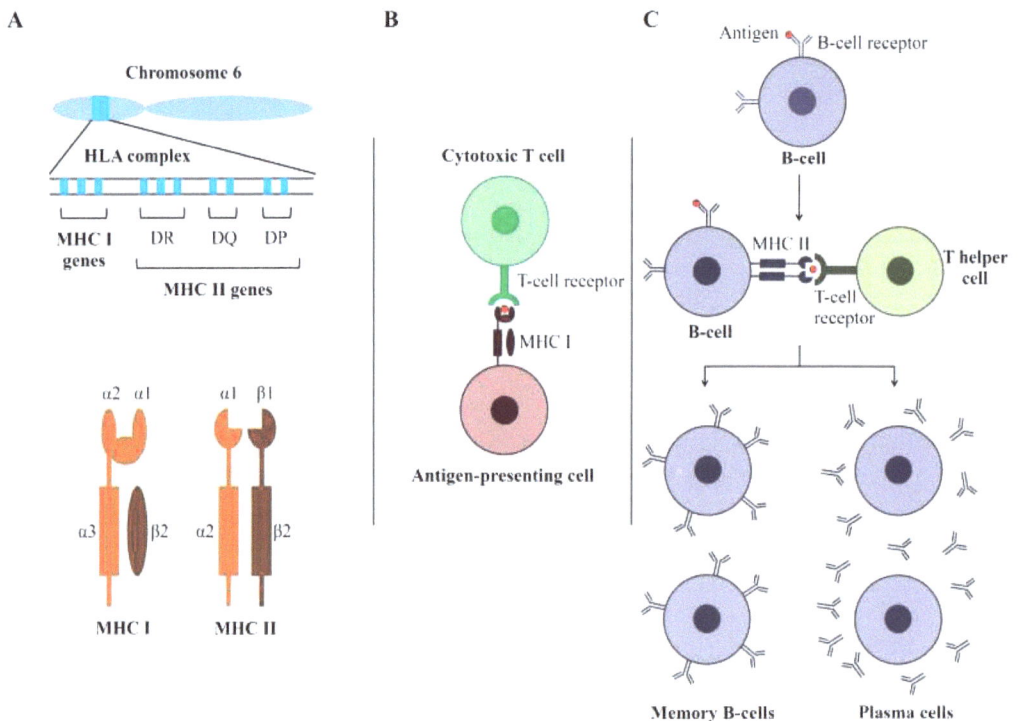

Fig. (1). Graphical illustrations of genes encoding MHC molecules and recognition of MHC-bound peptides by T cells. (**A**) The location of genes encoding MHC class I and MHC class II molecules on chromosome 6 in humans. (**B**) Recognition of peptides bound to MHC class I molecules on antigen-presenting cells by cytotoxic T cells. (**C**) The interaction of a foreign antigen with a B-cell receptor leads to the presentation of the foreign peptide to T helper cells through MHC class II molecules on the surface of B-cells. Activated B-cells proliferate and differentiate into memory B-cells and antibody-producing plasma cells.

2. IDENTIFICATION OF EPITOPES

Identification of epitopes in target antigens is important for practical purposes such as studying disease etiology, analyzing immune responses, designing epitope-based vaccines, developing immunodiagnostic assays, and producing therapeutic antibodies. Several experimental approaches have been developed for the identification of epitopes which can be classified as structural and functional methods. For the identification of B-cell epitopes, X-ray crystallography is the most preferred structural method owing to its accuracy. Through X-ray crystallography, the structure of antigen-antibody complexes can be solved to define structural epitopes. The analysis of antigen-antibody complexes has revealed that B-cell epitopes can be distinguished from non-epitopic regions on an antigen based on their structural, physicochemical, and geometrical properties [5]. Commonly used methods for identification of B-cell epitopes through functional methods involve the proteolysis of antigens and screening of the resulting peptides for antibody binding as well as evaluation of the reactivity of antibodies towards variants of antigens obtained through site-directed- or random mutagenesis [12]. Other methods such as display technologies and mimotope analysis are also employed [13, 14]. T-cell epitopes can be identified through binding assays which entail biochemical techniques such as size-exclusion chromatography which can be used to analyze the binding of synthetic peptides to MHC molecules [15]. Data generated from such studies can be found in freely available databases such as the Immune Epitope Database (IEDB) [16 - 18], an up-to-date database of experimentally characterized epitopes.

It is pertinent to note that most of the developed experimental techniques (structural and functional methods) for the identification of epitopes are costly, laborious, time-consuming, technically difficult, often do not lead to the identification of all epitopes, and not practicable for all antigens [12]. Recently, *in silico* epitope prediction methods have become increasingly available that can be used to select potential epitope candidates for experimental studies. For this purpose, candidate antigens are chosen at the first step (Fig. **2**). Then, *in silico* (immunoinformatics) tools are used to predict epitopes of B- and T-cells. Most of the epitope prediction tools have been developed based on neural networks that use artificial intelligence methods to predict B- and T- cell epitopes. Although the predictive performance of currently available epitope prediction tools has shown promising outcomes, there is still room for improvement. Nevertheless, many of these methods exhibit better accuracy than computational predictors used in other fields of biology [19, 20].

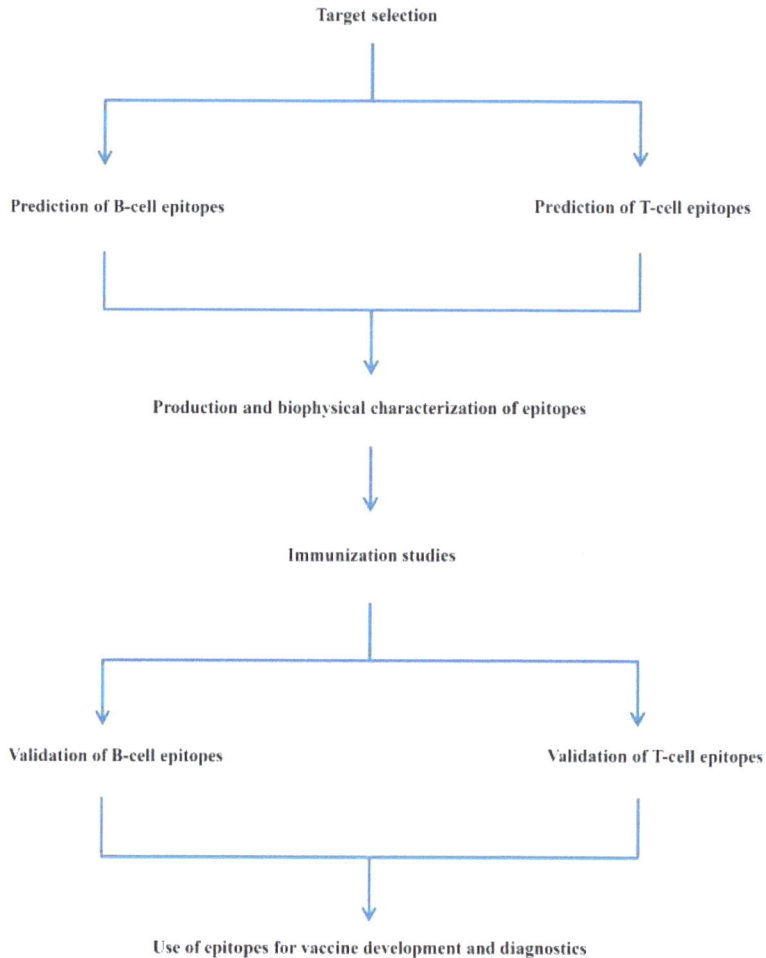

Fig. (2). Schematic representation of various steps involved in epitope prediction and validation on target proteins for vaccine development and diagnostics.

3. TARGET SELECTION FOR EPITOPE PREDICTION

Given that very few proteins (less than 0.05%) encoded by the genome of a pathogen can elicit an immune response [21], it is important to select a suitable (immunogenic) protein target for epitope prediction. Common features of immunogenic proteins are summarized here. Most immunogenic proteins are structural proteins that usually constitute the surface components such as the capsid (of viruses) and the cell membrane (of microorganisms). These proteins could be partially or completely exposed on the surface, are predominantly hydrophilic, contain flexible regions, and are accessible to the outer environment.

It has been found that most of the secreted proteins are antigenic. These are attractive protein targets for epitope prediction because such proteins are usually involved in pathogenicity, cell adherence to host cells, and induction of the immune response [22]. Moreover, signal peptide and transmembrane proteins of microorganisms are broadly antigenic and contain epitope density regions, also known as 'hot spots', which are in rich MHC class II binding regions. Identification of these 'hot spots' is important for epitope prediction [23]. Non-structural proteins or those present inside the pathogen are not attractive for epitope prediction as these proteins are not exposed on the surface and hence are generally not accessible to the immune system. It is important to note that the selected epitopes prioritized for the study should be antigenic, immunogenic, stable, and safe. They should also not bear any homology with the host proteome to avoid chances of allergenicity [24]. Such a dataset of non-homologous proteins can be obtained using subtractive proteomics approaches. It is also useful to consider the size of the protein before selecting it as a target for vaccine development. Due to the ease of expression and purification, small proteins (molecular weight less than 100 kDa) are more attractive as vaccine candidates. An associated advantage of selecting a low molecular weight protein is that the structure of the protein can be easily solved using structural biology techniques (for example, X-ray crystallography), which can yield useful structural information for the selection of epitopic regions on the target protein. Antigens encoded by genes acquired through horizontal gene transfer between microorganisms are more likely to produce a protective immune response. Proteins that are constitutively expressed by pathogens are attractive targets for vaccine development, and diagnostics as antibodies against such proteins are expected to be abundantly present in the host organism. Furthermore, it is generally regarded that antigenic epitopes originating from essential and virulent proteins exhibit better safety levels and yield improved efficacy when employed as vaccine candidates. Proteins which are implicated in virulence or other biological functions such as interaction with host receptors are usually highly immunogenic [25].

Reverse vaccinology, which entails the identification of antigens through the analysis of proteins encoded by the genome of an organism, has revolutionized the field of vaccine development [26]. Conceived in the year 2000, the reverse technology approach has been successfully applied for antigen discovery from the genome sequence of a variety of pathogens such as *Neisseria meningitidis* serogroup B strain [26], *Streptococcus pneumonia* [27], *Porphyromonas gingi-valis* [28], *Chlamydia pneumoniae* [29], *Bacillus anthracis* [30], *Staphy-lococcus aureus* [31, 32], extraintestinal pathogenic *Escherichia coli* [33], *Brucella meli-tensis* [34], *Rickettsia prowazekii* [35], *Echinococcus granulosus* [36], *Francisella tularensis* [37], *Theileria parva* [38], *Gallibacterium anatis* [39], *Leishmania spp*

[40], cytomegalovirus (CMV) [41], and influenza virus [42], to name a few. The workflow of a typical reverse vaccinology pipeline involves identification of open reading frames in the sequenced genome through bioinformatic analyses, *in silico* prediction of proteins that are either secreted or surface-exposed and exhibit similarities to virulence factors, analysis of antigen conservation across species for broad coverage, exclusion of proteins which either exhibit homology to host organism or are found in non-pathogenic organisms of the body, heterologous expression and purification of target proteins, immunization studies in model organisms, formulation of promising vaccine candidates, large-scale clinical trials, and vaccine licensing [25].

In the following sections, information on the commonly used computational tools for the prediction of B- and T-cell epitopes is provided.

4. TOOLS FOR PREDICTION OF LINEAR B-CELL EPITOPES

Currently available computational tools for the prediction of linear B-cell epitopes include ABCpred [43], BcePred [44], BepiPred [45], BCPred [46], FBCPred [47], COBEpro [48], LBtope [49], PREDITOP [50], and PEOPLE [51] (Table **1**). Among these tools, BepiPred [45] available at IEDB [16 - 18] has been widely used for the prediction of linear B-cell epitopes, frequently in combination with other methods which use hydrophilicity [52], flexibility [53], β-turns [54], solvent accessibility [55], and antigenicity [56] of proteins. The latest version, BepiPred-2.0, is trained on data (epitopes) derived from structures of antigen-antibody complexes determined through X-ray crystallography [57].

Table 1. Commonly used tools for the prediction of B- and T-cell epitopes.

B-Cell Epitopes	Tools	Website	References
Linear B-cell epitopes	ABCpred	https://webs.iiitd.edu.in/raghava/abcpred/	[43]
	BcePred	http://crdd.osdd.net/raghava/bcepred/	[44]
	BCPred	http://ailab-projects1.ist.psu.edu:8080/bcpred/	[46]
	BepiPred	http://www.cbs.dtu.dk/services/BepiPred/	[57]
	COBEpro	http://scratch.proteomics.ics.uci.edu/	[48]
	LBtope	http://crdd.osdd.net/raghava/lbtope/	[49]

(Table 1) cont.....

B-Cell Epitopes	Tools	Website	References
Conformational B-cell epitopes	BEpro (PEPITO)	http://pepito.proteomics.ics.uci.edu/	[62]
	CBTOPE	http://crdd.osdd.net/raghava/cbtope/	[61]
	CEP	http://bioinfo.ernet.in/cep.htm	[58]
	DiscoTope	http://www.cbs.dtu.dk/services/DiscoTope/	[59]
	Ellipro	http://tools.iedb.org/ellipro/	[60]
	EpiPred	http://opig.stats.ox.ac.uk/webapps/newsabdab/sabpred/epipred/	[65]
	EPISEARCH	http://curie.utmb.edu/episearch.html	[72]
	Epitopia	http://epitopia.tau.ac.il/	[68]
	IgPred	http://crdd.osdd.net/raghava/igpred/	[75]
	PEASE	www.ofranlab.org/PEASE	[64]
	PEPITOPE	http://pepitope.tau.ac.il/	[71]
	PEPMAPPER	http://informatics.nenu.edu.cn/PepMapper/	[74]
	SEPPA	http://www.badd-cao.net/seppa3/index.html	[63]
T-cell Epitopes	Tools	Website	References
MHC class I binders	CTLPred	http://crdd.osdd.net/raghava/ctlpred/	[76]
	NetCTL	http://www.cbs.dtu.dk/services/NetCTL/	[84]
	NetMHC	http://www.cbs.dtu.dk/services/NetMHC/	[81]
	NetMHCcons	http://www.cbs.dtu.dk/services/NetMHCcons/	[80]
	NetMHCpan	http://www.cbs.dtu.dk/services/NetMHCpan/	[82]
	ProPred-I	http://crdd.osdd.net/raghava/propred1/	[96]
	RANKPEP	http://imed.med.ucm.es/Tools/rankpep.html	[90]
	SVMHC	http://www-bs.informatik.uni-tuebingen.de/SVMHC/	[91]
	SYFPEITHI	http://www.syfpeithi.de/0-Home.htm	[86]
MHC class II binders	EpiDOCK	http://www.ddg-pharmfac.net/epidock/	[98]
	EpiTOP	http://www.ddg-pharmfac.net/EpiTOP3/	[99]
	NetMHCIIpan	http://www.cbs.dtu.dk/services/NetMHCIIpan/	[92]
	PREDIVAC	http://predivac.biosci.uq.edu.au/	[97]
	ProPred	http://crdd.osdd.net/raghava/propred/	[96]

5. TOOLS FOR PREDICTION OF CONFORMATIONAL B-CELL EPITOPES

Earlier methods such as Conformational Epitope Predictor (CEP) [58] developed for the prediction of conformational B-cell epitopes were based on predicting solvent-exposed regions in antigens. Other frequently used tools for the prediction of conformational B-cell epitopes include DiscoTope [59], Ellipro [60], CBTOPE [61], BEpro (also known as PEPITO) [62], and SEPPA [63]. Most of these methods predict conformational B-cell epitopes from the three-dimensional structure of antigens. Other tools, such as PEASE [64], require the availability of the sequence information of the antibody in addition to the structure of the target antigen, while certain other tools, such as EpiPred [65], require both antigens and antibody structures as inputs to predictors. Bepar [66] and ABepar [67] are sequence-based methods that only require sequences of the target antigen and antibody for predicting conformational epitopes. Some methods, such as Epitopia [68], can use either the sequence information or the structure of the antigen for predicting discontinuous epitopic regions. BEST [69] and CBTOPE [61] are examples of computational methods which only use antigen sequences for the prediction of conformational epitopes. Computational tools for prediction of conformational B-cell epitopes using mimotopes include MIMOX [70], PEPITOPE [71], EPISEARCH [72], MIMOPRO [73], and PEPMAPPER [74], while IgPred [75] is a tool of choice for predicting antibody-specific epitopes.

6. PREDICTION OF T-CELL EPITOPES

Little attention has been focused on developing tools such as CTLPred [76], which predict T-cell epitopes through direct methods, taking into account the pattern of epitopes recognized by T-cells. Most of the tools used for the prediction of T-cell epitopes are based on indirect methods which predict MHC binding peptides instead of peptides recognized by T-cells (Table **1**). Since binding of peptides in the groove of MHC molecules is a highly selective and more specific process than other steps in the MHC antigen processing and presentation pathway (proteasome cleavage and TAP (transporter associated with antigen processing) transport), MHC binding peptides are highly immunogenic epitopes that are efficiently recognized by T-cells [11, 77 - 79].

6.1. Tools for Prediction of MHC Class I Binding Peptides

A frequently used computational tool for the prediction of highly accurate MHC class I binding peptides is NetMHCcons 1.1 [80], which combines three programs (NetMHC [81], NetMHCpan [82], and PickPocket [83]) for the prediction of MHC class I binders. The NetMHCpan 2.0 server can be used to predict binding peptides for HLA-A and HLA-B as well as for MHC Class I molecules belonging to several animal models such as chimpanzee, rhesus macaque, gorilla, and mouse [82]. Examples of other widely used computational tools for prediction of MHC class I binders include NetCTL [84, 85], SYFPEITHI [86], MixMHCpred [87], MHCflurry [88], ProPred-I [89], RANKPEP [90], and SVMHC [91].

6.2. Tools for Prediction of MHC Class II Binding Peptides

The NetMHCIIpan method [92] is currently the leading method for the prediction of peptide binding to MHC class II. Other prediction tools for MHC class II binders include TEPITOPE [93], MultiRTA [94], MHCIIMulti [95], ProPred [96], and PREDIVAC [97]. For the prediction of promiscuous MHC class II binders, tools such as EpiDOCK [98] and EpiTOP [99] have been developed.

7. FURTHER CONSIDERATIONS

Before conducting immunization studies, it is important to analyze potential toxic effects or the allergic nature of predicted epitopes, and parts of predicted epitopes that are potentially toxic/allergic should be removed. For this purpose, computational tools such as Toxinpred [100], Algpred [101], AllerHunter [102], AllerTOP [103], AllergenFP [104], and PREAL [105] have been developed which can predict the toxic/allergic nature of peptides. Other side effects such as hemolysis or hypertension can be analyzed using tools such as Hemolytik [106] and AHTpin [107, 108], respectively.

Epitopes selected for vaccine development studies should be antigenic. The antigenicity profile of shortlisted B- and T- cell epitopes can be calculated using AntigenPro [109] and VaxiJen [110] tools.

After predicting B- and T-cell epitopes, it is pertinent to study the interaction of predicted epitopes with the binding partner (antibodies, MHC class I or MHC class II molecules) using computational tools such as Autodock Vina [111], Autodock 4 [112], PatchDock [113], ClusPro [114], HADDOCK [115], and GalaxyPepDock [116]. Further insights into the binding interaction can be obtained by performing molecular dynamics (MD) simulations using tools such as GROMACS [117] and YASARA [118].

8. LIMITATIONS OF EPITOPE PREDICTION TOOLS AND THE WAY FORWARD

Species-specific tools for the prediction of B-cell epitopes are needed to generate antibody responses in different animals using species-specific B-cell epitopes of a target protein. A few such computational tools (for example, VacSol [119] and PanRV [120]) have been developed. VacSol [119] uses the whole genome sequence of an organism and yields a list of epitopes that can be processed further for vaccine development studies. PanRV [120] is a pipeline that predicts core proteins from multiple strains/isolates of species and yields core genes. These core genes are further filtered to predict potential vaccine candidates. However, there is room for improvement in these tools to yield more promising results [119, 120].

In silico tools for predicting adjuvants for subunit vaccines are currently lacking. Though tools for the prediction of adjuvants for DNA-based vaccines have been developed [121], such tools need to be urgently developed for peptide-based vaccines so that subunit vaccines would be able to elicit a better immune response in the presence of suitable adjuvants. Currently, adjuvants are selected based on data mining of the literature (for example, B-defensin and Cholera Toxin B were added as adjuvants to multi-epitope vaccines [122, 123]). However, the adjuvant reported in the literature may not be suitable for use with epitopes other than the reported ones. Thus, reliable tools that can predict appropriate adjuvant for epitopes against a particular pathogen based on physiochemical properties of the epitope are urgently needed.

Currently available tools for the prediction of MHC binders primarily select the peptide based on their binding affinity for MHC molecules. However, it has been proposed that stability of the peptide bound to MHC molecules on the surface of APCs is more important than the binding affinity of the peptide to MHC molecules, as a stably bound peptide would be displayed on the surface for a long enough time to be recognized by a T-cell [124, 125]. Therefore, computational tools that can predict the stability of peptide-MHC complexes also need to be developed for the prediction of T-cell epitopes.

As the binding of peptides to MHC molecules is also affected by post-translational modifications such as phosphorylation, citrullination, and glycosylation of peptides [126 - 128], it is important to develop computational tools which can reliably predict the effect of post-translational modifications of peptides on the binding specificity to MHC molecules. Some of the tools that can predict phosphorylation and glycosylation in proteins are NetPhos [129], PhosphoPredict [130], and NetOGlyc [131].

Since the predictive performance of computational tools is related to the amount of data based on which such tools are developed, it is emphasized that more and more immunological data need to be generated. Likewise, new immunoin formatics tools with better accuracy need to be continuously developed for the prediction of B- and T-cells epitopes.

9. VALIDATION OF B- AND T-CELL EPITOPES

To evaluate the immunogenic potential of predicted epitopes, the immune response elicited by predicted epitopes is monitored by conducting immunization experiments in model systems. For this purpose, the selected epitope can be produced either through synthetic means or through recombinant approaches using a suitable host system. In the former case, solid-phase peptide synthesis is the method of choice for epitopes that are less than 50 amino acids long [132, 133]. For longer epitopes, recombinant methods could also be employed for the production of epitopes. In this case, the epitopic region is expressed either as a single protein or as virus-like particles when fused to the core protein of certain viruses (such as hepatitis B virus) at their major immunodominant regions, resulting in enhanced exposure of epitopic regions on the surface of virus-like particles [134] (Fig. **3**). Depending on the origin of the epitope, bacterial or eukaryotic expression systems (yeast, insect-, and mammalian cell lines) can be exploited for recombinant expression. Epitopes thus expressed can be purified using chromatographic techniques such as affinity chromatography, ion-exchange chromatography, and gel filtration. The purity and the integrity of purified epitopes can be analyzed through sodium dodecyl sulfate-polyacrylamide gel electrophoresis (SDS-PAGE), western blotting, and mass spectrometry. For conformational analysis of purified epitopes, biophysical techniques such as circular dichroism (CD) spectroscopy [135] and Fourier-transform infrared (FTIR) spectroscopy [136, 137] can be used to assess the secondary structure profile of epitopes. For further structural analysis, more sophisticated techniques such as nuclear magnetic resonance (NMR) spectroscopy [138] and X-ray crystallography [139] can be used. The immunogenic potential of properly folded epitopes can be evaluated by conducting immunization experiments using appropriate animal models. For this purpose, epitopes are either administered orally or injected into model organisms mostly in combination with suitable adjuvants [140]. Commonly used animal models for immunization studies include but are not limited to chickens, mice, rabbits, and chimpanzees [141, 142]. If an epitope is to be tested as a vaccine candidate in humans, its protective efficacy is first evaluated in animal models [143]. For vaccine development against animal disease, the model system used for vaccination trials is usually the same organism in which the pathogen causes the disease. Various immunological techniques have been developed for measuring humoral and cell-mediated immune responses evoked by B- and T-cell epitopes, respectively (Fig. **2**); (Table **2**). A brief description of commonly used immunological techniques for the detection of immune responses is given below.

A

B

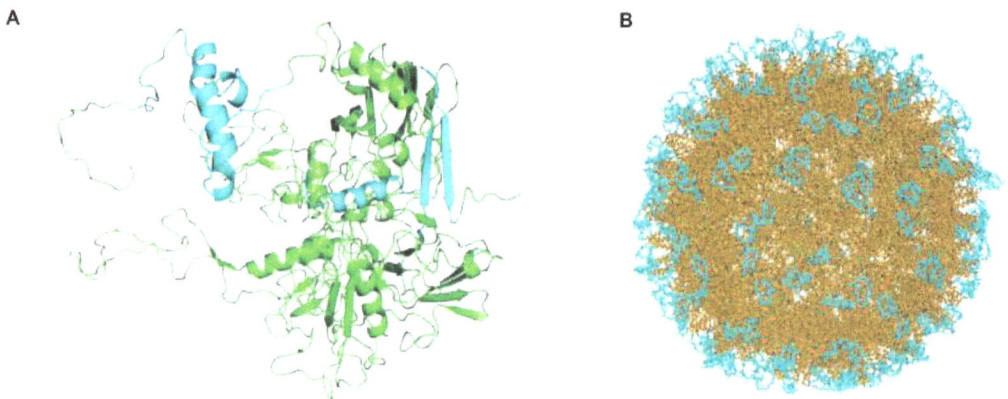

Fig. (3). Prediction of epitopic regions on the hexon capsid protein of Fowl adenovirus serotype 4 (FAdV-4) for use as vaccine candidates in the form of virus-like particles (VLPs). **(A)** Epitopic regions (colored cyan) located on the surface-exposed regions of FAdV-4 hexon (GenBank accession number: CAD30847.1; colored green) were predicted using various bioinformatics tools. Linear B-cell epitopes were predicted using BepiPred [45] and combining other criteria such as hydrophilicity [52], flexibility [53], β-turns [54], solvent accessibility [55], and antigenicity [56]. Conformational B-cell epitopes were predicted using Ellipro [60]. For the prediction of T-cell epitopes, MHC class I and MHC class II binders were predicted using the NetMHCcons 1.1 server [80] and the NetMHCII Pan 3.1 server [92], respectively. **(B)** The predicted epitopes (colored cyan) could be tested as vaccine candidates in the form of virus-like particles upon fusion with the core protein of hepatitis B virus (colored brown), for example.

Table 2. Commonly used tools for validation of B- and T-cell epitopes.

Epitope	Method	Reference(s)
B-cell epitopes	Enzyme-linked immunosorbent assay (ELISA)	[150 - 152]
	Immunoblotting	[156]
	Virus neutralization test (VNT)	[160]
	Lateral flow assay	[163]
T-cell epitopes	Mass spectrometry	[167, 168]
	Enzyme-linked immune absorbent spot (ELISpot)	[169]
	Intracellular cytokine staining (ICS) assay	[173, 174]
	Tetramer staining	[175, 176]

9.1. Tools for Validation of B-cell Epitopes

Epitope-based subunit vaccines generally mediate protection by inducing an immune response that leads to activation of B-cells and production of antibodies [144, 145]. Numerous immunoassays are available nowadays, which can be used to detect antibodies in serum or plasma samples [146, 147]. Most of these techniques are based on immobilization of specific surface-exposed recombinant antigens or their selected epitopes on a solid support and incubation with serum containing specific antibodies. It has been found that the inclusion of conformational B-cell epitopes in immunoassays yields better reactivity and specificity [148, 149]. A brief description of commonly used methods for the detection of antibodies in serum samples is given below:

9.1.1. Enzyme-Linked Immunosorbent Assay

The enzyme-linked immunosorbent assay (ELISA) is a typical serological method for the quantitative detection of antibodies. Serum or plasma can be used as test samples in ELISA. Although the basic principle of ELISA is the same, *i.e.*, the formation of an immune complex on a solid support, there are several formats available for performing the assay. Most commonly, indirect ELISA is performed for the detection of specific antibodies in a given serum sample. The test is based on the adsorption (coating) of the antigen or epitopes on solid support *e.g.*, the polystyrene surface. After that, the plates or tubes are blocked with a blocking agent such as bovine serum albumin or skimmed milk to fill the space not occupied by the protein. The test serum is then incubated, followed by washing the plates or tubes with an appropriate buffer (*e.g.*, Tris-buffer saline containing an anionic detergent such as Tween-20) and the addition of enzyme-labeled secondary antibodies which can specifically bind to antibodies of interest. After the unbound enzyme-labeled secondary antibodies are washed off, the development of a colored product upon addition of a chromogenic substrate of the enzyme confirms the presence of specific antibodies in the serum sample [150 - 152]. In most cases, enzymes conjugated to secondary antibodies include alkaline phosphatase, horseradish peroxidase, or glucose oxidase [150, 151, 153]. ELISA is a method of choice for the detection of antibodies in serum samples due to its high sensitivity, specificity, reproducibility, low cost, and feasibility of performance under various field conditions [154, 155].

9.1.2. Immunoblotting

Similar to ELISA, the formation of an immune complex in immunoblotting takes place on a solid surface. However, such a type of assay utilizes the capability of polyvinylidene fluoride (PVDF) or nitrocellulose membranes to bind proteins. Specific epitopes or antigens having a suitable concentration are added in the form of small dots, followed by drying the membranes. The treated membranes are added with a blocking buffer (containing *e.g.*, gelatin, ovalbumin, or milk proteins) to avoid non-specific adsorption. After blocking, the membranes are incubated with various dilutions of the serum sample. The membranes are washed to remove unbound antibodies and then incubated with secondary antibodies (conjugated with enzymes) generated against the antibodies of interest. Similar to the ELISA method, the antigen-antibody complex is visualized by the addition of the substrate, which produces a colored spot upon conversion by the antibody-bound enzyme. The spot intensity provides a semi-quantitative estimate of the number of specific serum antibodies present in the sample. In the advanced form of immunoblotting, termed Western blotting, the resolving power of electrophoresis and the discriminating power of an immunological reaction are combined. In this method, the identification of antibodies is achieved by electrophoretic separation of proteins (epitopes or antigens), which are transferred from the gel to the membrane. The remaining treatment is continued as described for dot blots. Finally, the presence of colored bands of a particular size of the protein (epitope or antigen) confirms the presence of antibodies [156]. Western blotting has been successfully employed for the detection of antibodies against the conformational or linear epitopes of the matrix protein [157], the immunodominant region of the VP2 protein of infectious bursal disease virus (IBDV) [158], and the fiber protein of fowl adenovirus serotype 8 (FAdV-8) [159], for example.

9.1.3. Virus Neutralization Test

An *in vitro* virus neutralization tests (VNT) are a serological method to assess the presence and quantitative estimation of functional systemic antibodies able to prevent virus infectivity. The viral growth and infectivity are inhibited when virus-specific neutralizing antibodies are transferred into host systems such as eggs, cell cultures, and animals in which viral replication and growth can take place. This principle forms the basis of VNT. The test indirectly provides an idea about the functionalization of antibodies and the protective efficacy of vaccines. VNT is a highly sensitive and specific assay to measure the titer of neutralizing antibodies post-vaccination or after-infection. In cell culture-based VNT, the antibody titer is determined based on the presence or the absence of cytopathic effects or by confirmation of the viral infection by immunoreactive techniques. Although the VNT assay is inexpensive and could be performed using standard laboratory equipment, however, the requirement of cell culture, more time, optimization, and technically skilled labor make it harder to conduct the VNT assay in comparison to other serological methods. The VNT assay is advantageous to assess the extent of serological cross-reactivity between vaccine antisera and variant viral strains leading to cross-protection in the host [160]. Hence, VNT is a better tool for *in vitro* evaluation of broadly active neutralizing antibodies effective against diverse strains of the pathogen [161]. VNT assays based on multiple epitopes and free of any vaccine material are the gold standards for unbiased evaluation of the protective efficacy of vaccine-induced antibodies

[162]. VNT assays can be used to completely characterize epitopic regions present on viral surfaces.

9.1.4. Lateral Flow Assay

Lateral flow assays (LFAs) have recently received considerable attention due to their excellent performance in terms of the detection of specific antibodies in a given serum sample. LFAs have several formats [163]. All LFA strips have four components: a conjugate pad, a sample pad, a nitrocellulose membrane, and an adsorbent pad. Monodisperse colloidal gold nanoparticles having a typical absorption spectrum (400–800 nm) conjugated to a specific protein (epitope or antigen) are generally used as reporters for colorimetric antibody detection in LFA kits [164]. The nitrocellulose membrane is printed with the protein (epitope or antigen) or antibody at the test line and the control line, depending on the test whether antibody or antigen is detected in the test sample. The assembly of these components in the form of kits enables the field workers to perform the test in 10-15 minutes without requiring any specialized equipment. LFAs have been successfully used for the detection of specific antibodies against several bacterial and viral diseases as well as their targeted epitopic regions (vaccines), *e.g.*, LFA based on the gold-conjugated surface capsid VP2 protein of IBDV was used to detect anti-IBDV antibodies in birds [165], and the nucleocapsid protein-gold conjugate-based LFA was used to investigate specific antibodies against SARS-CoV-2 [166].

9.2. Tools for Validation of T-Cell Epitopes

9.2.1. Mass spectrometry (MS)

The MHC-peptide complexes are purified from the cell lysate by immunoaffinity, the MHC-associated peptides are eluted and the sequence of the peptides is analyzed using mass spectrometry [167, 168].

9.2.2. Enzyme-linked Immune Absorbent Spot (ELISpot)

For measuring the immune response mediated by T-cells, ELISPOT can be employed. In this case, cytokines produced by cells are detected. For this purpose, cytokines released by stimulated cells are first captured by antibodies. The captured cytokines are then detected using cytokine-specific biotinylated antibodies and streptavidin-enzyme conjugates which catalyze the formation of insoluble/colored precipitates [169].

9.2.3. Intracellular Cytokine Staining (ICS) Assay

T-cell epitopes can also be characterized using ICS. In this case, cells are first activated with antigens in the presence of secretion inhibitors (monensin or brefeldin A, for example) [170, 171]. The cells are then permeabilized, and intracellular cytokines are stained with fluorescently labeled antibodies [172]. The stained cells are finally analyzed using flow cytometry [173, 174].

9.2.4. Tetramer Staining

For tetramer staining, four MHC molecules are stained with fluorophores specific for a peptide and analyzed using flow cytometry. Thus, a tetramer stain that specifically binds to a given MHC-peptide complex is needed to perform this technique [175, 176].

CONCLUDING REMARKS

There are several tools currently available for the prediction of epitopic regions on proteins. However, little attention has been directed to the development of tools for the prediction of adjuvants or the effects of post-translational modifications of peptides on MHC binding. Several complementary approaches are also available for validating B- and T-cell epitopes, and technologies with better throughput are also emerging.

CONSENT FOR PUBLICATION

Not applicable.

CONFLICT OF INTEREST

The author declares no conflict of interest, financial or otherwise.

ACKNOWLEDGEMENTS

This work was supported by the International Centre for Genetic Engineering and Biotechnology (ICGEB), Italy (Grant No. CRP/PAK10-01 awarded to Moazur Rahman), Agriculture Linkage Programme-Pakistan Agriculture Research Council (Grant No. AS015 awarded to Moazur Rahman), and the Higher Education Commission (HEC) of Pakistan (Grant Nos. TDF-03-006 and NRPU-7254 awarded to Moazur Rahman, and HEC-Indigenous Ph.D. and HEC- international IRSIP Fellowships awarded to Soban Tufail, Majid Ali Shah, and Maryam Zafar).

REFERENCES

[1]   Blum JS, Wearsch PA, Cresswell P. Pathways of antigen processing. Annu Rev Immunol 2013; 31: 443-73.
[http://dx.doi.org/10.1146/annurev-immunol-032712-095910] [PMID: 23298205]

[2]   Wieczorek M, Abualrous ET, Sticht J, *et al.* Major histocompatibility complex (MHC) class I and MHC class II proteins: Conformational plasticity in antigen presentation. Front Immunol 2017; 8: 292.
[http://dx.doi.org/10.3389/fimmu.2017.00292] [PMID: 28367149]

[3]   Getzoff ED, Tainer JA, Lerner RA, Geysen HM. The chemistry and mechanism of antibody binding to protein antigens. Adv Immunol 1988; 43: 1-98.
[http://dx.doi.org/10.1016/S0065-2776(08)60363-6] [PMID: 3055852]

[4]   Sela-Culang I, Ofran Y, Peters B. Antibody specific epitope prediction-emergence of a new paradigm. Curr Opin Virol 2015; 11: 98-102.
[http://dx.doi.org/10.1016/j.coviro.2015.03.012] [PMID: 25837466]

[5]   Rubinstein ND, Mayrose I, Halperin D, Yekutieli D, Gershoni JM, Pupko T. Computational characterization of B-cell epitopes. Mol Immunol 2008; 45(12): 3477-89.
[http://dx.doi.org/10.1016/j.molimm.2007.10.016] [PMID: 18023478]

[6]   Barlow DJ, Edwards MS, Thornton JM. Continuous and discontinuous protein antigenic determinants. Nature 1986; 322(6081): 747-8.
[http://dx.doi.org/10.1038/322747a0] [PMID: 2427953]

[7]   Langeveld JPM, Martinez-Torrecuadrada J, Boshuizen RS, Meloen RH, Ignacio Casal J. Characterisation of a protective linear B cell epitope against feline parvoviruses.Vaccine. Vaccine 2001; pp. 2352-60.
[http://dx.doi.org/10.1016/S0264-410X(00)00526-0]

[8]   Van Regenmortel MHV. What is a B-cell epitope? Methods Mol Biol 2009; 524: 3-20.
[http://dx.doi.org/10.1007/978-1-59745-450-6_1] [PMID: 19377933]

[9]   Walter G. Production and use of antibodies against synthetic peptides. J Immunol Methods 1986; 88(2): 149-61.
[http://dx.doi.org/10.1016/0022-1759(86)90001-3] [PMID: 2420900]

[10]   Paul W. Fundamental immunology. 7th ed., Philadelphia: Wolters Kluwer Health/Lippincott Williams & Wilkins 2013.

[11]   Peters B, Nielsen M, Sette A. T cell epitope predictions. Annu Rev Immunol 2020; 38: 123-45.
[http://dx.doi.org/10.1146/annurev-immunol-082119-124838] [PMID: 32045313]

[12]   Morris GE. Epitope mapping. Methods Mol Biol 2005; 295: 255-68.
[http://dx.doi.org/10.1385/1-59259-873-0:255] [PMID: 15596901]

[13]   Huang J, Ru B, Dai P. Bioinformatics resources and tools for phage display. Molecules 2011; 16(1): 694-709.
[http://dx.doi.org/10.3390/molecules16010694] [PMID: 21245805]

[14]   Moreau V, Granier C, Villard S, Laune D, Molina F. Discontinuous epitope prediction based on mimotope analysis. Bioinformatics 2006; 22(9): 1088-95.
[http://dx.doi.org/10.1093/bioinformatics/btl012] [PMID: 16434442]

[15]   Buus S, Sette A, Colon SM, Jenis DM, Grey HM. Isolation and characterization of antigen-Ia complexes involved in T cell recognition. Cell 1986; 47(6): 1071-7.
[http://dx.doi.org/10.1016/0092-8674(86)90822-6] [PMID: 3490919]

[16]   Vita R, Overton JA, Greenbaum JA, *et al.* The immune epitope database (IEDB) 3.0. Nucleic Acids Res 2015; 43(Database issue): D405-12.
[http://dx.doi.org/10.1093/nar/gku938] [PMID: 25300482]

[17]   Peters B, Sidney J, Bourne P, *et al.* The immune epitope database and analysis resource: from vision to blueprint. PLOS Biol. 2005; 3: p. (3)e91.
[http://dx.doi.org/10.1371/journal.pbio.0030091]

[18]   Zhang Q, Wang P, Kim Y, *et al.* Immune epitope database analysis resource (IEDB-AR). Nucleic Acids Res 2008; 36(Web Server issue): W513-8.
[http://dx.doi.org/10.1093/nar/gkn254] [PMID: 18515843]

[19]   Greenbaum JA, Andersen PH, Blythe M, *et al.* Towards a consensus on datasets and evaluation metrics for developing B-cell epitope prediction tools. J Mol Recognit 2007; 20(2): 75-82.
[http://dx.doi.org/10.1002/jmr.815] [PMID: 17205610]

[20]   El-Manzalawy Y, Honavar V. Recent advances in B-cell epitope prediction methods. Immunome Res 2010; 6 (Suppl. 2): S2.
[http://dx.doi.org/10.1186/1745-7580-6-S2-S2] [PMID: 21067544]

[21]   Yewdell JW, Bennink JR. Immunodominance in major histocompatibility complex class I-restricted T lymphocyte responses. Annu Rev Immunol 1999; 17: 51-88.
[http://dx.doi.org/10.1146/annurev.immunol.17.1.51] [PMID: 10358753]

[22]   Santos AR, Pereira VB, Barbosa E, *et al.* Mature Epitope Density--a strategy for target selection based on immunoinformatics and exported prokaryotic proteins. BMC Genomics 2013; 14 (Suppl. 6): S4.
[http://dx.doi.org/10.1186/1471-2164-14-S6-S4] [PMID: 24564223]

[23]   Kovjazin R, Volovitz I, Daon Y, *et al.* Signal peptides and trans-membrane regions are broadly immunogenic and have high CD8+ T cell epitope densities: Implications for vaccine development. Mol Immunol 2011; 48(8): 1009-18.
[http://dx.doi.org/10.1016/j.molimm.2011.01.006] [PMID: 21316766]

[24]   Dall'antonia F, Pavkov-Keller T, Zangger K, Keller W. Structure of allergens and structure based epitope predictions. Methods 2014; 66(1): 3-21.
[http://dx.doi.org/10.1016/j.ymeth.2013.07.024] [PMID: 23891546]

[25]   Seib KL, Zhao X, Rappuoli R. Developing vaccines in the era of genomics: a decade of reverse vaccinology. Clin Microbiol Infect 2012; 18 (Suppl. 5): 109-16.
[http://dx.doi.org/10.1111/j.1469-0691.2012.03939.x] [PMID: 22882709]

[26]   Pizza M, Scarlato V, Masignani V, *et al.* Identification of vaccine candidates against serogroup B meningococcus by whole-genome sequencing, Science 2000; 287(2000): 1816-20.
[http://dx.doi.org/10.1126/science.287.5459.1816]

[27]   Wizemann TM, Heinrichs JH, Adamou JE, *et al.* Use of a whole genome approach to identify vaccine molecules affording protection against *Streptococcus pneumoniae* infection. Infect Immun 2001; 69(3): 1593-8.
[http://dx.doi.org/10.1128/IAI.69.3.1593-1598.2001] [PMID: 11179332]

[28]   Ross BC, Czajkowski L, Hocking D, *et al.* Identification of vaccine candidate antigens from a genomic analysis of *Porphyromonas gingivalis*. Vaccine 2001; 19(30): 4135-42.
[http://dx.doi.org/10.1016/S0264-410X(01)00173-6] [PMID: 11457538]

[29]   Montigiani S, Falugi F, Scarselli M, *et al.* Genomic approach for analysis of surface proteins in *Chlamydia pneumoniae.* Infect Immun 2002; 70(1): 368-79.
[http://dx.doi.org/10.1128/IAI.70.1.368-379.2002] [PMID: 11748203]

[30]   Ariel N, Zvi A, Grosfeld H, *et al.* Search for potential vaccine candidate open reading frames in the *Bacillus anthracis* virulence plasmid pXO1: *in silico* and *in vitro* screening. Infect Immun 2002; 70(12): 6817-27.
[http://dx.doi.org/10.1128/IAI.70.12.6817-6827.2002] [PMID: 12438358]

[31]   Etz H, Minh DB, Henics T, *et al.* Identification of *in vivo* expressed vaccine candidate antigens from *Staphylococcus aureus.* Proc Natl Acad Sci USA 2002; 99(10): 6573-8.
[http://dx.doi.org/10.1073/pnas.092569199] [PMID: 11997460]

[32]   Bagnoli F, Fontana MR, Soldaini E, *et al.* Vaccine composition formulated with a novel TLR7-dependent adjuvant induces high and broad protection against *Staphylococcus aureus.* Proc Natl Acad Sci USA 2015; 112(12): 3680-5.
[http://dx.doi.org/10.1073/pnas.1424924112] [PMID: 25775551]

[33]   Moriel DG, Bertoldi I, Spagnuolo A, *et al.* Identification of protective and broadly conserved vaccine antigens from the genome of extraintestinal pathogenic *Escherichia coli.* Proc Natl Acad Sci USA 2010; 107(20): 9072-7.
[http://dx.doi.org/10.1073/pnas.0915077107] [PMID: 20439758]

[34]   Gomez G, Pei J, Mwangi W, Adams LG, Rice-Ficht A, Ficht TA. Immunogenic and invasive properties of *Brucella melitensis* 16M outer membrane protein vaccine candidates identified via a reverse vaccinology approach. PLoS One 2013; 8(3): e59751.
[http://dx.doi.org/10.1371/journal.pone.0059751] [PMID: 23533646]

[35]   Caro-Gomez E, Gazi M, Goez Y, Valbuena G. Discovery of novel cross-protective *Rickettsia prowazekii* T-cell antigens using a combined reverse vaccinology and *in vivo* screening approach. Vaccine 2014; 32(39): 4968-76.
[http://dx.doi.org/10.1016/j.vaccine.2014.06.089] [PMID: 25010827]

[36]   Gan W, Zhao G, Xu H, *et al.* Reverse vaccinology approach identify an *Echinococcus granulosus* tegumental membrane protein enolase as vaccine candidate. Parasitol Res 2010; 106(4): 873-82.
[http://dx.doi.org/10.1007/s00436-010-1729-x] [PMID: 20127115]

[37]   Chandler JC, Sutherland MD, Harton MR, *et al. Francisella tularensis* LVS surface and membrane proteins as targets of effective post-exposure immunization for tularemia. J Proteome Res 2015; 14(2): 664-75.
[http://dx.doi.org/10.1021/pr500628k] [PMID: 25494920]

[38]   Graham SP, Honda Y, Pellé R, *et al.* A novel strategy for the identification of antigens that are recognised by bovine MHC class I restricted cytotoxic T cells in a protozoan infection using reverse vaccinology. Immunome Res 2007; 3: 2.
[http://dx.doi.org/10.1186/1745-7580-3-2] [PMID: 17291333]

[39]   Bager RJ, Kudirkiene E, da Piedade I, *et al. In silico* prediction of *Gallibacterium anatis* pan-immunogens. Vet Res 2014; 45: 80.
[http://dx.doi.org/10.1186/s13567-014-0080-0] [PMID: 25223320]

[40]   John L, John GJ, Kholia T. A reverse vaccinology approach for the identification of potential vaccine candidates from Leishmania spp. Appl. Biochem. Biotechnol., Appl Biochem Biotechnol 2012; pp. 1340-50.
[http://dx.doi.org/10.1007/s12010-012-9649-0]

[41]   Fouts AE, Chan P, Stephan J-P, Vandlen R, Feierbach B. Antibodies against the gH/gL/UL128/UL130/UL131 complex comprise the majority of the anti-cytomegalovirus (anti-CMV) neutralizing antibody response in CMV hyperimmune globulin. J Virol 2012; 86(13): 7444-7.
[http://dx.doi.org/10.1128/JVI.00467-12] [PMID: 22532696]

[42]   Steel J, Lowen AC, Wang TT, *et al.* Influenza virus vaccine based on the conserved hemagglutinin stalk domain. MBio 2010; 1(1): e00018-10.
[http://dx.doi.org/10.1128/mBio.00018-10] [PMID: 20689752]

[43]   Saha S, Raghava GPS. Prediction of continuous B-cell epitopes in an antigen using recurrent neural network. Proteins 2006; 65(1): 40-8.
[http://dx.doi.org/10.1002/prot.21078] [PMID: 16894596]

[44]   Saha S, Raghava GPS. BcePred: Prediction of continuous B-cell epitopes in antigenic sequences using

physico-chemical properties, Lect. Notes Comput. Sci (Including Subser Lect Notes Artif Intell Lect Notes Bioinformatics). 2004; 3239: pp. 197-204.
[http://dx.doi.org/10.1007/978-3-540-30220-9_16]

[45]   Larsen JEP, Lund O, Nielsen M. Improved method for predicting linear B-cell epitopes. Immunome Res 2006; 2: 2.
[http://dx.doi.org/10.1186/1745-7580-2-2] [PMID: 16635264]

[46]   El-Manzalawy Y, Dobbs D, Honavar V. Predicting linear B-cell epitopes using string kernels. J Mol Recognit 2008; 21(4): 243-55.
[http://dx.doi.org/10.1002/jmr.893] [PMID: 18496882]

[47]   El-Manzalawy Y, Dobbs D, Honavar V. Predicting flexible length linear B-cell epitopes. Comput Syst Bioinformatics Conf 2008; 7: 121-32.
[http://dx.doi.org/10.1142/9781848162648_0011] [PMID: 19642274]

[48]   Sweredoski MJ, Baldi P. COBEpro: a novel system for predicting continuous B-cell epitopes. Protein Eng Des Sel 2009; 22(3): 113-20.
[http://dx.doi.org/10.1093/protein/gzn075] [PMID: 19074155]

[49]   Singh H, Ansari HR, Raghava GPS. Improved method for linear B-cell epitope prediction using antigen's primary sequence. PLoS One 2013; 8(5): e62216.
[http://dx.doi.org/10.1371/journal.pone.0062216] [PMID: 23667458]

[50]   Pellequer JL, Westhof E. PREDITOP: a program for antigenicity prediction. J Mol Graph 1993; 11(3): 204-210, 191-192.
[http://dx.doi.org/10.1016/0263-7855(93)80074-2] [PMID: 7509182]

[51]   Alix AJP. Predictive estimation of protein linear epitopes by using the program PEOPLE.Vaccine. Vaccine 1999; pp. 311-4.
[http://dx.doi.org/10.1016/S0264-410X(99)00329-1]

[52]   Parker JMR, Guo D, Hodges RS. New hydrophilicity scale derived from high-performance liquid chromatography peptide retention data: correlation of predicted surface residues with antigenicity and X-ray-derived accessible sites. Biochemistry 1986; 25(19): 5425-32.
[http://dx.doi.org/10.1021/bi00367a013] [PMID: 2430611]

[53]   Karplus PA, Schulz GE. Prediction of chain flexibility in proteins - A tool for the selection of peptide antigens. Naturwissenschaften 1985; 72: 212-3.
[http://dx.doi.org/10.1007/BF01195768]

[54]   Chou PY, Fasman GD. Empirical predictions of protein conformation. Annu Rev Biochem 1978; 47: 251-76.
[http://dx.doi.org/10.1146/annurev.bi.47.070178.001343] [PMID: 354496]

[55]   Emini EA, Hughes JV, Perlow DS, Boger J. Induction of hepatitis A virus-neutralizing antibody by a virus-specific synthetic peptide. J Virol 1985; 55(3): 836-9.
[http://dx.doi.org/10.1128/jvi.55.3.836-839.1985] [PMID: 2991600]

[56]   Kolaskar AS, Tongaonkar PC. A semi-empirical method for prediction of antigenic determinants on protein antigens. FEBS Lett 1990; 276(1-2): 172-4.
[http://dx.doi.org/10.1016/0014-5793(90)80535-Q] [PMID: 1702393]

[57]   Jespersen MC, Peters B, Nielsen M, Marcatili P. BepiPred-2.0: improving sequence-based B-cell epitope prediction using conformational epitopes. Nucleic Acids Res 2017; 45(W1): W24-9.
[http://dx.doi.org/10.1093/nar/gkx346] [PMID: 28472356]

[58]   Kulkarni-Kale U, Bhosle S, Kolaskar AS. CEP: a conformational epitope prediction server. Nucleic Acids Res 2005; 33(Web Server issue): W168-71.
[http://dx.doi.org/10.1093/nar/gki460] [PMID: 15980448]

[59]   Haste Andersen P, Nielsen M, Lund O. Prediction of residues in discontinuous B-cell epitopes using protein 3D structures. Protein Sci 2006; 15(11): 2558-67.

[http://dx.doi.org/10.1110/ps.062405906] [PMID: 17001032]

[60]   Ponomarenko J, Bui HH, Li W, *et al.* ElliPro: a new structure-based tool for the prediction of antibody epitopes. BMC Bioinformatics 2008; 9: 514.
[http://dx.doi.org/10.1186/1471-2105-9-514] [PMID: 19055730]

[61]   Ansari HR, Raghava GP. Identification of conformational B-cell epitopes in an antigen from its primary sequence. Immunome Res 2010; 6: 6.
[http://dx.doi.org/10.1186/1745-7580-6-6] [PMID: 20961417]

[62]   Sweredoski MJ, Baldi P. PEPITO: improved discontinuous B-cell epitope prediction using multiple distance thresholds and half sphere exposure. Bioinformatics 2008; 24(12): 1459-60.
[http://dx.doi.org/10.1093/bioinformatics/btn199] [PMID: 18443018]

[63]   Sun J, Wu D, Xu T, *et al.* SEPPA: a computational server for spatial epitope prediction of protein antigens. Nucleic Acids Res 2009; 37(Web Server issue): W612-6.
[http://dx.doi.org/10.1093/nar/gkp417] [PMID: 19465377]

[64]   Sela-Culang I, Ashkenazi S, Peters B, Ofran Y. PEASE: predicting B-cell epitopes utilizing antibody sequence. Bioinformatics 2015; 31(8): 1313-5.
[http://dx.doi.org/10.1093/bioinformatics/btu790] [PMID: 25432167]

[65]   Krawczyk K, Liu X, Baker T, Shi J, Deane CM. Improving B-cell epitope prediction and its application to global antibody-antigen docking. Bioinformatics 2014; 30(16): 2288-94.
[http://dx.doi.org/10.1093/bioinformatics/btu190] [PMID: 24753488]

[66]   Zhao L, Li J. Mining for the antibody-antigen interacting associations that predict the B cell epitopes. BMC Struct Biol 2010; 10 (Suppl. 1): S6.
[http://dx.doi.org/10.1186/1472-6807-10-S1-S6] [PMID: 20487513]

[67]   Zhao L, Wong L, Li J. Antibody-specified B-cell epitope prediction in line with the principle of context-awareness. IEEE/ACM Trans Comput Biol Bioinformatics 2011; 8(6): 1483-94.
[http://dx.doi.org/10.1109/TCBB.2011.49] [PMID: 21383422]

[68]   Rubinstein ND, Mayrose I, Martz E, Pupko T. Epitopia: a web-server for predicting B-cell epitopes. BMC Bioinformatics 2009; 10: 287.
[http://dx.doi.org/10.1186/1471-2105-10-287] [PMID: 19751513]

[69]   Gao J, Faraggi E, Zhou Y, Ruan J, Kurgan L. BEST: improved prediction of B-cell epitopes from antigen sequences. PLoS One 2012; 7(6): e40104.
[http://dx.doi.org/10.1371/journal.pone.0040104] [PMID: 22761950]

[70]   Huang J, Gutteridge A, Honda W, Kanehisa M. MIMOX: a web tool for phage display based epitope mapping. BMC Bioinformatics 2006; 7: 451.
[http://dx.doi.org/10.1186/1471-2105-7-451] [PMID: 17038191]

[71]   Mayrose I, Penn O, Erez E, *et al.* Pepitope: epitope mapping from affinity-selected peptides. Bioinformatics 2007; 23(23): 3244-6.
[http://dx.doi.org/10.1093/bioinformatics/btm493] [PMID: 17977889]

[72]   Negi SS, Braun W. Automated detection of conformational epitopes using phage display Peptide sequences. Bioinform Biol Insights 2009; 3: 71-81.
[http://dx.doi.org/10.4137/BBI.S2745] [PMID: 20140073]

[73]   Chen WH, Sun PP, Lu Y, Guo WW, Huang YX, Ma ZQ. MimoPro: a more efficient Web-based tool for epitope prediction using phage display libraries. BMC Bioinformatics 2011; 12: 199.
[http://dx.doi.org/10.1186/1471-2105-12-199] [PMID: 21609501]

[74]   Chen W, Guo WW, Huang Y, Ma Z. PepMapper: a collaborative web tool for mapping epitopes from affinity-selected peptides. PLoS One 2012; 7(5): e37869.
[http://dx.doi.org/10.1371/journal.pone.0037869] [PMID: 22701536]

[75]   Soni DK, Nath A, Dubey SK. Evaluation and use of *in silico* structure based epitope prediction for

listeriolysin O of Listeria monocytogenes. Ind J Biotechnol 2015; 14(2): 160-6.http://lifecenter.sgst.cn/seppa/

[76]   Bhasin M, Raghava GPS. Prediction of CTL epitopes using QM, SVM and ANN techniques. Vaccine 2004; 22(23-24): 3195-204.
[http://dx.doi.org/10.1016/j.vaccine.2004.02.005] [PMID: 15297074]

[77]   Koşaloğlu-Yalçın Z, Lanka M, Frentzen A, *et al.* Predicting T cell recognition of MHC class I restricted neoepitopes. OncoImmunology 2018; 7(11): e1492508.
[http://dx.doi.org/10.1080/2162402X.2018.1492508] [PMID: 30377561]

[78]   Bjerregaard AM, Nielsen M, Jurtz V, *et al.* An analysis of natural T cell responses to predicted tumor neoepitopes. Front Immunol 2017; 8: 1566.
[http://dx.doi.org/10.3389/fimmu.2017.01566] [PMID: 29187854]

[79]   Paul S, Croft NP, Purcell AW, *et al.* Benchmarking predictions of MHC class I restricted T cell epitopes in a comprehensively studied model system. PLOS Comput Biol 2020; 16(5): e1007757.
[http://dx.doi.org/10.1371/journal.pcbi.1007757] [PMID: 32453790]

[80]   Karosiene E, Lundegaard C, Lund O, Nielsen M. NetMHCcons: a consensus method for the major histocompatibility complex class I predictions. Immunogenetics 2012; 64(3): 177-86.
[http://dx.doi.org/10.1007/s00251-011-0579-8] [PMID: 22009319]

[81]   Lundegaard C, Lamberth K, Harndahl M, Buus S, Lund O, Nielsen M. NetMHC-3.0: accurate web accessible predictions of human, mouse and monkey MHC class I affinities for peptides of length 8-11. Nucleic Acids Res 2008; 36(Web Server issue): W509-12.
[http://dx.doi.org/10.1093/nar/gkn202] [PMID: 18463140]

[82]   Hoof I, Peters B, Sidney J, *et al.* NetMHCpan, a method for MHC class I binding prediction beyond humans. Immunogenetics 2009; 61(1): 1-13.
[http://dx.doi.org/10.1007/s00251-008-0341-z] [PMID: 19002680]

[83]   Zhang H, Lund O, Nielsen M. The PickPocket method for predicting binding specificities for receptors based on receptor pocket similarities: application to MHC-peptide binding. Bioinformatics 2009; 25(10): 1293-9.
[http://dx.doi.org/10.1093/bioinformatics/btp137] [PMID: 19297351]

[84]   Larsen MV, Lundegaard C, Lamberth K, Buus S, Lund O, Nielsen M. Large-scale validation of methods for cytotoxic T-lymphocyte epitope prediction. BMC Bioinformatics 2007; 8: 424.
[http://dx.doi.org/10.1186/1471-2105-8-424] [PMID: 17973982]

[85]   Larsen MV, Lundegaard C, Lamberth K, *et al.* An integrative approach to CTL epitope prediction: a combined algorithm integrating MHC class I binding, TAP transport efficiency, and proteasomal cleavage predictions. Eur J Immunol 2005; 35(8): 2295-303.
[http://dx.doi.org/10.1002/eji.200425811] [PMID: 15997466]

[86]   Rammensee H, Bachmann J, Emmerich NPN, Bachor OA, Stevanović S. SYFPEITHI: database for MHC ligands and peptide motifs. Immunogenetics 1999; 50(3-4): 213-9.
[http://dx.doi.org/10.1007/s002510050595] [PMID: 10602881]

[87]   Bassani-Sternberg M, Gfeller D. Unsupervised HLA peptidome deconvolution improves ligand prediction accuracy and predicts cooperative effects in peptide–HLA interactions. J Immunol 2016; 197(6): 2492-9.
[http://dx.doi.org/10.4049/jimmunol.1600808] [PMID: 27511729]

[88]   O'Donnell TJ, Rubinsteyn A, Bonsack M, Riemer AB, Laserson U, Hammerbacher J. Hammerbacher, MHCflurry: Open-source class I MHC binding affinity prediction, Cell Syst. 7 2018; 129-32.
[http://dx.doi.org/10.1016/j.cels.2018.05.014]

[89]   Singh H, Raghava GPS. ProPred1: prediction of promiscuous MHC Class-I binding sites. Bioinformatics 2003; 19(8): 1009-14.
[http://dx.doi.org/10.1093/bioinformatics/btg108] [PMID: 12761064]

[90]   Reche PA, Glutting JP, Reinherz EL. Prediction of MHC class I binding peptides using profile motifs. Hum Immunol 2002; 63(9): 701-9.
[http://dx.doi.org/10.1016/S0198-8859(02)00432-9] [PMID: 12175724]

[91]   Dönnes P, Kohlbacher O. SVMHC: a server for prediction of MHC-binding peptides. Nucleic Acids Res 2006; 34(Web Server issue): W194-7.
[http://dx.doi.org/10.1093/nar/gkl284] [PMID: 16844990]

[92]   Jensen KK, Andreatta M, Marcatili P, *et al.* Improved methods for predicting peptide binding affinity to MHC class II molecules. Immunology 2018; 154(3): 394-406.
[http://dx.doi.org/10.1111/imm.12889] [PMID: 29315598]

[93]   Sturniolo T, Bono E, Ding J, *et al.* Generation of tissue-specific and promiscuous HLA ligand databases using DNA microarrays and virtual HLA class II matrices. Nat Biotechnol 1999; 17(6): 555-61.
[http://dx.doi.org/10.1038/9858] [PMID: 10385319]

[94]   Bordner AJ, Mittelmann HD, Multi RTA. MultiRTA: a simple yet reliable method for predicting peptide binding affinities for multiple class II MHC allotypes. BMC Bioinformatics 2010; 11: 482.
[http://dx.doi.org/10.1186/1471-2105-11-482] [PMID: 20868497]

[95]   Pfeifer N, Kohlbacher O. Multiple instance learning allows MHC class II epitope predictions across alleles 2008.
[http://dx.doi.org/10.1007/978-3-540-87361-7_18]

[96]   Singh H, Raghava GPS. ProPred: prediction of HLA-DR binding sites. Bioinformatics 2001; 17(12): 1236-7.
[http://dx.doi.org/10.1093/bioinformatics/17.12.1236] [PMID: 11751237]

[97]   Oyarzún P, Ellis JJ, Bodén M, Kobe B. PREDIVAC: CD4+ T-cell epitope prediction for vaccine design that covers 95% of HLA class II DR protein diversity. BMC Bioinformatics 2013; 14: 52.
[http://dx.doi.org/10.1186/1471-2105-14-52] [PMID: 23409948]

[98]   Atanasova M, Patronov A, Dimitrov I, Flower DR, Doytchinova I. EpiDOCK: a molecular docking-based tool for MHC class II binding prediction. Protein Eng Des Sel 2013; 26(10): 631-4.
[http://dx.doi.org/10.1093/protein/gzt018] [PMID: 23661105]

[99]   Dimitrov I, Garnev P, Flower DR, Doytchinova I. EpiTOP--a proteochemometric tool for MHC class II binding prediction. Bioinformatics 2010; 26(16): 2066-8.
[http://dx.doi.org/10.1093/bioinformatics/btq324] [PMID: 20576624]

[100]   Gupta S, Kapoor P, Chaudhary K, Gautam A, Kumar R, Raghava GPS. *In silico* approach for predicting toxicity of peptides and proteins. PLoS One 2013; 8(9): e73957.
[http://dx.doi.org/10.1371/journal.pone.0073957] [PMID: 24058508]

[101]   Saha S, Raghava GPS. AlgPred: prediction of allergenic proteins and mapping of IgE epitopes. Nucleic Acids Res 2006; 34(Web Server issue): W202-9.
[http://dx.doi.org/10.1093/nar/gkl343] [PMID: 16844994]

[102]   Muh HC, Tong JC, Tammi MT. AllerHunter: a SVM-pairwise system for assessment of allergenicity and allergic cross-reactivity in proteins. PLoS One 2009; 4(6): e5861.
[http://dx.doi.org/10.1371/journal.pone.0005861] [PMID: 19516900]

[103]   Dimitrov I, Flower DR, Doytchinova I. AllerTOP--a server for *in silico* prediction of allergens. BMC Bioinformatics 2013; 14 (Suppl. 6): S4.
[http://dx.doi.org/10.1186/1471-2105-14-S6-S4] [PMID: 23735058]

[104]   Dimitrov I, Naneva L, Doytchinova I, Bangov I, Allergen FP. AllergenFP: allergenicity prediction by descriptor fingerprints. Bioinformatics 2014; 30(6): 846-51.
[http://dx.doi.org/10.1093/bioinformatics/btt619] [PMID: 24167156]

[105]   Wang J, Zhang D, Li J. PREAL: Prediction of allergenic protein by maximum Relevance Minimum

Redundancy (mRMR) feature selection. BMC Syst Biol 2013; 7 (Suppl. 5): S9.
[http://dx.doi.org/10.1186/1752-0509-7-S5-S9] [PMID: 24565053]

[106] Gautam A, Chaudhary K, Singh S, *et al.* Hemolytik: a database of experimentally determined hemolytic and non-hemolytic peptides. Nucleic Acids Res 2014; 42(Database issue): D444-9.
[http://dx.doi.org/10.1093/nar/gkt1008] [PMID: 24174543]

[107] Kumar R, Chaudhary K, Sharma M, *et al.* AHTPDB: a comprehensive platform for analysis and presentation of antihypertensive peptides. Nucleic Acids Res 2015; 43(Database issue): D956-62.
[http://dx.doi.org/10.1093/nar/gku1141] [PMID: 25392419]

[108] Kumar R, Chaudhary K, Singh Chauhan J, *et al.* An *in silico* platform for predicting, screening and designing of antihypertensive peptides. Sci Rep 2015; 5: 12512.
[http://dx.doi.org/10.1038/srep12512] [PMID: 26213115]

[109] Magnan CN, Zeller M, Kayala MA, *et al.* High-throughput prediction of protein antigenicity using protein microarray data. Bioinformatics 2010; 26(23): 2936-43.
[http://dx.doi.org/10.1093/bioinformatics/btq551] [PMID: 20934990]

[110] Doytchinova IA, Flower DR. VaxiJen: a server for prediction of protective antigens, tumour antigens and subunit vaccines. BMC Bioinformatics 2007; 8: 4.
[http://dx.doi.org/10.1186/1471-2105-8-4] [PMID: 17207271]

[111] Trott O, Olson AJ. AutoDock Vina: Improving the speed and accuracy of docking with a new scoring function, efficient optimization, and multithreading, J. Comput Chem 31 2009. NA-NA.
[http://dx.doi.org/10.1002/jcc.21334]

[112] Morris GM, Huey R, Lindstrom W, *et al.* AutoDock4 and AutoDockTools4: Automated docking with selective receptor flexibility. J Comput Chem 2009; 30(16): 2785-91.
[http://dx.doi.org/10.1002/jcc.21256] [PMID: 19399780]

[113] Schneidman-Duhovny D, Inbar Y, Nussinov R, Wolfson HJ. PatchDock and SymmDock: servers for rigid and symmetric docking. Nucleic Acids Res 2005; 33(Web Server issue): W363-7.
[http://dx.doi.org/10.1093/nar/gki481] [PMID: 15980490]

[114] Kozakov D, Hall DR, Xia B, *et al.* The ClusPro web server for protein-protein docking. Nat Protoc 2017; 12(2): 255-78.
[http://dx.doi.org/10.1038/nprot.2016.169] [PMID: 28079879]

[115] Dominguez C, Boelens R, Bonvin AMJJ. HADDOCK: a protein-protein docking approach based on biochemical or biophysical information. J Am Chem Soc 2003; 125(7): 1731-7.
[http://dx.doi.org/10.1021/ja026939x] [PMID: 12580598]

[116] Lee H, Heo L, Lee MS, Seok C. GalaxyPepDock: a protein-peptide docking tool based on interaction similarity and energy optimization. Nucleic Acids Res 2015; 43(W1): W431-5.
[http://dx.doi.org/10.1093/nar/gkv495] [PMID: 25969449]

[117] Abraham MJ, Murtola T, Schulz R, *et al.* Gromacs: High performance molecular simulations through multi-level parallelism from laptops to supercomputers. SoftwareX 2015; 1–2: 19-25.
[http://dx.doi.org/10.1016/j.softx.2015.06.001]

[118] Krieger E, Vriend G. YASARA View - molecular graphics for all devices - from smartphones to workstations. Bioinformatics 2014; 30(20): 2981-2.
[http://dx.doi.org/10.1093/bioinformatics/btu426] [PMID: 24996895]

[119] Rizwan M, Naz A, Ahmad J, *et al.* VacSol: a high throughput in silico pipeline to predict potential therapeutic targets in prokaryotic pathogens using subtractive reverse vaccinology. BMC Bioinformatics 2017; 18(1): 106.
[http://dx.doi.org/10.1186/s12859-017-1540-0] [PMID: 28193166]

[120] Naz K, Naz A, Ashraf ST, *et al.* PanRV: Pangenome-reverse vaccinology approach for identifications of potential vaccine candidates in microbial pangenome. BMC Bioinformatics 2019; 20(1): 123.
[http://dx.doi.org/10.1186/s12859-019-2713-9] [PMID: 30871454]

[121] Nagpal G, Gupta S, Chaudhary K, *et al.* VaccineDA: Prediction, design and genome-wide screening of oligodeoxynucleotide-based vaccine adjuvants. Sci Rep 2015; 5: 12478.
[http://dx.doi.org/10.1038/srep12478] [PMID: 26212482]

[122] Zaheer T, Waseem M, Waqar W, *et al.* Anti-COVID-19 multi-epitope vaccine designs employing global viral genome sequences. PeerJ 2020; 8: e9541.
[http://dx.doi.org/10.7717/peerj.9541] [PMID: 32832263]

[123] Dar HA, Zaheer T, Shehroz M, *et al.* Immunoinformatics-aided design and evaluation of a potential multi-epitope vaccine against *Klebsiella pneumoniae.* Vaccines (Basel) 2019; 7(3): E88.
[http://dx.doi.org/10.3390/vaccines7030088] [PMID: 31409021]

[124] van der Burg SH, Visseren MJ, Brandt RM, Kast WM, Melief CJ. Immunogenicity of peptides bound to MHC class I molecules depends on the MHC-peptide complex stability. J Immunol 1996; 156(9): 3308-14.
[PMID: 8617954]

[125] Harndahl M, Rasmussen M, Roder G, *et al.* Peptide-MHC class I stability is a better predictor than peptide affinity of CTL immunogenicity. Eur J Immunol 2012; 42(6): 1405-16.
[http://dx.doi.org/10.1002/eji.201141774] [PMID: 22678897]

[126] Sidney J, Becart S, Zhou M, *et al.* Citrullination only infrequently impacts peptide binding to HLA class II MHC. PLoS One 2017; 12(5): e0177140.
[http://dx.doi.org/10.1371/journal.pone.0177140] [PMID: 28481943]

[127] Zarling AL, Polefrone JM, Evans AM, *et al.* Identification of class I MHC-associated phosphopeptides as targets for cancer immunotherapy. Proc Natl Acad Sci USA 2006; 103(40): 14889-94.
[http://dx.doi.org/10.1073/pnas.0604045103] [PMID: 17001009]

[128] Andersen MH, Bonfill JE, Neisig A, *et al.* Phosphorylated peptides can be transported by TAP molecules, presented by class I MHC molecules, and recognized by phosphopeptide-specific CTL. J Immunol 1999; 163(7): 3812-8.
[PMID: 10490979]

[129] Blom N, Gammeltoft S, Brunak S. Sequence and structure-based prediction of eukaryotic protein phosphorylation sites. J Mol Biol 1999; 294(5): 1351-62.
[http://dx.doi.org/10.1006/jmbi.1999.3310] [PMID: 10600390]

[130] Song J, Wang H, Wang J, *et al.* PhosphoPredict: A bioinformatics tool for prediction of human kinase-specific phosphorylation substrates and sites by integrating heterogeneous feature selection. Sci Rep 2017; 7(1): 6862.
[http://dx.doi.org/10.1038/s41598-017-07199-4] [PMID: 28761071]

[131] Steentoft C, Vakhrushev SY, Joshi HJ, *et al.* Precision mapping of the human O-GalNAc glycoproteome through SimpleCell technology. EMBO J 2013; 32(10): 1478-88.
[http://dx.doi.org/10.1038/emboj.2013.79] [PMID: 23584533]

[132] Merrifield RB. Solid phase peptide synthesis. I. The synthesis of a tetrapeptide. J Am Chem Soc 1963; 85: 2149-54.
[http://dx.doi.org/10.1021/ja00897a025]

[133] Behrendt R, White P, Offer J. Advances in Fmoc solid-phase peptide synthesis. J Pept Sci 2016; 22(1): 4-27.
[http://dx.doi.org/10.1002/psc.2836] [PMID: 26785684]

[134] Roose K, De Baets S, Schepens B, Saelens X. Hepatitis B core-based virus-like particles to present heterologous epitopes. Expert Rev Vaccines 2013; 12(2): 183-98.
[http://dx.doi.org/10.1586/erv.12.150] [PMID: 23414409]

[135] Whitmore L, Wallace BA. Protein secondary structure analyses from circular dichroism spectroscopy: methods and reference databases. Biopolymers 2008; 89(5): 392-400.
[http://dx.doi.org/10.1002/bip.20853] [PMID: 17896349]

[136] Goormaghtigh E, Cabiaux V, Ruysschaert JM. Determination of soluble and membrane protein structure by Fourier transform infrared spectroscopy. I. Assignments and model compounds. Subcell Biochem 1994; 23: 329-62.
[http://dx.doi.org/10.1007/978-1-4615-1863-1_8] [PMID: 7855877]

[137] Goormaghtigh E, Cabiaux V, Ruysschaert JM. Determination of soluble and membrane protein structure by Fourier transform infrared spectroscopy. III. Secondary structures. Subcell Biochem 1994; 23: 405-50.
[http://dx.doi.org/10.1007/978-1-4615-1863-1_10] [PMID: 7855879]

[138] Markwick PRL, Malliavin T, Nilges M. Structural biology by NMR: structure, dynamics, and interactions. PLOS Comput Biol 2008; 4(9): e1000168.
[http://dx.doi.org/10.1371/journal.pcbi.1000168] [PMID: 18818721]

[139] Schiffer M. Structure determination of proteins by X-ray diffraction.Biophys Tech Photosynth. Kluwer Academic Publishers 2006; pp. 317-24.
[http://dx.doi.org/10.1007/0-306-47960-5_19]

[140] Yang H, Kim DS. Peptide immunotherapy in vaccine development: From epitope to adjuvant. Adv Protein Chem Struct Biol. Academic Press Inc. 2015; pp. 1-14.
[http://dx.doi.org/10.1016/bs.apcsb.2015.03.001]

[141] Kiros TG, Levast B, Auray G, Strom S, Van Kessel J, Gerdts V. The importance of animal models in the development of vaccines. Innov Vaccinol From Des Through to Deliv Test. Springer Netherlands 2012; pp. 251-64.
[http://dx.doi.org/10.1007/978-94-007-4543-8_11]

[142] Nath BM, Schumann KE, Boyer JD. The chimpanzee and other non-human-primate models in HIV-1 vaccine research. Trends Microbiol 2000; 8(9): 426-31.
[http://dx.doi.org/10.1016/S0966-842X(00)01816-3] [PMID: 10989311]

[143] Gerdts V, Littel-van den Hurk Sv, Griebel PJ, Babiuk LA. Use of animal models in the development of human vaccines. Future Microbiol 2007; 2(6): 667-75.
[http://dx.doi.org/10.2217/17460913.2.6.667] [PMID: 18041907]

[144] MacLennan IC, Gray D. Antigen-driven selection of virgin and memory B cells. Immunol Rev 1986; 91: 61-85.
[http://dx.doi.org/10.1111/j.1600-065X.1986.tb01484.x] [PMID: 3089914]

[145] Garlapati S. Do we know the Th1/Th2/Th17 determinants of vaccine response? Expert Rev Vaccines 2012; 11(11): 1307-10.
[http://dx.doi.org/10.1586/erv.12.111] [PMID: 23249229]

[146] Milena S, Vladimir P, Vladimir G, *et al.* Detection of antibodies against lumpy skin disease virus by virus neutralization test and ELISA methods. Acta Vet Brno 2019; 69: 47-60.
[http://dx.doi.org/10.2478/acve-2019-0003]

[147] Yang W, Chen W, Huang J, *et al.* Generation, identification, and functional analysis of monoclonal antibodies against porcine epidemic diarrhea virus nucleocapsid. Appl Microbiol Biotechnol 2019; 103(9): 3705-14.
[http://dx.doi.org/10.1007/s00253-019-09702-5] [PMID: 30877355]

[148] Dashprakash M, Venkatesan G, Kumar A, *et al.* Prokaryotic expression, purification and evaluation of goatpox virus ORF117 protein as a diagnostic antigen in indirect ELISA to detect goatpox. Arch Virol 2019; 164(4): 1049-58.
[http://dx.doi.org/10.1007/s00705-019-04170-8] [PMID: 30778744]

[149] Heegaard ED, Rasksen CJ, Christensen J. Detection of parvovirus B19 NS1-specific antibodies by ELISA and western blotting employing recombinant NS1 protein as antigen. J Med Virol 2002; 67(3): 375-83.
[http://dx.doi.org/10.1002/jmv.10079] [PMID: 12116031]

[150]   Engvall E, Perlmann P. Enzyme-linked immunosorbent assay (ELISA). Quantitative assay of immunoglobulin G. Immunochemistry 1971; 8(9): 871-4.
[http://dx.doi.org/10.1016/0019-2791(71)90454-X] [PMID: 5135623]

[151]   Engvall E, Perlmann P. Enzyme-linked immunosorbent assay, Elisa. 3. Quantitation of specific antibodies by enzyme-labeled anti-immunoglobulin in antigen-coated tubes. J Immunol 1972; 109(1): 129-35.
[PMID: 4113792]

[152]   Gan SD, Patel KR. Enzyme immunoassay and enzyme-linked immunosorbent assay. J Invest Dermatol 2013; 133(9): e12.
[http://dx.doi.org/10.1038/jid.2013.287] [PMID: 23949770]

[153]   Van Weemen BK, Schuurs AHWM. Immunoassay using antigen-enzyme conjugates. FEBS Lett 1971; 15(3): 232-6.
[http://dx.doi.org/10.1016/0014-5793(71)80319-8] [PMID: 11945853]

[154]   Quinn CP, Semenova VA, Elie CM, *et al.* Specific, sensitive, and quantitative enzyme-linked immunosorbent assay for human immunoglobulin G antibodies to anthrax toxin protective antigen. Emerg Infect Dis 2002; 8(10): 1103-10.
[http://dx.doi.org/10.3201/eid0810.020380] [PMID: 12396924]

[155]   Dhanze H, Bhilegaonkar KN, Rawat S, *et al.* Development of recombinant nonstructural 1 protein based indirect enzyme linked immunosorbent assay for sero-surveillance of Japanese encephalitis in swine. J Virol Methods 2019; 272: 113705.
[http://dx.doi.org/10.1016/j.jviromet.2019.113705] [PMID: 31351167]

[156]   Gallagher S, Winston SE, Fuller SA, Hurrell JGR. Immunoblotting and immunodetection, Curr. Protoc. Immunol 2008.
[http://dx.doi.org/10.1002/0471142735.im0810s83]

[157]   Bi Y, Jin Z, Wang Y, *et al.* Identification of two distinct linear B cell epitopes of the matrix protein of the Newcastle disease virus vaccine strain LaSota. Viral Immunol 2019; 32(5): 221-9.
[http://dx.doi.org/10.1089/vim.2019.0007] [PMID: 31094659]

[158]   Salem R, Assem SK, Omar OA, *et al.* Expressing the immunodominant projection domain of infectious bursal disease virus fused to the fragment crystallizable of chicken IgY in yellow maize for a prospective edible vaccine. Mol Immunol 2020; 118: 132-41.
[http://dx.doi.org/10.1016/j.molimm.2019.12.015] [PMID: 31881424]

[159]   Lu H, Shao H, Chen H, *et al.* Identification of novel B cell epitopes in the fiber protein of serotype 8 Fowl adenovirus. AMB Express 2019; 9(1): 172.
[http://dx.doi.org/10.1186/s13568-019-0895-1] [PMID: 31673824]

[160]   Gauger PC, Vincent AL. Serum virus neutralization assay for detection and quantitation of serum neutralizing antibodies to influenza A virus in swine. Methods Mol Biol. Humana Press Inc. 2020; pp. 321-33.
[http://dx.doi.org/10.1007/978-1-0716-0346-8_23]

[161]   Xu K, Acharya P, Kong R, *et al.* Epitope-based vaccine design yields fusion peptide-directed antibodies that neutralize diverse strains of HIV-1. Nat Med 2018; 24(6): 857-67.
[http://dx.doi.org/10.1038/s41591-018-0042-6] [PMID: 29867235]

[162]   Dessy FJ, Giannini SL, Bougelet CA, *et al.* Correlation between direct ELISA, single epitope-based inhibition ELISA and pseudovirion-based neutralization assay for measuring anti-HPV-16 and anti-HPV-18 antibody response after vaccination with the AS04-adjuvanted HPV-16/18 cervical cancer vaccine. Hum Vaccin 2008; 4(6): 425-34.
[http://dx.doi.org/10.4161/hv.4.6.6912] [PMID: 18948732]

[163]   Sajid M, Kawde AN, Daud M. Designs, formats and applications of lateral flow assay: A literature review. J Saudi Chem Soc 2015; 19: 689-705.

[http://dx.doi.org/10.1016/j.jscs.2014.09.001]

[164] Liu J, Wang J, Li Z, *et al.* A lateral flow assay for the determination of human tetanus antibody in whole blood by using gold nanoparticle labeled tetanus antigen. Mikrochim Acta 2018; 185(2): 110.
[http://dx.doi.org/10.1007/s00604-017-2657-6] [PMID: 29594594]

[165] Zafar M, Shah MA, Shehzad A, *et al.* Characterization of the highly immunogenic VP2 protrusion domain as a diagnostic antigen for members of *Birnaviridae* family. Appl Microbiol Biotechnol 2020; 104(8): 3391-402.
[http://dx.doi.org/10.1007/s00253-020-10458-6] [PMID: 32088761]

[166] Wen T, Huang C, Shi FJ, *et al.* Development of a lateral flow immunoassay strip for rapid detection of IgG antibody against SARS-CoV-2 virus. Analyst (Lond) 2020; 145(15): 5345-52.
[http://dx.doi.org/10.1039/D0AN00629G] [PMID: 32568341]

[167] Hunt DF, Henderson RA, Shabanowitz J, *et al.* , Characterization of peptides bound to the class I MHC molecule HLA-A2.1 by mass spectrometry, Science (80-.). 1992; 255(1992): 1216-63.
[http://dx.doi.org/10.1126/science.1546328]

[168] Caron E, Kowalewski DJ, Chiek Koh C, Sturm T, Schuster H, Aebersold R. Analysis of major histocompatibility complex (MHC) immunopeptidomes using mass spectrometry. Mol Cell Proteomics 2015; 14(12): 3105-17.
[http://dx.doi.org/10.1074/mcp.O115.052431] [PMID: 26628741]

[169] Czerkinsky C, Andersson G, Ekre HP, Nilsson LÅ, Klareskog L, Ouchterlony O. Reverse ELISPOT assay for clonal analysis of cytokine production. I. Enumeration of gamma-interferon-secreting cells. J Immunol Methods 1988; 110(1): 29-36.
[http://dx.doi.org/10.1016/0022-1759(88)90079-8] [PMID: 3131436]

[170] Picker LJ, Singh MK, Zdraveski Z, *et al.* Direct demonstration of cytokine synthesis heterogeneity among human memory/effector T cells by flow cytometry. Blood 1995; 86(4): 1408-19.
[http://dx.doi.org/10.1182/blood.V86.4.1408.bloodjournal8641408] [PMID: 7632949]

[171] Nylander S, Kalies I, Brefeldin A. Brefeldin A, but not monensin, completely blocks CD69 expression on mouse lymphocytes: efficacy of inhibitors of protein secretion in protocols for intracellular cytokine staining by flow cytometry. J Immunol Methods 1999; 224(1-2): 69-76.
[http://dx.doi.org/10.1016/S0022-1759(99)00010-1] [PMID: 10357208]

[172] Prussin C, Metcalfe DD. Detection of intracytoplasmic cytokine using flow cytometry and directly conjugated anti-cytokine antibodies. J Immunol Methods 1995; 188(1): 117-28.
[http://dx.doi.org/10.1016/0022-1759(95)00209-X] [PMID: 8551029]

[173] Foster B, Prussin C, Liu F, Whitmire JK, Whitton JL. Detection of intracellular cytokines by flow cytometry. Curr. Protoc. Immunol 2007.
[http://dx.doi.org/10.1002/0471142735.im0624s78]

[174] Suni MA, Picker LJ, Maino VC. Detection of antigen-specific T cell cytokine expression in whole blood by flow cytometry. J Immunol Methods 1998; 212(1): 89-98.
[http://dx.doi.org/10.1016/S0022-1759(98)00004-0] [PMID: 9671156]

[175] Altman JD, Moss PAH, Goulder PJR, *et al.* Phenotypic analysis of antigen-specific T lymphocytes, Science. 1996; 274(1996): 94-6.
[http://dx.doi.org/10.1126/science.274.5284.94]

[176] Skinner PJ, Haase AT. *In situ* tetramer staining. J Immunol Methods 2002; 268(1): 29-34.
[http://dx.doi.org/10.1016/S0022-1759(02)00197-7] [PMID: 12213340]

CHAPTER 2

Immunoinformatics and its Role in Vaccine Development

Iqra Mehmood[1], Amna Bari[2], Sajjad Ahmad[3], Anam Naz[4], Farah Shahid[1], Usman Ali Ashfaq[1], Kishver Tusleem[5] and Muhammad Tahir ul Qamar[6,*]

[1] *Department of Bioinformatics and Biotechnology, Government College University Faisalabad, Faisalabad, Pakistan*

[2] *Hubei Key Laboratory of Agricultural Bioinformatics, College of Informatics, Huazhong Agricultural University, Wuhan, P. R. China*

[3] *Department of Health and Biological Sciences, Abasyn University, Peshawar, Pakistan*

[4] *Institute of Molecular Biology and Biotechnology, The University of Lahore, Lahore, Pakistan*

[5] *Sir Ganga Ram Hospital, Fatima Jinnah Medical University, Lahore, Pakistan*

[6] *College of Life Science and Technology, Guangxi University, Nanning, P. R. China*

Abstract: Immunoinformatics is currently an emerging field that has accelerated immunological research to a great extent. It is playing a significant role in antigen identification, immunodiagnostic development, and vaccine design. The arrival of genome sequencing with recent advancements in immunoinformatics has provided a lot of data that can be annotated using databases and tools to reduce the cost required for antibody and vaccine development, ultimately saving time, cost, and resources. The selection and identification of immunogenic regions from the pathogen genomes by computational methods play an important role in devising new hypotheses by a comprehensive examination of immunologic data composite, which is otherwise impossible to achieve by using traditional methods alone. Presently, many epitope-based vaccines, especially multi-epitope vaccines designed employing immunoinformatics approaches, are successfully trailed and being developed against pathogens. In this chapter, we provide an outline of the recent progress in the field of vaccinology and immunoinformatics, enlisted recent tools and databases available for epitopes prediction, validation, and vaccine design, and give a brief description of the role of immunoinformatics in vaccine design against recent COVID-19.

Keywords: Computationally designed vaccines, Immunoinformatics, Reverse vaccinology, Subtractive genomics.

* **Corresponding author Muhammad Tahir ul Qamar:** College of Life Science and Technology, Guangxi University, Nanning, P. R. China; Tel: +8615271915965; ORCID: 0000-0003-4832-4250; E-mail: m.tahirulqamar@hotmail.com

1. INTRODUCTION

Vaccination is an important approach that involves the designation of vaccines and its administration in the host to protect them from various diseases [1]. The immunization journey started with the discovery of a vaccine against smallpox by Edward Jenner in 1798 [2]. Later on, many vaccines have been developed, and some are still in the process against emerging diseases. Various conventional methods being used for vaccine development generally involved inactivated or live attenuated vaccines [3], nucleic acid vaccines [4], subunit vaccines [5], and virus-like particles [6], which are discussed separately in a later section of this chapter. All these methods are generally *in vitro* and have many complications and limitations associated with them. For example, inactivated vaccines, if not properly designed, can reactivate in the host and cause diseases. Additionally, some vaccines require boosters for working effectively at later stages because they are unable to generate a robust immune response in the first attempt [2, 7, 8].

In the recent decade, whole-genome sequencing of humans and other organism resulted in a huge amount of functional, epidemiological, and clinical data [9, 10]. This assembled information available in specialist repositories and databases enables the researchers to get deep insights into the mechanisms of human diseases and to understand host immune responses. Computational immunology or immunoinformatics deals with such rapidly growing immunological data [11 - 13]. Immunoinformatics is now a crucial constituent of modern immunology research and correlates with experimental immunology and computer science. It utilizes computational resources and methods to understand, generate, process, and propagate immunological information. Immunoinformatics came into being 90 years ago with the hypothetical demonstration of malarial epidemiology [11, 12]. This emerging field includes various databases and tools that manage the observation of immunologic records produced by tentative researchers and assist in introducing and representing new therapeutic targets [14, 15]. It seems to provide the platform to develop and progress immunological research in less time. Vaccine informatics successfully utilized various bioinformatics methods and applications to accommodate different locations of the preclinical, clinical, and licensed vaccine activities. Also, the advancement in immunology and molecular biology facilitates the improvement of epitope-based vaccines, which has become a way for further research of molecular vaccines [15 - 18]. Furthermore, reverse vaccinology can expand vaccine protocols, design, and production by predicting various protein-vaccine candidates within genomes using computational approaches [19]. Immunoinformatics allow the potential B and T cell epitope prediction within selected candidate proteins, which can elicit a proper immune response against the pathogen . It has provided ease in the analysis, vaccine design, immunization modeling, and evaluation of vaccine efficacy and safety [18, 20, 21].

In silico vaccine design, immunoproteomics, immunogenomics, and epitope prediction are some areas of Immunoinformatics. Recently, system biology methodologies have also been used to understand the various aspects and variable behavior of the complex immune system [22]. It constitutes the utilization of computational resources and methods to understand immunological evidence. It is a recent advancement in bioinformatics that uses numerous computational approaches to identify, understand, and predict a wide range of interactions among antigens and host immune system's receptors, including major histocompatibility complex (MHC) receptors also B and T cell receptors [23]. Immunoinformatics not merely aids in handling enormous data but also plays a crucial role in explaining new hypotheses associated with immune responses [24]. It is a potential and advantageous approach since conventional methods for vaccine development needs viruses to cultivate for understanding their binding patterns and to extract their antigenic proteins [25]. The term "Immunome" is used for all the data of proteins and genes in the immune system, excluding proteins and genes present in other cells except immune cells [26]. Immunoreactions that result from reactions between host cell proteins and antigens are referred to as immunome reactions, and the study of these reactions is called Immunomics [27]. Immunomics is a branch of knowledge that deals with the use of high throughput methods and techniques to determine and understand mechanisms involved in the immune system [28, 29]. In this chapter, we will briefly discuss the transformation of traditional vaccinology into modern vaccinology, recent advancements in immunoinformatics, important tools, and databases for prediction and validation of epitopes for vaccine design, and the role of immunoinformatics in COVID-19 vaccine development. It is anticipated that the information provided in this chapter will help researchers in choosing the most suitable tools and approaches for vaccine designs and immunological diagnostics development.

2. TRANSFORMATION OF VACCINOLOGY FROM CONVENTIONAL TO MODERN ERA

Previously, in the absence of effective therapies and preventive methods; infectious diseases such as measles, smallpox, diphtheria, rubella, chickenpox, and influenza were top listed child killers. Fortunately, these devastating diseases have diminished in various developed countries due to the widespread distribution and evolution of effective, safe, and affordable vaccines [30]. Vaccines can stop pathogens before they cause life-threatening damage, and millions of lives have been saved [31]. The immunization and vaccination story begins with Edward Jenner, who performed the world's first vaccination in 1798 by taking into account the pus cells obtained from cowpox lesions on the hands of milkmaid [32]. Edward Jenner published his work which became a definitive text in medicinal history: *Inquiry into the Causes and Effects of the Variolae Vaccine*. His statement "that the cow-pox defends the human structure from the smallpox infection" laid the basis of modern vaccinology [33]. With the start of the 19th Century, Jenner's vaccination formula quickly spread globally, supported by various governments due to its favorable effects and its ability to diminish the overwhelming effects of epidemics on the human population. These remarkable features show that our current ideas regarding vaccine development and therapy are strongly incorporated and deeply rooted in the science of immunology.

Later, vaccines against Anthrax and Swine erysipelas were developed in 1880 and 1881, respectively. Louis Pasteur developed a vaccine against rabies in 1885. Pasteur produced antitoxin against rabies by expanding the incubation period of rabies pathogen which acted as a post-infection antidote [34]. Pasteur defined vaccine as "suspension of live attenuated or inactivated microorganisms that when administered, induce immunity and help to prevent infectious diseases" [35]. In 1890, French veterinarian Henri Toussaint developed a vaccine against Fowl Cholera. In 1894, Serum therapy was introduced against diphtheria in Germany and France by Emile Roux and Emil von Behring, respectively. In 1895, serum therapy was used by Sclavo and Marchoux against Anthrax. Similarly, Friedrich Loffler utilized serum therapy to treat foot and mouth disease in cattle, and, later on, it was applied on a larger scale in Denmark [36]. Trials for vaccination against tuberculosis started in 1882 by Robert Koch and continued up to 1933. Due to allergic reactions and severe immune response, the use of vaccination against tuberculosis stopped in 1954, yet it gives no beneficial outcomes in animals and humans under trials [36]. Gaston Ramon introduced adjuvants in 1924 with the development of an anti-tetanus vaccine, also known as anatoxin. Anatoxin contained tetanus toxin and was preserved with formaldehyde or heat [37]. The efficacy of vaccines can be enhanced when used in combination with adjuvants of immunity such as aluminium hydroxide. In this way, Ramon created the first adjuvant vaccine which provided a potential benefit by enhancing the effect of the vaccine against targeted disease [36]. Thus, with the discovery of the vaccine by Edward Jenner, this journey continued to protect humans and animals from various pathogenic diseases. Still, many vaccines are under trial against previously and newly identified diseases.

Until the 20th century, many vaccines were developed using these traditional methods, but new approaches were needed to overcome the remaining pathogens and the limitations of traditional methods. Hence, significant progress was made with the development of new technologies, *e.g.*, recombinant DNA technology with the whole genome sequencing of microbes [38]. Recent advancements allowed researchers to move beyond Pasteur's rules by utilizing *in silico* approaches and genomic information for realistic vaccine development without culturing the specific organism under lab conditions. This new approach was termed "reverse vaccinology" [39]. The *Meningococcus B (MenB)* was the first pathogen subjected to the reverse vaccinology approach. This pathogen is responsible for 50% of meningococcal meningitis globally [40]. Meanwhile, reverse vaccinology has been utilized for a variety of pathogens. Eight genomes of *B. streptococcus* have been studied, directing/focusing on the expression of 312 surface proteins. However, vaccine development was based on 4 proteins which were conserved in all serotypes [41]. Protein-based vaccines were developed against antibiotic-resistant *Streptococcus pneumonia* and *Staphylococcus aureus* by utilizing a genome-based approach. By using the same approach, vaccine has also been formulated against *Chlamydia*. In recent times, genome power is supplemented with the capability to cross-examine the whole antigenic catalogue. This was accomplished by directly examining the bacterial surface antigens by estimating their amount and by determining their presence; or by utilizing the information of expressed antigens and by determining the ability of an antigen to induce an immune response in the course of infection, which is referred to as the "antigenome" [42]. Former technology earns benefits from the usage of mass spectrometry to quantify and identify bacterial surface antigens [43]. Partial digestion of bacterial surface proteins performed by using proteases and resulting peptides are analyzed by a spectrometer, which provided useful information regarding exposed surface proteins in terms of their quantity. Reverse vaccinology not only provides useful information for effective vaccine development but also made it possible to study antigen functions to understand the biology of pathogens. Considerable examples are the detection of *meningococcus* factor H binding protein and pili detection in gram-positive bacteria. The selection of defensive antigens followed by reverse vaccinology confirmed that defensive antigen was a part of high molecular weight pilus in group B *streptococcus* and mediate them in adhesion while colonization [44]. This initial observation led to the discovery of a pilus in cluster A *pneumococcus* and *streptococcus* revealed a distinctive pathogenesis mechanism in human pathogens. *Meningococcus* antigen GNA1870 was found to be bind with the human complement regulator factor H which confirmed that *meningococcus* downregulates the various pathways involved in complement activation and can survive in human blood. However, *meningococcus* was unable to grow in the blood of animal models such as rats and mice as the same protein was incapable to bind factor H [45, 46]. Furthermore, to screen out the best vaccine candidates reverse vaccinology utilizes the whole proteome of each pathogen. In this way, effective and successful vaccines can be developed, which is otherwise impossible or a difficult task. Besides, the comparison between conventional vaccinology and reverse vaccinology is depicted in Table **1**.

Table 1. Comparison of conventional and reverse vaccinology [47].

-	Conventional Vaccinology	Reverse Vaccinology
Antigens Available	10-25 detected by genetic and biochemical tools.	All genome encoding antigens are available.
Property of Antigens	Most plentiful antigens and the most immunogenic during disease, only from cultivable microorganisms.	All antigens are available whether they are immunogenic or not. Antigens of non-cultured microbes can also be determined.
Immunology of the Antigens	Extremely immunogenic antigens, often variable in sequence, because of immune selective pressure. Some can promote autoimmunity as they contain domains which are copy of self-proteins.	The most conserved protective antigens can be identified. Usually these are not the most immunogenic during infection. Also involves the comparison of newly found antigens with human genome and homologous proteins are removed.
Polysaccharide Antigens	A major target of traditional bacterial vaccines.	Reverse vaccinology cannot identify them but such operons that are involved in the biosynthesis of polysaccharides can be found. In this way, novel carbohydrates encoding antigens can be determined.
T –cell Epitope	Known epitopes limited to the known antigens.	Nearly all T cell epitopes are available. Overlapping peptides can be used to screen overall T cell immunity.

3. SUBTRACTIVE GENOMICS AND REVERSE VACCINOLOGY

The subtractive genomics approach examines the whole proteome/genome of pathogens by excluding the host homologous proteins for different purposes, for example, for the prediction of vaccine and drug targets, particularly for pathogens that are difficult to grow in the laboratory [48, 49]. In recent times, many studies have utilized the subtractive genomics approach on numerous pathogenic strains and successfully identified species-specific unique therapeutic targets [50 - 53]. Subtractive genomics aids in filtering the pan-genome of pathogens. In pan-genome filtration core genes, shared genes, and singletons can be identified. Core genes are those genes that are found in all strains of a species while shared genes are only present in some pathogenic strains. On the other hand, singletons are strain-specific and found only in one strain *i.e.*, each unique strain will have its unique singleton sequence [54]. Besides, subtractive genomics also identifies essential genes of pathogens. For this purpose, the database of essential genes (DEG) is usually used. This database contains all essential genes of bacterial and eukaryotic genomes [55]. This platform provides information regarding non-coding RNA and proteins produced from essential genes of prokaryotes and eukaryotes. For the prediction of their subcellular localization SurfG+, PSORTb or CELLO2GO can be used [56 - 58]. These tools and databases help to identify proteins localizing within cytoplasmic membranes, extracellular proteins, cell walls, and in the cytoplasm [58]. Cytoplasmic proteins are deliberated as potential drug targets due to the presence of hydrophobic pockets for drug binding and are involved in key metabolic pathways responsible for the survival of micro-organisms [59]. However, those proteins which are present on the membrane, putatively exposed to the surface (PSE) and secretory proteins can induce an immune response when coming in contact with host immune cells [60]. Paralogous/ redundant proteins can also be removed with subtractive genomics. Paralogous proteins share significant similarities and are removed to generate a set of non-paralogous pathogenic proteins. BLASTp and CD-HIT database can be used to remove redundant proteins [61]. Further tools and software for all the above-mentioned purposes are summarized in Table **2**.

Table 2. Tools, software and databases use in subtractive genomics approach.

To Remove Redundant/Paralogous Proteins	
BLASTp	https://blast.ncbi.nlm.nih.gov/Blast.cgi?PAGE=Proteins
MCL clustering algorithm	https://micans.org/mcl/
CD-HIT database	http://weizhong-lab.ucsd.edu/cdhit_suite/cgi-bin/index.cgi
For Filtering Pan-genome	
In-house scripts	https://github.com/eco-gaoshaom/in-house-scripts
PanRV	https://sourceforge.net/projects/panrv2/
BPGA	https://sourceforge.net/projects/bpgatool/
PGAWeb	http://pgaweb.vlcc.cn/
Panseq	https://lfz.corefacility.ca/panseq/
PanCGHweb	http://bamics2.cmbi.ru.nl/websoftware/pancgh/
ppsPCP	http://cbi.hzau.edu.cn/ppsPCP/
PANNOTATOR	http://pannotator.facom.ufu.br/
PanCake	https://drops.dagstuhl.de/opus/volltexte/2013/4231/
To Remove Host Homologues Proteins	
BLASTp	https://blast.ncbi.nlm.nih.gov/Blast.cgi?PAGE=Proteins
OrthoFinder	https://github.com/davidemms/OrthoFinder
To Identify Essential Genes of Pathogens	

(Table 2) cont.....

Database of Essential Genes (DEG)	http://www.essentialgene.org/
BLASTp	https://blast.ncbi.nlm.nih.gov/Blast.cgi?PAGE=Proteins
MP3	http://metagenomics.iiserb.ac.in/mp3/index.php
VirulentPred	http://www.bioinfo.icgeb.res.in/virulent/
VICMpred	http://crdd.osdd.net/raghava/vicmpred/
To Identify Protein Localization	
SurfG+	https://mulcyber.toulouse.inra.fr/projects/surfgplus
PSORTb v.3.0	https://www.psort.org/psortb/
CELLO v2.5	http://cello.life.nctu.edu.tw/
TMHMM v0.2 server	http://www.cbs.dtu.dk/services/TMHMM/
Mature epitope density server (MED)	https://med.compbio.sdu.dk/
Pathway Analysis	
KEGG (Kyoto encyclopedia of gene and genome)	https://www.genome.jp/kegg/
KAAS server	https://www.genome.jp/kegg/kaas/
Virulence Factor Determination	
VFDB (virulence factor database)	http://www.mgc.ac.cn/cgi-bin/VFs/v5/main.cgi
EVIRDB	https://ijpbs.net/abstract.php?article=NDQzOQ==
VFDB	http://www.mgc.ac.cn/VFs/main.htm
IMG4	http://img.jgi.doe.gov/w
The Pathogenicity Island Database (PAIDB)	http://www.paidb.re.kr/about_paidb.php
Virulence finder	http://cge.cbs.dtu.dk/services/VirulenceFinder/
Antibiotic resistance gene pipeline (ARG)	http://smile.hku.hk/SARGs
ResFinder	https://cge.cbs.dtu.dk/services/ResFinder/
Molecular Weight Analysis of Proteins	
ExPASy server	https://www.expasy.org/
Biochemistry-online	http://vitalonic.narod.ru/biochem/index_en.html
Peptide molecular weight calculator	https://www.genscript.com/tools/peptide-molecular-weight-calculator
Protein Information Resources (PIR)	https://proteininformationresource.org/pirwww/search/comp_mw.shtml

(Table 2) cont.....

Sequence Manipulation Suite (SMS)	http://www.bioinformatics.org/sms2/protein_mw.html
Protein Molecular weight	https://www.bioinformatics.org/sms/prot_mw.html
Peptide and Protein molecular weight calculator	https://www.aatbio.com/tools/calculate-peptide-and-protein-molecular-weight-mw

Subtractive genomics and reverse vaccinology (RV) are integrated in terms of vaccine design. The availability of pathogens genome and a variety of vigorous sequence examination tools allow computational prediction of vaccine candidates on a genome level [62]. Initially, RV was formulated to overcome main restrictions in common experimental methods. In the current post-genomic era, RV has become the usual method for vaccine development. Instead of the identification of defensive antigens in conventional approaches, RV utilizes the whole potential antigenic spectrum. In this way, vaccinologists got a chance to obtain a variety of vaccine candidates including antigens that failed to indicate either due to impossibility of *in vitro* culturing the pathogen or absent/poor expression. Pipelines involved in vaccine development are depicted in Fig. (**1**) starting from reverse vaccinology.

3.1. Epitope Based Vaccine Design

In general, an artificial functional vaccine may comprise of non-proteinaceous danger-signals and one or more B and T-cell epitopes. Moreover, it can be a natural antigen or an artificially designed poly-epitope vaccine followed by an adjuvant. However, functional vaccines are readily captured by the immune system and effectors involved in the immune system. Understanding the vigorous immune response against the pathogenic organism is an important consideration if, we have to deal with them for fruitful design and development of vaccines. Computational approaches are highly convenient to design effective vaccines, for both the improvement and discovery of existing and new vaccines respectively. Identification of undiscriminating MHC (major histocompatibility complex) receptors is important to design appropriate subunit vaccines. In this era, the preliminary studies by Rammensee *et al.* [63] and Parker *et al.* [64] have much importance in providing opportunities for further investigations. Based on experimental data, they introduced a workflow to identify the binding patterns of non-peptides to MHC I molecules and their relative strengths [64]. They also delivered the first catalogue of peptide motifs and MHC ligands [63]. Afterward, many methods were introduced to identify the MHC I [65, 66] and MHC II receptors [67, 68]. Furthermore, many comprehensive methods have been developed to study a variety of components, which play part in antigen processing *e.g.* TAP binders [69 - 71] and cleavage site [72, 73] prediction.

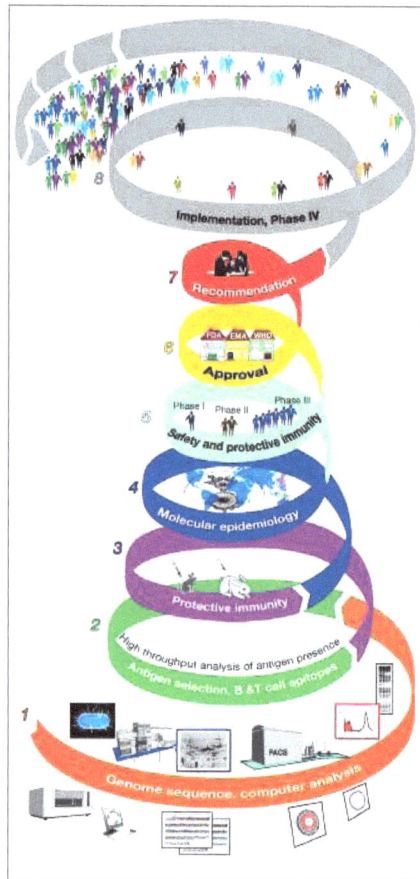

Fig. (1). Schematic diagram depicting pipelines involved in vaccine development starting from reverse vaccinology [47].

Various tools and databases have been developed that can predict continuous and conformational epitopes based on their physicochemical characterization. For example, BEPITOPE is directed to predict the patterns in proteins and the position of continuous B cell epitopes [74]. CEP (conformational epitope prediction) server is a conformational epitope prediction algorithm. It also predicts sequential epitopes and antigenic determinants [75]. Already existing methods are not good at predicting B-cell epitopes [76, 77]. Therefore, new tools are continuously being developed including machine-learning techniques [78].

In vaccine design, the discovery of discontinuous B-cell epitopes is a major challenge. Structure-based platforms provide improved information regarding discontinuous B cell epitopes and help in predicting these epitopes as compared to sequence-based methods which solely rely on sequences. Results of these methods help to map epitopes in both developments of diagnostic tools and in the rational vaccine design, which might play a central role in efficient epitopes prediction [79].

Cancer immunotherapy depends on the identification of peptide epitopes, which are usually obtained from self-antigenic proteins of the tumor. Thus structure based methods that can predict the association of antigenic peptides with MHC molecules can be helpful in vaccine design against cancer [80]. For T cell epitope-based vaccine design, combination of structure-based modeling and immunoinformatics was used [81]. The evaluation of epitope maintenance is extremely important. A high level of maintenance during vaccine design would facilitate wider protection through various pathogenic strains. During disease monitoring or diagnostic analysis, with the observation of an epitope, a specific marker is achieved with low protection. Under both conditions, an epitope maintenance evaluation tool is essential [82].

3.2. Multiple Epitope Sub-unit Vaccines

Immune response has a critical role in the host's protection against viral infections as well as tumors. Antigenic epitope provokes both humoral and cellular immune responses. Therefore, a multi-epitope based vaccine approach is a perfect idea for the successful prevention and treatment of viral infections (Fig. **2**) [50, 83 - 85]. Multi-epitope based vaccines consist of overlapping peptides and have a unique design as compared to single-epitope and conventional vaccines [86 - 88]. Various multi-epitope-based vaccines are being intended against a variety of pathogens. For example, *Staphylococcus aureus showed resistance to a variety of antibiotics including vancomycin and methicillin.* Methicillin-resistant *S. aureus* (MRSA) strains are associated with community-acquired and nosocomial infections [89]. Recently, *S. aureus* strains appeared from less to extreme vancomycin resistant which was considered as the last line treatment [90, 91]. For effective vaccine design against various *S. aureus strains,* multiple antigens should be selected. Moreover, for effective immune responses, the vaccine is usually accompanied by an appropriate adjuvant [92]. So, a combination of several interferon-gamma inducing, B cell and T cell epitopes, was designed as a novel multi-epitope vaccine to initiate vigorous innate, humoral, and cellular immune responses against the pathogenic *S. aureus* strains which caused antibiotic-resistant infections [93].

Similarly, in the case of cancer, multi-epitope based vaccine is a considerable therapeutic methodregarding immune system suppression and tumor escape. In general, two mechanisms are involved in improving anti-tumor immunity: increase in the elements of the immune system, which facilitate effective CD8+ cytolytic T lymphocytes (CTLs) and CD4+ helper T lymphocytes (HTLs) responses; or by preventing elements (such as T_{reg}) which overwhelm the immune response [95]. The key benefit of immunotherapy as compared to other cancer treatments is less undesirable and non-specific toxicities as well as specific suppression of cancerous cells [96]. Generally, cancer immunotherapies are divided into three groups: (1) Active immunotherapy, which contains vaccines that estimates immune response against tumor-associated antigens (TAAs)/ tumor-specific antigens (TSAs), (2) passive/ adaptive immunotherapy, which involves *ex vivo* administration of monoclonal antibodies for targeted degradation of cancer cell, and (3) non-specific immunotherapy, which induce immune response commonly by interleukins, interferons and cytokines administration.

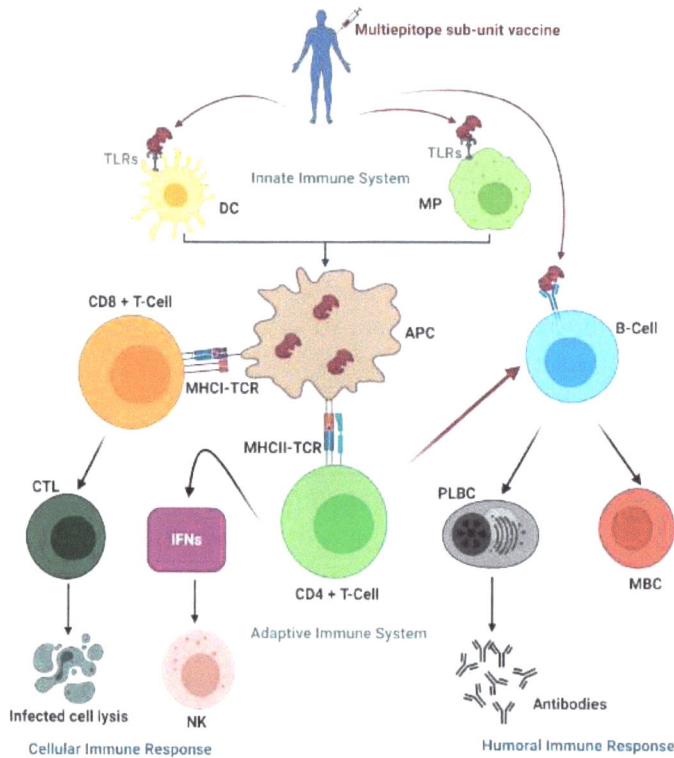

Fig. (2). Illustration of host immune response to multiple epitope sub-unit vaccine [94].

Existing limitations in the way of multi-epitope based vaccine design are a selection of suitable candidate antigens and their epitopes which induce immune response as well as the advancement in the effective delivery system. The estimation of appropriate antigenic epitopes of a directed protein by immunoinformatics methods is particularly essential to design a multi-epitope based vaccine [97, 98]. Multi-epitope based vaccines can be contemplated as an optimistic approach against viral infections and tumors. Immunoinformatics methods can facilitate the effective prediction and successful screening of suitable epitopes for effective multi-epitope based vaccine design. Both nanoparticles as well as VLPs (virus-like particles) can be used for the successful delivery of a multi-epitope vaccine which could increase its immunogenicity.

3.3. Antigen Based Vaccine Design

For the effective designation of vaccines against infective microbial agents, adhesins are vital components [99]. Against whooping cough whose causative agent is *Bordetella pertussis*, an acellular pertussis vaccine has been approved. One of the components of this vaccine is filamentous adhesin known as haemagglutinin [100]. Adhesins, as well as adhesin like proteins, can be identified experimentally but this process is laborious. Therefore, bioinformatics is the perfect master plan for the identification of adhesins. Homology-based methods are suitable for quick identification of homologous sequences, but their accuracy can decrease in the case of diverse sequences. To solve this problem, molecular composition and machine-learning techniques are being utilized to improve numerous algorithms for specified predictions and are dependent on the properties of sequences and do not rely on the homology of sequences. For instance, SPAAN (a software program for the prediction of adhesins and adhesin-like proteins using a neural network) [101] relies on neural networks and highly curated datasets which were ideally directed for compositional features. The final probability of an adhesin protein is obtained from the average of different likelihoods obtained from every five accurate set-ups. SPAAN has identified several adhesins of *Mycobacterium tuberculosis*, further confirmed by functional evidence *e.g.*, Rv1818c displayed adhesin-like properties [102, 103] and is further examined now for vaccine design [104]. Consequently, candidate adhesin proteins prediction considers as an auspicious approach for the identification of novel vaccine candidates. Examples of these vaccines and their targeted organisms are discussed in Table **3**.

4. RECENT ADVANCEMENTS IN IMMUNOINFORMATICS

With advancements in immunoinformatics, many tools and databases are now available to study interactions among host cell receptors and antigens. Host cell receptors involve B cell receptors, T cell receptors, and major histocompatibility complex (MHC) receptors. Immunoinformatics also consists of the use of various *in silico* approaches to design immunogenic proteins and multi epitopes.

Table 3. Immunoinformatics based vaccine design against bacteria, viruses, and other targets.

Immunoinformatics Based Vaccine design	Targeted Species	References
Immunoinformatics Based Vaccine Design Against Bacteria		
Epitope based vaccine	*Streptococcus pneumoniae, Neisseria meningitidis*, and *Haemophilus influenzae* Type b	[105]
Multi-epitope peptide vaccine	*Vibrio cholerae*	[106]
Multi-epitope oral vaccine	*Helicobacter pylori*	[107]
Multi-epitope peptide vaccine	*Staphylococcus aureus*	[93]
Multi-epitope vaccine	*Klebsiella pneumoniae*	[108]
Multi-epitope based vaccine	*Leptospirosis*	[109]
Multi-epitope based vaccine	*Elizabethkingia anophelis*	[110]
Peptide based vaccine	Multiple drug resistant bacteria *Providencia stuartii*	[111]
Multi-epitope vaccine	Enterotoxigenic *Bacteroides fragilis*	[112]
Epitope based vaccine	*Nocardia asteroides*	[113]
Multi-epitope subunit vaccine	*Mycobacterium tuberclosis*	[114]
Epitope based vaccine	*Campylobacter jejuni*	[115]
Multi-epitope based vaccine	*Shigella spp.*	[116]
Multi-epitope vaccine	*Brucellosis*	[117]
Multi-epitope driven subunit vaccine	*Fasciola gigantica*	[118]
Immunoinformatics Based Vaccine Design Against Viruses		
Multi-epitope based vaccine	Yellow fever virus	[119]
Multi-epitope peptide vaccine	Human norovirus (Norwalk virus)	[120]
Multi-epitope based vaccine	Kaposi Sarcoma	[121]
Peptide based vaccine	African horse sickness virus	[122]
Peptide vaccine	*Chikungunya* virus	[123]
Epitope based vaccine	Shrimp white spot syndrome virus (WSSV)	[124]
Multi-epitope vaccine	*Mayaro* virus	[125]
Epitope based peptide vaccine	*Chikungunya* Virus	[126]
Predication and *in silico* modeling of epitope-based peptide vaccine	virulent Newcastle disease viruses	[127]
Conserved subunit multi-epitope vaccine	HCV	[128]

(Table 3) cont.....

Immunoinformatics Based Vaccine design	Targeted Species	References
Epitope-based peptide vaccine design and active site prediction	*Oropouche* virus	[129]
Epitope-based peptide vaccine	*Ebola* virus	[130]
Multi-epitope subunit vaccine	HIV	[131]
Peptide vaccines	*Zika* virus	[132]
Multi-epitope based subunit vaccine	Respiratory Syncytial Virus	[133]
Epitope based peptide vaccine	Middle East Respiratory Syndrome Coronavirus (MERS-CoV)	[134]
Immunoinformatics Based Vaccine Design Against other Targets (Fungi, Parasites and Protozoans)		
Epitope-based peptide vaccine	*Madurella mycetomatis*	[135]
Multi-epitope peptide vaccine	*Leishmania infantum*	[136]
T cell epitope-based	*Coccidioidomycosis*	[137]
T Lymphocyte epitope-based	*Coccidioides*	[138]
Epitope-based peptide vaccine	*Candida Glabrata*	[139]
Epitope-based subunit vaccine	*Trypanosoma cruzi*	[140]
Multi-epitope based vaccine candidate	*Theileria parasites*	[141]
Multi-epitope vaccine candidate	*Toxoplasma gondii*	[142]
Epitope ensemble vaccine candidate	*Visceral leishmaniasis*	[143]
Epitope-based vaccine ensemble	*Trypanosomiasis*	[144]
Multi-epitope based vaccine design	*Sarcoptes scabiei paramyosin*	[145]

4.1. B and T Cell Epitope Prediction Tools and Databases

B cell epitope prediction plays a vital role in antibody production, vaccine development, and immunodiagnostic studies. The probability of discontinuous B-cell epitopes is 90%. However, they contain peptide chains of linear amino acids which come closer in 3D structure [146]. Epitome database is comprised of all known structures of an antigen-antibody complex. Semi-automated tools predict the probable antigenic interactions among all identified structures of an antigen-antibody complex. These tools collect information related to antigen-antibody interactions in the Epitome database [24]. Epitome gets updated twice a year with the updating of SCOP [147]. The Bicep is another database that contains information regarding B cell epitopes and also has integrated tools that allow the mapping of epitopes on known antigenic sequences [148]. Conformational epitope database (CED) contains the literature information regarding conformational epitopes determined by various methods *e.g* NMR, overlapping peptides, phage display, X-ray diffraction, and scanning mutagenesis [149].

Effective T cell response subjected to interaction among MHC ligand and T cell receptors and MHC binding peptides. Recent findings involve the identification and T cell epitopes mapping which leads to the formulation of effective vaccines [24, 150]. IEDB contains >88,382 protein- based epitopes and has specifically designed to predict biochemical, intrinsic, and chemical information regarding immune epitopes as well as the mechanism that how they interact with the other molecules of the immune system [151]. IEDB beta version has different tools for B and T cell epitope prediction [152]. FRED (framework for T cell epitope prediction) has methods to compare the efficacy of prediction methods which compare various distinct values and to process huge data [153]. The international ImMuno GeneTics information system (IMGT) has an adequate collection of T cell receptor (TR), Immunoglobulin (IG), and major histocompatibility complex (MHC) also other proteins of vertebrates and humans immune system. It has 15 online tools and 5 databases for 3D structure, sequences, and genome analysis [154]. The IMGT/HLA Database provides an expert database that has 1,612 HLA class II and 5,518 HLA class I alleles. It also shares the international IMGT project [155]. Tools and databases available for T and B cell predictions are depicted in Table **4**. These tools are categorized into two major groups which identify patterns based on sequence similarity and interactions with MHC molecules by using 3D models and structures. Computational methods have employed to predict MHC binding peptides include sequence motifs, binding matrices, decision trees, the structure of peptides, Artificial Neural Networks (ANNs), Hidden Markov Models (HMM), and Support Vector Machines (SVMs) [146].

4.2. Allergens Informatics

Allergy is an irritation caused by allergens (a common problem in all age groups). Allergens are glycoproteins that are recognized by immunoglobulin E (IgE) in the immune system. Allergens upon binding with IgE (antibodies present on the surface of basophils and mast cells) release inflammatory mediators that cause inflammation and allergy symptoms. Bioinformatics based tools and databases determine physiochemical as well as structural properties of allergen proteins [156]. These databases and tools can be utilized to find out the out cross-reactivity among known allergens. Both internal and external factors usually cause allergy. Specific allergens that evoke IgE antibodies usually lead to Type I hypersensitive reactions [157]. With the use of therapeutics and genetically modified (GM) food, it became necessary to predict allergen proteins. Food and Agriculture Organization's (FAO) and World Health Organization (WHO) rules suggested that upon comparing with known allergens, a protein with a window of 80 amino acids and which has at least six same contiguous amino acids is considered an allergen.

Allergens do not share mutual structural features thus allergen-based databases are utilized for allergenicity evaluation and to find out sequence similarity among various allergens [158]. In the allergens, T and B cell localization may not co-occur. Within an antigen, a difference exists in both kinds of epitopes. For example, B cell epitopes are conformational/linear and acknowledged by IgE antibodies and are present on the surface of the molecule accessible to antibodies. On the other hand, T cell epitopes are linear and scattered throughout the primary structure of the allergen. Furthermore, it is a difficult task to predict allergenicity in the case of B cell epitopes based on already predicted epitopes. Allergen Nomenclature database of the International Union of Immunological Societies (IUIS) has an allergen database (http://www.allergen.org/), last updated in October 2009 [159]. AllergenPro database comprises 2,434 allergens, for instance, allergens in rice microbes, plants, and animals [160]. The web server Allergome 4.0 provides a comprehensive depository of IgE-binding compound data and has 1,736 allergen sources. The real-time monitoring of the IgE sensitization module allows uploading of raw data from both *in vitro* and *in vivo* experiments [161]. The structural database of Allergenic Proteins (SDAP) is a web server that imparts cross-referenced access to the structure and sequence of allergenic proteins and has 1,478 isoallergens and allergens. This server works by considering the conserved properties of the side chains of amino acids [162]. Another allergen prediction technique was devised by Kong *et al.* [163] which determines the combination of two allergen motifs within a protein sequence. For the allergen dataset, they consider 575 proteins and for non-allergen, 700 sequences were considered from the given reference sequences [164]. A database was developed which contain all probable arrangements of two motifs from catalogue of allergen motifs by utilizing the 500 motifs each was 35 amino acids long. Zorzet *et al.* [165] also presented a computational method to classify the amino acid sequences into allergens and non-allergens. They recognized 91 food allergens from the SWALL database (SWISSPROT and TrEMBL) and numerous particular public sources of food allergy. AlgPred [166] uses a similarity-based method and structural vector machines (SVM) to scan and analyze 183 IgE epitopes against all proteins of the dataset. The server enables prediction by a hybrid mechanism to predict allergen using a combined approach (SVMc, MAST, gE epitope, and ARPs BLAST) [166]. In another study, Stadler *et al.* [158] used the MEME motif discovery tool for recognition of appropriate motif in allergen sequence. A sequence is considered to be an allergen sequence if the E value became 10^{-8} in pair-wise alignment or if the query sequence recognizes an allergen motif [158]. Furthermore, these sequences are compared with the WHO/FAO guidelines for allergenicity prediction of sequence in SWISSPROT. ALLERMATCH is a web tool that uses a sliding window approach to determine the potential allergenicity of proteins [167]. It is performed according to the recommendations of the WHO and FAO experts, but this method generates false-negative and false-positive results, so the WHO recommends that the results should be compared with other allergenicity assessment methods. APPEL (Allergen Protein Prediction E-Lab) tool is based on a statistical method and utilizes SVM (Structural Vector Machine) to determine novel allergen proteins. This tool accurately classified 99.9% of 6717 non-allergens and 93% of 229 allergens [168]. EVALLER is another web that uses the filtered length-adjusted allergen peptides (DFLAP) method to determine the potential allergen proteins. DFLAP takes out allergen sequence fragments of variable length and then employs SVM [169]. APPEL and EVALLER servers allocated all calmodulin-like/ calmodulins proteins as likely non-allergens [168, 169]. But conventional approaches are favored to find sequence similarities between reference proteins and known allergens and put the above-mentioned proteins in the allergen category. These proteins are considered as non-allergenic homologues to the polcalcin family. Tools containing structural and physical characteristics are suitable to determine cross-reacting proteins that otherwise cannot be detected by sequence similarity-based methods. Tools for allergen prediction are given in Table **4**.

5. IMMUNOINFORMATICS QUEST AGAINST SARS-COV-2

Coronavirus disease (COVID-19) is a deadly disease, caused by severe acute respiratory syndrome coronavirus 2 (SARS-CoV-2) and affected millions of people worldwide [170, 171]. COVID-19 is the worst infectious disease of recent decades and immunoinformatics is playing a vital role in the development of its vaccine. Being a time-saving technology, it is used to discover the candidate epitopes for efficient vaccine development against SARS-CoV-2 [172]. Recently, Sohail *et al.* compared the 65 different studies which reported candidate epitopes using different approaches to design an effective vaccine against COVID-19 [173]. They stated that the majority (>95%) of the experimentally determined epitopes for HLA-A*02:01, were identical to the epitopes predicted by immunoinformatics studies, which highlighted the importance and accuracy of immunoinformatics methods [173]. Immunoinformatics approaches are also being used in combination with the subtractive genomics approach to identify the potential epitopes from most virulent proteins of SARS-CoV-2 to develop multi - epitope subunit vaccines [174]. Besides, Immunoinformatics approaches coupled with additional *in silico* approaches were also used to propose a stable mRNA vaccine against COVID-19 [175]. Several other reviewed/published studies reporting vaccine constructs against SARS-CoV-2 using various immunoinformatics approaches are enlisted in Table **5**.

Table 4. Tools for T cell, B cell epitope and allergens prediction [24, 146].

Epitope	Tool name	URL
B cell Epitope Prediction	ABCpred	http://www.imtech.res.in/raghava/abcpred
	Bepipred	http://www.cbs.dtu.dk/services/BepiPred
	IMGT	http://www.imgt.org
	Bcepred	http://www.imtech.res.in/raghava/bcepred/
	DiscoTope	http://www.cbs.dtu.dk/services/DiscoTope/
	CEP	http://www.115.111.37.205/cgi-bin/cep.pl
	AgAbDb	http://www.115.111.37.206:8080/agabdb2/home.jsp
	MIMOP	franck.molina@cpbs.univ-montp1.fr
	MIMOX	http://www.immunet.cn/mimox/
	Pepitope	http://www.pepitope.tau.ac.il/
	3DEX	http://www.schreiber-abc.com/3dex/
	AntiJen	http://www.jenner.ac.uk/AntiJen
	CED	http://web.kuicr.kyoto-u.ac.jp/~ced
	ePitope	http://www.epitope-informatics.com/
	Mapitope	gershoni@tauex.tau.ac.il
	MEPS	http://www.caspur.it/meps
	Vaxijen	http://www.jenner.ac.uk/VaxiJen
	IEDB	http://www.immuneepitope.org
	COBEpro	http://www.scartch.proteomics.uci.edu
	FIMM	http://research.i2r.a-star.edu.sg/fimm
	SYFPEITHI	http://www.syfpeithi.de
	JenPep	http://www.jenner.ac.uk/jenpep
	EPIMHC	http://bio.dfci.harvard.edu/epimhc
	MHCBN	http://www.imtech.res.in/raghava/mhcbn
	Epitome	http://www.rostlab.org/services/epitome

(Table 4) cont.....

T cell Epitope Prediction	MMBPred	http://www.imtech.res.in/raghava/mmbpred/
	NetCTL	http://www.cbs.dtu.dk/services/NetCTL/
	NetMHC 3.0	http://www.cbs.dtu.dk/services/NetMHC/
	TAPPred	http://www.imtech.res.in/raghava/tappred/
	Pcleavage	http://www.imtech.res.in/raghava/pcleavage/
	ElliPro	http://www.tools.immuneepitope.org/tools/ElliPro
	MHCPred	http://www.ddg-pharmfac.net/mhcpred/MHCPred/
	Propred	http://www.imtech.res.in/raghava/propred1/
	EpiToolKit	http://www.epitoolkit.org
	Syfpeithi	http://www.syfpeithi.de
	IMGT	http://www.imgt.org
	IEDB	http://www.immuneepitope.org/
	EpiJen v 1.0	http://www.ddgharmfac.net/epijen/EpiJen/EpiJen.htm
	BLAST	http://www.ncbi.nlm.nih.gov/blast/Blast.cgi
	EpiMatrix	http://www.epivax.com/
	HLABIND	http;//www-bimas.cit.nih.gov/molbio/hla_bind
	Hepitope	http://www.hiv.lanl.gov/content/immunology/hepitopes/index.html
	ELF	http://www.hiv.lanl.gov/content/sequence/ELF/epitope_analyzer.html
	IMTECH	http://imtech.res.in/raghava/mhc
	Lib Score	http://libscore.ddbj.nig.ac.jp/cgi-bin/libscore/request.rb?lang=E
	MEME	http://meme.sdsc.edu/
	MHCBench	http://www.imtech.res.in/raghava/mhcbench
	Multipred	http://research.i2r.a-star.edu.sg/multipred
	NetMHCPan	http://www.cbs.dtu.dk/services/NetMHCpan
	nHLAPred	http://www.imtech.res.in/raghava/nhlapred
	PepDist	http://www.pepdist.cs.huji.ac.il/
	PREDEP	http://margalit.huji.ac.il/Teppred/mhc-bind/index.html
	RankPep	http://bio.dfci.harvard.edu/Tools/rankpep.html
	SVMHC	http://www-bs.informatik.uni-tuebingen.de/Services/SVMHC
	CED	http://www.immunet.cn/ced/log.html
	Epitome	http://www.rostlab.org/services/epitome/
	Bcipep	http://www.imtech.res.in/raghava/bcipep

(Table 4) cont.....

Allergens prediction	AlgPred	http://www.imtech.res.in/raghava/algpred
	Allermatch	http://www.allermatch.org
	APPEL	http://www.jing.cz3.nus.edu.sg/cgi-bin/APPEL
	EVALLER	http://www.slv.se/en-gb/Group1/ Food-Safety/e-Testing-of-protein-allergenicity/
	ALLERDB	http://sdmc.i2r.a-star.edu.sg/Templar/DB/Allergen
	Allergome	http://www.allergome.org
	WebAllergen	http://weballergen.bii.a-star.edu.sg/
	Database of IUIS	http://www.allergen.org
	SDAP	http://www.fermi.utmb.edu/SDAP/

Table 5. Vaccine design against SARS-CoV-2 using immunoinformatics approaches.

Study References	Target Proteins	Target Human Toll-like Receptors
Tahir ul Qamar *et al* **[176]**	Structural Proteins	TLR3
Ojha *et al* **[177]**	Non-structural Proteins	TLR3
Tahir ul Qamar *et al* **[94]**	ORF10, E, M, ORF6, ORF7a, ORF8, N	TLR3, TLR8
Ahmad *et al* **[178]**	Nsp8, 3CL^{pro}, S	TLR3, TLR4
Enayatkhani *et al* **[179]**	N, ORF3a, M	TLR4
Naz *et al* **[180]**	S	TLR2, TLR4
Dong *et al* **[181]**	ORF7a, ORF8, nsp9, nsp6, nsp3, endoRNAse, ORF3a, M, N	TLR3
Jakhar *et al* **[182]**	3CL^{pro}	TLR3
Chauhan *et al* **[183]**	Whole Proteome	TLR3
Sarkar *et al* **[184]**	N, M, ORF3a, S	TLR8
Kalita *et al* **[185]**	N, M, S	TLR3

6. LIMITATIONS OF IMMUNOINFORMATICS

There are some challenges in the way of immunoinformatics due to its multidisciplinary nature. The positive points of each of these disciplines therefore need to be integrated and combined to create flawless assimilation of workflow that would help in mathematical modeling of host and virus interactions [186]. Another drawback of immunoinformatics is that it does not address the influences of post-translational modifications (PTMs) *e.g.* glycosylation, methylation, and acetylation, *etc.* These PTMs play a vital role in recognizing T and B cell epitopes [187 - 191]. There is a need to work in immunoinformatics regarding PTMs as many viruses are known to modify the protein's function by post-translational modifications (PTMs) events that are important to cause significant effects [192]. Moreover, immunogenicity and antigenicity of v accine candidates is another considerable issue [193, 194]. In actual words, these limitations explain why artificially designed vaccines are far away from becoming a reality [195]. Besides, causes of vaccine failure are highly diverse ranging from variations in the hosts and pathogenic population. Next-generation sequencing (NGS) technology has altered the vaccinology era to diminish adverse reactions of vaccines and their failures by empowering pre-vaccination screenings of whole genome [196, 197].

7. FUTURE PERSPECTIVE AND CONCLUSIONS

By understanding the overall advantages of immunoinformatics, further evolution in this field requires effective benefits to find, proceed, and elucidate the data which allows its ease of access to researchers. In the next few years, interest in clustered distributed and computing systems will be enhanced for effective screening and analysis of a huge amount of data. With the outburst of variety, number, and refinement of tools and algorithms, the challenge is the integration of strengths and efficacies of approaches rather than their weak points. Various computational approaches have been described, which allow the modeling of the human immune system with different aspects. A variety of bioinformatics, immunological, and mathematical modeling systems together with effective computational methods permit the establishment of new models and allow systemic understanding of dynamics and structure at cellular and organism levels. Computational tools and databases help in the B cell and T cell epitopes prediction and allow us to study and determine antigenic and antibody interactions thus helpful in computer-based drug designing. By using these computational approaches, successful vaccine candidates can be determined, and many of them are in the preclinical and clinical phase against various life-threatening diseases. Immunoinformatics based approaches allow the study of interactions between an infective agent and host antibodies, thus help in the prediction of successful drug targets. In the case of SARS-CoV-2, immunoinformatics is playing an important part to determine highly antigenic viral protein (structural proteins). The epitopes of these antigenic peptides can be determined by using *in silico* approaches and candidate vaccines can be suggested. Moreover, the interaction of these vaccines with host cell receptors (TLRs and MHC class I and II molecules) can also be studied. Moreover, the effectiveness of adaptive and innate immune response elicited by a vaccine can also be determined, Hence, immunoinformatics is a time saving and cost-effective approach for vaccine design and discovery.

CONSENT FOR PUBLICATION

Not applicable.

CONFLICT OF INTEREST

The author declares no conflict of interest, financial or otherwise.

ACKNOWLEDGMENTS

Authors would like to acknowledge Guangxi University, Nanning, China for providing facilities during this study.

REFERENCES

[1]   Ada G. Overview of vaccines and vaccination. Mol Biotechnol 2005; 29(3): 255-72.
        [http://dx.doi.org/10.1385/MB:29:3:255] [PMID: 15767703]

[2]   Plotkin SA. History of vaccine development Springer Science & Business Media 2011.
        [http://dx.doi.org/10.1007/978-1-4419-1339-5]

[3]   Shan C, Muruato AE, Jagger BW, *et al.* A single-dose live-attenuated vaccine prevents Zika virus
        pregnancy transmission and testis damage. Nat Commun 2017; 8(1): 676.
        [http://dx.doi.org/10.1038/s41467-017-00737-8] [PMID: 28939807]

[4]   McKenzie BS, Corbett AJ, Brady JL, *et al.* Nucleic acid vaccines: tasks and tactics. Immunol Res
        2001; 24(3): 225-44.
        [http://dx.doi.org/10.1385/IR:24:3:225] [PMID: 11817323]

[5]   Moyle PM, Toth I. Modern subunit vaccines: development, components, and research opportunities.
        Chem Med Chem 2013; 8(3): 360-76.
        [http://dx.doi.org/10.1002/cmdc.201200487] [PMID: 23316023]

[6]   Noad R, Roy P. Virus-like particles as immunogens. Trends Microbiol 2003; 11(9): 438-44.
        [http://dx.doi.org/10.1016/S0966-842X(03)00208-7] [PMID: 13678860]

[7]   Gao Q, Bao L, Mao H, *et al.* Development of an inactivated vaccine candidate for SARS-CoV-2.
        Science 2020; 369(6499): 77-81.
        [http://dx.doi.org/10.1126/science.abc1932] [PMID: 32376603]

[8]   Karimkhanilouyi S, Ghorbian S. Nucleic acid vaccines for hepatitis B and C virus. Infect Genet Evol
        2019; 75: 103968.
        [http://dx.doi.org/10.1016/j.meegid.2019.103968] [PMID: 31325609]

[9]   Kazi A, Chuah C, Majeed ABA, Leow CH, Lim BH, Leow CY. Current progress of
        immunoinformatics approach harnessed for cellular- and antibody-dependent vaccine design. Pathog
        Glob Health 2018; 112(3): 123-31.
        [http://dx.doi.org/10.1080/20477724.2018.1446773] [PMID: 29528265]

[10]   Bahrami AA, Payandeh Z, Khalili S, Zakeri A, Bandehpour M. Immunoinformatics: *in silico*
        approaches and computational design of a multi-epitope, immunogenic protein. Int Rev Immunol
        2019; 38(6): 307-22.
        [http://dx.doi.org/10.1080/08830185.2019.1657426] [PMID: 31478759]

[11]   Tomar N, De RK. Immunoinformatics: an integrated scenario. Immunology 2010; 131(2): 153-68.
        [http://dx.doi.org/10.1111/j.1365-2567.2010.03330.x] [PMID: 20722763]

[12]   Korber B, LaBute M, Yusim K. Immunoinformatics comes of age. PLOS Comput Biol 2006; 2(6):
        e71.

[http://dx.doi.org/10.1371/journal.pcbi.0020071] [PMID: 16846250]

[13]　Naz A, Obaid A, Shahid F, Dar HA, Naz K, Ullah N, *et al*. Reverse vaccinology and drug target identification through pan-genomics. Pan-genomics: Applications, challenges, and future prospects. Elsevier 2020; pp. 317-33.
[http://dx.doi.org/10.1016/B978-0-12-817076-2.00016-0]

[14]　Backert L, Kohlbacher O. Immunoinformatics and epitope prediction in the age of genomic medicine. Genome Med 2015; 7(1): 119.
[http://dx.doi.org/10.1186/s13073-015-0245-0] [PMID: 26589500]

[15]　Jespersen MC, Peters B, Nielsen M, Marcatili P. BepiPred-2.0: improving sequence-based B-cell epitope prediction using conformational epitopes. Nucleic Acids Res 2017; 45(W1): W24-9.
[http://dx.doi.org/10.1093/nar/gkx346] [PMID: 28472356]

[16]　Moutaftsi M, Peters B, Pasquetto V, *et al*. A consensus epitope prediction approach identifies the breadth of murine T(CD8+)-cell responses to vaccinia virus. Nat Biotechnol 2006; 24(7): 817-9.
[http://dx.doi.org/10.1038/nbt1215] [PMID: 16767078]

[17]　Peters B, Nielsen M, Sette A. T cell epitope predictions. Annu Rev Immunol 2020; 38: 123-45.
[http://dx.doi.org/10.1146/annurev-immunol-082119-124838] [PMID: 32045313]

[18]　Sanchez-Trincado JL, Gomez-Perosanz M, Reche PA. Fundamentals and methods for T-and B-cell epitope prediction. J Immunol Res. 2017.

[19]　Vivona S, Gardy JL, Ramachandran S, *et al*. Computer-aided biotechnology: from immuno-informatics to reverse vaccinology. Trends Biotechnol 2008; 26(4): 190-200.
[http://dx.doi.org/10.1016/j.tibtech.2007.12.006] [PMID: 18291542]

[20]　Potocnakova L, Bhide M, Pulzova LB. An introduction to B-cell epitope mapping and *in silico* epitope prediction. J Immunol Res 2016.

[21]　El-Manzalawy Y, Honavar V. Recent advances in B-cell epitope prediction methods. Immunome Res 2010; 6(2) (Suppl. 2): S2.
[http://dx.doi.org/10.1186/1745-7580-6-S2-S2] [PMID: 21067544]

[22]　Gardy JL, Lynn DJ, Brinkman FS, Hancock RE. Enabling a systems biology approach to immunology: focus on innate immunity. Trends Immunol 2009; 30(6): 249-62.
[http://dx.doi.org/10.1016/j.it.2009.03.009] [PMID: 19428301]

[23]　Bock GR, Goode JA. Immunoinformatics: bioinformatic strategies for better understanding of immune function. . John Wiley & Sons 2004; pp. 23-55.

[24]　Tomar N, De RK. Immunoinformatics: a brief review. Immunoinformatics. Springer 2014; pp. 23-55.
[http://dx.doi.org/10.1007/978-1-4939-1115-8_3]

[25]　Naz A, Awan FM, Obaid A, *et al*. Identification of putative vaccine candidates against Helicobacter pylori exploiting exoproteome and secretome: a reverse vaccinology based approach. Infect Genet Evol 2015; 32: 280-91.
[http://dx.doi.org/10.1016/j.meegid.2015.03.027] [PMID: 25818402]

[26]　Ortutay C, Vihinen M. Immunome knowledge base (IKB): an integrated service for immunome research. BMC Immunol 2009; 10(1): 3.
[http://dx.doi.org/10.1186/1471-2172-10-3] [PMID: 19134210]

[27]　Sette A, Fleri W, Peters B, Sathiamurthy M, Bui H-H, Wilson S. A roadmap for the immunomics of category A-C pathogens. Immunity 2005; 22(2): 155-61.
[http://dx.doi.org/10.1016/j.immuni.2005.01.009] [PMID: 15773067]

[28]　Grainger DJ. Immunomics: principles and practice. IRTL 2004; 2: 1-6.

[29]　De Groot AS. Immunomics: discovering new targets for vaccines and therapeutics. Drug Discov Today 2006; 11(5-6): 203-9.
[http://dx.doi.org/10.1016/S1359-6446(05)03720-7] [PMID: 16580597]

[30]   Stern AM, Markel H. The history of vaccines and immunization: familiar patterns, new challenges. Health Aff (Millwood) 2005; 24(3): 611-21.
[http://dx.doi.org/10.1377/hlthaff.24.3.611] [PMID: 15886151]

[31]   Markel H. Taking shots: The modern miracle of vaccines. Medscape Pediatrics 2004; p. 6.

[32]   Baxby D. Jenner's smallpox vaccine: the riddle of vaccinia virus and its origin. Heinemann Educational Publishers 1981.

[33]   Jenner E. An inquiry into the causes and effects of the variolae vaccinae, a disease discovered in some of the western counties of England, particularly Gloucestershire, and known by the name of the cow pox: author; 1800. 1800.

[34]   Hansen B. America's first medical breakthrough: how popular excitement about a French rabies cure in 1885 raised new expectations for medical progress. Am Hist Rev 1998; 103(2): 373-418.
[http://dx.doi.org/10.2307/2649773] [PMID: 11620083]

[35]   Pickering LK, Baker CJ, Freed GL, *et al.* Immunization programs for infants, children, adolescents, and adults: clinical practice guidelines by the Infectious Diseases Society of America. Clin Infect Dis 2009; 49(6): 817-40.
[http://dx.doi.org/10.1086/605430] [PMID: 19659433]

[36]   Lombard M, Pastoret P-P, Moulin AM. A brief history of vaccines and vaccination. Rev Sci Tech 2007; 26(1): 29-48.
[http://dx.doi.org/10.20506/rst.26.1.1724] [PMID: 17633292]

[37]   Ramon G. Sur la toxine et sur l'anatoxine diphtheriques. Ann Inst Pasteur (Paris) 1924; 38(1).

[38]   Fleischmann RD, Adams MD, White O, *et al.* Whole-genome random sequencing and assembly of Haemophilus influenzae Rd. Science 1995; 269(5223): 496-512.
[http://dx.doi.org/10.1126/science.7542800] [PMID: 7542800]

[39]   Rappuoli R. Reverse vaccinology. Curr Opin Microbiol 2000; 3(5): 445-50.
[http://dx.doi.org/10.1016/S1369-5274(00)00119-3] [PMID: 11050440]

[40]   Masignani V, Pizza M, Moxon ER. The development of a vaccine against meningococcus B using reverse vaccinology. Front Immunol 2019; 10: 751.
[http://dx.doi.org/10.3389/fimmu.2019.00751] [PMID: 31040844]

[41]   Maione D, Margarit I, Rinaudo CD, *et al.* Identification of a universal Group B streptococcus vaccine by multiple genome screen. Science 2005; 309(5731): 148-50.
[http://dx.doi.org/10.1126/science.1109869] [PMID: 15994562]

[42]   Giefing C, Meinke AL, Hanner M, *et al.* Discovery of a novel class of highly conserved vaccine antigens using genomic scale antigenic fingerprinting of pneumococcus with human antibodies. J Exp Med 2008; 205(1): 117-31.
[http://dx.doi.org/10.1084/jem.20071168] [PMID: 18166586]

[43]   Rodríguez-Ortega MJ, Norais N, Bensi G, *et al.* Characterization and identification of vaccine candidate proteins through analysis of the group A Streptococcus surface proteome. Nat Biotechnol 2006; 24(2): 191-7.
[http://dx.doi.org/10.1038/nbt1179] [PMID: 16415855]

[44]   Lauer P, Rinaudo CD, Soriani M, *et al.* Genome analysis reveals pili in Group B Streptococcus. Science 2005; 309(5731): 105.
[http://dx.doi.org/10.1126/science.1111563] [PMID: 15994549]

[45]   Madico G, Welsch JA, Lewis LA, *et al.* The meningococcal vaccine candidate GNA1870 binds the complement regulatory protein factor H and enhances serum resistance. J Immunol 2006; 177(1): 501-10.
[http://dx.doi.org/10.4049/jimmunol.177.1.501] [PMID: 16785547]

[46]   Schneider MC, Prosser BE, Caesar JJ, *et al.* Neisseria meningitidis recruits factor H using protein

mimicry of host carbohydrates. Nature 2009; 458(7240): 890-3.
[http://dx.doi.org/10.1038/nature07769] [PMID: 19225461]

[47]   Sette A, Rappuoli R. Reverse vaccinology: developing vaccines in the era of genomics. Immunity 2010; 33(4): 530-41.
[http://dx.doi.org/10.1016/j.immuni.2010.09.017] [PMID: 21029963]

[48]   Sachdev K, Gupta MK. A comprehensive review of computational techniques for the prediction of drug side effects. Drug Dev Res 2020; 81(6): 650-70.
[http://dx.doi.org/10.1002/ddr.21669] [PMID: 32314424]

[49]   Chakrabarty RP, Alam ASMRU, Shill DK, Rahman A. Identification and qualitative characterization of new therapeutic targets in Stenotrophomonas maltophilia through *in silico* proteome exploration. Microb Pathog 2020; 149: 104293.
[http://dx.doi.org/10.1016/j.micpath.2020.104293] [PMID: 32531498]

[50]   Mondal SI, Ferdous S, Jewel NA, *et al.* Identification of potential drug targets by subtractive genome analysis of Escherichia coli O157:H7: an *in silico* approach. Adv Appl Bioinform Chem 2015; 8: 49-63.
[http://dx.doi.org/10.2147/AABC.S88522] [PMID: 26677339]

[51]   Hossain T, Kamruzzaman M, Choudhury TZ, Mahmood HN, Nabi A, Hosen M. Application of the subtractive genomics and molecular docking analysis for the identification of novel putative drug targets against *Salmonella enterica* subsp. *enterica serovar* Poona. BioMed Res Int 2017 . 2017; 3783714.

[52]   Sarangi AN, Aggarwal R, Rahman Q, Trivedi N. Subtractive genomics approach for *in silico* identification and characterization of novel drug targets in Neisseria Meningitides Serogroup B. J Comput Sci Syst Biol 2009; 2(5): 255-8.
[http://dx.doi.org/10.4172/jcsb.1000038]

[53]   Uddin R, Jamil F. Prioritization of potential drug targets against P. aeruginosa by core proteomic analysis using computational subtractive genomics and Protein-Protein interaction network. Comput Biol Chem 2018; 74: 115-22.
[http://dx.doi.org/10.1016/j.compbiolchem.2018.02.017] [PMID: 29587180]

[54]   Emms DM, Kelly S. OrthoFinder: solving fundamental biases in whole genome comparisons dramatically improves orthogroup inference accuracy. Genome Biol 2015; 16(1): 157.
[http://dx.doi.org/10.1186/s13059-015-0721-2] [PMID: 26243257]

[55]   Zhang R, Ou HY, Zhang CT. DEG: a database of essential genes. Nucleic Acids Res 2004; 32(Database issue) (Suppl. 1): D271-2.
[http://dx.doi.org/10.1093/nar/gkh024] [PMID: 14681410]

[56]   Pacheco LG, Slade SE, Seyffert N, *et al.* A combined approach for comparative exoproteome analysis of Corynebacterium pseudotuberculosis. BMC Microbiol 2011; 11(1): 12.
[http://dx.doi.org/10.1186/1471-2180-11-12] [PMID: 21241507]

[57]   Yu C-S, Cheng C-W, Su W-C, *et al.* CELLO2GO: a web server for protein subcellular Localization prediction with functional gene ontology annotation. PLoS One 2014; 9(6): e99368.
[http://dx.doi.org/10.1371/journal.pone.0099368] [PMID: 24911789]

[58]   Yu NY, Wagner JR, Laird MR, *et al.* PSORTb 3.0: improved protein subcellular localization prediction with refined localization subcategories and predictive capabilities for all prokaryotes. Bioinformatics 2010; 26(13): 1608-15.
[http://dx.doi.org/10.1093/bioinformatics/btq249] [PMID: 20472543]

[59]   Yang D-Q, Halaby M-J, Li Y, Hibma JC, Burn P. Cytoplasmic ATM protein kinase: an emerging therapeutic target for diabetes, cancer and neuronal degeneration. Drug Discov Today 2011; 16(7-8): 332-8.
[http://dx.doi.org/10.1016/j.drudis.2011.02.001] [PMID: 21315178]

[60] Barinov A, Loux V, Hammani A, *et al.* Prediction of surface exposed proteins in Streptococcus pyogenes, with a potential application to other Gram-positive bacteria. Proteomics 2009; 9(1): 61-73. [http://dx.doi.org/10.1002/pmic.200800195] [PMID: 19053137]

[61] Huang Y, Niu B, Gao Y, Fu L, Li W. CD-HIT Suite: a web server for clustering and comparing biological sequences. Bioinformatics 2010; 26(5): 680-2. [http://dx.doi.org/10.1093/bioinformatics/btq003] [PMID: 20053844]

[62] Rappuoli R, Bottomley MJ, D'Oro U, Finco O, De Gregorio E. Reverse vaccinology 2.0: Human immunology instructs vaccine antigen design. J Exp Med 2016; 213(4): 469-81. [http://dx.doi.org/10.1084/jem.20151960] [PMID: 27022144]

[63] Rammensee H-G, Friede T, Stevanoviíc S. MHC ligands and peptide motifs: first listing. Immunogenetics 1995; 41(4): 178-228. [http://dx.doi.org/10.1007/BF00172063] [PMID: 7890324]

[64] Parker KC, Bednarek MA, Coligan JE. Scheme for ranking potential HLA-A2 binding peptides based on independent binding of individual peptide side-chains. J Immunol 1994; 152(1): 163-75. [PMID: 8254189]

[65] Bhasin M, Raghava GP. A hybrid approach for predicting promiscuous MHC class I restricted T cell epitopes. J Biosci 2007; 32(1): 31-42. [http://dx.doi.org/10.1007/s12038-007-0004-5] [PMID: 17426378]

[66] Singh H, Raghava GP. ProPred1: prediction of promiscuous MHC Class-I binding sites. Bioinformatics 2003; 19(8): 1009-14. [http://dx.doi.org/10.1093/bioinformatics/btg108] [PMID: 12761064]

[67] Singh H, Raghava GP. ProPred: prediction of HLA-DR binding sites. Bioinformatics 2001; 17(12): 1236-7. [http://dx.doi.org/10.1093/bioinformatics/17.12.1236] [PMID: 11751237]

[68] Sturniolo T, Bono E, Ding J, *et al.* Generation of tissue-specific and promiscuous HLA ligand databases using DNA microarrays and virtual HLA class II matrices. Nat Biotechnol 1999; 17(6): 555-61. [http://dx.doi.org/10.1038/9858] [PMID: 10385319]

[69] Bhasin M, Raghava GP. Analysis and prediction of affinity of TAP binding peptides using cascade SVM. Protein Sci 2004; 13(3): 596-607. [http://dx.doi.org/10.1110/ps.03373104] [PMID: 14978300]

[70] Doytchinova IA, Flower DR. Class I T-cell epitope prediction: improvements using a combination of proteasome cleavage, TAP affinity, and MHC binding. Mol Immunol 2006; 43(13): 2037-44. [http://dx.doi.org/10.1016/j.molimm.2005.12.013] [PMID: 16524630]

[71] Tenzer S, Peters B, Bulik S, *et al.* Modeling the MHC class I pathway by combining predictions of proteasomal cleavage, TAP transport and MHC class I binding. Cell Mol Life Sci 2005; 62(9): 1025-37. [http://dx.doi.org/10.1007/s00018-005-4528-2] [PMID: 15868101]

[72] Bhasin M, Raghava GP. Pcleavage: an SVM based method for prediction of constitutive proteasome and immunoproteasome cleavage sites in antigenic sequences. Nucleic Acids Res 2005; 33(Web Server issue) (Suppl. 2): W202-7. [http://dx.doi.org/10.1093/nar/gki587] [PMID: 15988831]

[73] Keşmir C, Nussbaum AK, Schild H, Detours V, Brunak S. Prediction of proteasome cleavage motifs by neural networks. Protein Eng 2002; 15(4): 287-96. [http://dx.doi.org/10.1093/protein/15.4.287] [PMID: 11983929]

[74] Odorico M, Pellequer JL. BEPITOPE: predicting the location of continuous epitopes and patterns in proteins. J Mol Recognit 2003; 16(1): 20-2. [http://dx.doi.org/10.1002/jmr.602] [PMID: 12557235]

[75]   Kulkarni-Kale U, Bhosle S, Kolaskar AS. CEP: a conformational epitope prediction server. Nucleic Acids Res 2005; 33(Web Server issue) (Suppl. 2): W168-71.
[PMID: 15980448]

[76]   Saha S, Raghava GPS, Eds. BcePred: prediction of continuous B-cell epitopes in antigenic sequences using physico-chemical properties. International Conference on Artificial Immune Systems.
[http://dx.doi.org/10.1007/978-3-540-30220-9_16]

[77]   Blythe MJ, Flower DR. Benchmarking B cell epitope prediction: underperformance of existing methods. Protein Sci 2005; 14(1): 246-8.
[http://dx.doi.org/10.1110/ps.041059505] [PMID: 15576553]

[78]   Saha S, Raghava GPS. Prediction of continuous B-cell epitopes in an antigen using recurrent neural network. Proteins 2006; 65(1): 40-8.
[http://dx.doi.org/10.1002/prot.21078] [PMID: 16894596]

[79]   Haste Andersen P, Nielsen M, Lund O. Prediction of residues in discontinuous B-cell epitopes using protein 3D structures. Protein Sci 2006; 15(11): 2558-67.
[http://dx.doi.org/10.1110/ps.062405906] [PMID: 17001032]

[80]   Schiewe AJ, Haworth IS. Structure-based prediction of MHC-peptide association: algorithm comparison and application to cancer vaccine design. J Mol Graph Model 2007; 26(3): 667-75.
[http://dx.doi.org/10.1016/j.jmgm.2007.03.017] [PMID: 17493854]

[81]   Todman SJ, Halling-Brown MD, Davies MN, Flower DR, Kayikci M, Moss DS. Toward the atomistic simulation of T cell epitopes automated construction of MHC: peptide structures for free energy calculations. J Mol Graph Model 2008; 26(6): 957-61.
[http://dx.doi.org/10.1016/j.jmgm.2007.07.005] [PMID: 17766153]

[82]   Bui H-H, Sidney J, Li W, Fusseder N, Sette A. Development of an epitope conservancy analysis tool to facilitate the design of epitope-based diagnostics and vaccines. BMC Bioinformatics 2007; 8(1): 361.
[http://dx.doi.org/10.1186/1471-2105-8-361] [PMID: 17897458]

[83]   Buonaguro L, Consortium H. Developments in cancer vaccines for hepatocellular carcinoma. Cancer Immunol Immunother 2016; 65(1): 93-9.
[http://dx.doi.org/10.1007/s00262-015-1728-y] [PMID: 26093657]

[84]   Lu I-N, Farinelle S, Sausy A, Muller CP. Identification of a CD4 T-cell epitope in the hemagglutinin stalk domain of pandemic H1N1 influenza virus and its antigen-driven TCR usage signature in BALB/c mice. Cell Mol Immunol 2017; 14(6): 511-20.
[http://dx.doi.org/10.1038/cmi.2016.20] [PMID: 27157498]

[85]   Kuo T, Wang C, Badakhshan T, Chilukuri S, BenMohamed L. The challenges and opportunities for the development of a T-cell epitope-based herpes simplex vaccine. Vaccine 2014; 32(50): 6733-45.
[http://dx.doi.org/10.1016/j.vaccine.2014.10.002] [PMID: 25446827]

[86]   Jiang P, Cai Y, Chen J, *et al.* Evaluation of tandem Chlamydia trachomatis MOMP multi-epitopes vaccine in BALB/c mice model. Vaccine 2017; 35(23): 3096-103.
[http://dx.doi.org/10.1016/j.vaccine.2017.04.031] [PMID: 28456528]

[87]   Lennerz V, Gross S, Gallerani E, *et al.* Immunologic response to the survivin-derived multi-epitope vaccine EMD640744 in patients with advanced solid tumors. Cancer Immunol Immunother 2014; 63(4): 381-94.
[http://dx.doi.org/10.1007/s00262-013-1516-5] [PMID: 24487961]

[88]   Zhu S, Feng Y, Rao P, *et al.* Hepatitis B virus surface antigen as delivery vector can enhance Chlamydia trachomatis MOMP multi-epitope immune response in mice. Appl Microbiol Biotechnol 2014; 98(9): 4107-17.
[http://dx.doi.org/10.1007/s00253-014-5517-x] [PMID: 24458565]

[89]   Schaffer AC, Lee JC. Vaccination and passive immunisation against *Staphylococcus aureus*. Int J

Antimicrob Agents 2008; 32 (Suppl. 1): S71-8.
[http://dx.doi.org/10.1016/j.ijantimicag.2008.06.009] [PMID: 18757184]

[90]   Shekarabi M, Hajikhani B, Salimi Chirani A, Fazeli M, Goudarzi M. Molecular characterization of vancomycin-resistant *Staphylococcus aureus* strains isolated from clinical samples: A three year study in Tehran, Iran. PLoS One 2017; 12(8): e0183607.
[http://dx.doi.org/10.1371/journal.pone.0183607] [PMID: 28854219]

[91]   Cong Y, Yang S, Rao X. Vancomycin resistant *Staphylococcus aureus* infections: A review of case updating and clinical features. J Adv Res 2019; 21: 169-76.
[http://dx.doi.org/10.1016/j.jare.2019.10.005] [PMID: 32071785]

[92]   Adamczyk-Poplawska M, Markowicz S, Jagusztyn-Krynicka EK. Proteomics for development of vaccine. J Proteomics 2011; 74(12): 2596-616.
[http://dx.doi.org/10.1016/j.jprot.2011.01.019] [PMID: 21310271]

[93]   Hajighahramani N, Nezafat N, Eslami M, Negahdaripour M, Rahmatabadi SS, Ghasemi Y. Immunoinformatics analysis and *in silico* designing of a novel multi-epitope peptide vaccine against Staphylococcus aureus. Infect Genet Evol 2017; 48: 83-94.
[http://dx.doi.org/10.1016/j.meegid.2016.12.010] [PMID: 27989662]

[94]   Qamar M Tahir ul, A Rehman, K Tusleem, UA Ashfaq, M Qasim, X Zhu. Designing of a next generation multiepitope based vaccine (MEV) against SARS-COV-2: Immunoinformatics and *in silico* approaches. PLOS ONE 2020; 15(12)

[95]   Mellman I, Coukos G, Dranoff G. Cancer immunotherapy comes of age. Nature 2011; 480(7378): 480-9.
[http://dx.doi.org/10.1038/nature10673] [PMID: 22193102]

[96]   Nezafat N, Ghasemi Y, Javadi G, Khoshnoud MJ, Omidinia E. A novel multi-epitope peptide vaccine against cancer: an *in silico* approach. J Theor Biol 2014; 349: 121-34.
[http://dx.doi.org/10.1016/j.jtbi.2014.01.018] [PMID: 24512916]

[97]   Cherryholmes GA, Stanton SE, Disis ML. Current methods of epitope identification for cancer vaccine design. Vaccine 2015; 33(51): 7408-14.
[http://dx.doi.org/10.1016/j.vaccine.2015.06.116] [PMID: 26238725]

[98]   Yin D, Li L, Song X, *et al.* A novel multi-epitope recombined protein for diagnosis of human brucellosis. BMC Infect Dis 2016; 16(1): 219.
[http://dx.doi.org/10.1186/s12879-016-1552-9] [PMID: 27206475]

[99]   Wizemann TM, Adamou JE, Langermann S. Adhesins as targets for vaccine development. Emerg Infect Dis 1999; 5(3): 395-403.
[http://dx.doi.org/10.3201/eid0503.990310] [PMID: 10341176]

[100]   Colombi D, Oliveira ML, Campos IB, Monedero V, Pérez-Martinez G, Ho PL. Haemagglutination induced by Bordetella pertussis filamentous haemagglutinin adhesin (FHA) is inhibited by antibodies produced against FHA(430-873) fragment expressed in Lactobacillus casei. Curr Microbiol 2006; 53(6): 462-6.
[http://dx.doi.org/10.1007/s00284-005-0388-0] [PMID: 17106803]

[101]   Sachdeva G, Kumar K, Jain P, Ramachandran S. SPAAN: a software program for prediction of adhesins and adhesin-like proteins using neural networks. Bioinformatics 2005; 21(4): 483-91.
[http://dx.doi.org/10.1093/bioinformatics/bti028] [PMID: 15374866]

[102]   Delogu G, Pusceddu C, Bua A, Fadda G, Brennan MJ, Zanetti S. Rv1818c-encoded PE_PGRS protein of Mycobacterium tuberculosis is surface exposed and influences bacterial cell structure. Mol Microbiol 2004; 52(3): 725-33.
[http://dx.doi.org/10.1111/j.1365-2958.2004.04007.x] [PMID: 15101979]

[103]   Brennan MJ, Delogu G, Chen Y, *et al.* Evidence that mycobacterial PE_PGRS proteins are cell surface constituents that influence interactions with other cells. Infect Immun 2001; 69(12): 7326-33.

[http://dx.doi.org/10.1128/IAI.69.12.7326-7333.2001] [PMID: 11705904]

[104]   Chaitra MG, Shaila MS, Nayak R. Evaluation of T-cell responses to peptides with MHC class I-binding motifs derived from PE_PGRS 33 protein of Mycobacterium tuberculosis. J Med Microbiol 2007; 56(Pt 4): 466-74.
[http://dx.doi.org/10.1099/jmm.0.46928-0] [PMID: 17374885]

[105]   Zahroh H, Ma'rup A, Tambunan USF, Parikesit AA. Immunoinformatics approach in designing epitope-based vaccine against meningitis-inducing bacteria (Streptococcus pneumoniae, Neisseria meningitidis, and Haemophilus influenzae type b). Drug target insights 2016.

[106]   Nezafat N, Karimi Z, Eslami M, Mohkam M, Zandian S, Ghasemi Y. Designing an efficient multi-epitope peptide vaccine against *Vibrio choleraevia* combined immunoinformatics and protein interaction based approaches. Comput Biol Chem 2016; 62: 82-95.
[http://dx.doi.org/10.1016/j.compbiolchem.2016.04.006] [PMID: 27107181]

[107]   Nezafat N, Eslami M, Negahdaripour M, Rahbar MR, Ghasemi Y. Designing an efficient multi-epitope oral vaccine against *Helicobacter pylori* using immunoinformatics and structural vaccinology approaches. Mol Biosyst 2017; 13(4): 699-713.
[http://dx.doi.org/10.1039/C6MB00772D] [PMID: 28194462]

[108]   Dar HA, Zaheer T, Shehroz M, *et al.* Immunoinformatics-aided design and evaluation of a potential multi-epitope vaccine against *klebsiella pneumoniae.* Vaccines (basel) 2019; 7(3): 88.
[http://dx.doi.org/10.3390/vaccines7030088] [PMID: 31409021]

[109]   Validi M, Karkhah A, Prajapati VK, Nouri HR. Immuno-informatics based approaches to design a novel multi epitope-based vaccine for immune response reinforcement against Leptospirosis. Mol Immunol 2018; 104: 128-38.
[http://dx.doi.org/10.1016/j.molimm.2018.11.005] [PMID: 30448609]

[110]   Nain Z, Abdulla F, Rahman MM, *et al.* Proteome-wide screening for designing a multi-epitope vaccine against emerging pathogen *Elizabethkingia anophelis* using immunoinformatic approaches. J Biomol Struct Dyn 2020; 38(16): 4850-67.
[http://dx.doi.org/10.1080/07391102.2019.1692072] [PMID: 31709929]

[111]   Asad Y, Ahmad S, Rungrotmongkol T, Ranaghan KE, Azam SS. Immuno-informatics driven proteome-wide investigation revealed novel peptide-based vaccine targets against emerging multiple drug resistant Providencia stuartii. J Mol Graph Model 2018; 80: 238-50.
[http://dx.doi.org/10.1016/j.jmgm.2018.01.010] [PMID: 29414043]

[112]   Majid M, Andleeb S. Designing a multi-epitopic vaccine against the enterotoxigenic Bacteroides fragilis based on immunoinformatics approach. Sci Rep 2019; 9(1): 19780.
[http://dx.doi.org/10.1038/s41598-019-55613-w] [PMID: 31874963]

[113]   Patra P, Mondal N, Patra BC, Bhattacharya M. Epitope-based vaccine designing of nocardia asteroides targeting the virulence factor mce-family protein by immunoinformatics approach. Int J Pept Res Ther 2020; 26(2): 1165-76.
[http://dx.doi.org/10.1007/s10989-019-09921-4] [PMID: 32435172]

[114]   Chatterjee N, Ojha R, Khatoon N, Prajapati VK. Scrutinizing Mycobacterium tuberculosis membrane and secretory proteins to formulate multiepitope subunit vaccine against pulmonary tuberculosis by utilizing immunoinformatic approaches. Int J Biol Macromol 2018; 118(Pt A): 180-8.
[http://dx.doi.org/10.1016/j.ijbiomac.2018.06.080] [PMID: 29920369]

[115]   Gupta N, Khan F, Kumar A. Exploring highly antigenic protein of Campylobacter jejuni for designing epitope based vaccine: immunoinformatics approach. Int J Pept Res Ther 2019; 25(3): 1159-72.
[http://dx.doi.org/10.1007/s10989-018-9764-z]

[116]   Farhani I, Nezafat N, Mahmoodi S. Designing a novel multi-epitope peptide vaccine against pathogenic Shigella spp. based immunoinformatics approaches. Int J Pept Res Ther 2019; 25(2): 541-53.
[http://dx.doi.org/10.1007/s10989-018-9698-5]

[117] Saadi M, Karkhah A, Nouri HR. Development of a multi-epitope peptide vaccine inducing robust T cell responses against brucellosis using immunoinformatics based approaches. Infect Genet Evol 2017; 51: 227-34.
[http://dx.doi.org/10.1016/j.meegid.2017.04.009] [PMID: 28411163]

[118] Kalita P, Lyngdoh DL, Padhi AK, Shukla H, Tripathi T. Development of multi-epitope driven subunit vaccine against Fasciola gigantica using immunoinformatics approach. Int J Biol Macromol 2019; 138: 224-33.
[http://dx.doi.org/10.1016/j.ijbiomac.2019.07.024] [PMID: 31279880]

[119] Multi-epitope based vaccine against yellow fever virus applying immunoinformatics approaches. J Biomol Struct Dyn 2020; 1-17.
[PMID: 31854239]

[120] Azim KF, Hasan M, Hossain MN, *et al.* Immunoinformatics approaches for designing a novel multi epitope peptide vaccine against human norovirus (Norwalk virus). Infect Genet Evol 2019; 74: 103936.
[http://dx.doi.org/10.1016/j.meegid.2019.103936] [PMID: 31233780]

[121] Chauhan V, Rungta T, Goyal K, Singh MP. Designing a multi-epitope based vaccine to combat Kaposi Sarcoma utilizing immunoinformatics approach. Sci Rep 2019; 9(1): 2517.
[http://dx.doi.org/10.1038/s41598-019-39299-8] [PMID: 30792446]

[122] Abdelbagi M, Hassan T, Shihabeldin M, Bashir S, Ahmed E, Mohamed E, *et al.* Immunoinformatics prediction of peptide-based vaccine against african horse sickness virus. Immunome Res 2017; 13(135): 2.
[http://dx.doi.org/10.4172/1745-7580.1000135]

[123] Qamar M Tahir ul, A Bari, MM Adeel, A Maryam, UA Ashfaq, X Du , *et al.* Peptide vaccine against chikungunya virus: immuno-informatics combined with molecular docking approach. J Transl Med 2018; 16(1): 298.

[124] Momtaz F, Foysal J, Rahman M, Fotedar R. Design of epitope based vaccine against shrimp white spot syndrome virus (WSSV) by targeting the envelope proteins: an immunoinformatic approach. Turk J Fish Aquat Sci 2019; 19(2): 59-69.
[http://dx.doi.org/10.4194/1303-2712-v19_2_07]

[125] Khan S, Khan A, Rehman AU, *et al.* Immunoinformatics and structural vaccinology driven prediction of multi-epitope vaccine against Mayaro virus and validation through in-silico expression. Infect Genet Evol 2019; 73: 390-400.
[http://dx.doi.org/10.1016/j.meegid.2019.06.006] [PMID: 31173935]

[126] Anwar S, Mourosi JT, Khan MF, Hosen MJ. Prediction of epitope-based peptide vaccine against the chikungunya virus by immuno-informatics approach. Curr pharm biotechnol 2020; 21(4): 325-40.
[http://dx.doi.org/10.2174/1389201020666191112161743] [PMID: 31721709]

[127] Badawi MM, Alla AF, Alam SS, *et al.* Immunoinformatics predication and *in silico* modeling of epitope-based peptide vaccine against virulent Newcastle disease viruses. Am J Infect Dis Microbiol 2016; 4(3): 61-71.

[128] Ikram A, Zaheer T, Awan FM, *et al.* Exploring NS3/4A, NS5A and NS5B proteins to design conserved subunit multi-epitope vaccine against HCV utilizing immunoinformatics approaches. Sci Rep 2018; 8(1): 16107.
[http://dx.doi.org/10.1038/s41598-018-34254-5] [PMID: 30382118]

[129] Adhikari UK, Tayebi M, Rahman MM. Immunoinformatics approach for epitope-based peptide vaccine design and active site prediction against polyprotein of emerging oropouche virus. Journal of immunology research 2018.
[http://dx.doi.org/10.1155/2018/6718083]

[130] Khan MA, Hossain MU, Rakib-Uz-Zaman SM, Morshed MN. Epitope-based peptide vaccine design

and target site depiction against Ebola viruses: an immunoinformatics study. Scand J Immunol 2015; 82(1): 25-34.
[http://dx.doi.org/10.1111/sji.12302] [PMID: 25857850]

[131]   Pandey RK, Ojha R, Aathmanathan VS, Krishnan M, Prajapati VK. Immunoinformatics approaches to design a novel multi-epitope subunit vaccine against HIV infection. Vaccine 2018; 36(17): 2262-72.
[http://dx.doi.org/10.1016/j.vaccine.2018.03.042] [PMID: 29571972]

[132]   Mirza MU, Rafique S, Ali A, *et al.* Towards peptide vaccines against Zika virus: Immunoinformatics combined with molecular dynamics simulations to predict antigenic epitopes of Zika viral proteins. Sci Rep 2016; 6: 37313.
[http://dx.doi.org/10.1038/srep37313] [PMID: 27934901]

[133]   Qamar M Tahir ul, Z Shokat, I Muneer, UA Ashfaq, H Javed. Multiepitope-based subunit vaccine design and evaluation against respiratory syncytial virus using reverse vaccinology approach. Vaccines 2020; 8(2): 288.

[134]   Tahir Ul Qamar M, Saleem S, Ashfaq UA, Bari A, Anwar F, Alqahtani S. Epitope-based peptide vaccine design and target site depiction against Middle East Respiratory Syndrome Coronavirus: an immune-informatics study. J Transl Med 2019; 17(1): 362.
[http://dx.doi.org/10.1186/s12967-019-2116-8] [PMID: 31703698]

[135]   Mohammed AA, ALnaby AM, Sabeel SM, *et al.* Epitope-based peptide vaccine against fructose-bisphosphate aldolase of *madurella mycetomatis* using immunoinformatics approaches. Bioinform biol insights 2018; 12: 1177932218809703.
[http://dx.doi.org/10.1177/1177932218809703] [PMID: 30542244]

[136]   Vakili B, Eslami M, Hatam GR, *et al.* Immunoinformatics-aided design of a potential multi-epitope peptide vaccine against Leishmania infantum. Int J Biol Macromol 2018; 120(Pt A): 1127-39.
[http://dx.doi.org/10.1016/j.ijbiomac.2018.08.125] [PMID: 30172806]

[137]   Hurtgen BJ, Hung C-Y, Ostroff GR, Levitz SM, Cole GT. Construction and evaluation of a novel recombinant T cell epitope-based vaccine against Coccidioidomycosis. Infect Immun 2012; 80(11): 3960-74.
[http://dx.doi.org/10.1128/IAI.00566-12] [PMID: 22949556]

[138]   Hurtgen BJ, Hung C-Y. Rational design of T lymphocyte epitope-based vaccines against Coccidioides infection. Vaccines for Invasive Fungal Infections. Springer 2017; pp. 45-64.

[139]   Elhasan LM, Hassan MB, Elhassan RM, Abdelrhman F, Salih EA, Hassan AI, *et al.* Epitope-based peptide vaccine design against fructose bisphosphate aldolase of candida glabrata: an immunomics approach bioRxiv 2020.
[http://dx.doi.org/10.1101/2020.07.03.180430]

[140]   Khatoon N, Ojha R, Mishra A, Prajapati VK. Examination of antigenic proteins of *Trypanosoma cruzi* to fabricate an epitope-based subunit vaccine by exploiting epitope mapping mechanism. Vaccine 2018; 36(42): 6290-300.
[http://dx.doi.org/10.1016/j.vaccine.2018.09.004] [PMID: 30217522]

[141]   Kar PP, Srivastava A. Immuno-informatics analysis to identify novel vaccine candidates and design of a multi-epitope based vaccine candidate against Theileria parasites. Front Immunol 2018; 9: 2213.
[http://dx.doi.org/10.3389/fimmu.2018.02213] [PMID: 30374343]

[142]   Hajissa K, Zakaria R, Suppian R, Mohamed Z. Immunogenicity of multiepitope vaccine candidate against *toxoplasma gondii* infection in balb/c mice. Iran j parasitol 2018; 13(2): 215-24.
[PMID: 30069205]

[143]   Singh G, Pritam M, Banerjee M, Singh AK, Singh SP. Genome based screening of epitope ensemble vaccine candidates against dreadful visceral leishmaniasis using immunoinformatics approach. Microb Pathog 2019; 136: 103704.
[http://dx.doi.org/10.1016/j.micpath.2019.103704] [PMID: 31479726]

[144]   Michel-Todó L, Bigey P, Reche PA, Pinazo M-J, Gascón J, Alonso-Padilla J. Design of an epitope-based vaccine ensemble for animal trypanosomiasis by computational methods. Vaccines (basel) 2020; 8(1): 130.
[http://dx.doi.org/10.3390/vaccines8010130] [PMID: 32188062]

[145]   Naz S, Ahmad S, Walton S, Abbasi SW. Multi-epitope based vaccine design against Sarcoptes scabiei paramyosin using immunoinformatics approach. J Mol Liq 2020; 319: 114105.
[http://dx.doi.org/10.1016/j.molliq.2020.114105]

[146]   Evans MC. Recent advances in immunoinformatics: application of *in silico* tools to drug development. Curr Opin Drug Discov Devel 2008; 11(2): 233-41.
[PMID: 18283611]

[147]   Schlessinger A, Ofran Y, Yachdav G, Rost B. Epitome: database of structure-inferred antigenic epitopes. Nucleic Acids Res 2006; 34(Database issue) (Suppl. 1): D777-80.
[http://dx.doi.org/10.1093/nar/gkj053] [PMID: 16381978]

[148]   Saha S, Bhasin M, Raghava GP. Bcipep: a database of B-cell epitopes. BMC Genomics 2005; 6(1): 79.
[http://dx.doi.org/10.1186/1471-2164-6-79] [PMID: 15921533]

[149]   Huang J, Honda W. CED: a conformational epitope database. BMC Immunol 2006; 7(1): 7.
[http://dx.doi.org/10.1186/1471-2172-7-7] [PMID: 16603068]

[150]   Rammensee H, Bachmann J, Emmerich NPN, Bachor OA, Stevanović S. SYFPEITHI: database for MHC ligands and peptide motifs. Immunogenetics 1999; 50(3-4): 213-9.
[http://dx.doi.org/10.1007/s002510050595] [PMID: 10602881]

[151]   Sathiamurthy M, Peters B, Bui H-H, *et al.* An ontology for immune epitopes: application to the design of a broad scope database of immune reactivities. Immunome Res 2005; 1(1): 2.
[http://dx.doi.org/10.1186/1745-7580-1-2] [PMID: 16305755]

[152]   Peters B, Sidney J, Bourne P, *et al.* The immune epitope database and analysis resource: from vision to blueprint. PLoS Biol 2005; 3(3): e91.
[http://dx.doi.org/10.1371/journal.pbio.0030091] [PMID: 15760272]

[153]   Feldhahn M, Dönnes P, Thiel P, Kohlbacher O. FRED--a framework for T-cell epitope detection. Bioinformatics 2009; 25(20): 2758-9.
[http://dx.doi.org/10.1093/bioinformatics/btp409] [PMID: 19578173]

[154]   Lefranc M-P, Giudicelli V, Ginestoux C, *et al.* IMGT, the international ImMunoGeneTics information system. Nucleic Acids Res 2009; 37(Database issue) (Suppl. 1): D1006-12.
[http://dx.doi.org/10.1093/nar/gkn838] [PMID: 18978023]

[155]   Robinson J, Halliwell JA, McWilliam H, Lopez R, Parham P, Marsh SG. The IMGT/HLA database. Nucleic Acids Res 2013; 41(Database issue): D1222-7.
[PMID: 23080122]

[156]   Pomés A. Relevant B cell epitopes in allergic disease. Int Arch Allergy Immunol 2010; 152(1): 1-11.
[http://dx.doi.org/10.1159/000260078] [PMID: 19940500]

[157]   Kuby J, Goldsby RA, Kindt TJ, Osborne BA. Kuby immunology. WH Freeman 2006.

[158]   Stadler MB, Stadler BM. Allergenicity prediction by protein sequence. FASEB J 2003; 17(9): 1141-3.
[http://dx.doi.org/10.1096/fj.02-1052fje] [PMID: 12709401]

[159]   Jappe U. Nomenklatur der Allergene. Hautarzt 2018; 69(1): 90-1.
[http://dx.doi.org/10.1007/s00105-017-4103-4] [PMID: 29260247]

[160]   Kim C, Kwon S, Lee G, *et al.* A database for allergenic proteins and tools for allergenicity prediction. Bioinformation 2009; 3(8): 344-5.
[http://dx.doi.org/10.6026/97320630003344] [PMID: 19707297]

[161] Mari A, Scala E, Palazzo P, Ridolfi S, Zennaro D, Carabella G. Bioinformatics applied to allergy: allergen databases, from collecting sequence information to data integration. The Allergome platform as a model. Cell Immunol 2006; 244(2): 97-100.
[http://dx.doi.org/10.1016/j.cellimm.2007.02.012] [PMID: 17434469]

[162] Ivanciuc O, Schein CH, Braun W. SDAP: database and computational tools for allergenic proteins. Nucleic Acids Res 2003; 31(1): 359-62.
[http://dx.doi.org/10.1093/nar/gkg010] [PMID: 12520022]

[163] Kong W, Tan TS, Tham L, Choo KW. Improved prediction of allergenicity by combination of multiple sequence motifs. In Silico Biol 2007; 7(1): 77-86.
[PMID: 17688432]

[164] Björklund ÅK, Soeria-Atmadja D, Zorzet A, Hammerling U, Gustafsson MG. Supervised identification of allergen-representative peptides for *in silico* detection of potentially allergenic proteins. Bioinformatics 2005; 21(1): 39-50.
[http://dx.doi.org/10.1093/bioinformatics/bth477] [PMID: 15319257]

[165] Zorzet A, Gustafsson M, Hammerling U. Prediction of food protein allergenicity: a bioinformatic learning systems approach. In Silico Biol 2002; 2(4): 525-34.
[PMID: 12611632]

[166] Saha S, Raghava GP. AlgPred: prediction of allergenic proteins and mapping of IgE epitopes. Nucleic Acids Res 2006; 34(Web Server issue) (Suppl. 2): W202-9.
[http://dx.doi.org/10.1093/nar/gkl343] [PMID: 16844994]

[167] Fiers MW, Kleter GA, Nijland H, Peijnenburg AA, Nap JP, van Ham RC. Allermatch, a webtool for the prediction of potential allergenicity according to current FAO/WHO Codex alimentarius guidelines. BMC Bioinformatics 2004; 5(1): 133.
[http://dx.doi.org/10.1186/1471-2105-5-133] [PMID: 15373946]

[168] Martinez Barrio A, Soeria-Atmadja D, Nistér A, Gustafsson MG, Hammerling U, Bongcam-Rudloff E. EVALLER: a web server for *in silico* assessment of potential protein allergenicity. Nucleic Acids Res 2007; 35(Web Server issue) (Suppl. 2): W694-700.
[PMID: 17537818]

[169] Soeria-Atmadja D, Lundell T, Gustafsson MG, Hammerling U. Computational detection of allergenic proteins attains a new level of accuracy with *in silico* variable-length peptide extraction and machine learning. Nucleic Acids Res 2006; 34(13): 3779-93.
[http://dx.doi.org/10.1093/nar/gkl467] [PMID: 16977698]

[170] Alamri MA. ul Qamar MT, Mirza MU, Alqahtani SM, Froeyen M, Chen L-L. Discovery of human coronaviruses pan-papain-like protease inhibitors using computational approaches. J Pharm Anal 2020.
[http://dx.doi.org/10.1016/j.jpha.2020.08.012]

[171] Muhseen ZT, Hameed AR, Al-Hasani HMH, Tahir Ul Qamar M, Li G. Promising terpenes as SARS-CoV-2 spike receptor-binding domain (RBD) attachment inhibitors to the human ACE2 receptor: Integrated computational approach. J Mol Liq 2020; 320: 114493.
[http://dx.doi.org/10.1016/j.molliq.2020.114493] [PMID: 33041407]

[172] Rehman HM, Mirza MU, Ahmad MA, *et al.* A putative prophylactic solution for COVID-19: Development of novel multiepitope vaccine candidate against SARS-COV-2 by comprehensive immunoinformatic and molecular modelling approach. Biology (Basel) 2020; 9(9): 296.
[http://dx.doi.org/10.3390/biology9090296] [PMID: 32962156]

[173] Sohail MS, Ahmed SF, Quadeer AA, McKay M. *In silico* T cell epitope identification for SARS-Co--2: Progress and perspectives. SSRN 2020.
[http://dx.doi.org/10.2139/ssrn.3720371]

[174] Abdelmageed MI, Abdelmoneim AH, Mustafa MI, Elfadol NM, Murshed NS, Shantier SW, *et al.*

Design of a multiepitope-based peptide vaccine against the e protein of human covid-19: an immunoinformatics approach. BioMed Research International 2020.

[175] Ahammad I, Lira SS. Designing a novel mRNA vaccine against SARS-CoV-2: An immunoinformatics approach. Int J Biol Macromol 2020; 162: 820-37.
[http://dx.doi.org/10.1016/j.ijbiomac.2020.06.213] [PMID: 32599237]

[176] Qamar M Tahir ul, F Shahid, S Aslam, UA Ashfaq, S Aslam, I Fatima, *et al.* Reverse vaccinology assisted designing of multiepitope-based subunit vaccine against SARS-CoV-2. Infectious diseases of poverty 2020; 9(1): 1-14.

[177] Ojha R, Gupta N, Naik B, *et al.* High throughput and comprehensive approach to develop multiepitope vaccine against minacious COVID-19. Eur J Pharm Sci 2020; 151: 105375.
[http://dx.doi.org/10.1016/j.ejps.2020.105375] [PMID: 32417398]

[178] Ahmad S, Navid A, Farid R, *et al.* Design of a novel multi epitope-based vaccine for pandemic coronavirus disease (covid-19) by vaccinomics and probable prevention strategy against avenging zoonotics. Eur j pharm sci 2020; 151: 105387.
[http://dx.doi.org/10.1016/j.ejps.2020.105387] [PMID: 32454128]

[179] Enayatkhani M, Hasaniazad M, Faezi S, Guklani H, Davoodian P, Ahmadi N, *et al.* Reverse vaccinology approach to design a novel multi-epitope vaccine candidate against COVID-19: an *in silico* study. J Biomol Struct Dyn 2020; 39(8): 1-16.
[http://dx.doi.org/10.1080/07391102.2020.1857843] [PMID: 32295479]

[180] Naz A, Shahid F, Butt TT, Awan FM, Ali A, Malik A. Designing multi-epitope vaccines to combat emerging coronavirus disease 2019 (COVID-19) by employing immuno-informatics approach. Front Immunol 2020; 11: 1663.
[http://dx.doi.org/10.3389/fimmu.2020.01663] [PMID: 32754160]

[181] Dong R, Chu Z, Yu F, Zha Y. Contriving multi-epitope subunit of vaccine for COVID-19: immunoinformatics approaches. Front Immunol 2020; 11: 1784.
[http://dx.doi.org/10.3389/fimmu.2020.01784] [PMID: 32849643]

[182] Jakhar R, Kaushik S, Gakhar SK. 3CL hydrolase-based multiepitope peptide vaccine against SARS-CoV-2 using immunoinformatics. J Med Virol 2020; 92(10): 2114-23.
[http://dx.doi.org/10.1002/jmv.25993] [PMID: 32379348]

[183] Chauhan V, Rungta T, Rawat M, Goyal K, Gupta Y, Singh MP. Excavating SARS-coronavirus 2 genome for epitope-based subunit vaccine synthesis using immunoinformatics approach. J Cell Physiol 2021; 236(2): 1131-47.
[http://dx.doi.org/10.1002/jcp.29923] [PMID: 32643158]

[184] Sarkar B, Ullah MA, Johora FT, Taniya MA, Araf Y. Immunoinformatics-guided designing of epitope-based subunit vaccines against the SARS Coronavirus-2 (SARS-CoV-2). Immunobiology 2020; 225(3): 151955.
[http://dx.doi.org/10.1016/j.imbio.2020.151955] [PMID: 32517882]

[185] Kalita P, Padhi AK, Zhang KYJ, Tripathi T. Design of a peptide-based subunit vaccine against novel coronavirus SARS-CoV-2. Microb Pathog 2020; 145: 104236.
[http://dx.doi.org/10.1016/j.micpath.2020.104236] [PMID: 32376359]

[186] Pappalardo F, Zhang P, Halling-Brown M, *et al.* Computational simulations of the immune system for personalized medicine: state of the art and challenges. Curr Pharmacogenomics Person Med 2008; 6(4): 260-71. [Formerly Current Pharmacogenomics].
[http://dx.doi.org/10.2174/187569208786733839]

[187] Hudrisier D, Riond J, Mazarguil H, Gairin JE. Pleiotropic effects of post-translational modifications on the fate of viral glycopeptides as cytotoxic T cell epitopes. J Biol Chem 2001; 276(41): 38255-60.
[http://dx.doi.org/10.1074/jbc.M105974200] [PMID: 11479317]

[188] Haque MA, Hawes JW, Blum JS. Cysteinylation of MHC class II ligands: peptide endocytosis and

reduction within APC influences T cell recognition. J Immunol 2001; 166(7): 4543-51.
[http://dx.doi.org/10.4049/jimmunol.166.7.4543] [PMID: 11254711]

[189] Lisowska E. The role of glycosylation in protein antigenic properties. Cell Mol Life Sci 2002; 59(3): 445-55.
[http://dx.doi.org/10.1007/s00018-002-8437-3] [PMID: 11964123]

[190] Li H, Xu C-F, Blais S, *et al*. Proximal glycans outside of the epitopes regulate the presentation of HIV-1 envelope gp120 helper epitopes. J Immunol 2009; 182(10): 6369-78.
[http://dx.doi.org/10.4049/jimmunol.0804287] [PMID: 19414790]

[191] Petersen J, Purcell AW, Rossjohn J. Post-translationally modified T cell epitopes: immune recognition and immunotherapy. J Mol Med (Berl) 2009; 87(11): 1045-51.
[http://dx.doi.org/10.1007/s00109-009-0526-4] [PMID: 19763524]

[192] Chattopadhyay S, Bagchi P, Dutta D, Mukherjee A, Kobayashi N, Chawlasarkar M. Computational identification of post-translational modification sites and functional families reveal possible moonlighting role of rotaviral proteins. Bioinformation 2010; 4(10): 448-51.
[http://dx.doi.org/10.6026/97320630004448] [PMID: 20975908]

[193] Wang W, Lu B, Zhou H, *et al*. Glycosylation at 158N of the hemagglutinin protein and receptor binding specificity synergistically affect the antigenicity and immunogenicity of a live attenuated H5N1 A/Vietnam/1203/2004 vaccine virus in ferrets. J Virol 2010; 84(13): 6570-7.
[http://dx.doi.org/10.1128/JVI.00221-10] [PMID: 20427525]

[194] Flower DR, Doytchinova IA. Immunoinformatics and the prediction of immunogenicity. Appl Bioinformatics 2002; 1(4): 167-76.
[PMID: 15130835]

[195] Van Regenmortel MH. What is a B-cell epitope? Epitope Mapping Protocols. Springer 2009; pp. 3-20.
[http://dx.doi.org/10.1007/978-1-59745-450-6_1]

[196] Nelson CS, Herold BC, Permar SR. A new era in cytomegalovirus vaccinology: considerations for rational design of next-generation vaccines to prevent congenital cytomegalovirus infection. NPJ Vaccines 2018; 3(1): 38.
[http://dx.doi.org/10.1038/s41541-018-0074-4] [PMID: 30275984]

[197] Dhiman N, Smith DI, Poland GA. Next-generation sequencing: a transformative tool for vaccinology. Expert Rev Vaccines 2009; 8(8): 963-7.
[http://dx.doi.org/10.1586/erv.09.67] [PMID: 19627178]

CHAPTER 3

Computational Toolbox for Analysis of Protein Thermostability

Syed Farhat Ali[1,*]

[1] *School of Life Sciences, Forman Christian College (A Chartered University), Lahore, Pakistan*

Abstract: Thermostable proteins have many applications. They have utility in food and beverages, paper and pulp industries, animal feed production, laundry and detergent, and molecular biology and diagnostics. Many factors contribute to protein thermostability. These include covalent and non-covalent interactions, protein folding and conformation, and other thermodynamic factors. Although the available protein structures have been increasing over time, the increase in available protein sequences is overwhelmingly enormous. Also, structure determination can be a challenging job and many proteins are difficult to crystallize. This has resulted in a sequence-structure gap. The use of computer-assisted structure prediction has helped in filling this gap. There are many *in silico* strategies and methods available to predict and analyze protein structure and properties. These can be used to determine protein stability and are quite useful for *in silico* protein analysis to improve function and thermostability.

Keywords: *In silico*, Molecular dynamics, Mutagenesis, Protein thermostability, Rational design.

1. INTRODUCTION

Proteins are remarkable biomolecules. By utilizing a limited set of monomers (amino acids), the resulting variety of structure and function is extraordinary. Proteins are central to cellular processes, including catalysis, formation of cellular structures, storage, transport, communication, energy metabolism, movement, and flow of biological information.

Thermophiles are organisms that grow at temperatures exceeding 50°C. Hyperthermophiles grow at even higher temperatures. Many of them grow at temperatures close to the boiling point of water. Many industrial processes involving enzymatic reactions take place at high temperatures. So, enzymes from

* **Corresponding author Syed Farhat Ali:** School of Life Sciences, Forman Christian College (A Chartered University), Lahore, Pakistan; Tel: +92-42-99231581; E-mail: farhatali@fccollege.edu.pk

Muhammad Sarwar Khan (Ed)

mesophiles are not suitable for these high temperature-requiring processes. In order to improve the efficiency of industrial processes, there is a continuing need for enzymes with better properties. Thermophilic enzymes meet this requirement as they are robust and can tolerate high temperatures [1]. Enzymes have applications in foods, detergents, textiles, leather, starch, and other industries. Moreover, protein engineering strategies have been useful in improving the properties of existing enzymes and tailoring them to improve process efficacy [2]. Some important applications of enzymes for industrial processes are shown in Table **1** [2, 3].

Table 1. Applications of some important enzymes.

Enzymes	Application
Amylase	Baking, paper and pulp, starch processing
Starch debranching enzymes	Starch processing, food industries
Cellulase	Plant biomass treatment, paper and pulp, biofuel production
Invertase	Food industry
Lactase	Food industry, lactose-free dairy products
Lipase and esterase	Food industry, detergents
Pectinase	Food, juice production
Peroxidase	Textile
Protease	Meat processing, detergents, leather and textile, dairy products
DNA polymerase	Polymerase chain reaction (PCR), gene cloning, diagnosis, molecular biology applications
Chitinase	Pharmaceutical products
Laccase	Bioremediation, detoxification
Amidases	Pharmaceutical products
Phytase	Feed and food industries

2. BASIS OF PROTEIN THERMOSTABILITY

During past years, much work has been done to understand protein thermostability. Many thermophilic proteins have been characterized and their structures determined. The proteins from thermophiles are not very different from their mesophilic counterparts in terms of their function. It has been reported that the thermostability of these proteins depends on various types of interactions, including hydrogen bonds, hydrophobic interaction, van der Waals interactions, and ionic interactions [4].

2.1. Electrostatic Interactions

A greater number of charged residues are found on the surface of thermostable proteins [4]. Electrostatic interactions are important for protein stability. Optimization of electrostatic interactions can result in protein stabilization while maintaining its activity [5]. Displacement of water molecules from the enzyme active site can also result in enhancing its activity. Glucanases have water molecules on their active sites, which are removed upon substrate binding. Engineering glucanases to displace water molecules during catalysis can enhance enzyme activity [6].

Electrostatic interactions of charged amino acid residues and the dielectric response of the protein are important factors in governing its thermostability. Proteins from thermophiles have been found to contain a greater number of charged residues on their surface as compared to those from mesophiles. This increasing trend of dielectric constant from mesophilic to thermophilic proteins modulates the thermal stability [7]. It has also been shown that optimizing electrostatic interactions by increasing the number of salt bridges is an important factor for high-temperature tolerance of proteins. The glutamate dehydrogenase from *Pyrococcus furiosus* has a large network of ion pairs formed by 18 charged residues [8]. Similarly, the enzyme lumazine synthase from *Aquifex* has a greater number of ions pairs as compared to corresponding mesophilic enzymes [9]. Also, the DNA polymerase from *Pyrobaculum calidifontis* (*Pca*) was shown to contain a greater number of salt bridges as compared to the ones found in *E. coli* (*Eco*) DNA polymerase [10]. *Pca* DNA polymerase contains a total of 242 charged amino acids. On the other hand, this number in *Eco* DNA polymerase is 193. A comparison of structures of these DNA polymerases is given in Fig. (1), demonstrating the association of electrostatic interactions with protein thermostability.

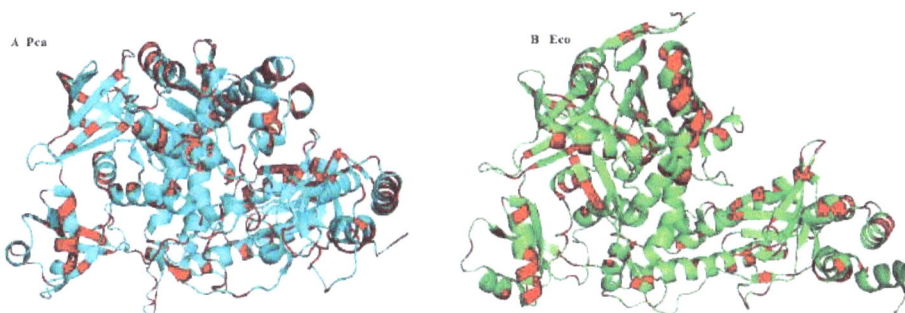

Fig. (1). Structures of **A)** *Pca* DNA polymerase – cyan (PDB ID: 5MDN) and **B)** *Eco* DNA Polymerase – green (PDB ID: 1Q8I). Charged residues are shown in red color in both structures. *Pca* DNA polymerase has a greater number of charged residues, resulting in more salt bridges, and is a thermostable enzyme. (The figure is generated in PyMol by using indicated PDB structures).

2.2. Hydrogen Bonds

Hydrogen bonds are an important type of non-covalent interaction found in proteins. Hydrogen bonds are formed between an electronegative atom and a polar hydrogen atom. These interactions occur not only within a protein but also with water molecules around it. Hydrogen bonds have been shown to contribute to the stability of a mutated recombinant lipase from *Geobacillus zalihae*. A mutation resulting in an increase of hydrogen bonds increased the melting temperature of the protein [11]. Similarly, the contribution of hydrogen bonds for protein thermostability has been highlighted by using proline racemase from *Thermococcus litoralis* (*Tli*) [12]. The native enzyme contains a tyrosine residue at position 171. It forms hydrogen bonds with Cys251 and Thr253. This interaction contributes to the thermostability of this enzyme. When Tyr171 was mutated to Phe, these hydrogen bonds were abolished, as shown in Fig. (**2**).

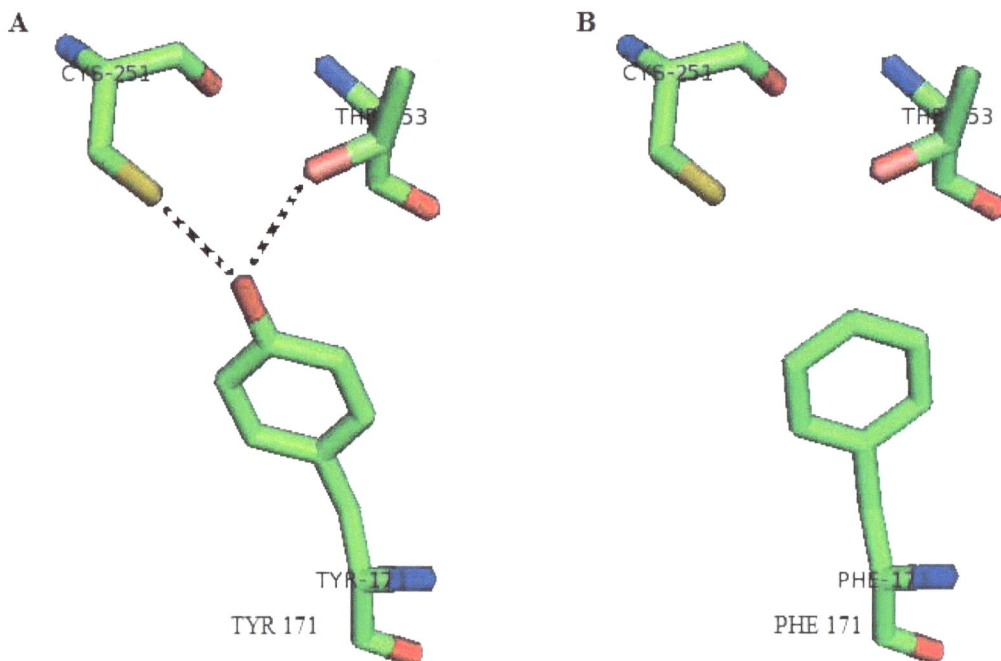

Fig. (2). A) Hydrogen bond formation by Y171 in *Thermococcus litoralis* (*Tli*) proline racemase (PDB ID: 6J7C) contributes to the thermostability of the enzyme. **B)** Y171F mutation abolishes hydrogen bond formation. (The figure is generated in PyMol by using indicated PDB structure).

2.3. Hydrophobic Interaction

Hydrophobic interaction is indeed a major determinant of structural stability.

There is a direct correlation between hydrophobicity and the thermostability of a protein. Various proteins have been described to show improved thermostability by increasing hydrophobic interaction [13]. It was reported that mutating a flexible region of trypsin resulted in increasing hydrophobic interaction hence enhancing its thermostability [14]. Similarly, hydrophobic core and surface salt bridges were shown to impart thermostability [15].

2.4. Disulfide Bonds

Disulfide bonds are formed between cysteine residues by oxidation of their sulfhydryl groups. The disulfide bridges help to stabilize proteins through the entropic effect. Through site-directed mutagenesis, when cysteine residues were introduced in subtilisin E, the half-life and *Tm* of the mutant enzyme was significantly higher than those of the wild-type enzyme [13].

2.5. Protein Rigidity and Flexibility

It has been postulated that thermophilic proteins have a more rigid structure with less structural flexibility. This, along with other molecular interactions, contributes to making them thermostable. However, both rigidity and flexibility of structure can synergistically affect the thermal stability of proteins. This effect is mediated by a fine balance of entropic and enthalpic changes, which result in a shift of unfolding temperature. Indeed, protein conformational flexibility is required for processes like ligand binding and release and enzyme activity [16, 17]. A recent study was done for comparative molecular dynamics simulations on thermophilic and mesophilic subtilisin-like serine protease. It was observed that global fluctuations of structure in the thermophilic enzyme were less than those observed in its mesophilic counterpart [18].

2.6. Amino Acid Composition

Amino acid composition of a protein has been demonstrated as a key determinant regarding thermostability. It has been shown that amino acids like serine, cysteine, glutamine and threonine are less frequent in thermophiles as compared to their abundance in mesophiles. Among these cysteine, asparagine and glutamine are thermolabile amino acids. Also, the distribution of proline has been found to be greater in thermophilic proteins as compared to their occurrence in mesophilic ones. Proline has limited conformational states and thus leads to lower conformational entropy. The presence of charged residues at the "weak points" in protein structure also provides stability by forming salt bridges [19]. Likewise, a thermostable laminarinase was shown to contain an increased number of arginine and glutamate residues as compared to other members of this enzyme family. An abundance of charged and aromatic residues is correlated to the thermophilic nature of this protein [20].

3. TOOLS AND APPROACHES FOR *IN SILICO* ANALYSIS OF PROTEIN THERMOSTABILITY

Despite technological advances, the ability to experimentally determine protein structures is still limited. This has led to emergence of alternate computer-based methods to meet this gap. Various approaches have been developed for *in silico* analysis of evaluating protein thermostability. Some of these approaches include methods based on force fields and energy potentials, machine learning, graph theory, support vector machine, neural networks and other modelling-based approaches.

These methods have been quite useful for studying protein thermophilic behavior and computer-aided design to improve their thermostability. Conventionally, the following approaches have been used for improving thermostability, namely protein-directed evolution, ancestral or consensus approach, and rational protein design guided by structure. Although with different levels of success rates of these methods, they have their own practical limitations. Hence there is a need for faster and powerful approaches for rendering proteins thermostable. This led to the development of methods for accessing protein thermostability and predicting the effects of mutations and changes on it [21 - 23]. Some of these important methods and tools are summarized in Table **2** and briefly described in this section.

Table 2. Some important tools for protein thermostability analysis.

Name	Description	References
FireProt	Prediction of multiple point mutants centered on energy- and evolution-based approaches	[24]
ScooP	Temperature-dependent stability curve of protein-based on two-state folding transition	[25]
KStable	Sequence-based approach; includes temperature and pH values to predict thermostability changes	[26]
I-Mutant2.0	Support vector machine-based tool; predicts the extent of the effect on thermostability by mutation	[27]
CUPSAT	Based on structural environment-specific atom potential and torsion angle potential	[28]
PROTS	Fragment-based thermostability potential to predict stability upon mutation	[29]

(Table 2) cont.....

Name	Description	References
FoldX	Force filed algorithm to calculate the total energy and free energy changes	[23]
RankProt	Heuristic model based on analytical hierarchical process to rank protein feature predictors	[21]
iStable2.0	Integrated prediction tool based on machine learning approach for sequence-structure prediction	[30]
ProTstab	Machine learning method; uses gradient boosting of regression trees, based on limited proteolysis and mass spectrometry	[31]
Rosetta Design	Based on Monte Carlo optimization for protein stability analysis *via* single and multiple mutations	[19]
SDM	Based on conformationally constrained environment-specific substitution tables	[33]
PoPMuSIC	Predicts thermodynamic stability changes using a combination of statistical potentials	[34]

3.1. FireProt

FireProt presents a computational strategy for the prediction of stable point mutations based on a combination of energy-based and evolution-based approaches with filtering ability to identify stabilizing mutants in an additive manner. This was successfully applied to design a thermostable haloalkane dehalogenase and a thermostable γ-hexachlorocyclohexane dehydrochlorinase. This energy-based approach removed conserved and correlated positions and was useful in reducing the number of possible mutants for analysis, thus facilitating the computational process. Evolution-based approach filtered function-related evolutionary constraints, thereby improving process efficacy [24].

3.2. ScooP

The folding transition of proteins is based on many factors, including amino acid interaction and their energetic balance is influenced by factors like temperature, pH, and ionic strength. This method is based on a stability curve predictor and takes into account features including the temperature of the environment and solvent accessible area. It is a convenient method to analyze the thermal stability of proteins on a large scale, especially from thermophilic extremophiles [25].

3.3. KStable

Methods based only on sequence have a limitation to accurately predict structural information. This could be improved by incorporating other available information regarding protein stability. KStable is a computational tool based on sequence information for prediction of protein thermostability changes upon mutation. It includes basis factors (location of mutation, nature of amino acids, temperature, and pH) and minimal redundancy –maximal relevance features [26].

3.4. CUPSAT

This tool analyzes protein stability changes upon mutations. It is based on environment-specific atom potentials and torsion angle potentials to predict the difference in free energy of unfolding between wild-type and mutant proteins. It classifies mutations and mean force potentials into different structural regions. The output also includes information about solvent accessibility, torsion angles, secondary structures, and comprehensive information about saturating mutagenesis [28].

3.5. PROTS

Thermophiles grow at high temperatures, so their proteins are intrinsically stable at higher temperature as compared to mesophilic proteins. PROTS is based on these differences in structures of thermophilic proteins as compared to mesophilic ones. It works on a fragment-based protein thermostability potential. It relies on a non-redundant representative collection of thermophilic and mesophilic protein structures and experimentally evaluated melting temperatures of point mutations. Protein thermostability changes are experimentally measured by monitoring melting temperature and free energy changes of unfolding upon mutation. PROTS uses these features from experimentally determined data [29].

3.6. FoldX

FoldX is based on a force filed algorithm. The package has various features. This includes side-chain rearrangement for minimizing energy, the introduction of mutations and optimizations of mutant protein structure, calculation of total energy and free energy change at oligomer interfaces. FoldX can also be applied to study energy changes of DNA-protein interactions [23].

3.7. RankProt

This method works on a strategy to achieve protein thermostability by predicting site-specific mutations. It uses 17 biophysical features of protein to predict thermostabilizing mutations by ranking intra-protein interactions using an

Analytical Hierarchical Process. The process involves the following steps: structuring complexity into clusters of factors, representing criteria weights with numbers, using these numbers for determining the priorities of criteria, and lastly, determining the most important alternative (mutant) [21].

3.8. iStable2.0

iStable2.0 is an integrated prediction tool utilizing a machine learning approach. It integrates 11 sequence-based and structure-based prediction tools and adds sequence information as features. The sequence-based methods consider point mutations, up- and downstream amino acid residues, their features and physiochemical properties. Structure-based methods utilize protein 3D structure information, including secondary structures, chemical composition and interaction between residue atoms. In this manner, iStable integrates tools developed by various groups based on different approaches [30].

3.9. Rosetta Design

It is based on Monte Carlo optimization to find sequences of amino acids which can fit into a given protein structure. This consists of optimal packing, hydrogen bonding and burial of hydrophobic residues. Rosetta Design can also estimate protein stability changes upon single and multiple mutations and can be used for improving protein thermostability [19].

3.10. Comparative Analysis of Tools for Thermostability Prediction

Several reports have been published demonstrating a comparison among various tools and approaches used for protein stability prediction. Khan and Vihinen compared the performance of 11 online stability predictors. The analysis included CUPSAT, Dmutant, FoldX, I-Mutant 2.0 and 3.0, MultiMutate, MUpro, SCide, Scpred and SRide. In a set of 80 proteins, 1784 single mutations (different from the training data) were used. Various parameters of these mutants were evaluated, which included the location of the mutation, volume at the residue site, and charge interactions. Among all, I-Mutant 3.0, Dmutant and FoldX were found to be the most reliable predictors [32].

SDM webserver uses environment-specific amino acid substitution tables for analyzing protein stability. It takes into account residue conformation and interaction parameters, packing density, and residue depth. SDM was compared with other methods of prediction by using a set of 350 proteins. It turned out to be one of the top-performing methods [33].

In addition, a recent work compared the performance of nine prediction tools to predict the production of soluble protein. 51 previously published mutants for β-glucosidase were used in this study. It was found that Rosetta, FoldX, PoPMuSIC and SDM were able to predict the possibility of obtaining a mutant as a soluble protein [22].

4. RELEVANT DATABASES

Although protein sequence and structure databases have been in place for a while now, during recent years, attempts have been made to come up with protein stability databases as well. Table **3** summarizes some of these databases.

Table 3. List of databases relevant to protein stability.

Database	Description	References
Protein Databank	Contains information about 3D structures of biomolecules, including proteins	[35]
ProTherm	Database of experimentally determined thermodynamic parameters regarding protein stability	[36]
ProtData Therm	Database for protein thermostability analysis and engineering based on thermal stability and protein family	[37]
ProTherm* (curated ProTherm)	Selection of entries from ProTherm database containing mutations with experimental ΔΔG values	[38]
PGTdb	Prokaryotic growth temperature database. Provides information about growth temperatures and thermostability	[39]

5. APPLICATIONS OF *IN SILICO* APPROACHES FOR PROTEIN THERMOSTABILITY ANALYSIS AND IMPROVEMENT

In silico approaches have been instrumental in screening the effect of mutations on protein stability and design variants with increased thermostability. Many reports published during past years have demonstrated the success of these methods for achieving desirable characteristics in proteins.

One such mechanism is to stabilize protein structure by the addition of disulfide bonds. This covalent interaction, in turn, results in improving protein thermostability. Through a rational design approach, Liu *et al*. introduced seven pairs of cysteine residues in alkaline α-amylase from *Alkalimonas amylolytica*. Three of these single disulfide bond mutants showed a significant increase in their thermostability. This was reflected by *in vitro* experimental analysis done on these mutants. Also, this resulted in an improvement of other biochemical properties of the mutant variants as compared to wild-type protein [40]. Similarly, in another study done on endoglucanase from *Penicillium verruculosum,* two variants with disulfide bonds were produced through *in silico* design. These mutants were produced through site-directed mutagenesis, and their properties were analyzed biochemically. The engineered variants showed improvement in specific activity and thermostability [41].

In silico protein design can potentially facilitate crystallography research by producing more stable mutants. This was demonstrated by enhancing the thermostability of a G-protein coupled receptor CXCR1-T4 lysozyme complex through computer-aided engineering. Through homology modelling and molecular dynamics simulations, a mutant with improved thermostability was produced. Experimental analysis of mutant through circular dichroism (CD) spectroscopy also indicated an increase in thermostability [42].

Another mechanism to improve thermostability is remodeling protein flexible loops. This has been demonstrated in the case of *E. coli* transketolase. The mutational design in this study included computational design based on $\Delta\Delta G$ using Rosetta. Several mutants were generated and experimentally characterized. This resulted in producing variants with greater thermostability as compared to the wild-type enzyme [43].

Intra-protein interactions are required for the greater stability of proteins. Mutations that lead to a stable set of interactions tend to improve the thermostability of protein. This was also demonstrated in a report describing the design of a thermostable fungal laccase. Followed by homology modelling, potential thermostable mutants were generated *in silico* based on ionic interactions, salt bridges and disulfide bonds. This resulted in variants with greater tolerance to high temperatures without affecting their active site [44].

Orthologous proteins from species that evolved at various temperatures have shown temperature-dependent variation in structural stability. A study of malate dehydrogenase from marine molluscs adapted to various temperatures was reported. The adaptive variations in thermal stability were analyzed by site-directed mutagenesis and *in silico* mutagenesis by using molecular dynamics simulations. The findings suggest that protein surface regions other than the active site can help in enzyme's thermal responses and evolutionary adaptation of function [45].

In silico saturating mutagenesis is a useful method to explore possible effects of amino acid substitution. In this method, a given amino acid is substituted with all other 19 amino acids. This strategy was utilized for producing thermostable glucose oxidase from *Penicillium*. Selected mutants were analyzed by *in vitro* enzyme assay and found to show improvement in their thermostability [46].

CONCLUSION

In recent years, many methods and approaches for protein stability analysis have emerged. With improvement in methods, training data, and computational power, the accuracy of prediction has also increased. *In silico* computational methods provide a powerful means to analyze proteins and predict changes therein. This is not only helpful in improving the attributes of existing proteins but also leads to design proteins from scratch with required properties.

CONSENT FOR PUBLICATION

Not applicable.

CONFLICT OF INTEREST

The author declares no conflict of interest, financial or otherwise.

ACKNOWLEDGMENTS

Declared none.

REFERENCES

[1]   Atalah J, Cáceres-Moreno P, Espina G, Blamey JM. Thermophiles and the applications of their enzymes as new biocatalysts. Bioresour Technol 2019; 280: 478-88.
[http://dx.doi.org/10.1016/j.biortech.2019.02.008] [PMID: 30826176]

[2]   Rigoldi F, Donini S, Redaelli A, Parisini E, Gautieri A. Review: Engineering of thermostable enzymes for industrial applications. APL Bioeng 2018; 2(1): 011501.
[http://dx.doi.org/10.1063/1.4997367] [PMID: 31069285]

[3]   Cabrera MA, Blamey JM. Biotechnological applications of archaeal enzymes from extreme environments. Biol Res 2018; 51(1): 37.
[http://dx.doi.org/10.1186/s40659-018-0186-3] [PMID: 30290805]

[4]   Kohli I, Joshi NC, Mohapatra S, Varma A. Extremophile - an adaptive strategy for extreme conditions and applications. Curr genomics 2020; 21(2): 96-110.
[http://dx.doi.org/10.2174/1389202921666200401105908] [PMID: 32655304]

[5]   Siddiqui KS. Defying the activity-stability trade-off in enzymes: taking advantage of entropy to enhance activity and thermostability. Crit Rev Biotechnol 2017; 37(3): 309-22.
[http://dx.doi.org/10.3109/07388551.2016.1144045] [PMID: 26940154]

[6]   Guerriero G, Hausman JF, Strauss J, Ertan H, Siddiqui KS. Destructuring plant biomass: focus on fungal and extremophilic cell wall hydrolases. Plant Sci 2015; 234: 180-93.
[http://dx.doi.org/10.1016/j.plantsci.2015.02.010] [PMID: 25804821]

[7]   Dominy BN, Minoux H, Brooks CL III. An electrostatic basis for the stability of thermophilic proteins. Proteins 2004; 57(1): 128-41.

[http://dx.doi.org/10.1002/prot.20190] [PMID: 15326599]

[8]   Karshikoff A, Ladenstein R. Ion pairs and the thermotolerance of proteins from hyperthermophiles: a "traffic rule" for hot roads. Trends Biochem Sci 2001; 26(9): 550-6.
[http://dx.doi.org/10.1016/S0968-0004(01)01918-1] [PMID: 11551792]

[9]   Zhang X, Meining W, Fischer M, Bacher A, Ladenstein R. X-ray structure analysis and crystallographic refinement of lumazine synthase from the hyperthermophile *Aquifex aeolicus* at 1.6 A resolution: determinants of thermostability revealed from structural comparisons. J Mol Biol 2001; 306(5): 1099-114.
[http://dx.doi.org/10.1006/jmbi.2000.4435] [PMID: 11237620]

[10]   Guo J, Zhang W, Coker AR, *et al.* Structure of the family B DNA polymerase from the hyperthermophilic archaeon *Pyrobaculum calidifontis*. Acta Crystallogr D Struct Biol 2017; 73(Pt 5): 420-7.
[http://dx.doi.org/10.1107/S2059798317004090] [PMID: 28471366]

[11]   Ishak SNH, Kamarudin NHA, Ali MSM, Leow ATC, Rahman RNZRA. Ion-pair interaction and hydrogen bonds as main features of protein thermostability in mutated T1 recombinant lipase originating from Geobacillus zalihae. Molecules 2020; 25(15): 3430.
[http://dx.doi.org/10.3390/molecules25153430] [PMID: 32731607]

[12]   Watanabe Y, Watanabe S, Itoh Y, Watanabe Y. Crystal structure of substrate-bound bifunctional proline racemase/hydroxyproline epimerase from a hyperthermophilic archaeon. Biochem Biophys Res Commun 2019; 511(1): 135-40.
[http://dx.doi.org/10.1016/j.bbrc.2019.01.141] [PMID: 30773259]

[13]   Li WF, Zhou XX, Lu P. Structural features of thermozymes. Biotechnol Adv 2005; 23(4): 271-81.
[http://dx.doi.org/10.1016/j.biotechadv.2005.01.002] [PMID: 15848038]

[14]   Liu L, Yu H, Du K, Wang Z, Gan Y, Huang H. Enhanced trypsin thermostability in *Pichia pastoris* through truncating the flexible region. Microb Cell Fact 2018; 17(1): 165.
[http://dx.doi.org/10.1186/s12934-018-1012-x] [PMID: 30359279]

[15]   Nguyen C, Young JT, Slade GG, Oliveira RJ, McCully ME. A dynamic hydrophobic core and surface salt bridges thermostabilize a designed three-helix bundle. Biophys J 2019; 116(4): 621-32.
[http://dx.doi.org/10.1016/j.bpj.2019.01.012] [PMID: 30704856]

[16]   Radestock S, Gohlke H. Protein rigidity and thermophilic adaptation. Proteins 2011; 79(4): 1089-108.
[http://dx.doi.org/10.1002/prot.22946] [PMID: 21246632]

[17]   Karshikoff A, Nilsson L, Ladenstein R. Rigidity *versus* flexibility: the dilemma of understanding protein thermal stability. FEBS J 2015; 282(20): 3899-917.
[http://dx.doi.org/10.1111/febs.13343] [PMID: 26074325]

[18]   Sang P, Liu SQ, Yang LQ. New insight into mechanisms of protein adaptation to high temperatures: a comparative molecular dynamics simulation study of thermophilic and mesophilic subtilisin-like serine proteases. Int J Mol Sci 2020; 21: 3128.
[http://dx.doi.org/10.3390/ijms21093128]

[19]   Modarres HP, Mofradab MR, Sanati-Nezhad A. Protein thermostability engineering. RCS Advances 2016; 6: 115252-70.

[20]   Bleicher L, Prates ET, Gomes TCF, *et al.* Molecular basis of the thermostability and thermophilicity of laminarinases: X-ray structure of the hyperthermostable laminarinase from *Rhodothermus marinus* and molecular dynamics simulations. J Phys Chem B 2011; 115(24): 7940-9.
[http://dx.doi.org/10.1021/jp200330z] [PMID: 21619042]

[21]   Chakravorty D, Patra S. RankProt: A multi criteria-ranking platform to attain protein thermostabilizing mutations and its *in vitro* applications - Attribute based prediction method on the principles of Analytical Hierarchical Process. PLoS One 2018; 13(10): e0203036.
[http://dx.doi.org/10.1371/journal.pone.0203036] [PMID: 30286107]

[22] Huang P, Chu SKS, Frizzo HN, Connolly MP, Caster RW, Siegel JB. Evaluating protein engineering thermostability prediction tools using an independently generated dataset. ACS Omega 2020; 5(12): 6487-93.
[http://dx.doi.org/10.1021/acsomega.9b04105] [PMID: 32258884]

[23] Buss O, Rudat J, Ochsenreither K. FoldX as protein engineering tool: better than random based approaches Comput Struct Biotechnol J 2018; 16: 25-33.
[http://dx.doi.org/10.1016/j.csbj.2018.01.002] [PMID: 30275935]

[24] Bednar D, Beerens K, Sebestova E, *et al.* FireProt: Energy- and evolution-based computational design of thermostable multiple-point mutants. PLOS Comput Biol 2015; 11(11): e1004556.
[http://dx.doi.org/10.1371/journal.pcbi.1004556] [PMID: 26529612]

[25] Pucci F, Kwasigroch JM, Rooman M. SCooP: an accurate and fast predictor of protein stability curves as a function of temperature. Bioinformatics 2017; 33(21): 3415-22.
[http://dx.doi.org/10.1093/bioinformatics/btx417] [PMID: 29036273]

[26] Chen CW, Chang KP, Ho CW, Chang HP, Chu YW. KStable: A computational method for predicting protein thermal stability changes by K-Star with regular-mRMR feature selection. Entropy (Basel) 2018; 20: 988.
[http://dx.doi.org/10.3390/e20120988]

[27] Capriotti E, Fariselli P, Casadio R. I-Mutant2.0: predicting stability changes upon mutation from the protein sequence or structure. Nucleic Acids Res 2005; 33(Web Server issue): W306-10.
[http://dx.doi.org/10.1093/nar/gki375] [PMID: 15980478]

[28] Parthiban V, Gromiha MM, Schomburg D. CUPSAT: prediction of protein stability upon point mutations. Nucleic Acids Res 2006; 34(Web Server issue): W239-42.
[http://dx.doi.org/10.1093/nar/gkl190] [PMID: 16845001]

[29] Li Y, Zhang J, Tai D, Middaugh CR, Zhang Y, Fang J. PROTS: a fragment based protein thermo-stability potential. Proteins 2012; 80(1): 81-92.
[http://dx.doi.org/10.1002/prot.23163] [PMID: 21976375]

[30] Chen CW, Lin MH, Liao CC, Chang HP, Chu YW. iStable 2.0: Predicting protein thermal stability changes by integrating various characteristic modules. Comput Struct Biotechnol J 2020; 18: 622-30.
[http://dx.doi.org/10.1016/j.csbj.2020.02.021] [PMID: 32226595]

[31] Yang Y, Ding X, Zhu G, Niroula A, Lv Q, Vihinen M. ProTstab - predictor for cellular protein stability. BMC Genomics 2019; 20(1): 804.
[http://dx.doi.org/10.1186/s12864-019-6138-7] [PMID: 31684883]

[32] Khan S, Vihinen M. Performance of protein stability predictors. Hum Mutat 2010; 31(6): 675-84.
[http://dx.doi.org/10.1002/humu.21242] [PMID: 20232415]

[33] Pandurangan AP, Ochoa-Montaño B, Ascher DB, Blundell TL. SDM: a server for predicting effects of mutations on protein stability. Nucleic Acids Res 2017; 45(W1): W229-35.
[http://dx.doi.org/10.1093/nar/gkx439] [PMID: 28525590]

[34] Dehouck Y, Kwasigroch JM, Gilis D, Rooman M. PoPMuSiC 2.1: a web server for the estimation of protein stability changes upon mutation and sequence optimality. BMC Bioinformatics 2011.

[35] Berman HM, Westbrook J, Feng Z, Gilliland Z, Bhat TN. The protein data bank. 2000; Nucleic Acids Res 28: 235-42.

[36] Kumar MD, Bava KA, Gromiha MM, Prabakaran P, Kitajima K. ProTherm and ProNIT: thermodynamic databases for proteins and protein-nucleic acids interactions. Nucleic Acids Res 2006; 34: 204-6.
[http://dx.doi.org/10.1093/nar/gkj103]

[37] Modarres HP, Mofrad MR, Sanati-Nezhad A. ProtDataTherm: A database for thermostability analysis and engineering of proteins. PLoS One 2018; 13: e0191222.

[http://dx.doi.org/10.1371/journal.pone.0191222] [PMID: 29377907]

[38] Frenz B, Lewis SM, King I, DiMaio F, Park H, Song Y. Prediction of protein mutational free energy: benchmark and sampling improvements increase classification accuracy. Front Bioeng Biotechnol 2020; 8: 558247.
[http://dx.doi.org/10.3389/fbioe.2020.558247] [PMID: 33134287]

[39] Huang SL, Wu LC, Liang HK, Pan KT, Horng JT, Ko MT. PGTdb: a database providing growth temperatures of prokaryotes. Bioinformatics 2004; 20(2): 276-8.
[http://dx.doi.org/10.1093/bioinformatics/btg403] [PMID: 14734322]

[40] Liu L, Deng Z, Yang H, *et al. In silico* rational design and systems engineering of disulfide bridges in the catalytic domain of an alkaline α-amylase from *Alkalimonas amylolytica* to improve thermostability. Appl Environ Microbiol 2014; 80(3): 798-807.
[http://dx.doi.org/10.1128/AEM.03045-13] [PMID: 24212581]

[41] Bashirova A, Pramanik S, Volkov P, *et al.* Disulfide bond engineering of an endoglucanase from *Penicillium verruculosum* to improve its thermostability. 2019. Int J Mol Sci 1602; 20.

[42] Wang Y, Park JH, Lupala CS, *et al.* Computer aided protein engineering to enhance the thermo-stability of CXCR1- T4 lysozyme complex. Sci Rep 2019; 9(1): 5317.
[http://dx.doi.org/10.1038/s41598-019-41838-2] [PMID: 30926935]

[43] Yu H, Yan Y, Zhang C, Dalby PA. Two strategies to engineer flexible loops for improved enzyme thermostability. Sci Rep 2017; 7: 41212.
[http://dx.doi.org/10.1038/srep41212] [PMID: 28145457]

[44] Díaz R, Díaz-Godínez G, Anducho-Reyes MA, Mercado-Flores Y, Herrera-Zúñiga LD. *In silico* design of laccase thermostable mutants from Lacc 6 of *Pleurotus ostreatus.* Front Microbiol 2018; 9: 2743.
[http://dx.doi.org/10.3389/fmicb.2018.02743] [PMID: 30487785]

[45] Liao ML, Somero GN, Dong YW. Comparing mutagenesis and simulations as tools for identifying functionally important sequence changes for protein thermal adaptation. Proc Natl Acad Sci USA 2019; 116(2): 679-88.
[http://dx.doi.org/10.1073/pnas.1817455116] [PMID: 30584112]

[46] Ning X, Zhang Y, Yuan T, *et al.* Enhanced thermostability of glucose oxidase through computer-aided molecular design. Int J Mol Sci 2018; 19(2): 425.
[http://dx.doi.org/10.3390/ijms19020425] [PMID: 29385094]

CHAPTER 4

Pan-Proteomics to Analyze the Functional Complexity of Organisms

Muhammad Tahir ul Qamar[1,2]**, Barira Zahid**[3]**, Fatima Khalid**[3]**, Anam Naz**[4]**, Jia-Ming Song**[1,2]**, Sajjad Ahmad**[5]**, Xitong Zhu**[2]**, Feng Xing**[6]**, Muhammad Sarwar Khan**[7]** and Ling-Ling Chen**[1,2,*]

[1] *State Key Laboratory for Conservation and Utilization of Subtropical Agro-bioresources, College of Life Science and Technology, Guangxi University, Nanning, P.R. China*

[2] *Hubei Key Laboratory of Agricultural Bioinformatics, College of Informatics, Huazhong Agricultural University, Wuhan, P.R. China*

[3] *Key Laboratory of Horticultural Plant Biology (Ministry of Education), College of Horticulture and Forestry, Huazhong Agricultural University, Wuhan, P.R. China*

[4] *Institute of Molecular Biology and Biotechnology (IMBB), The University of Lahore, Lahore, Pakistan*

[5] *Department of Health and Biological Sciences, Abasyn University, Peshawar, Pakistan*

[6] *6College of Life Science, Xinyang Normal University, Xinyang 464000, P.R. China*

[7] *7Center of Agricultural Biochemistry and Biotechnology, University of Agriculture, Faisalabad, Pakistan*

Abstract: Proteomics is rapidly expanding with the advent of high throughput technologies and offers a greater understanding of the complexity of life and the process of evolution. Protein profile comparison between genetically heterogeneous individuals may provide important insights into physiological diversity and function. A new term, pan-proteomics, has also been introduced that permits the qualitative and quantitative comparison of proteomes of genetically heterogeneous organisms. Here in this chapter, various aspects of pan-proteomics along with its basic methodology and applications have been discussed.

Keywords: Bioinformatics, Comparative proteomics, Proteomics, Pan-Proteomics.

* **Corresponding author Ling-Ling Chen:** State Key Laboratory for Conservation and Utilization of Subtropical Agro-bioresources, College of Life Science and Technology, Guangxi University, Nanning, P. R. China; Tel: +8618971629380; E-mail: llchen@gxu.edu.cn; ORCID: https://orcid.org/0000-0002-3005-526X

Muhammad Sarwar Khan (Ed)

1. INTRODUCTION

Proteins are a quintessential part of all living organisms, with different functions essential for life. The word "proteome" was first coined in 1994 by Mark Wilkins. Proteomics is the large-scale protein structure and function analysis [1]. Major advances in molecular biology techniques providing new insights into the nature of genes and proteins are due to the staggering number of genomic projects performed over the past decade. However, many types of information cannot be obtained, and it is impossible to interpret the mechanism of disease and the effect of the environment solely by these results. In this context, to characterize the protein modifications, large-scale study of proteins is inevitable. Proteomics is a large-scale analysis of the whole set of proteins expressed by a cell, tissue, or organism, resulting in an information-rich landscape of expressed proteins and their modulations under a defined set of conditions. Unlike the previous studies of individual proteins or simple macromolecular complexes, proteomics offers a more comprehensive and systematic understanding of biological systems [2, 3].

Proteomics allows evaluating different cellular processes by providing a comprehensive qualitative and quantitative description of protein contents in a cell. Quantitative proteomics provided several valuable insights into proteomes, and it can be further classified in global proteome analysis and relative quantification [4]. Global proteomics strategies are often used for the absolute quantification of one or more proteins in a given sample. On the other hand, relative quantification of two samples with two different physiological states provides insights into the cellular and molecular mechanisms involved in a biological process. Quantification of mRNA content can confirm the presence of protein and its quantity in a cell [5, 6]. This concept allows the comparative analysis of protein measurements on a whole-genome level. So, the protein's detected differences can enable new insights into the processes accountable for a detected phenotype [7].

Recent technological advances in proteomics have greatly propelled our knowledge about drug development, host-pathogen interaction, and human or animal health. Proteomics offers the ability to interpret the physiological diversity and functions in a complex system. Many of the proteomes of genetically heterozygous organisms may fail to account for underlying genetic variance when compared. Therefore, recent knowledge concerning the functional heterogeneity of individuals at the protein level may be inaccurate or incomplete. To address this, there is a need to consider the impact of genetic heterogeneity on proteome comparisons more significantly. Pan-proteomics is the possible solution here that allows the qualitative and quantitative comparison between genetically heterogeneous organisms [7]. The aim of this chapter is to briefly explain the concept of pan-proteomics and its applications in different areas of basic life sciences. In addition, proteomics strategies and bioinformatics approaches developed to increase pan-proteomic power have also been discussed.

2. CONCEPT OF PAN-PROTEOMICS

Pan-proteomics is an emerging analytical approach to compare proteome variability within genetically heterogenous organisms across species of interest [7]. Like the pan-genome, the pan-proteome also refers to the full set of non-redundant proteins thought to be expressed within members of the same species. Similarly, it can be categorized into core proteome (proteins present in all members) and variable or dispensable proteome (proteins present only in some members). The pan-proteomics complements both pan-genomics [8, 9] and pan-transcriptomics [10, 11], which helps to identify genetic variants and the presence of homologous sequences of an organism at the protein level [12, 13]. Comparative proteomic analysis of different individuals of the same species can reveal relationships and genetic variations among individuals without assessing the genomic data [14, 15].

3. APPROACHES AND SOFTWARE USED FOR PAN-PROTEOMICS

Proteomics relies on three fundamental technological cornerstones that include, Mass Spectrometry (MS) to obtain the data essential for individual protein identification; a method to separate complex proteins; and bioinformatics to analyze and assemble the MS data [16]. Just like proteomics, some steps are considered important in pan-proteomics as well, including (i) sample preparation – considering the type of protein fraction and selection of extraction buffer based on different physical-chemical properties of proteins; (ii) identification method – implementation of high-performance methodologies like MudPIT for maximum proteome coverage and dynamic range; (iii) protein sequence database – with the proteome of all species utilized in the study and curated protein sequences. So, the pan-database will contain reference proteome and unique sequences that are not present in the reference proteome [17, 18]. Once a pan-database has been constructed, sequence homology clusters can be inferred from it [7]. In this regard, different methods of homology cluster inference are available or have been reported, including: BLASTclust [19], orthoMCL [20], PipeAlign [21], OrthoFinder [22], CD-HIT [23] and OrthoVenn [24]. In addition, GROMACS [25] can be used for protein's structural clustering confirmations analysis [26]. Once inferred, clusters can be further classified into core, dispensable and specie-specific categories. Proteomic data is usually explored with the help of gene ontology (GO) and pathway analysis [7]. Several platforms are available for such analyses, including UniProt [27], GO annotations with the Gene Ontology Consortium [28] and KEGG [29]. The best way to study gene/protein ontology is by using BLAST2GO [30]. From every protein cluster, some representative sequences can be used as input and in turn, the GO annotations screened for every protein cluster representative are based on the sequence similarity with already annotated sequences. The GO annotation resulted from the BLAST2GO evaluation can be further easily employed to GO terms using statistical approaches [7].

To further enhance the analytical power of proteomic studies, many advanced methodologies have been developed that generate a huge amount of MS data. In addition, a number of databases, bioinformatics software, and computational programs are developed, with the ability to store and analyze these high-throughput proteomics methodologies data [31]. Some of the software can be utilized in pan-proteomics as well (Table 1). However, no simple, rapid and comprehensive tool is available yet, that can process proteomics data and define pan-proteomes directly. Therefore, it is crucial to develop robust software capable of analyzing, processing, and transforming the raw data into qualitative/quantitative data, to increase the accessibility of pan-proteomics to researchers globally.

Table 1. List of available tools which can be utilized for pan-proteome analysis.

Resources for Protein Quantification Analysis	
MASCOT	https://www.matrixscience.com/
PEAKS Q	https://www.bioinfor.com/quantification/
ProteinPilot	https://sciex.com/products/software/proteinpilot-software
Progenesis	http://www.nonlinear.com/progenesis/qi-for-proteomics/
MassMatrix	https://massmatrix.bio/
TopPIC	http://proteomics.informatics.iupui.edu/software/toppic/
PEAKS DB	https://www.bioinfor.com/peaksdb/
SEQUEST	https://proteomicsresource.washington.edu/protocols06/sequest.php
Resources for Protein Annotation	
PSAT	http://psat.llnl.gov/psat/
PGTools	http://qcmg.org/bioinformatics/PGTools
PFAM	https://pfam.xfam.org/
PROSITE	https://prosite.expasy.org/
customProDB	https://bioconductor.org/packages/release/bioc/html/customProDB.html
STRAP	http://www.bumc.bu.edu/cardiovascularproteomics/cpctools/strap/

(Table 1) cont.....

Resources for Protein Quantification Analysis	
MAKER	http://www.yandell-lab.org/software/maker.html
Resources for Protein Homologous Analysis	
OrthoVenn2	https://orthovenn2.bioinfotoolkits.net/home
BLAST	https://blast.ncbi.nlm.nih.gov/Blast.cgi
kClust	http://toolkit.lmb.uni-muenchen.de/pub/kClust/
OrthoFinder	https://github.com/davidemms/OrthoFinder
SPIDER	https://www.bioinfor.com/peptide-mutations-homology-searching/
COBALT	https://www.ncbi.nlm.nih.gov/tools/cobalt/re_cobalt.cgi
HHPred	https://toolkit.tuebingen.mpg.de/tools/hhpred#/tools/hhpred
OrthoMCL	https://orthomcl.org/orthomcl/
Resources for Pan-Genome Investigations	
GET_HOMOLOGUES	http://www.eead.csic.es/compbio/soft/gethoms.php
Roary	https://sanger-pathogens.github.io/Roary/
BPGA	https://iicb.res.in/bpga/
ppsPCP	http://cbi.hzau.edu.cn/ppsPCP/
PanRV	https://sourceforge.net/projects/panrv2/
PanTools	https://github.com/sheikhizadeh/pantools
Panseq	https://lfz.corefacility.ca/panseq/
PanCGHweb	http://bamics2.cmbi.ru.nl/websoftware/pancgh/pancgh_start.php
PanCake	https://bitbucket.org/CorinnaErnst/pancake/wiki/Home
panX	http://pangenome.de/
Resources for Protein Functional Analysis	
EggNOG	http://eggnog5.embl.de/#/app/home
GO	http://geneontology.org/
STRING	https://string-db.org/
KEGG	https://www.genome.jp/kegg/
PANTHER	http://www.pantherdb.org/
BAR 3.0	http://bar.biocomp.unibo.it/bar3
InterPro	https://www.ebi.ac.uk/interpro/

4. APPLICATIONS AND EXPERIMENTAL DESIGN OF PAN-PROTEOMICS IN PROKARYOTES RESEARCH

Conventional proteomic analysis methods, no doubt, can provide knowledge about function and diversity, which are the key basis of bacterial strain virulence.

However, such analysis often fails to explain the genetic variation among the strains, which ultimately provides incomplete or false insights into the knowledge about the diversity of function at the protein level. At this point, the science of pan-proteomics comes to the rescue, which gives greater importance to the genetic heterogeneity on proteome comparisons [7]. In prokaryotic research, proteomic studies have played an essential role in enhancing our understanding of the responses against environmental stress, the interaction between the host and pathogen, and microbial meta-proteomics; besides this, it also has assisted us with the concept of prokaryotic pan-proteomics. The comparison of an unlimited number of bacterial strains is now possible with pan-proteomics, which can provide scientists a deeper understanding of the biological pathways existing within them [32]. While dealing with closer phylogenetic relatives of bacteria, pan-proteomics can mostly play an important role in proteome investigation of those particular bacterial strains [33].

Characterization of various bacterial pathogens has been done with the aid of pan-proteomics. Four strains of *Salmonella paratyphi* A were investigated by doing their proteomic and pan-proteomic analyses under laboratory conditions, provided the knowledge that the strains' protein expression had limited genetic diversity among them [14]. When quantitative proteomic analysis of seven enterotoxigenic *Escherichia coli* (ETEC) strains was done, it indicated the introduction of proteins related to the processes of acquisition of iron, metabolism of iron and resistance to acids [34]. Ocean bacteria *Roseobacters* were analyzed based on their pan-exoproteome, which gave us an insight into their adaptation in the sea [35]. Two zoonotic infection causative agent *Brucella* bacterial strains of human brucellosis share a higher degree of genomic similarities. Their pan-proteomic analysis was also utilized to determine the protein expression level difference, which provided an in-depth understanding of the differences in bimolecular transport and protein synthesis mechanisms between these two strains [13]. Pan-proteomic analysis of a global commodity of the aquaculture sector known as *Streptococcus agalactiae* (GBS), a major pathogen of *Nile tilapia* evaluated the protein expression in its main genotypes. The study concluded conversion in the core proteome and had a high similarity percentage, which put forward the idea that a monovalent vaccine can prove very useful in this current scenario [36]. When pathogens are concerned, comparative analysis between avirulent, virulent, and clinical strains gave us a clear-cut indication about the mechanism of their virulence and drug resistance, which provides us with the knowledge to dive deep into the specific biology of a particular pathogen [37]. Tavora *et al.*, recently evaluated the contrasting virulence between two isolates of *Xanthomonas campestris* pv. *campestris* (Xcc) - Xcc51 (more virulent) and XccY21 (less virulent) by determining their pan-proteome profiles [38]. Their results shed light on the complex Xcc pathogenicity mechanisms and point out a set of proteins related to the higher virulence of Xcc51. This information is essential for the development of more efficient strategies aiming at the control of black rot disease [38]. In the bacterial pathogenesis cycle, secreted and surface proteins have an important role. To understand these secreted proteins, proteomic research was conducted on the secretome of enterohemorrhagic *Escherichia coli* (EHEC) [39]. *Acinetobacter baumanii* (*A. baumannii*) has proven its resistance to a lot of antibiotics and is a major nosocomial pathogen. Vaccination can be an effective alternative strategy to tackle *A. baumannii*. A study tried to determine the vaccination strategy by exploring its pan-genome and pan-proteome. The effectiveness of this strategy due to the global expansion of the sequencing data has the potential to produce vaccine against *A. baumannii* and other bacterial pathogens [40].

Besides pathogens, pan-proteomics analytical studies also highlight the critical factors of the non-pathogenic strains related to their adaptive and physiological processes. Four strains of the *Lactococcus lactis* (*L. Lactis*) were analyzed based on their proteome that distinguished the proteins that have a role in the stress response of the strains. Moreover, those proteins were also involved in the probiotic characteristic of *L. lactis* [41]. With the aid of continuously increasing knowledge of pan-proteomics, scientists are reaching closer and closer to finding out the absolute essential proteins required for the survival of bacteria [42].

Pan-proteome analysis has great potential to interpret different strain types, identify drug and vaccine targets, and discern drug activity mechanisms. Regarding vaccine development, those candidate targets are usually preferred that represents: (i) high sequence homology of said antigen; (ii) require a minimum of ubiquitous existence across an appropriate strain; (iii) present in excessive abundance to initiate an immune response; and (iv) granted immune system accessibility. From that perspective, pan-proteomics provide information regarding protein presence/ absence, abundance, and subcellular-localization of prospective vaccine candidates within a pathogenic species, which is highly valuable in vaccine development [43, 44]. Recent research has been published to find out effective drug and vaccine targets against *Vibrio parahaemolyticus,* a highly reported pathogenic bacteria of the aquatic environment, using a comprehensive genome-based analysis [45]. Here are the steps that can be followed to identify novel drug targets as well as potential vaccine candidates using pan-proteome. The complete workflow of screening potential drug and vaccine targets has been shown in Fig. (**1**).

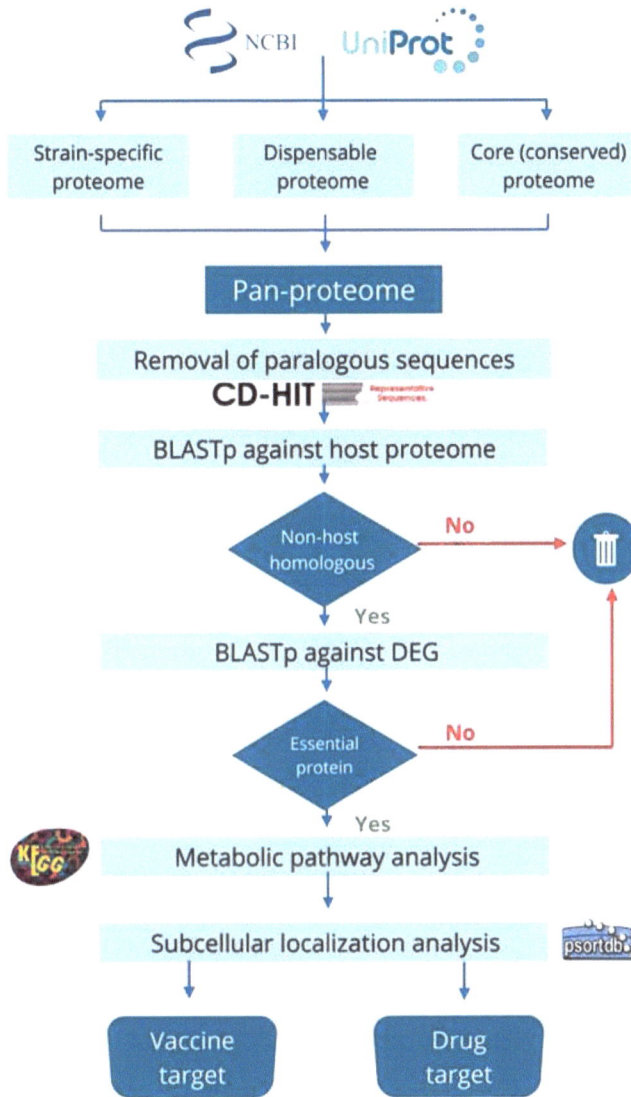

Fig. (1). Workflow for screening drug and vaccine targets using pan-proteome [48].

4.1. Proteome Retrieval and Removal of Duplicate Sequences

The whole proteome sequences of a specific strain and all other strains are mainly retrieved from the National Center for Biotechnology Information (NCBI) [46] and Universal Protein (UniProt) databases [47], which are the most extensive and informative protein databases. These protein sequences together make the whole pan-proteome [48]. The rapid expansion in the use of next-generation sequencing (NGS) technology has led to an unrivaled explosion in the amount of biological sequencing data, and data redundancy elimination has become one of the significant challenges to upcoming bioinformatics analyses [48]. To date, CD-HIT has been widely used to cluster and eliminate duplicate sequences having similarities more than the expected threshold values [49].

4.2. Searching of essential, non-homologous proteins

The selection of non-homologous sequences to those in the host is a critical step of the process. If the target proteins have similarities to the host protein sequences, the designed drug may lead to nonspecific interactions with the host protein, resulting in negative effects [48]. A basic local alignment search tool (BLAST) is the best choice to perform a similarity search that helps to compare the target non-paralogous proteins with the entire host proteome [50, 51]. The next critical step in the process is the selection of essential proteins. These proteins are necessary for the proper functioning of bacteria and important for their survival. Therefore, therapeutic effects can be improved by inhibiting the activity of essential genes. The Database of Essential Genes (DEG) provides comprehensive information on essential genes in bacteria. An essentiality test is conducted on the selected target proteins; homology with the sequences found in the DEG database is the basis of the essentiality of non-homologous proteins [52, 53].

4.3. Metabolic pathway analysis

KEGG database is a resource that contains a network of metabolic pathways present in living organisms with their complete annotation [54]. That gives the advantage to perform metabolic pathway analysis on the non-homologous essential proteins *via* Automatic Annotation Server (KAAS) [55, 56]. Comparative pathway analysis of bacteria is the next step; through that, unique metabolic pathways of non-homologous essential proteins can be mapped, which can further be used as key targets for the treatment of diseases.

4.4. Drug-Ability Analysis

A 'druggable' target is the one that has the potential to bind to the drugs with high affinity. So, the screening of selected non-homologous essential proteins is done through DrugBank 5.1.0 [57], using default parameters to identify both novel therapeutic targets and druggable proteins.

4.5. Prediction of Subcellular Localization

It is the most critical step for the identification of target proteins that also provides information about the function of the protein. Moreover, it has been demonstrated that cytoplasmic proteins can be used as effective drug targets [58], while proteins distributed on extracellular-membranes are considered effective vaccine candidates [59]. At this stage, shortlisted pathogen-specific essential proteins are subjected to subcellular localization analysis to predict effective targets using PSORTb server [60], CELLO [61], SignalIP [62], and ngLOC [63], to achieve more precise targets.

5. APPLICATION OF PAN-PROTEOMICS IN EUKARYOTES RESEARCH

5.1. Utilization in Plant's Research

As we move towards eukaryotic organisms from prokaryotic organisms, it becomes apparent that they have a vast complexity compared to prokaryotic organisms. If plants are considered first, there are now different types of cells, organs, and tissues that interact with each other and make the whole living system sustainable. During the research, various organs are especially studied for different types of analysis, *i.e.*, if the research is related to the mechanism of photosynthesis samples, it will be collected from the leaves of the plant for experimentation [64]. Different proteomic analysis techniques have been in use for a long time and are continuously evolving. All the simple proteomic analysis had their limitations and were affecting the clarity of experimental results and their interpretations. With the advancement of the proteomic sciences, there was a new beginning in plant proteomics research. New techniques and technologies were evolved that had a higher chance of precision and accuracy than conventional procedures. Recent studies also reported that alternative splicing might not be the key to proteome complexity, and proteomics techniques can detect the expression of stable alternative splice isoforms on a genome-wide scale [65, 66]. This situation ultimately leads to advancements in the field of proteomics which caused the evolution of pan-proteomics [64]. The pan-proteomic techniques have been used for some studies related to different aspects like development, defense, *etc*. The major use of such type of pan-proteomic analysis is to determine the relative abundance of protein expression under different circumstances and development stages. This comparative study provides detailed knowledge about protein components and interaction between the protein components to form a complete protein network or routes [67]. Pan-proteomic analysis was used in a study that analyzed the response of the *Arabidopsis thaliana* to the hormone of strigolactone that stimulates development in seed [68]. This method provided deep insights into the whole molecular mechanism of this hormone in the *A. thaliana*. Similarly, different other aspects including plant's stress response can also be understood in detail using the pan-proteomics [69]. A pan-plant proteome complex reveals deep molecular conversion and novel assemblies using co-fractionation mass spectrometry to find the association between approximately two million proteins from thirteen genetically variant and scientifically/agriculturally important plant species [12]. In previous studies, these protein complexes were only inferred in plants by the gene content. With the proteomics advancements, especially with the concept of pan-proteomics, now the protein complexes are more understood with their subunits and their entire organization. Previously unknown interactions between protein complexes came into light following this study. Moreover, plant analogs of animal complexes with distinct molecular assemblies, including a megadalton-scale tRNA multi-synthetase complex, were observed [12]. James *et al.* recently presented that they are applying the pan-proteome approach to wheat for trait classification of diverse wheat strains, prediction of time-to-flowering and yield improvement [70]. They identified wheat pan-proteome from transcriptome-driven protein sequences. They also identify individual proteins and their network proponents that show a close association with time-to-flower measurements [70].

5.2. Utilization in Animals and Human Research

In the case of animals, correlating protein samples collected from different tissues is challenging. Likewise, for animals, pan-proteomics has proved its importance and helped to understand the biological system and mechanisms in many organisms by using the protein samples from the organs and secretions [71]. Trapp *et al.* performed a pan-proteome analysis of ovaries from five *Senticaudata* amphipods. They defined the core-proteome of *Senticaudata* female reproductive systems and reported that proteins are conserved among all amphipods [72]. Several approaches have been employed to understand the disease origin or status, to study the animal's reproduction, and to characterize the animal's products. Later, a comprehensive syndicate was established for the applications of proteomics approaches to animal production and health, involving 31 countries' researchers [69, 73]. A human proteome study has been established to investigate the expression profile of ~20,000 genes to get deep insights into human biology for health and disease management [74]. In an *in silico* study of healthy human skin, proteome produced data of about 3000 proteins, correlated through the help of an automated literature review, gives an idea about the possible application of pan-proteomics [69, 75]. Many scientists around the globe are working on specific human organs. For example, there is a project called EyeOme that focuses on the proteins found in the human eye organ and the interaction between them. Around 16000 proteins have been determined from the ciliary body, retina, iris, retrobulbar optic nerve, and sclera [76].

5.3. Utilization in Cancer Studies

In the cancer research field, pan-proteomics is playing a very important role by making a correlation of data and emphasizing the determination of proteins that can help prognostic and therapeutic of the functional proteome [69]. For example, for breast cancer, Mucin-1 was determined as a biomarker protein used to monitor the patients' progression with metastasis [77]. In lung cancer, higher expression of Stanniocalcin-2 (STC2) than normal cells has confirmed STC2 as a promising biomarker for cancer [78]. Plasma proteins that serve in the diagnosis of many diseases have high complexity due to the extensive heterogeneity among the diseases, and this complexity becomes problematic in the research field [69]. Serum protein samples were collected from patients with gastric cancer biomarkers SERPINA1 and ENOSF1 were identified through MS which are related to gastric cancer [79]. The aim of the pan-cancer analysis is to highlight the differences and similarities present among different types of tumors as it focuses on the cellular alteration and genomic sequence manipulations. In 2018, the Cancer Genome Atlas (TCGA) Research Network used exome, transcriptome, and DNA methylome data so that an integrated picture of commonalities, differences, and emergent themes across tumor types can be developed [80, 81].

CONCLUSIONS AND FUTURE PERSPECTIVES

The concepts of proteomics have been applied to many of the knowledge disciplines to understand the basic principles and mechanisms of biological systems. Initially, techniques like two-dimensional gel electrophoresis (2-DE) and liquid chromatography coupled to mass spectrometry (LC-MS) were in service, which were later evolved into advanced techniques and technologies like multidimensional protein identification technology (MudPIT) and pan-proteomics. Pan-proteomics offer an insight into the system biology and the treatments of diseases of plants, humans and animals. In a given organism/species, the combination of pan-proteomics with the latest knowledge disciplines like bioinformatics provides a better understanding of genetic variation. The relation of pan-proteomics with genomics, transcriptomics, metabolomics, and interactomics serves to gain knowledge about the gene function and regulation. In the future, pan-proteomic may expand investigations of genetically heterogeneous organisms and aid in understanding their functional complexity.

CONSENT FOR PUBLICATION

Not applicable.

CONFLICT OF INTEREST

The author declares no conflict of interest, financial or otherwise.

ACKNOWLEDGMENTS:

This work was supported by the starting research grant for High-level Talents from Guangxi University, China, and Postdoctoral Project from Guangxi University, China.

REFERENCES

[1] Wilkins MR, Sanchez JC, Gooley AA, *et al.* Progress with proteome projects: why all proteins expressed by a genome should be identified and how to do it. Biotechnol Genet Eng Rev 1996; 13: 19-50.
[http://dx.doi.org/10.1080/02648725.1996.10647923] [PMID: 8948108]

[2] Graves PR, Haystead TAJ. Molecular biologist's guide to proteomics. Microbiol mol biol rev 2002; 66(1): 39-63.
[http://dx.doi.org/10.1128/MMBR.66.1.39-63.2002] [PMID: 11875127]

[3] Shah TR, Misra A. 8 - Proteomics. In: Misra A, Ed. Challenges in Delivery of Therapeutic Genomics and Proteomics. London: Elsevier 2011; pp. 387-427.
[http://dx.doi.org/10.1016/B978-0-12-384964-9.00008-6]

[4] Chao T-C, Hansmeier N. The current state of microbial proteomics: where we are and where we want to go. Proteomics 2012; 12(4-5): 638-50.
[http://dx.doi.org/10.1002/pmic.201100381] [PMID: 22246737]

[5] Ankney JA, Muneer A, Chen X. Relative and absolute quantitation in mass spectrometry-based proteomics. annu rev anal chem (palo alto, calif) 2018; 11(1): 49-77.
[http://dx.doi.org/10.1146/annurev-anchem-061516-045357] [PMID: 29894226]

[6] Deracinois B, Flahaut C, Duban-Deweer S, Karamanos Y. Comparative and quantitative global proteomics approaches: an overview. Proteomes 2013; 1(3): 180-218.
[http://dx.doi.org/10.3390/proteomes1030180] [PMID: 28250403]

[7] Broadbent JA, Broszczak DA, Tennakoon IU, Huygens F. Pan-proteomics, a concept for unifying quantitative proteome measurements when comparing closely-related bacterial strains. Expert Rev Proteomics 2016; 13(4): 355-65.
[http://dx.doi.org/10.1586/14789450.2016.1155986] [PMID: 26889693]

[8] Tahir Ul Qamar M, Zhu X, Khan MS, Xing F, Chen LL. Pan-genome: A promising resource for noncoding RNA discovery in plants. Plant Genome 2020; 13(3): e20046.
[http://dx.doi.org/10.1002/tpg2.20046] [PMID: 33217199]

[9] Tahir Ul Qamar M, Zhu X, Xing F, Chen L-L. ppsPCP: a plant presence/absence variants scanner and pan-genome construction pipeline. Bioinformatics 2019; 35(20): 4156-8.
[http://dx.doi.org/10.1093/bioinformatics/btz168] [PMID: 30851098]

[10] Bhatti A, Shah FS, Azhar J, Ahmad S, John P. Pan-transcriptomics and its applications.Pan-genomics: Applications, Challenges, and Future Prospects. Academic Press 2020; pp. 343-56.
[http://dx.doi.org/10.1016/B978-0-12-817076-2.00018-4]

[11]   Hirsch CN, Foerster JM, Johnson JM, *et al.* Insights into the maize pan-genome and pan-transcriptome. Plant Cell 2014; 26(1): 121-35.
[http://dx.doi.org/10.1105/tpc.113.119982] [PMID: 24488960]

[12]   McWhite CD, Papoulas O, Drew K, *et al.* A pan-plant protein complex map reveals deep conservation and novel assemblies. Cell 2020; 181(2): 460-474.e14.
[http://dx.doi.org/10.1016/j.cell.2020.02.049] [PMID: 32191846]

[13]   Murugaiyan J, Eravci M, Weise C, *et al.* Pan-proteomic analysis and elucidation of protein abundance among the closely related *brucella* species, *brucella abortus* and *brucella melitensis.* Biomolecules 2020; 10(6): E836.
[http://dx.doi.org/10.3390/biom10060836] [PMID: 32486122]

[14]   Zhang L, Xiao D, Pang B, *et al.* The core proteome and pan proteome of Salmonella Paratyphi A epidemic strains. PLoS One 2014; 9(2): e89197.
[http://dx.doi.org/10.1371/journal.pone.0089197] [PMID: 24586590]

[15]   Tahir Ul Qamar M, Faryad A, Bari A, Zahid B, Zhu X, Chen L-L. Effectiveness of conventional crop improvement strategies vs omics environment, climate, plant and vegetation growth 1. Springer 2020; pp. 253-84.
[http://dx.doi.org/10.1007/978-3-030-49732-3_11]

[16]   Yu L-R, Stewart NA, Veenstra TD. Proteomics: the deciphering of the functional genome. In: Ginsburg GS, Willard HF, Eds. Essentials of genomic and personalized medicine. San Diego: Academic Press 2010; pp. 89-96.
[http://dx.doi.org/10.1016/B978-0-12-374934-5.00008-8]

[17]   Zhang Z, Wu S, Stenoien DL, Paša-Tolić L. High-throughput proteomics. Annu Rev Anal Chem (Palo Alto, Calif) 2014; 7(1): 427-54.
[http://dx.doi.org/10.1146/annurev-anchem-071213-020216] [PMID: 25014346]

[18]   R Apweiler, A Bairoch, CH Wu. UniProt: the universal protein knowledgebase. Nucleic Acids Res 2017; 45(D1): D158-69.
[http://dx.doi.org/10.1093/nar/gkw1099] [PMID: 27899622]

[19]   Altschul SF, Madden TL, Schäffer AA, *et al.* Gapped BLAST and PSI-BLAST: a new generation of protein database search programs. Nucleic Acids Res 1997; 25(17): 3389-402.
[http://dx.doi.org/10.1093/nar/25.17.3389] [PMID: 9254694]

[20]   Li L, Stoeckert CJ Jr, Roos DS. OrthoMCL: identification of ortholog groups for eukaryotic genomes. Genome Res 2003; 13(9): 2178-89.
[http://dx.doi.org/10.1101/gr.1224503] [PMID: 12952885]

[21]   Plewniak F, Bianchetti L, Brelivet Y, *et al.* PipeAlign: A new toolkit for protein family analysis. Nucleic Acids Res 2003; 31(13): 3829-32.
[http://dx.doi.org/10.1093/nar/gkg518] [PMID: 12824430]

[22]   Emms DM, Kelly S. OrthoFinder: phylogenetic orthology inference for comparative genomics. Genome Biol 2019; 20(1): 238.
[http://dx.doi.org/10.1186/s13059-019-1832-y] [PMID: 31727128]

[23]   Fu L, Niu B, Zhu Z, Wu S, Li W. CD-HIT: accelerated for clustering the next-generation sequencing data. Bioinformatics 2012; 28(23): 3150-2.
[http://dx.doi.org/10.1093/bioinformatics/bts565] [PMID: 23060610]

[24]   Wang Y, Coleman-Derr D, Chen G, Gu YQ. OrthoVenn: a web server for genome wide comparison and annotation of orthologous clusters across multiple species. Nucleic Acids Res 2015; 43(W1): W78-84.
[http://dx.doi.org/10.1093/nar/gkv487] [PMID: 25964301]

[25]   Van Der Spoel D, Lindahl E, Hess B, Groenhof G, Mark AE, Berendsen HJ. GROMACS: fast, flexible, and free. J Comput Chem 2005; 26(16): 1701-18.

[http://dx.doi.org/10.1002/jcc.20291] [PMID: 16211538]

[26] Papaleo E, Mereghetti P, Fantucci P, Grandori R, De Gioia L. Free-energy landscape, principal component analysis, and structural clustering to identify representative conformations from molecular dynamics simulations: the myoglobin case. J Mol Graph Model 2009; 27(8): 889-99.
[http://dx.doi.org/10.1016/j.jmgm.2009.01.006] [PMID: 19264523]

[27] Consortium U. UniProt: a hub for protein information. Nucleic Acids Res 2015; 43(Database issue): D204-12.
[http://dx.doi.org/10.1093/nar/gku989] [PMID: 25348405]

[28] Consortium GO. Gene Ontology Consortium: going forward. Nucleic Acids Res 2015; 43(Database issue): D1049-56.
[http://dx.doi.org/10.1093/nar/gku1179] [PMID: 25428369]

[29] Kanehisa M, Araki M, Goto S, *et al.* KEGG for linking genomes to life and the environment. Nucleic Acids Res 2008; 36(Database issue) (Suppl. 1): D480-4.
[PMID: 18077471]

[30] Conesa A, Götz S, García-Gómez JM, Terol J, Talón M, Robles M. Blast2GO: a universal tool for annotation, visualization and analysis in functional genomics research. Bioinformatics 2005; 21(18): 3674-6.
[http://dx.doi.org/10.1093/bioinformatics/bti610] [PMID: 16081474]

[31] Keerthikumar S. An Introduction to Proteome Bioinformatics. Methods Mol Biol 2017; 1549: 1-3.
[http://dx.doi.org/10.1007/978-1-4939-6740-7_1] [PMID: 27975279]

[32] Cho WC. Proteomics technologies and challenges. Genomics Proteomics Bioinformatics 2007; 5(2): 77-85.
[http://dx.doi.org/10.1016/S1672-0229(07)60018-7] [PMID: 17893073]

[33] Gouveia D, Grenga L, Pible O, Armengaud J. Quick microbial molecular phenotyping by differential shotgun proteomics. Environ Microbiol 2020; 22(8): 2996-3004.
[http://dx.doi.org/10.1111/1462-2920.14975] [PMID: 32133743]

[34] Vedøy OB, Hanevik K, Sakkestad ST, Sommerfelt H, Steinsland H. Proliferation of enterotoxigenic *Escherichia coli* strain TW11681 in stools of experimentally infected human volunteers. Gut Pathog 2018; 10: 46.
[http://dx.doi.org/10.1186/s13099-018-0273-6] [PMID: 30349586]

[35] Christie-Oleza JA, Piña-Villalonga JM, Bosch R, Nogales B, Armengaud J. Comparative proteogenomics of twelve Roseobacter exoproteomes reveals different adaptive strategies among these marine bacteria. Mol Cell Proteomics 2012; 11(2)

[36] Tavares GC, Pereira FL, Barony GM, *et al.* Delineation of the pan-proteome of fish-pathogenic Streptococcus agalactiae strains using a label-free shotgun approach. BMC Genomics 2019; 20(1): 11.
[http://dx.doi.org/10.1186/s12864-018-5423-1] [PMID: 30616502]

[37] Jhingan GD, Kumari S, Jamwal SV, *et al.* Comparative proteomic analyses of avirulent, virulent, and clinical strains of mycobacterium tuberculosis identify strain-specific patterns. j biol chem 2016; 291(27): 14257-73.
[http://dx.doi.org/10.1074/jbc.M115.666123] [PMID: 27151218]

[38] Távora FTPK, Santos C, Maximiano MR, *et al.* Pan Proteome of Xanthomonas campestris pv. campestris Isolates Contrasting in Virulence. Proteomics 2019; 19(13): e1900082.
[http://dx.doi.org/10.1002/pmic.201900082] [PMID: 31050381]

[39] Nirujogi RS, Muthusamy B, Kim MS, *et al.* Secretome analysis of diarrhea-inducing strains of Escherichia coli. Proteomics 2017; 17(6): 10.
[http://dx.doi.org/10.1002/pmic.201770040] [PMID: 28070933]

[40] Hassan A, Naz A, Obaid A, *et al.* Pangenome and immuno-proteomics analysis of Acinetobacter baumannii strains revealed the core peptide vaccine targets. BMC Genomics 2016; 17(1): 732.

[http://dx.doi.org/10.1186/s12864-016-2951-4] [PMID: 27634541]

[41]   Silva WM, Sousa CS, Oliveira LC, *et al.* Comparative proteomic analysis of four biotechnological strains *Lactococcus lactis* through label-free quantitative proteomics. Microb Biotechnol 2019; 12(2): 265-74.
[http://dx.doi.org/10.1111/1751-7915.13305] [PMID: 30341804]

[42]   Mihăşan M, Babii C, Aslebagh R, Channaveerappa D, Dupree EJ, Darie CC. Exploration of Nicotine Metabolism in Paenarthrobacter nicotinovorans pAO1 by Microbial Proteomics. Adv Exp Med Biol 2019; 1140: 515-29.
[http://dx.doi.org/10.1007/978-3-030-15950-4_30] [PMID: 31347068]

[43]   Ziebandt AK, Kusch H, Degner M, *et al.* Proteomics uncovers extreme heterogeneity in the Staphylococcus aureus exoproteome due to genomic plasticity and variant gene regulation. Proteomics 2010; 10(8): 1634-44.
[http://dx.doi.org/10.1002/pmic.200900313] [PMID: 20186749]

[44]   Naz K, Naz A, Ashraf ST, *et al.* PanRV: Pangenome-reverse vaccinology approach for identifications of potential vaccine candidates in microbial pangenome. BMC Bioinformatics 2019; 20(1): 123.
[http://dx.doi.org/10.1186/s12859-019-2713-9] [PMID: 30871454]

[45]   Hasan M, Azim KF, Imran MAS, *et al.* Comprehensive genome based analysis of *Vibrio parahaemolyticus* for identifying novel drug and vaccine molecules: Subtractive proteomics and vaccinomics approach. PLoS One 2020; 15(8): e0237181.
[http://dx.doi.org/10.1371/journal.pone.0237181] [PMID: 32813697]

[46]   Sayers EW, Barrett T, Benson DA, *et al.* Database resources of the national center for biotechnology information. nucleic acids res 2011; 39(Database issue): D38-51.
[http://dx.doi.org/10.1093/nar/gkq1172] [PMID: 21097890]

[47]   Consortium U. Activities at the Universal Protein Resource (UniProt). Nucleic Acids Res 2014; 42(Database issue): D191-8.
[PMID: 24253303]

[48]   Yan F, Gao F. A systematic strategy for the investigation of vaccines and drugs targeting bacteria. Comput Struct Biotechnol J 2020; 18: 1525-38.
[http://dx.doi.org/10.1016/j.csbj.2020.06.008] [PMID: 32637049]

[49]   Huang Y, Niu B, Gao Y, Fu L, Li W. CD-HIT Suite: a web server for clustering and comparing biological sequences. Bioinformatics 2010; 26(5): 680-2.
[http://dx.doi.org/10.1093/bioinformatics/btq003] [PMID: 20053844]

[50]   Hossain MU, Khan MA, Hashem A, *et al.* Finding Potential Therapeutic Targets against *Shigella flexneri* through Proteome Exploration. Front Microbiol 2016; 7: 1817.
[http://dx.doi.org/10.3389/fmicb.2016.01817] [PMID: 27920755]

[51]   Sivashanmugam M, Nagarajan H, Vetrivel U, Ramasubban G, Therese KL, Narahari MH. In silico analysis and prioritization of drug targets in *Fusarium solani*. Med Hypotheses 2015; 84(2): 81-4.
[http://dx.doi.org/10.1016/j.mehy.2014.12.015] [PMID: 25555413]

[52]   Zhang R, Ou HY, Zhang CT. DEG: a database of essential genes. Nucleic Acids Res 2004; 32(Database issue): D271-2.
[http://dx.doi.org/10.1093/nar/gkh024] [PMID: 14681410]

[53]   Peng C, Lin Y, Luo H, Gao F. A comprehensive overview of online resources to identify and predict bacterial essential genes. Front microbiol 2017; 8: 2331.
[http://dx.doi.org/10.3389/fmicb.2017.02331] [PMID: 29230204]

[54]   Kanehisa M, Goto S. KEGG: kyoto encyclopedia of genes and genomes. Nucleic Acids Res 2000; 28(1): 27-30.
[http://dx.doi.org/10.1093/nar/28.1.27] [PMID: 10592173]

[55]   Schilling CH, Schuster S, Palsson BO, Heinrich R. Metabolic pathway analysis: basic concepts and

scientific applications in the post-genomic era. Biotechnol Prog 1999; 15(3): 296-303.
[http://dx.doi.org/10.1021/bp990048k] [PMID: 10356246]

[56] Moriya Y, Itoh M, Okuda S, Yoshizawa AC, Kanehisa M. an automatic genome annotation and pathway reconstruction server. Nucleic Acids Res 2007; 35: 182-5. (Web Server issue):
[http://dx.doi.org/10.1093/nar/gkm321]

[57] Wishart DS, Feunang YD, Guo AC, *et al.* DrugBank 5.0: a major update to the DrugBank database for 2018. Nucleic Acids Res 2018; 46(D1): D1074-82.
[http://dx.doi.org/10.1093/nar/gkx1037] [PMID: 29126136]

[58] Bakheet TM, Doig AJ. Properties and identification of antibiotic drug targets. BMC Bioinformatics 2010; 11: 195.
[http://dx.doi.org/10.1186/1471-2105-11-195] [PMID: 20406434]

[59] Zagursky RJ, Olmsted SB, Russell DP, Wooters JL. Bioinformatics: how it is being used to identify bacterial vaccine candidates. Expert Rev Vaccines 2003; 2(3): 417-36.
[http://dx.doi.org/10.1586/14760584.2.3.417] [PMID: 12903807]

[60] Yu NY, Wagner JR, Laird MR, *et al.* PSORTb 3.0: improved protein subcellular localization prediction with refined localization subcategories and predictive capabilities for all prokaryotes. Bioinformatics 2010; 26(13): 1608-15.
[http://dx.doi.org/10.1093/bioinformatics/btq249] [PMID: 20472543]

[61] Yu CS, Chen YC, Lu CH, Hwang JK. Prediction of protein subcellular localization. Proteins 2006; 64(3): 643-51.
[http://dx.doi.org/10.1002/prot.21018] [PMID: 16752418]

[62] Petersen TN, Brunak S, von Heijne G, Nielsen H. SignalP 4.0: discriminating signal peptides from transmembrane regions. Nat Methods 2011; 8(10): 785-6.
[http://dx.doi.org/10.1038/nmeth.1701] [PMID: 21959131]

[63] King BR, Guda C. nGLOC: an n-gram-based Bayesian method for estimating the subcellular proteomes of eukaryotes. Genome Biol 2007; 8(5): R68.
[http://dx.doi.org/10.1186/gb-2007-8-5-r68] [PMID: 17472741]

[64] Hochholdinger F, Sauer M, Dembinsky D, *et al.* Proteomic dissection of plant development. Proteomics 2006; 6(14): 4076-83.
[http://dx.doi.org/10.1002/pmic.200500851] [PMID: 16786485]

[65] Tress ML, Abascal F, Valencia A. Alternative splicing may not be the key to proteome complexity. Trends Biochem Sci 2017; 42(2): 98-110.
[http://dx.doi.org/10.1016/j.tibs.2016.08.008] [PMID: 27712956]

[66] Tress ML, Bodenmiller B, Aebersold R, Valencia A. Proteomics studies confirm the presence of alternative protein isoforms on a large scale. Genome Biol 2008; 9(11): R162.
[http://dx.doi.org/10.1186/gb-2008-9-11-r162] [PMID: 19017398]

[67] Roe MR, Griffin TJ. Gel-free mass spectrometry-based high throughput proteomics: tools for studying biological response of proteins and proteomes. Proteomics 2006; 6(17): 4678-87.
[http://dx.doi.org/10.1002/pmic.200500876] [PMID: 16888762]

[68] Li Z, Czarnecki O, Chourey K, *et al.* Strigolactone-regulated proteins revealed by iTRAQ-based quantitative proteomics in Arabidopsis. J Proteome Res 2014; 13(3): 1359-72.
[http://dx.doi.org/10.1021/pr400925t] [PMID: 24559214]

[69] da Silva WM, Seyffert N. Pan-proteomics: Technologies, applications, and challenges Pan-genomics: Applications, Challenges, and Future Prospects. Elsevier 2020; pp. 357-69.
[http://dx.doi.org/10.1016/B978-0-12-817076-2.00019-6]

[70] Broadbent J, Stockwell S, Byrne K, Bose U, Dillon S, Ramm K, Eds. Wheat pan-proteomics: Unifying data-independent LC-MS proteome measurements across diverse genetic backgrounds for trait screening and classification ASMS. Atlanta, GA, USA: ASMS 2019.

[71] Boehmer JL, Ward JL, Peters RR, Shefcheck KJ, McFarland MA, Bannerman DD. Proteomic analysis of the temporal expression of bovine milk proteins during coliform mastitis and label-free relative quantification. J Dairy Sci 2010; 93(2): 593-603.
[http://dx.doi.org/10.3168/jds.2009-2526] [PMID: 20105531]

[72] Trapp J, Almunia C, Gaillard J-C, *et al.* Proteogenomic insights into the core-proteome of female reproductive tissues from crustacean amphipods. J Proteomics 2016; 135: 51-61.
[http://dx.doi.org/10.1016/j.jprot.2015.06.017] [PMID: 26170043]

[73] Almeida AM, Bassols A, Bendixen E, *et al.* Animal board invited review: advances in proteomics for animal and food sciences. Animal 2015; 9(1): 1-17.
[http://dx.doi.org/10.1017/S1751731114002602] [PMID: 25359324]

[74] Thul PJ, Lindskog C. The human protein atlas: A spatial map of the human proteome. Protein Sci 2018; 27(1): 233-44.
[http://dx.doi.org/10.1002/pro.3307] [PMID: 28940711]

[75] Hibbert SA, Ozols M, Griffiths CEM, Watson REB, Bell M, Sherratt MJ. Defining tissue proteomes by systematic literature review. Sci Rep 2018; 8(1): 546.
[http://dx.doi.org/10.1038/s41598-017-18699-8] [PMID: 29323144]

[76] Semba RD, Enghild JJ, Venkatraman V, Dyrlund TF, Van Eyk JE. The Human Eye Proteome Project: perspectives on an emerging proteome. Proteomics 2013; 13(16): 2500-11.
[http://dx.doi.org/10.1002/pmic.201300075] [PMID: 23749747]

[77] Liu Y, Liao Y, Xiang L, *et al.* A panel of autoantibodies as potential early diagnostic serum biomarkers in patients with breast cancer. Int J Clin Oncol 2017; 22(2): 291-6.
[http://dx.doi.org/10.1007/s10147-016-1047-0] [PMID: 27778118]

[78] Na SS, Aldonza MB, Sung HJ, *et al.* Stanniocalcin-2 (STC2): A potential lung cancer biomarker promotes lung cancer metastasis and progression. Biochim Biophys Acta 2015; 1854(6): 668-76.
[http://dx.doi.org/10.1016/j.bbapap.2014.11.002] [PMID: 25463045]

[79] Yang J, Xiong X, Wang X, Guo B, He K, Huang C. Identification of peptide regions of SERPINA1 and ENOSF1 and their protein expression as potential serum biomarkers for gastric cancer. Tumour Biol 2015; 36(7): 5109-18.
[http://dx.doi.org/10.1007/s13277-015-3163-2] [PMID: 25677901]

[80] Weinstein JN, Collisson EA, Mills GB, *et al.* The cancer genome atlas pan-cancer analysis project. Nat Genet 2013; 45(10): 1113-20.
[http://dx.doi.org/10.1038/ng.2764] [PMID: 24071849]

[81] Tomczak K, Czerwińska P, Wiznerowicz M. The Cancer Genome Atlas (TCGA): an immeasurable source of knowledge. Contemp Oncol (Pozn) 2015; 19(1A): A68-77.
[http://dx.doi.org/10.5114/wo.2014.47136] [PMID: 25691825]

CHAPTER 5

Functional Characterization of Proteins and Peptides Using Computational Approaches

Zeshan Haider[1,2] and **Adnan K. Niazi**[1,*]

[1] *Centre of Agricultural Biochemistry and Biotechnology (CABB), University of Agriculture Faisalabad, Faisalabad, 38040, Pakistan*

[2] *State Key Laboratory of Grassland Agro-Ecosystems, College of Pastoral Agriculture Science and Technology, Lanzhou University, Lanzhou, China*

Abstract: Bioinformatics tools have produced enormous data in different repositories over the past decade, which is of great interest today for *in-silico* analysis. Proteins contain one or more peptides that are essential molecules having biological and biomedical functions. Instead of working with sequences, proteomics, and peptidomics, researchers now concentrate more on molecules' processes and metabolic interactions in which proteins or peptides are involved. As a preliminary assessment of possible biological roles, bioinformatics methods are an essential step and greatly reduce experimental verification time and cost. This chapter offers a brief overview of computational methods for predicting the biological features of peptides and proteins. Algorithms using structural, evolutionary, or statistical patterns and strategies based on molecular docking are considered based on machine learning techniques, which are the most common today. The protein and bioactive peptide databases are reported, providing the knowledge required to develop new algorithms. The biological functions for forecasting, the features of proteins, and peptides should be considered, based on the possibility of concluding their natural role. The report includes a list of online resources that researchers can use to evaluate possible protein function and peptides.

Keywords: Bioinformatics, Biological functions, Databases, Machine learning, Molecular docking, Peptides and protein.

1. INTRODUCTION

Proteomics, including peptidomics, is one of the most rapidly developing areas of biochemistry nowadays. These science branches study the totality of proteins and protein fragments (peptides), their functions, and interactions in living organisms [1 - 3]. Mass spectrometry, various electrophoretic approaches, and liquid chro-

---

* **Corresponding author Adnan K. Niazi:** Centre of Agricultural Biochemistry and Biotechnology (CABB), University of Agriculture Faisalabad, Faisalabad, 38040, Pakistan; E-mail: niazi@uaf.edu.pk, adnan1753@yahoo.com

Muhammad Sarwar Khan (Ed)

matography are the primary study methods in proteomics and peptidomics [3 - 5]. To classify and further analyze peptides revealed by mass spectrometry, special algorithms are commonly used. The data obtained allow peptide abundance analysis in different samples (comparative peptidomics), determine the position of a peptide in a protein sequence, elucidate the proteases involved in peptide generation, and predict the structures and functions of the peptide [6]. One of such studies' potential objectives is to look for and evaluate bioactive peptides that, by binding to particular receptors or other targets, modulate physiological functions [7]. In higher organisms like a human, bioactive peptides can affect almost all systems. Antithrombotic peptides play a significant role in inhibiting platelet aggregation and fibrinogen binding, hypotensive peptides that inhibit angiotensin-transforming enzymes and antioxidant peptides that can scavenge free radicals (produced as by-products of cell oxidative metabolism) and inhibit lipid peroxidation are influenced by the cardiovascular system [8, 9]. Such peptides bind the cell surfaces of bacteria and thus inhibit the work of membranes or kill bacterial biofilms. In addition, certain peptides have an antiviral activity or antifungal activity [10].

Technological advances in DNA sequencing have made whole-genome sequencing a hot topic of current research. However, scientists' key challenges regarding sequence data entry are to elucidate the role of proteins and large-scale analysis of entire proteomes (the protein component of genomes) rapidly and accurately. Many local databases exist, *i.e.*, UniProtKB, GenBank, RefSeq and TPA, and Swissport, PIR, PRF, and PDB that record the processed protein sequences [11]. To equate a target sequence to those of known functions, scientists typically use sequence similarity searches. Still, this approach has its drawbacks and depends on the precision of the remaining data offered to date. Protein signatures are used in alternative strategies for the classification of proteins. Some common databases that establish diagnostic protein signatures have emerged for recognized protein families or domains. Each of these databases has its methodology, signature generation parameters, and methods.

As a consequence, it also has strengths and shortcomings of its own [12]. *In-silico* draws on the techniques of bioinformatics to classify proteins and peptides. It offers references to protein and peptide databases. It works on strategies to develop algorithms to study proteins and peptides that would be of interest to specialists working in the field.

2. *IN-SILICO* ANALYSIS OF PEPTIDES

Peptides are biochemical compounds present in nature. In all living species, peptides are present naturally and play a key role in all biological activities, and are synthesized by DNA transcription, like proteins [13 - 15].

There are only 20 amino acids that can be mixed into a lot of diverse molecules. If the molecule is 2 to 50 amino acids in length, is a peptide, or is a longer chain of more than 50 amino acids, it is commonly referred to as a protein. For peptide toxicity, *in-silico* based databases and peptide sequence analysis tools have been developed., which have revolutionized peptidomics.

2.1. Classification and Databases of Peptides

Peptides are deemed to be appropriate resources in different biological fields. They can be used mainly for the rational design of molecules that are bioactive. In the manufacture of targeted drugs and diagnostics, vaccine production, or agriculture, they may be used as ligands. You can classify peptides into two broad structural classes: linear peptides and cyclic peptides. A special type of macrocyclic ring polypeptides, monocyclic peptides, exhibit advantages such as more selective binding and target receptor absorption and greater efficiency and stability compared to rectilinear forms [16]. There are many different online databases and tools available to retrieve the different types of peptides sequences. Such databases and tools are listed below in Table **1**.

Table 1. Different types of peptides-based databases and functions.

Sr.no.	Database name	Biological Type of Peptide	Online Link	Kind of Data to be Stored
1	APD	**Antimicrobial and Antiviral Peptides**	http://aps.unmc.edu/AP/	Antimicrobial peptides and their activity
2	ParaPep		http://crdd.osdd.net/raghava/parapep/	Antiparasitic peptides and their structures
3	CAMPR3		http://www.camp3.bicnirrh.res.in/	Antimicrobial peptides
4	DBAASP		https://dbaasp.org/	Antimicrobial peptides, their activities, and structures
5	AVPdb		http://crdd.osdd.net/servers/avpdb/	Antiviral peptides
6	AVPpred		http://crdd.osdd.net/servers/avppred/	antiviral and NOT antiviral peptides
7	BACTIBASE 2.0		http://bactibase.pfba-lab-tun.org.	Bacteriocins, proteins and peptides produced by bacteria suppressing other strains of the same or related species
8	DADP		http://split4.pmfst.hr/dadp/	Protective peptides of 167 species of tailless amphibians
9	BaAMPs		BaAMPs - Home	Peptides destroying bacterial biofilms
10	MilkAMP		milkampdb.org	milk antimicrobial peptides
11	CancerPPD		CancerPPD (osdd.net)	Anticancer peptides
12	TumorHoPe		Home Page of TumorHoPe a Database of Tumor Homing Peptides (osdd.net)	Tumour-migrating peptides
13	Allergome	**Immunology Peptides**	http://www.allergome.org/	Allergens
14	AgAbDb		Antigen-Antibody Interaction Database	Protein and peptide antigens
15	HPVdb		Human Short Peptide Variation Database (bioinformatics.nl)	Epitopes of human MHC and T cells
16	AHTPDB	**Hypotensive Peptides**	AHTPDB: Anti-Hypertensive Inhibiting peptide database (osdd.net)	Hypotensive peptides
17	ACEpepDB		www.cftri.com	Food-derived hypotensive peptides

(Table 1) cont.....

Sr.no.	Database name	Biological Type of Peptide	Online Link	Kind of Data to be Stored
18	ArachnoServer 2.0	**Toxins**	Page Not Found | ArachnoServer	Spider toxins
19	ATDB		湖南师范大学 (hunnu.edu.cn)	Animal toxins
20	ConoServer		ConoServer	Conopeptides, peptides of marine carnivorous cone-shaped snails
21	DBETH		DBETH (iicb.res.in)	Bacteria toxins
22	CPPsite 2.0	**Cell-penetrating Peptides**	CPPsite 2.0: A Database of Cell-Penetrating Peptides (osdd.net)	Cell-penetrating peptides, their secondary and tertiary structures

2.2. Algorithm for Prediction of Peptides and their Function

Algorithms can be divided into two broad classes to search for peptides and predict their functions, generated either using machine learning techniques or without machine learning (historical ones). Each of these methods focuses on their work: looking for similar fragments, searching for evolutionary reasons, and searching for statistical patterns.

2.2.1. Search for Similar Fragments

Based on the search for related fragments in peptide amino acid sequences, the first method remains the simplest and most widely used one. Linked pieces are expected to perform similar functions. The probability that the proteins will have the same procedure will be over 90 percent if the protein amino acid sequences have 70 percent similarity [17].

Besides, short linear motifs, for instance, SLiMs, are responsible for protein-peptide interactions in eukaryotes cells [18, 19]. In general, the minimum length of the amino acid sequence is needed to involve in protein-protein business and immune interaction [19]. It is highly likely to perform similar functions for a long time with the same arrangements. The following are the scan for related sequences; algorithms are most commonly used: NCBI BLAST for local alignment, TCoffee or Clustal for multiple alignments, and HMMER to look for poor patterns of homology or series. For instance, these techniques have been used to scan for new peptides in crustaceans [20 - 24].

2.2.2. Search for Evolutionary Conservation

Evolutionary preservation of its sequence can be the second attribute that demonstrates the bioactivity of a certain peptide. For example, in the search for peptides controlling platelet function, orthologs were searched among vertebrates [25]. About 47 peptides with transmembrane domains were chosen from almost three thousand proteins expressed in platelets. Ten consecutive amino acid sequences near the membrane in the cytoplasm and orthologs among vertebrates have been selected to find functional peptides. The peptides are proven to be bioactive experimentally. Work can also refer to this category by Michael and co-authors, who used evolutionary conservation as an additional criterion to check for bioactive peptides using motif resemblance [26].

2.2.3. Search for Statistical Patterns

The third category of algorithms that do not use machine learning is focused on statistical pattern search. These patterns are mainly searched among the sequences of amino acids and physicochemical properties, as the latter describe the roles of bioactive peptides. As the effect of most pharmacologic peptides is based on the ability to inhibit the transforming angiotensin enzyme. The pharmacologic peptides are considered to have an obligation to have aromatic or necessary N-terminal and C-terminal amino acids, positive and hydrophobic amino acids [27]. To bind the enzyme requires this structure. Because they interact with the anionic cell wall and bacterial membrane, most antimicrobial peptides have an amphiphilic system [28, 29].

2.3. Prediction Features for Prediction of Peptides Model

For model growth, three types of peptide characteristics are used in this analysis, *i.e.*, the structure of amino acids, atomic composition, and descriptors of chemistry. Here is a concise description of the following characteristics:

2.3.1. Amino Acid Composition

The composition of amino acids in proteins or peptides is preserved from species to species and from various protein or peptide groups, so the design of amino acids can be distinguished the characteristics of two types of peptides [30]. The composition of amino acids reflects the fraction in a peptide. Each amino acid is characterized by a trajectory of 20 amino acids after each of the amino acids of the twenty amino acids. The structure of amino acids was computed using the following equation (equation 1):

$$\text{Composition of an amino acid } (a.a) = \frac{\text{Frequency of amino acid } (a.a)}{\text{Length of the peptides}} \tag{1}$$

Where *a.a* can be any natural amino acid.

2.3.2. Atomic Composition of Amino Acids

The frequency of each atom in the amino acid composition is measured by evaluating each amino acid's chemical formula. Five bits are made up of natural amino acids (C, H, N, O, S), so we first use these five atom frequencies as prediction features. Since peptides consist of 20 types of amino acids, we have calculated them. The number and type of atoms in each amino, as shown in Table **2**, for each amino.

Table 2. The atomic composition of twenty amino acids.

Sr.No.	A.A.	Types of Atoms					Chemical Formula	Single Bonds (-)	Double Bonds (=)
		Carbon	Hydrogen	Nitrogen	Oxygen	Sulphur			
1	Ala	Three	Seven	One	Two	Zero	C3H7N1O2	Eleven	One
2	Cys	Three	Seven	One	Two	One	C3H7N1O2S1	Twelve	One
3	Asp	Four	Seven	One	Four	Zero	C4H7N1O4	Thirteen	Two
4	Glu	Five	Nine	One	Four	Zero	C5H9N1O4	Sixteen	Two
5	Phe	Nine	Eleven	One	Two	Zero	C9H11N1O2	Nineteen	Four
6	Gly	Two	Five	One	Two	Zero	C2H5N1O2	Eight	One
7	His	Six	Nine	Three	Two	Zero	C6H9N3O2	Seventeen	Three
8	Ile	Six	Thirteen	One	Two	Zero	C6H13N1O2	Twenty	One
9	Lys	Six	Fourteen	Two	Two	Zero	C6H14N2O2	Twenty-two	One
10	Leu	Six	Thirteen	One	Two	Zero	C6H13N1O2	Twenty	One
11	Met	Five	Eleven	One	Two	One	C5H11N1O2S1	Eighteen	One
12	Asn	Four	Eight	Two	Three	Zero	C4H8N2O3	Fourteen	Two
13	Pro	Five	Nine	One	Two	Zero	C5H9N1O2	Sixteen	One
14	Gln	Five	Ten	Two	Three	Zero	C5H10N2O3	Seventeen	Two
15	Arg	Six	Fourteen	Four	Two	Zero	C6H14N4O2	Twenty-three	Two

(Table 2) cont.....

Sr.No.	A.A.	Types of Atoms					Chemical Formula	Single Bonds (-)	Double Bonds (=)
		Carbon	Hydrogen	Nitrogen	Oxygen	Sulphur			
16	Ser	Three	Seven	One	Three	Zero	C3H7N1O3	Twelve	One
17	The	Four	Nine	One	Three	Zero	C4H9N1O3	Fifteen	One
18	Val	Five	Eleven	One	Two	Zero	C5H11N1O2	Seventeen	One
19	Trp	Eleven	Twelve	Two	Two	Zero	C11H12N2O2	Twenty-three	Five
20	Tyr	Nine	Eleven	One	Three	Zero	C9H11N1O3	Twenty	Four

2.3.3. Chemical Descriptors

Chemical descriptors play a significant role in the biological determination of any chemical molecule's behaviour. Therefore, they were utilized as crucial characteristics in designing QSAR models [31]. The PaDEL, open-source software to measure various types of the descriptor. By using PaDEL programme quantified 15,537 descriptor types, involving 1D, 2D, 3D and ten distinct binary fingerprint types [32]. In the past, it has been shown that biological activity does not correspond with all chemical descriptors [33]. So, it is easier to eliminate unnecessary descriptors that in the model may produce noise or over-fit. First, we choose the required set of descriptors to create QSAR models using Weka software at default parameters (maximum variance percentage, M= 99). For the delete descriptors that either do not differ or vary too much at the maximum variance percentage, the 'Remove Useless' function was added [34]. Another element evaluator, 'Cfs Subset Eval', was applied with 'Best First' as a search tool with default forward direction settings parameters (lookup size, D=1 and backtracking quantity, N=5) Weka software. In the third step, the F-stepping technique was applied to decrease further the descriptors that are not meaningful. All the good descriptors were retained from the "BestFirst" algorithm along with F-stepping and evaluated the output by extracting each descriptor separately (This is also referred to as the descriptor selection reverse direction). If the output is improved or unchanged by eliminating a descriptor, the descriptor is constantly removed. On the other side, they put back the descriptor if the output is diminished by removing a descriptor, and this step has been repeated for each descriptor. Finally, using the leave-one-out cross-validation method, finally selected descriptors were used as input for training and testing SVM-based QSAR models [35].

3. *IN-SILICO* ANALYSIS OF PROTEINS

3.1. Protein Databases for Sequence Retrieval

Several new protein analysis technologies allow numerous proteins of complex shapes to be quickly recognized, protein-protein interactions to be mapped in the cellular context, their cell location to be confirmed, and their biological behaviour to be analyzed. As a central repository for storing and publishing the data from these science efforts, protein sequence databases play an important role. In the conventional sense, large-scale experimental data is often not released but is stored in a database. The proteins databases are the most extensive collection of knowledge about proteins that scientists can access sequences the protein sequence from databases.

It is important to distinguish between them and understand the types of data they contain to get the most out of the various tools. Universal protein databases protect all species' proteins, whereas specialized data sets have or are associated with a particular protein family or category of proteins. Besides, universal protein sequence databases may be divided into master sequence data files or sequence repositories where data is processed with minimal or no manual data.

Many databases of protein sequences serve as online platforms for the retrieval of protein sequences. These databases add nothing to the flow records they hold or no additional data and usually do not seek to construct a non-redundant row for users. GenBank, or GenPept, a gene product database created by the National Center for Biotechnology Information, is an example of this (NCBI) (https://www.ncbi.nlm.nih.gov/) [36]. Database entries are derived from translations of nucleotide database sequences maintained together by DDBJ [37], EMBL Nucleotide Sequence Database [38], and GenBank [39] and comprised limited annotation primarily obtained by the entrance of the corresponding nucleotide. The records are devoid of additional annotations, and no proteins collected from amino acid sequencing are included in the database. It offers a redundant view of the world of proteins, which means that several records represent each protein, and no effort is made to combine these records into a single document in the database. Another example of a sequence repository is NCBI's Entrez Protein database [36].

As many records contain additional information, the database differs from GenPept, but none of the annotated information includes additional information. New data is extracted from chosen databases and is not added to a word that cannot be found in other records. As in GenPept, the set of sequences is redundant. The RefSeq provided by the NCBI is a more optimistic initiative [40]. For key research activities, including genomic DNA, transcription (RNA), and protein products, the project aims to provide a fully integrated and non-redundant set of sequences. NCBI employees and collaborators ensure permanent storage, and the status of verification is indicated on each paper.

3.1.1. Organism-specific and Protein Family Based-databases

Several specialized databases are accessible to the life sciences community and the universal protein sequence databases. Others concentrate on the basic characterization of proteins or protein groups, families, or a particular organism, while others aim to combine available tools to make full use of them. The databases vary in the size and type of information that they contain. PDB, which stores three-dimensional structural structures, is an example of a database from the previous category. Genetic ontology (GO) is a dynamically regulated language which while gathering and refining information about the role of genes and proteins in cells, can be applied to all species [41]. The Gene Ontology project annotation [41], annotation of proteins to GO terms. IntAct stores information on protein interaction [42].

3.1.2. Protein Family-Based Databases

Protein family databases contain information about a specific family or group of proteins. Experts generally manage these databases in the field and, due to the limited nature of the data they contain, can provide a finer level of granularity than is usually possible. An example of such a database is MEROPS, a knowledge resource on peptidases and their inhibitors [43]. A summary page that describes the classification and nomenclature of proteins is given for each peptidase and includes links to additional pages that list identifiers of sequences, known structures, and bibliographic references. Proteins are classified using a hierarchical structure method in which each peptidase is assigned to a family based on statistically significant amino acid sequence similarity and families considered homologous are grouped together.

3.1.3. Organism Specified Protein Databases

The root of illnesses and biological processes in humans is a valuable tool for understanding model organisms. Various model systems of organisms provide the quantities of genetic, phenotypic, and protein-related information, and many databases that are directly applicable to the biology of these species have been created to collect this information and provide it to the scientific community. The FlyBase database [44] covering the genetics of Drosophila species, the Mouse

Genome Database [45] containing laboratory mouse-related information, the Worm Base [46] surrounding Caenorhabditis elegans, the Saccharomyces Genome Database (SGD) [47], the Saccharomyces protein, and the Arabidopsis Information Resource (TAIR) [48] are examples. These databases play an important role in providing integrated access to the data available about these organisms.

3.2. Classification of Protein Sequences

The grouping of similar items together is one of the simplest ways to understand very broad datasets, and the protein sequences may not vary. Dividing proteins into families decreases the total problem space and predicts the activity of proteins. Similar functions should be given to proteins with identical sequences. Thus, the group's uncharacterized sequences may be extracted from the feature when parallel lines are grouped, and one or more of these sequences have a known characteristic. Numerous methods have been developed to classify protein sequences, but perhaps the best methods are protein signatures, which mathematically describe protein sequences.

3.2.1. Methods for Classification of Protein Sequences

For the production of the protein signature, various related proteins are used. Sequences are expected there. It is impossible to recognize common characteristics of a family of proteins or domains from a single sequence. Still, it is possible to use the alignment of similar sequences to achieve consensus on a family of proteins or to classify domains. Or residues that have been preserved. The general function of such sequences, such as highly conserved residual triads, which may include an active site, is likely to contain highly conserved regions. These derived regions, which diagnose a family, domain, or place of functional proteins, can create protein signatures using various approaches. They are then used to scan the protein sequence requested to categories the family to which the protein belongs.

The simplest protein signature method uses regular expressions to map patterns of stored amino acid residues. The standard expression indicates which amino acids may appear or may not appear at each site. This pattern centre is tested against the set of sequences commented on and is optimized until only the right sequences in the test set are reached [49]. For the identification of highly conserved active and binding sites, regular expressions are helpful. Still, their lack of universality in identifying individual residual variants restricts their capacity to find more distant sequences.

Profiling is another commonly used protein signature technique. Profiles are produced based on different sequence alignments and arrays of positional based amino acid weights and spacing values or templates that specify the probability of an amino acid being present at a specific position in a sequence [50]. The numbers (scores) are used to estimate the similarity scores for a particular orientation between the profile and the sequence. A threshold score is calculated for any sequences to decide if the query sequence relates to the original set of matched sequences. A complementary profile-driven approach is the Hidden Markov Model (HMM), a predictive profile based on probabilities rather than predictions [51, 52]. In Sean Eddy's HMMER package, HMM is the most widely used method that allows a user to create an HMM from a sequence alignment and search for an HMM in a sequence database without understanding the functioning of the HMMs. Profiling and HMM are good protein signature strategies and compensate for regex shortcomings, usually covering wider sequence regions and differentiate between divergent family members [24].

3.2.2. Signatures Databases for Protein

A variety of public domain protein signature databases are used to generate analytic signatures for protein families and protein domains using the methods mentioned above, including sequence-based PROSITE [53], PRINTS [54], Pfam [55], SMART [56], TIGRFAM [57], PIRSF [58] and PANTHER [59], and structure-based SUPERFAMILY [60] and Gene3D [61]. Databases using sequence clustering and alignment methods are also available, such as ProDom [62], which clusters all of the database's proteins into families.

3.2.3. Super Integrated Signatures Databases for Proteins

In the protein families and domains-based databases, all the databases mentioned above have major overlaps. If consistency is beyond your coverage area, it would be inefficient to use only one of the important databases to test query consistency and expose you to limitations that the chosen database may have. Instead, at the same time, you are trying to use them together, but in essence. When trying to rationalize the various outcomes obtained at each place, location can create uncertainty. This InterPro database combines all the databases based on signatures of protein into a single database. PROSITE, PRINTS, Pfam, ProDom, SMART, TIGRFAM, PIRSF, SUPERFAMILY, Gene3D, and PANTHER signatures represent the same domain, family, repeat, active site, binding site, or post-translational modification are grouped with unique accession numbers into single InterPro entries. InterPro records are related (family and subfamily) as parent/child or are found in relationships (domain composition) that match a subset of proteins that match other records.

4. MOLECULAR DOCKING-BASED PEPTIDE PREDICTION

A separate method class is defined by molecular modelling methods based on knowledge on molecules' spatial structure, allowing the most likely orientation of a bioactive peptide to be predicted. And the target of its protein upon contact. When the action of a peptide has already been revealed or expected, these techniques are used. On the other hand, many peptides can be studied when the molecular structure is the target, and those that selectively bind the target can be revealed. Molecular docking can also be used for bioactive peptide hunting.

Unlike small molecules for which conformation space is limited, the docking process is more complicated for larger, more versatile peptides because their conformation can change significantly when protein interactions occur. Most algorithms built for molecular docking are, therefore, not applicable to peptides [63]. The peptide bioactivity analysis algorithm by docking usually includes the following steps: determining the most probable conformation of the peptide, checking for the binding site of the protein-peptide (may be omitted if) the interaction with a known position is being investigated), and determining the complex structure and evaluating the complex energy of formation.

The simplest services allow a protein fragment that is involved in protein-peptide interaction to be predicted. The PepSite online service (http://pepsite2. russelllab.org/), for example, belongs to this type; it uses weight matrices of spatial location. Calculated based on a broad array of identified peptide and protein complexes [64]. The service uses peptide amino acid sequence and protein spatial structure data on the protein surface to predict the peptide-binding site. Similar knowledge can be obtained using the PEP-SiteFinder algorithm, which indicates potential peptide conformations, carries out binding energy minimization docking, and assesses each amino acid residue's contribution to the interaction [65]. PEP-SiteFinder works much faster because molecules needless calculations. These are perceived as rigid bodies.

The Peptiderive (https://rosie.rosettacommons.org/peptiderive) algorithm is also an important one. It enables the search in protein-protein complexes for a linear peptide area of interaction, which can design a drug that inhibits the interaction between protein and protein [66]. The configuration of the receptor and the interaction partner complex requires input data. A preset sliding frame travels along the amino acid sequence of the protein partner chain and, with each resulting peptide, evaluates the binding energy. The peptide with the highest attraction energy is the product of the algorithm. Then by adding disulfide bonds to the peptide under analysis, the algorithm verifies whether the complex could be made any stronger. The division of the target sequence into segments is another way to predict peptide structure. Analysis of short fragments continues quicker and is similar to the gradual binding mechanism between a disordered protein.

The tertiary structure of a protein with a spatial structure is stable. However, it can also be used for peptide studies. The peptide-protein interaction analysis protocol comprises the following steps: a peptide overview with a collection of short fragments similar amino acid sequences (short binding motifs); fragment docking as rigid bodies; and structure optimization of the fragment complex assembled into a peptide using the Rosetta Flex-PepDock algorithm that can consider receptor and peptide conformational changes. In the research on 27 complexes, 52 percent of the structures were identified by the system [63, 67-68].

Also, as in the GalaxyTBM algorithm [69], a peptide's spatial structure can be determined by its resemblance to an existing structure. In the first stage of the algorithm, a prototype complex is searched for in the PepBind database based on the protein structure and peptide amino acid sequence [70]. A model minimizing the energy of interaction is constructed at the next level. The algorithm was evaluated on 22 complexes, 17 of which were correctly predicted.

CONCLUSION

The bioinformatics/*in-silico* approaches are being developed to make the process more effective in analyzing the biological roles of proteins and peptides. To identify potentially biologically significant peptides from proteins using machine learning techniques, algorithms can be developed. However, it also focused primarily on the study of different peptide properties: amino acid composition, physicochemical properties, and functional motifs. While there are methods by which a peptide can be functionally determined, most algorithms allow you to search for peptides that perform a particular function. Some algorithms that have been developed in practice to predict certain functions are more flexible than expected, which would probably only allow algorithms to identify a wide range of functional behaviour in the future.

Secondly, this chapter's tools and services are already aimed at reducing the load on the data stream of raw sequencing. In making all known sequences publicly available and, in the latter case, adding value to that data, sequence repositories and certain databases are vital. Databases of protein sequences such as UniProtKB serve as consolidated protein information resources, while prospective databases such as UniRef have smaller data subsets that can be better handled for analysis. The UniProtKB / TrEMBL. InterPro serves as a protein sequence classification system with its built-in protein signatures that enable the automatic prediction of functions.

Continuous research moves from a single gene approach to a whole-genome method, and bottom-up, hypothesis-based approaches are replaced by top-down analysis. Proteins and peptides are analyzed in a cell's sense to recognize which molecules or segments they associate with and the biological systems they play a role in. This approach to systems biology is much more relevant to real life. It is more likely to provide a deeper understanding of organisms' molecular biology than a single gene approach. Researchers need to access large, complex data sets containing as much accessible information as possible in order to do this. It is expected that a combination of tools such as those discussed here by biologists would help shed light on the biological function of newly discovered proteins and peptides. Only then can the data be used optimally in medical or industrial applications.

CONSENT FOR PUBLICATION

Not applicable.

CONFLICT OF INTEREST

The author declares no conflict of interest, financial or otherwise.

ACKNOWLEDGEMENTS

Declared none.

REFERENCES

[1]   Aebersold R, Mann M. Mass-spectrometric exploration of proteome structure and function. Nature 2016; 537(7620): 347-55.
[http://dx.doi.org/10.1038/nature19949] [PMID: 27629641]

[2]   Schrader M, Schulz-Knappe P, Fricker LD. Historical perspective of peptidomics. EuPA Open Proteom 2014; 3: 171-82.
[http://dx.doi.org/10.1016/j.euprot.2014.02.014]

[3]   Ivanov VT, Yatskin ON. Peptidomics: a logical sequel to proteomics. Expert Rev Proteomics 2005; 2(4): 463-73.
[http://dx.doi.org/10.1586/14789450.2.4.463] [PMID: 16097881]

[4]   Mann M, Kulak NA, Nagaraj N, Cox J. The coming age of complete, accurate, and ubiquitous proteomes.   Mol   Cell   2013;   49(4): 583-90.http://www.sciencedirect.com/science/article/pii/S1097276513000932 [Internet].
[http://dx.doi.org/10.1016/j.molcel.2013.01.029] [PMID: 23438854]

[5]   Mann M. The rise of mass spectrometry and the fall of Edman degradation. Clin Chem 2016; 62(1): 293-4.
[http://dx.doi.org/10.1373/clinchem.2014.237271] [PMID: 26430071]

[6]   Kalmykova SD, Arapidi GP, Urban AS, Osetrova MS, Gordeeva VD, Ivanov VT, *et al. In silico* analysis of peptide potential biological functions. russ j bioorganic chem 2018; 44(4): 367-85. [Internet].
[http://dx.doi.org/10.1134/S106816201804009X]

[7]   Fitzgerald RJ, Murray BA. Bioactive peptides and lactic fermentations. Int J Dairy Technol 2006; 59(2): 118-25.
[http://dx.doi.org/10.1111/j.1471-0307.2006.00250.x]

[8]   Fitzgerald C, Aluko RE, Hossain M, Rai DK, Hayes M. Potential of a renin inhibitory peptide from the red seaweed Palmaria palmata as a functional food ingredient following confirmation and characterization of a hypotensive effect in spontaneously hypertensive rats. J Agric Food Chem 2014; 62(33): 8352-6.
[http://dx.doi.org/10.1021/jf500983n] [PMID: 25062358]

[9]   Singh BP, Vij S, Hati S. Functional significance of bioactive peptides derived from soybean. Peptides 2014; 54: 171-9.
[http://dx.doi.org/10.1016/j.peptides.2014.01.022] [PMID: 24508378]

[10]   El-Fattah AA, Sakr S, El-Dieb S, Elkashef H. Angiotensin-converting enzyme inhibition and antioxidant activity of commercial dairy starter cultures. Food Sci Biotechnol 2016; 25(6): 1745-51.
[http://dx.doi.org/10.1007/s10068-016-0266-5] [PMID: 30263470]

[11]   Consortium U. The universal protein resource (UniProt). Nucleic Acids Res 2008; 36(Database issue) (Suppl. 1): D190-5.
[PMID: 18045787]

[12]   Mulder NJ, Apweiler R, Attwood TK, *et al.* New developments in the InterPro database. Nucleic Acids Res 2007; 35(Database issue) (Suppl. 1): D224-8.
[http://dx.doi.org/10.1093/nar/gkl841] [PMID: 17202162]

[13]   Mor A. Peptides: Biological Activities of Small Peptides. eLS 2001.
[http://dx.doi.org/10.1038/npg.els.0001329]

[14]   Vanhoof G, Goossens F, De Meester I, Hendriks D, Scharpé S. Proline motifs in peptides and their biological processing. FASEB J 1995; 9(9): 736-44.
[http://dx.doi.org/10.1096/fasebj.9.9.7601338] [PMID: 7601338]

[15]   Papo N, Shai Y. Can we predict biological activity of antimicrobial peptides from their interactions with model phospholipid membranes? Peptides 2003; 24(11): 1693-703.
[http://dx.doi.org/10.1016/j.peptides.2003.09.013] [PMID: 15019200]

[16]   Graf von Roedern E, Lohof E, Hessler G, Hoffmann M, Kessler H. Synthesis and conformational analysis of linear and cyclic peptides containing sugar amino acids. J Am Chem Soc 1996; 118(42): 10156-67.
[http://dx.doi.org/10.1021/ja961068a]

[17]   Joshi T, Xu D. Quantitative assessment of relationship between sequence similarity and function similarity. BMC Genomics 2007; 8(1): 222.
[http://dx.doi.org/10.1186/1471-2164-8-222] [PMID: 17620139]

[18]   Dinkel H, Van Roey K, Michael S, *et al.* The eukaryotic linear motif resource ELM: 10 years and counting. Nucleic Acids Res 2014; 42(Database issue): D259-66.
[http://dx.doi.org/10.1093/nar/gkt1047] [PMID: 24214962]

[19]   Lucchese G, Stufano A, Trost B, Kusalik A, Kanduc D. Peptidology: short amino acid modules in cell biology and immunology. Amino Acids 2007; 33(4): 703-7.
[http://dx.doi.org/10.1007/s00726-006-0458-z] [PMID: 17077961]

[20]   Camacho C, Coulouris G, Avagyan V, *et al.* BLAST+: architecture and applications. BMC Bioinformatics 2009; 10(1): 421.
[http://dx.doi.org/10.1186/1471-2105-10-421] [PMID: 20003500]

[21] Notredame C, Higgins DG, Heringa J. T-Coffee: A novel method for fast and accurate multiple sequence alignment. J Mol Biol 2000; 302(1): 205-17.
[http://dx.doi.org/10.1006/jmbi.2000.4042] [PMID: 10964570]

[22] Sievers F, Wilm A, Dineen D, *et al.* Fast, scalable generation of high-quality protein multiple sequence alignments using Clustal Omega. Mol Syst Biol 2011; 7(1): 539.
[http://dx.doi.org/10.1038/msb.2011.75] [PMID: 21988835]

[23] Eddy SR. Profile hidden Markov models. Bioinformatics 1998; 14(9): 755-63.
[http://dx.doi.org/10.1093/bioinformatics/14.9.755] [PMID: 9918945]

[24] Christie AE. Prediction of the peptidomes of Tigriopus californicus and Lepeophtheirus salmonis (Copepoda, Crustacea). Gen Comp Endocrinol 2014; 201: 87-106.
[http://dx.doi.org/10.1016/j.ygcen.2014.02.015] [PMID: 24613138]

[25] Edwards RJ, Moran N, Devocelle M, *et al.* Bioinformatic discovery of novel bioactive peptides. Nat Chem Biol 2007; 3(2): 108-12.
[http://dx.doi.org/10.1038/nchembio854] [PMID: 17220901]

[26] Michael S, Travé G, Ramu C, Chica C, Gibson TJ. Discovery of candidate KEN-box motifs using cell cycle keyword enrichment combined with native disorder prediction and motif conservation. Bioinformatics 2008; 24(4): 453-7.
[http://dx.doi.org/10.1093/bioinformatics/btm624] [PMID: 18184688]

[27] Li Y, Yu J. Research progress in structure-activity relationship of bioactive peptides. J Med Food 2015; 18(2): 147-56.
[http://dx.doi.org/10.1089/jmf.2014.0028] [PMID: 25137594]

[28] Pimenta AMC, De Lima ME. Small peptides, big world: biotechnological potential in neglected bioactive peptides from arthropod venoms. J Pept Sci 2005; 11(11): 670-6.
[http://dx.doi.org/10.1002/psc.701] [PMID: 16103988]

[29] Lee EY, Fulan BM, Wong GCL, Ferguson AL. Mapping membrane activity in undiscovered peptide sequence space using machine learning. Proc Natl Acad Sci USA 2016; 113(48): 13588-93.
[http://dx.doi.org/10.1073/pnas.1609893113] [PMID: 27849600]

[30] Kumar R, Raghava GPS. Hybrid approach for predicting coreceptor used by HIV-1 from its V3 loop amino acid sequence. PLoS One 2013; 8(4): e61437.
[http://dx.doi.org/10.1371/journal.pone.0061437] [PMID: 23596523]

[31] Eriksson L, Johansson E. Multivariate design and modeling in QSAR. Chemom Intell Lab Syst 1996; 34(1): 1-19.
[http://dx.doi.org/10.1016/0169-7439(96)00023-8]

[32] Yap CW. PaDEL-descriptor: an open source software to calculate molecular descriptors and fingerprints. J Comput Chem 2011; 32(7): 1466-74.
[http://dx.doi.org/10.1002/jcc.21707] [PMID: 21425294]

[33] Shahlaei M. Descriptor selection methods in quantitative structure-activity relationship studies: a review study. Chem Rev 2013; 113(10): 8093-103.
[http://dx.doi.org/10.1021/cr3004339] [PMID: 23822589]

[34] Frank E, Hall M, Trigg L, Holmes G, Witten IH. Data mining in bioinformatics using Weka. Bioinformatics 2004; 20(15): 2479-81.
[http://dx.doi.org/10.1093/bioinformatics/bth261] [PMID: 15073010]

[35] Garg A, Tewari R, Raghava GPS. KiDoQ: using docking based energy scores to develop ligand based model for predicting antibacterials. BMC Bioinformatics 2010; 11(1): 125.
[http://dx.doi.org/10.1186/1471-2105-11-125] [PMID: 20222969]

[36] Wheeler DL, Barrett T, Benson DA, *et al.* Database resources of the national center for biotechnology information. Nucleic Acids Res 2008; 36(Database issue) (Suppl. 1): D13-21.

[PMID: 18045790]

[37] Okubo K, Sugawara H, Gojobori T, Tateno Y. DDBJ in preparation for overview of research activities behind data submissions. Nucleic Acids Res 2006; 34(Database issue) (Suppl. 1): D6-9.
[http://dx.doi.org/10.1093/nar/gkj111] [PMID: 16381940]

[38] Kulikova T, Akhtar R, Aldebert P, *et al.* EMBL nucleotide sequence database in 2006. Nucleic Acids Res 2007; 35(Database issue) (Suppl. 1): D16-20.
[http://dx.doi.org/10.1093/nar/gkl913] [PMID: 17148479]

[39] Benson DA, Karsch-Mizrachi I, Lipman DJ, Ostell J, Wheeler DL. GenBank. Nucleic Acids Res 2007; 35(Database issue) (Suppl. 1): D21-5.
[http://dx.doi.org/10.1093/nar/gkl986] [PMID: 17202161]

[40] Pruitt KD, Tatusova T, Maglott DR. NCBI reference sequences (RefSeq): a curated non-redundant sequence database of genomes, transcripts and proteins. Nucleic Acids Res 2007; 35: D61-5.
[http://dx.doi.org/10.1093/nar/gkl842] [PMID: 17130148]

[41] Camon E, Magrane M, Barrell D, *et al.* The gene ontology annotation (goa) database: sharing knowledge in uniprot with gene ontology. Nucleic Acids Res 2004; 32: D262-6.
[http://dx.doi.org/10.1093/nar/gkh021] [PMID: 14681408]

[42] Kerrien S, Alam-Faruque Y, Aranda B, *et al.* IntAct--open source resource for molecular interaction data. Nucleic Acids Res 2007; 35: D561-5.
[http://dx.doi.org/10.1093/nar/gkl958] [PMID: 17145710]

[43] Rawlings ND, Morton FR, Barrett AJ. MEROPS: the peptidase database. Nucleic Acids Res 2006; 34: D270-2.
[http://dx.doi.org/10.1093/nar/gkj089] [PMID: 16381862]

[44] Crosby MA, Goodman JL, Strelets VB, Zhang P, Gelbart WM, Consortium F. FlyBase: genomes by the dozen. Nucleic Acids Res 2007; 35: D486-91.
[http://dx.doi.org/10.1093/nar/gkl827] [PMID: 17099233]

[45] Eppig JT, Blake JA, Bult CJ, Kadin JA, Richardson JE, Group MGD. The mouse genome database (MGD): new features facilitating a model system. Nucleic Acids Res 2007; 35: D630-7.
[http://dx.doi.org/10.1093/nar/gkl940] [PMID: 17135206]

[46] Bieri T, Blasiar D, Ozersky P, *et al.* WormBase: new content and better access. Nucleic Acids Res 2007; 35: D506-10.
[http://dx.doi.org/10.1093/nar/gkl818] [PMID: 17099234]

[47] Cherry JM, Hong EL, Amundsen C, *et al.* Saccharomyces Genome Database: the genomics resource of budding yeast. Nucleic Acids Res 2012; 40: D700-5.
[http://dx.doi.org/10.1093/nar/gkr1029] [PMID: 22110037]

[48] Rhee SY, Beavis W, Berardini TZ, *et al.* The Arabidopsis Information Resource (TAIR): a model organism database providing a centralized, curated gateway to Arabidopsis biology, research materials and community. Nucleic Acids Res 2003; 31(1): 224-8.
[http://dx.doi.org/10.1093/nar/gkg076] [PMID: 12519987]

[49] Sigrist CJA, Cerutti L, Hulo N, *et al.* PROSITE: a documented database using patterns and profiles as motif descriptors. Brief Bioinform 2002; 3(3): 265-74.
[http://dx.doi.org/10.1093/bib/3.3.265] [PMID: 12230035]

[50] Lüthy R, Xenarios I, Bucher P. Improving the sensitivity of the sequence profile method. Protein Sci 1994; 3(1): 139-46.
[http://dx.doi.org/10.1002/pro.5560030118] [PMID: 7511453]

[51] Krogh A, Brown M, Mian IS, Sjölander K, Haussler D. Hidden Markov models in computational biology. Applications to protein modeling. J Mol Biol 1994; 235(5): 1501-31.
[http://dx.doi.org/10.1006/jmbi.1994.1104] [PMID: 8107089]

[52]   Durbin R, Eddy SR, Krogh A, Mitchison G. Biological sequence analysis: probabilistic models of proteins and nucleic acids. Cambridge university press 1998.
[http://dx.doi.org/10.1017/CBO9780511790492]

[53]   de Castro E, Sigrist CJA, Gattiker A, *et al.* ScanProsite: detection of PROSITE signature matches and ProRule-associated functional and structural residues in proteins. Nucleic Acids Res 2006; 34(Web Server issue) (Suppl. 2): W362-5.
[http://dx.doi.org/10.1093/nar/gkl124] [PMID: 16845026]

[54]   Attwood TK, Bradley P, Flower DR, *et al.* PRINTS and its automatic supplement, prePRINTS. Nucleic Acids Res 2003; 31(1): 400-2.
[http://dx.doi.org/10.1093/nar/gkg030] [PMID: 12520033]

[55]   Finn RD, Mistry J, Schuster-Böckler B, *et al.* Pfam: clans, web tools and services. Nucleic Acids Res 2006; 34(Database issue) (Suppl. 1): D247-51.
[http://dx.doi.org/10.1093/nar/gkj149] [PMID: 16381856]

[56]   Letunic I, Copley RR, Pils B, Pinkert S, Schultz J, Bork P. SMART 5: domains in the context of genomes and networks. Nucleic Acids Res 2006; 34(Database issue) (Suppl. 1): D257-60.
[http://dx.doi.org/10.1093/nar/gkj079] [PMID: 16381859]

[57]   Selengut JD, Haft DH, Davidsen T, *et al.* TIGRFAMs and Genome Properties: tools for the assignment of molecular function and biological process in prokaryotic genomes. Nucleic Acids Res 2007; 35(Database issue) (Suppl. 1): D260-4.
[http://dx.doi.org/10.1093/nar/gkl1043] [PMID: 17151080]

[58]   Mulder NJ, Apweiler R, Attwood TK, *et al.* InterPro, progress and status in 2005. Nucleic Acids Res 2005; 33(Database issue) (Suppl. 1): D201-5.
[http://dx.doi.org/10.1093/nar/gki106] [PMID: 15608177]

[59]   Mi H, Guo N, Kejariwal A, Thomas PD. PANTHER version 6: protein sequence and function evolution data with expanded representation of biological pathways. Nucleic Acids Res 2007; 35(Database issue) (Suppl. 1): D247-52.
[http://dx.doi.org/10.1093/nar/gkl869] [PMID: 17130144]

[60]   Andreeva A, Howorth D, Chandonia J-M, *et al.* Data growth and its impact on the SCOP database: new developments. Nucleic Acids Res 2008; 36(Database issue) (Suppl. 1): D419-25.
[PMID: 18000004]

[61]   Yeats C, Maibaum M, Marsden R, *et al.* Gene3D: modelling protein structure, function and evolution. Nucleic Acids Res 2006; 34(Database issue) (Suppl. 1): D281-4.
[http://dx.doi.org/10.1093/nar/gkj057] [PMID: 16381865]

[62]   Bru C, Courcelle E, Carrère S, Beausse Y, Dalmar S, Kahn D. The ProDom database of protein domain families: more emphasis on 3D. Nucleic Acids Res 2005; 33(Database issue) (Suppl. 1): D212-5.
[http://dx.doi.org/10.1093/nar/gki034] [PMID: 15608179]

[63]   Chandrasekaran P, Doss CGP, Nisha J, Sethumadhavan R, Shanthi V, Ramanathan K, *et al. In silico* analysis of detrimental mutations in ADD domain of chromatin remodeling protein ATRX that cause ATR-X syndrome: X-linked disorder. Netw Model Anal Health Inform Bioinform 2013; 2(3): 123-35.
[http://dx.doi.org/10.1007/s13721-013-0031-0]

[64]   Lee H, Heo L, Lee MS, Seok C. GalaxyPepDock: a protein-peptide docking tool based on interaction similarity and energy optimization. Nucleic Acids Res 2015; 43(W1): W431-5.
[http://dx.doi.org/10.1093/nar/gkv495] [PMID: 25969449]

[65]   Lee H, Seok C. Template-based prediction of protein-peptide interactions by using GalaxyPepDock.Modeling Peptide-Protein Interactions. Springer 2017; pp. 37-47.
[http://dx.doi.org/10.1007/978-1-4939-6798-8_4]

[66]   Sedan Y, Marcu O, Lyskov S, Schueler-Furman O. Peptiderive server: derive peptide inhibitors from

protein-protein interactions. Nucleic Acids Res 2016; 44(W1): W536-41.
[http://dx.doi.org/10.1093/nar/gkw385] [PMID: 27141963]

[67]   Peterson LX, Roy A, Christoffer C, Terashi G, Kihara D. Modeling disordered protein interactions from biophysical principles. PLOS Comput Biol 2017; 13(4): e1005485.
[http://dx.doi.org/10.1371/journal.pcbi.1005485] [PMID: 28394890]

[68]   Wang J, Alekseenko A, Kozakov D, Miao Y. Improved Modeling of Peptide-Protein Binding Through Global Docking and Accelerated Molecular Dynamics Simulations. Front Mol Biosci 2019; 6: 112.
[http://dx.doi.org/10.3389/fmolb.2019.00112] [PMID: 31737642]

[69]   Aldonza MBD, Ku J, Hong J-Y, *et al.* Prior acquired resistance to paclitaxel relays diverse EGFR-targeted therapy persistence mechanisms. Sci Adv 2020; 6(6): eaav7416.
[http://dx.doi.org/10.1126/sciadv.aav7416] [PMID: 32083171]

[70]   Das AA, Sharma OP, Kumar MS, Krishna R, Mathur PP. PepBind: a comprehensive database and computational tool for analysis of protein-peptide interactions. Genomics Proteomics Bioinformatics 2013; 11(4): 241-6.
[http://dx.doi.org/10.1016/j.gpb.2013.03.002] [PMID: 23896518]

SECTION II: Molecular Pharming for Human Beings

CHAPTER 6

Molecular Pharming: Research, Developments and Future Perspective

Muhammad Sarwar Khan[1,*], **Ghulam Mustafa**[1] and **Faiz Ahmad Joyia**[1]

[1] *Centre of Agricultural Biochemistry & Biotechnology (CABB), University of Agriculture, Faisalabad, Pakistan*

Abstract: Plants are tamed to function as production factories of pharmaceuticals. Recently, several pharmaceuticals, including therapeutics, drugs, vaccines, vitamins, antibiotics, nutraceuticals, and diagnostic molecules, have been produced through these green factories. Compared with conventional systems, for example, bacterial, yeast, fungal, and mammalian cell cultures, plants are accepted as a cost-effective source of pharmaceuticals products. Considering plants as a versatile, cost-effective, and robust production platform, the system could be exploited in different ways like plant cell culture, transient expression and harvesting, and stable transgenics. This chapter highlights the importance and potential of molecular pharming with special emphasis on methodological aspects, proving the suitability of plants as the most appropriate biopharmaceutical production platform with recent interventions.

Keywords: Biopharmaceuticals, Cell culture, Cost-effective, Plant expression system, Recent innovations.

1. INTRODUCTION

Biopharmaceutics are biomolecules produced using biotechnological processes. These are protein or nucleic acid-based substances used for therapeutic or diagnostic purposes [1]. Pharmaceutical technology will continue to provide breakthroughs in medical research, leading to the effective treatment of noxious diseases including AIDS, cancer, asthma, Parkinson's, and Alzheimer's diseases [2]. Plant-derived pharmaceuticals have drawn special attention in this context owing to certain salient advantages of this production system as compared with others.

---

* **Corresponding author Faiz Ahmad Joyia:** Centre of Agricultural Biochemistry & Biotechnology (CABB), University of Agriculture, Faisalabad, Pakistan; E-mail: faizahmad1980@gmail.com

Muhammad Sarwar Khan (Ed)

The pharmaceutical market of recombinant proteins is growing steadily, with almost all major pharmaceutical firms reporting a rising share of these products. These pharmaceuticals are usually produced using mammalian or bacterial cell-based processes that are complex to operate and potentially susceptible to human pathogen contamination [3]. Current good manufacturing practice (cGMP) compliant manufacturing facilities require tremendous capital investment and include substantial financial risk. Plant biotechnology can solve some of these drawbacks, but before this emerging industry can compete successfully in the pharmaceutical market, many obstacles remain to be addressed. A variety of different technologies are now operated by plant biotechnology for recombinant protein expression. These developments constitute opportunities in the sector for current and future commercial enterprises [4].

The idea of using plants for the production of recombinant pharmaceutical proteins, known as plant molecular pharming (PMF) or pharmacy (PMP), is not new. The 1st plant-derived recombinant therapeutic protein was human growth hormone, initially developed in tobacco and sunflower in 1986 [5]. The HBs Ag (hepatitis B surface antigen) was later expressed in transgenic tobacco. Physically and antigenically, this plant-derived antigen was identical to HBs Ag obtained from recombinant yeast and human serum. HBs Ag derived from yeast is clinically used for vaccination against HBV. It has been reported that above 100 recombinant proteins were expressed and produced through a plant-based expression system [6].

The basic term 'ATMPs (advanced therapy medicinal products)' is used by the European Medicine Agency (EMA) to refer to human medicines that are based on tissue, cells, and gene engineering. Cell therapy products (CTPs) are cells/tissues containing biomedicines that have been manipulated to modify their biological properties and can be used to cure, prevent, or diagnose diseases. The common benefits of these systems are ease of manipulation, low cost, speed, and high protein yield.

The developments in functional genomics and recombinant DNA technology have been merged to provide vast opportunities for the cost-effective production of commercial-scale recombinant protein. Enzymes are now widely used in industry for bio-catalytic reactions, allowing them to be used in anything from cultivation and bioremediation to medicine and food preparation. The rise of the biopharmaceutical industry has been brought about by further exploitation and modifications in the recombinant protein structure and function for therapeutic uses [7]. Any of the therapeutic proteins, including growth hormones, pancreatic enzymes, can now be produced from alternative sources, thus available to the ultimate consumer in natural form. Further, advancements in recombinant DNA technology have refined it by making possible the isolation and expression of protein-coding genes in the transgenic host cells. Transgenic sources of these recombinant proteins pave the way to the production of low-cost proteins with increased protection and efficiency [8].

2. HISTORY OF THE BIOPHARMACEUTICAL INDUSTRY

Proteins of mammalian origin have been produced in plants since the late 1980s. Since then, 'molecular plant farming' has widely been exploited to use plants as protein factories. Fisher introduced the concept of molecular pharming (biopharming) and highlighted that any mammalian protein, including vaccines, blood proteins, antibodies, and medicinal therapeutics, can effectively be produced in plants [9]. Over the past century, vaccine production has progressed tremendously. The burden of many life-threatening diseases has been minimized by traditional vaccines. Alternative approaches have been employed for the development of novel vaccines with the ability to effectively protect against new diseases [10, 11].

Three key groups had been striving to ascertain the notion of vaccine production in plants under the leadership of Charles Arntzen, Roy Curtsis, and Hilary Koprowski. Arntzen's research contributed first peer-reviewed publication in the area of plant molecular farming [12]. It was suggested that an edible vaccine is a fruit or vegetable with a particular antigen that can be delivered cost-effectively with increased efficacy. This proved the worth of plants as an alternative source of vaccine having the ability to compete with the existing market. Thus, using plants as edible vaccines could be an effective alternative to treat infectious diseases. The beauty of this concept was that recombinant plants could be grown near to the target population, facilitating the availability of inexpensive and effective vaccines. The reduced cost of downstream processes, transportation, and purification would result in reducing the cost as compared with traditional vaccines. Despite several advantages, there are certain limitations as well, including optimization of dosage and contamination of the food chain by producing recombinant plants [13].

Dow Agro Sciences LLC announced (on 31st January 2006) to be the global leader and obtained the World's first regulatory approval from USDA. The approved plant-based vaccine could combat the ND (Newcastle Disease). The vaccine antigen was expressed in tobacco plant cell lines through suspension culture in a traditional bioreactor system. The resulting cells were extracted and minimally processed for the final formulation of the vaccine [14, 15].

3. VARIOUS PRODUCTION SYSTEMS FOR BIOPHARMACEUTICALS

Different expression systems have been used for the commercial-scale production of biopharmaceuticals. While selecting an expression system, the important parameters are quality and quantity of proteins, output rate, and the expected cost. Various biopharming platforms are discussed in the following sections. Each platform offers several advantages along with certain limitations; hence, for every protein of interest, the most suitable production system should be selected carefully [16].

3.1. Using Microbes for Biopharming

Bacterial expression system has widespread applications in biopharming. The first recombinant protein produced in *E. coli* was human insulin which was approved in 1982. Nevertheless, the complex proteins of eukaryotic origin, when produced by *E. coli*, further need post-translational modifications (PTM) like denaturation and refolding *in vitro*. However, such PTM can be achieved in mammalian cells. Such cells are difficult to cultivate and manage than *E. coli*. In *E. coli,* protein targeting is required, which can direct recombinant proteins to the periplasm, where mammalian proteins can fold correctly due to a favorable environment for disulfide bond formation [17].

Another versatile and potential microbes-based expression system for biopharming is yeast. Just like bacteria, yeast needs less expensive and simpler growth media. On the other hand, yeast cells have been found much more competent in expressing eukaryotic proteins as they can fold and assemble complex human proteins better than bacterial cells. *Saccharomyces cerevisiae* was the first and the most commonly used yeast for biopharming. There are some limitations to the yeast expression system. It includes poor plasmid stability, lower expression level, hyperglycosylated glycoproteins, and inefficient secretion of proteins [18]. Similarly, fungal mycelia can also be used for the production of human therapeutic proteins with their high product secretion capacity [19].

3.2. Using Mammalian Cell Lines for Biopharming

Mammalian cell lines have been used predominantly for biopharming due to several advantages like proper folding and glycosylation of proteins, making them biologically active. If proteins are expressed in yeast and fungi, they exhibit variable glycosylation than that of native human proteins. Hence, regardless of the maintenance problems and higher costs, human and mammalian cell lines are more suitable for biopharming. This is the reason that almost half of the Federal

Drug Administration (FDA) approved biopharming products (vaccines, therapeutics, and diagnostics) have been produced through such cell lines. However, there are certain limitations associated with this expression system, including pathogenic contamination, higher cost of production, and downstream processing [20, 21].

3.3. Using GM Animals for Biopharming

The whole animals can be made genetically modified, and biopharmaceuticals can be expressed in their body fluids which can be obtained without killing the animal. Hence, biopharmaceuticals will be available continuously and in ample quantity. Most of the time, recombinant proteins are expressed in mammary glands under mammary-gland-specific promoters so that the proteins can be secreted in milk. Similarly, recombinant proteins can be secreted in other body fluids, including seminal fluid, urine, and blood serum. Various animals with large milk/meat production abilities, including goats, sheep, rabbits, cows, and pigs, are being used for this purpose [22]. Although GM animals are promising in terms of protein quality and quantity, yet their maintenance is time-consuming and expensive [23].

Similarly, eggs also appear to be a good biopharming platform for the production of therapeutic proteins. A single chicken may lay up to 300 eggs in a year, with each egg containing approximately 6 g protein. Mere production of 100 mg of recombinant protein may lead to the production of a huge quantity of protein as a flock of 5000 chickens may produce up to 150 kg of pharmaceutical protein within one hatching cycle. Moreover, rearing chickens is more economical and cost-effective as compared with the rest of the aforementioned animals [24].

3.4. Using GM Crops for Biopharming

Genetically modified plants have also been used for biopharming. The important features include lack of human pathogens, low cost, higher expression levels, ease of cultivation, and maintenance. Hence, plants have attracted significant interest as a more competitive alternative to produce high quality, cost-effective, and safer as well as biologically active pharmaceuticals. Similarly, the much lower upscale costs for commercial-scale production of biopharmaceuticals in plants may instigate cultivation over vast areas [25]. Furthermore, encapsulation of recombinant proteins within chloroplasts or seeds makes them an ideal candidate for oral administration [26 - 28]. Hence, a variety of options in plant-based biopharming platforms, including stable genetic transformation in nucleus and chloroplasts, transient expression system, and plant suspension cultures, have been developed for commercial-scale production [1].

4. TYPES OF PLANT-BASED BIOPHARMACEUTICAL PRODUCTS

Plant biotechnology contributes to the commercial-scale development of pharmacologically significant proteins, in many instances, almost similar to mammalian counterparts and functional. The use of agricultural biotechnology for the development of biologically active molecules and hormone started almost 20 years ago, with the expressions of functional antibodies in the plant, showing that plants are an economic platform for the production of pharmaceutical/therapeutic proteins. However, the bacterial system of protein production is more economical and certain proteins that require specific post-translational modifications are reported to be better expressed in the eukaryotic expression system, including plants [29].

4.1. Antibodies

Monoclonal antibody (mAb) production has been an important advancement in biotechnology for both diagnostic and therapeutic purposes. Traditional monoclonal therapeutic antibodies have been produced from mice. The human immune system has readily recognized this foreign protein, limiting the usefulness of antibodies for therapeutic use, particularly with repeated dosage [30]. The presence of neutralizing antibodies inactivating the drugs also precluded further clinical use, although in the absence of serum sickness or anaphylaxis. Recombinant developments, however, have made it possible to substitute murine antibodies with partly chimeric or humanized antibody and now it is possible to express the fully functional human antibody. After might be derived from the mice having the genes of human immunoglobulins or using yeasts or another expression system. To attain high-affinity binding (affinity maturation), recombinant technology could be used to evolve the antibody genes. Thus, the current antibody exhibit decreases immunogenicity and increases biological activity compared to earlier monoclonal antibodies [31]. There are currently more than a dozen approved mAbs by FDA, and more than 700 therapeutics antibodies are under development. As a potentially limitless supply of mAb, some referred to as 'plantibody', the plant has now been explored as a potential candidate for antibody expression. Tobacco, being the model plant for transformation, has extensively been used for these sorts of studies. Likewise, other plant species, including soybeans, potatoes, rice, alfalfa, and maize, have also been engineered for the expression of various types of antibodies [32].

4.2. Vaccines

For the development of an edible vaccine with a low cost, there has been a significant interest in the scientific community and pharmaceutical companies. As for polio, conventional edible vaccines use whole, partially purified material or attenuated species for stimulating both Ig-G-mediated (systemic immunity) and Ig-A-mediated (local membrane immunity) [33]. The significance of the plant expression system is that it uses only the DNA encoding the desirable antigenic sequence from the virus, pathogenic bacteria, and parasite. In plant tissues, immunogenic protein or antigenic sequence may be produced and later consumed as an edible antiviral vaccine. The immune system of the mucosa could trigger a protective immune response to pathogen or toxin and might be useful in inducing tolerance to antigens inhaled. In a mucosal region, the production of sIg-A (secretory Ig-A) and provocation of the specific immune lymphocyte can occur, and these regions are of particular importance for the production of the edible vaccine [34].

Plant-based vaccines deliver a range of specific advantages, in addition to intrinsically low production costs, including improved protection, stability, flexibility, and efficacy [35]. Where appropriate, vaccines from plants could be grown locally, reducing the storage and cost of transport. Naturally, the related antigen is preserved in the plant tissues. The oral vaccine can be administered directly in food products in which it is produced, reducing the expense of purification. Immunotoxicity and other adverse effects can be minimized by plants bred for expressing only selected antigenic portion of related pathogens, and the plant-derived vaccine is free from mammalian virus contaminations [36].

4.3. Other Therapeutic Agents

A higher range of other therapeutics drugs, including enzymes, hormones (somatotropin), interferon (IFN), interleukins, and HSA (human serum albumin), have been expressed in plants [31]. Related biotherapeutic agents from bacterial and mammalian cell systems have also been expressed. There is a huge market for HSA, and the processing of plants will offer the benefit of being free from the contamination of human pathogenic viruses. In emphysema and hepatic diseases, modified rice plant is capable of producing the human alpha-1-antitrypsin. It is a protein that recognizes the therapeutic potential. Originally extracted from the leech, hirudin is the blood anticoagulant that could be used to express in the oilseed rapes, mustard, and tobacco. A transgenic plant of potato may encode human INF for at least two subtypes, both of which may moderate the role of viral agents in certain cancers and diseases. In a transgenic plant of tobacco,

Erythropoietin (EPO) (glycoproteins used in the treatment of anemia and marketed about 20 years ago in mammalian systems) has also been expressed. Blood replacements, *e.g.*, human hemoglobin, have long been sought, and transgenic tobacco-derived human hemoglobin is being tested for ensuring the functions and efficiency of carrying oxygen of the molecules [37].

5. TRANSGENIC PLANTS IN THE BIOPHARMACEUTICAL MARKET

It is important to regard the first 10 years of commercial experience with GM (genetically modified) crops as a success. Adoptions have been remarkable: in 1996 1st commercially planted, worldwide planting of GM crops in 2020 crossed 190 million hectares. However, large-scale adoption in only a few countries has been limited to a few crops. Currently, GM crops are all focused on only agronomic traits that confer tolerance to herbicides and insect pasts [38, 39].

The development of a transgenic crop depends upon the ability of foreign DNA to get incorporated into a plant cell in such a way that new proteins can then be expressed. The insertion of single genes conferring resistance against herbicide or insect has been achieved in the first generation of GM crops. However, a genetic transformation technique could be used for the incorporation of DNA, which codes for the expression of the novel protein of direct interest, into plants. It enables the host plants to be used in the 'molecular farming,' whereby high-valued recombinant protein and organic compound which could be used for manufacturing of pharmaceutical compound, recombinant enzyme, and polymer with the industrial application is intended harvest products [40].

Given the ability to beat the growing demand for drugs and pharmaceuticals, biotechnological innovations are of particular interest. The concept is that a suitably designed GM plant can be competitive because there are well-known disadvantages of alternative development platforms in the biotechnology industry. Microbial fermentation, for example, is widespread, but effectiveness is restricted by the requirement of downstream processing because yeast and bacteria are unable to perform post-translation modifications (*e.g.*, glycosylation), which are essential for the biological activity of complex protein with pharmaceutical value [41]. Most of these problems are solved by recombinant mammalian cell lines cultivated in bioreactors and currently constitute the site of choice for the large and complex protein. This form of bio-manufacturing is expensive. A transgenic plant, since it possesses a mechanism of protein synthesis like eukaryote nearly similar to animals, offers a desirable alternative in theory. In essence, suitably converted GM host plants serve as a bioreactor and could manufacture the recombinant protein more effectively than mammalian cell lines [42].

6.   METHODOLOGICAL   ASPECTS   OF   PLANT-BASED BIOPHARMACEUTICALS

Agriculture biotechnology can contribute to the commercial development of pharmacologically significant proteins, in several instances, completely functional and almost the same as a mammalian counterpart. The use of agricultural biotechnology for the production of the hormone or other biologically active molecule started almost twenty years, with the expression of functional antibodies in the plant being a critical advance, showing that plants can produce complex protein having therapeutic significance [43]. However, bacteria is an inexpensive and suitable system for the production of several proteins (*e.g.*, human insulin). In more complex multi-component proteins, e.g., antibody, post-translational changes, and assembly step required for biological activity. Plant exhibit an efficient pathway of eukaryotes protein synthesis, and plants can easily generate tonnes of protein by merging the currently available system for expression of genes with sufficient land area [44].

In plants, biopharmaceutical development needs a series of careful decisions about three crucial areas: (i) the mechanism for the expression of a gene, (ii) the position inside the plant of gene expression (iii) the type of plants to be used. Here are a variety of methods for the expression of a gene that could be used in plants for generating particular proteins. Gene sequence is introduced into the cells of a plant with transient expression (TE) using plant virus, gene-gun, or other methods, without the introduction into the plant chromosome of the latest genetic material. TE system could be deployed quickly and generate large quantities of protein, but since the mitosis or meiosis phase does not copy non-chromosomal DNA, expression of a gene is neither permanent and nor heritable [32]. Although the TE system is very useful for development and research and might be useful for the production of pharmaceuticals, yet they involve the freshly transformed plants at each planting stage and might be less appealing to the production of long-term proteins.

Alternatively, the main plant chromosome may be changed to allow a specific protein to be permanently and heritably expressed, *i.e.*, to create plants that grow seeds that carry the desired variation. It could be achieved through *Agrobacterium tumefaciens*, a plant pathogen that passes genetic materials to plant in nature, a chromosome in plants [45]. Through modifying the genetic material of Agrobacterium, it is possible to easily insert the desired genes into several types of plants, especially soybeans. Small metallic particles can also be coated with genetic material and inserted into a cell by the 'gene-gun' method. This latter device is useful for the higher range of plant species. Although permanent alteration of a plant's genome is more expensive and laborious, with repeated planting alone, it provides the simple advantage of secure, ongoing protein production [46].

In the future production of biopharmaceuticals, plant chloroplast can play a critical role. This small energy-producing organelle seems to have an advantage over the transformation in the nucleus, particularly given that hundreds of such organelles can be carried by each cell, which results in the capacity to retain a large number of copies of a functional gene. For example, chloroplast transformed tobacco could generate human somatotropins at a protein level more than 100 times higher in contrast to nuclear transgenic equivalent, with somatotropins and Bt insecticidal proteins output representing seven percent and forty-five percent of the total production of plant proteins. Analysis, cost, protection, and production factors affect the selection of the plants' expression systems [32].

6.1. Upstream Processing

Pharmacologically active proteins, obtained in a plant using it as a bioreactor to ensure good health, require a precise regulatory setting. Working techniques are primarily aimed at acquiring economically important crops resistant to herbicides and pathogens (sweet potatoes, corns, sugar canes, potatoes, pineapples, citrus, tomatoes, papayas, coffees, banana, and rice) [47]. Particularly in comparison to bacterial and mammalian processes, plants have become more and more attractive sources to produce high-value proteins, providing the potential for a highly scalable system, alleviating many safety issues (*e.g.*, no animal pathogens), and acquiring the ability to assemble multi-subunit proteins with suitable post-translational modifications. Research conducted by Hiat and Barta emphasized that a plant-based production system is most advantageous and desirable [48]. The PMF sector is in a process, with major knowledge strides resulting in the development of more effective multifunctional vaccines, therapeutics, and pharmaceuticals. The ability of the plant to fulfill proper post-translational modifications by properly folding RPs and retaining the functional structure of protein predicts the potential of PMF applications. Compared with pharmaceutical production platforms (mammalian bacteria and yeast), plants are taken to be more efficient and economical. These prevailing factors are the result of an increase in biopharmaceutical production [49].

6.1.1. Plant Transformation

To generate recombinant pharmaceuticals in plants, two transformation methods are widely used. The first one involves Agrobacterium, gene gun, or other transformation methods to generate stably transformed transgenic plants. The second approach is to infect the non-transgenic plant with a recombinant virus which, during its replication in the host, express transgenes [50].

6.1.1.1. Stable Expression vs. Transient Expression

6.1.1.1.1. Nuclear Transformation:

Whether the transgene of interest is stably integrated into the host nuclear genome or not is of pivotal importance to harvest the above-mentioned features of plant-based production of therapeutics. Stable transgene integration is very important to express its encoded protein, thereby creating a new genetic germline of the host organism that was not present in the original untransformed host. Initially, the SNT approach was the most common way to produce RPs. The opportunity to accumulate RPs in the seeds of cereals facilitates its storage at room temperature for long periods without thinking about protein degradation is one benefit of the SNT process. As cereal crops can be grown relatively anywhere worldwide, the system allows for high scale-up ability. Yet, certain disadvantages also prevail, including a long processing cycle and risk of cross-contamination wild-type indigenous species [50].

6.1.1.1.2. Plastid Transformation:

The plastid expression system provides an alternative to the nuclear transformation with certain valuable advantages, including transgene containment through maternal inheritance. Plastids are maternally inherited in most of the plant species. Hence, pollens are contained, making the transgenes non-transferrable, helping to ease some public scrutiny. Though plastid transformation has been extended to numerous dicot and monocot plants still certain limitations are there [51]. Since several research groups have published the expression of therapeutics, nutraceuticals, antibiotics, and vaccines, *etc.*, using plastid expression systems [52].

6.1.1.1.3. Transient Expression System

Transient expression was primarily used to testify whether recombinant genes and vectors are performing or not. As the transgene expression is not heritable, one has to reproduce all of the processes [53]. With the advancements in research, scientists started to use this concept for the economical production of valuable proteins as well as scaled-up into stable transgene expression. Various alternative systems were proposed when the production of edible vaccines using some part or whole of the plant. Transient gene expression is an important technique for generating large quantities of a recombinant protein that can save time and is widely accepted. Since the last few decades, the benefits of this system have been well established, and this sort of gene expression is easy to verify quickly (three to six days) [54].

For transient gene expression, the process can be divided into different phases, including the pre-inoculation and post-inoculation phases. Optimization of culture conditions include nutrient composition, the temperature of the environment, light regime, and vacuum at the time of agro-infiltration and post infiltration dehydration [55]. It has been observed that a higher amount of nitrate-containing nutrients results in higher expression of recombinant protein in transient expression. Hence, the important factors include nitrate-enriched fertilizers, plant growth regulators, quick post-Agro-infiltration dehydration, and reasonably lesser plant density [56]. Some other studies elucidated that plant density and the location of a leaf on the plant body have a significant impact on the biomass harvested and hence, are considered critical factors in achieving economically viable production. It was found that yield of recombinant hemagglutinin (HA) protein was impaired substantially by dehydration of bacterial suspension. Moreover, the expression differences in detached leaves as compared with that of intact leaves [57].

6.1.1.1.4. Virus Infection Method and Magnification Technology

The platform for virus infection relies on viruses like the potato virus X (PVX), the bean yellow dwarf virus (BeYDV), or the tobacco mosaic virus (TMV) and their ability to transmit transgenes to plant hosts as vectors without integration into the host chromosome. These methods of virus expression infect tobacco plants, transiently releasing their target protein in the tissue of the plant leaf. As with most 'modern' plant-based platforms, the RPs produced must be collected and processed immediately (or frozen) to avoid necrosis of plant tissue and protein instability, one major downside to the system. Large Scale Biology Company is the first to implement this platform to manufacture vaccines for the treatment of B-cell non-Hodgkin's lymphoma, despite this slight disadvantage [58].

To achieve high-level co-expression of more extensive polypeptides that are needed for hetero-oligomeric protein synthesis, agro-infiltration and viral vector techniques have limited potential. A robust transient expression system, called MagnICON technology, was developed by Icon Genetics, where *A. tumefaciens* is the principal method of transmission and infection. The coat protein of the noncompeting virus strains necessary for systemic movement is separated from this method, and the vector is transmitted to the whole plant using *A. tumefaciens* have enhanced infectivity and better amplification. Ultimately, the method results in high-level co-expression of several different polypeptides, resulting in up to a 100-fold increase in the assembly of hetero-oligomeric proteins. A major autoantigen involved in autoimmune type 1- diabetes, the human autoantigen GAD65, is an RP developed with exceptional yields by this platform [59].

6.1.2. Media Hydration

In suspension culture, rice is the most commonly used for establishing cell lines. By expressing the genes under the promoter of the alpha-amylase gene from rice, the problem of low gene expression levels in rice suspension cultures can be resolved. The alpha-amylase 3D promoter is among the most commonly employed promoters for metabolite-regulation. It is triggered by sugar starvation. It has been employed in suspension cultures of rice to produce numerous therapeutic proteins such as hGM-CSF, human VEGF165, FimA mAb, bovine trypsin, and human pepsinogen C [60].

Plant cell cultures provide a cultivation system with added advantages of a controlled fermenter-based process. While using whole plants for therapeutic production, following good manufacturing practices (GMP) becomes a significant problem as it becomes problematic to limit the entire supply chain to a sterile room. On the other hand, the GMP requirements for clinical products manufactured in bacterial or mammalian cell lines can be extended to plant cells if cultured in a bioreactor [61]. Another intervention is the addition of tags at suitable positions in the expression vectors to improve the isolation and subsequent purification of recombinant proteins. Such tags include molecules like elastin-like polypeptides, zein-derived peptides, and hydrophobins. The key features of such tags are the stability and accumulation enhancement of the fusion protein in distinct storage organelles. As a result, the final yield of the recombinant protein is enhanced significantly. In a study, a two-fold increase in expression of green fluorescent protein (GFP) was observed when fused with hydrophobin and expressed transiently in *N. benthamiana* leaves. In another study, a 3 fold increase in expression with reasonable enhancement in purity and up to 60 percent recovery was observed when GFP-HFB fusion protein was expressed in suspension culture tobacco BY-2 cells [59].

6.1.3. Cell Culture (Bioreactor)

Suspension culture systems, which have been used since 1902, have a long history. They are commonly used as a suitable production platform for recombinant proteins, secondary metabolites, and numerous other types of pharmaceutical molecules. Moreover, the secretion of target compounds into the media permits to streamline of the production as well as purification processes and reduces the cost, thus offering major advantages over approaches focused on isolation and purification of final products from plant cells. Moreover, conditions can be regulated easily, which brings about a more consistent yield [62]. Many studies of plant suspension culture have been carried due to these advantages; proteins developed using this method include the Newcastle disease virus HN protein, human glucocerebrosidase, recombinant alpha-galactosidase-A, and anti-tumor necrosis factor antibodies. Choice of the host plant species, the culture type (*e.g.*, callus or hairy root), media ingredients, and the design and operation of the bioreactor must be considered to optimize important factors affecting the suspension community. These considerations are close to those for well-characterized microbial or mammalian cell-dependent production systems [63].

6.2. Harvesting

Many of the harvesting methods available are not appropriate for the complete dewatering of plant therapeutic products and must be used to achieve the cell concentrations required for the production of drug feedstock in conjunction with other methods. This has been referred to as either primary concentration and secondary concentration or thickening and bulk harvesting. Although the language is different, there are the same procedures. To raise the concentration of solids, a secondary or thickening phase is needed for most downstream processing. Although primary harvesting remains a major challenge for biomass generated by photoautotrophy (due to large volumes at low density), processes that bypass the secondary concentration stage of higher energy could provide an advantage for the commercial production of biopharmaceutical products [64].

6.2.1. Harvesting from Plant Material (Centrifugation or Filtration)

To extract plant biomass from a dilute solution, centrifugation is a common and efficient process. It is versatile and can be continuously worked to accommodate enormous amounts on a scale. Besides, plants containing high lipids are harder to centrifuge and thus take extra energy to recover by centrifugation [65].

In a continuous and large-scale method, the use of a centrifuge for secondary concentration with enhanced and reliable centrifuges will most likely be useful for drug production. Hydro cyclones are low-cost, continuous centrifuges. With reduced hydro cyclone size and higher flow speeds, their productivity improves, which may be a challenge for industrial scale-up. An example of this problem is a study of the efficiency of solid separation of hydro cylones of 5 and 18 cm diameter that yielded 34 to 29 percent removal, respectively [66].

6.3. Downstream Processing of the Proteins of Pharmaceutical Value

In downstream processing, several steps are used for the extraction of highly purified products from biological matrices. Downstream processing (DSP) is a critical component of overall processes for obtaining exceptionally pure and highly regulated biopharmaceutical. In the beginning era of genetic engineering, scientists mainly focused on the production of recombinant proteins from plants. Thus, they neglect the importance of purification and extractions methods development. However, currently, downstream processing in plants for the production of recombinant proteins is the most costly part and it is estimated that it accounts for more than 80% of the total cost [67]. The development of the DSPs is also a critical part to ensure GMPs compliance. For obtaining pure and homogeneous products, contaminations that are present in the product need to be lowered or removed to acceptable levels. Steps include in downstream processing are generally tissue harvesting, extraction of proteins, purifications, and formulations [68]. Schematic sketch so various steps of DSP is given in Fig. (**1**). The first step emphasizes that biological product processing should be under the production platforms whereas other steps should ensure isolation of a stable protein of interest [69]. Its intentions are for maximizing purity and that goal is attained through increasing the quantity of the recovered recombinant proteins which reduced time and cost for completing the process [70].

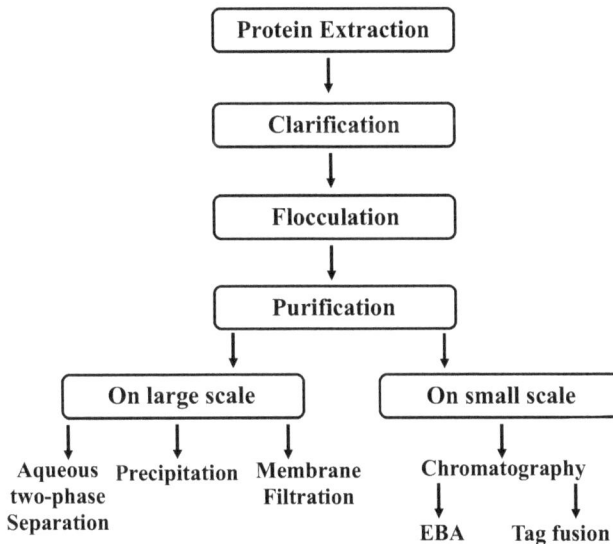

Fig. (1). Schematic sketch showing various steps of downstream processes for the purification of recombinant proteins.

6.3.1. Protein Extraction

During the extraction process, target proteins are set free from the biological matrix. Protein extraction is a critical step in defining the quality of target proteins, concentrations, and total volume of the extract. Additionally, extraction stages are mostly responsible for the types and quantities of the impurity. This method mainly includes harvesting of plant material and its maceration. In an expression system of seed-based, the first step includes milling. Leafy biomass or milled seed homogenized through pressing or blade-based homogenizer. Buffers are generally added to the received biomasses for making the recovery of protein easier. Additionally, it provides enough salinity and pH, maintain solubility, and adds antioxidants. In that step, a protease inhibitor is generally added because of the threat of the proteolysis of the target proteins. The type of buffers depends on the extracted protein's type. Buffers are widely different from each other. But aqueous buffer is most common. Sometimes another substance *e.g.* ionic and supercritical fluid is also used. An organic solvent such as phenol or hexane is used for the non-protein target, membrane protein, and removal of fat from biomass [71].

6.3.2. Clarification

Disruption of the tissues can heavily contaminate the protein extract through releasing of particles burden. Such impurities generally consist of plants' RNA and DNA, alkaloid, polysaccharide, soluble protein, and chlorophyll [72]. The presence of polyphenol in the plant extract is troublesome for the efficient purification of protein. These are responsible for structural and conformational changes in protein and the formation of aggregates of polyphenol-protein (higher molecular size of polyphenols, greater the ability for the formation of a complex with protein) [73]. Preliminary, polyphenol elimination substantially increases microfiltration rates. Sensitive filter and chromatographic apparatus need high extract purity; thus, before the process of purification, even the cell culture supernatants and hydroponic medium need to be clarified [74]. Filtrations and centrifugations are generally used together for the extract clarification. Centrifugations are frequently used for the small-scale procedure. Such as, special centrifugation is used for obtaining the apoplast-targeted protein directly from the plant cell, without disruption. The filter is easy to handle, effective, and generally does not require cleaning as it is for single-use that further reduced the contamination risk. So, filter accounts for 25% cost of the downstream processing. Clarification can be scaled up easily with filters as compared with centrifugation. Hence, the process of serial filtration has become a method of choice for manufacturing, though filter capacity is a critical factor from an operational point of view. Heavily particulate feed stream could reduce the capacity of the filter. In such a situation, the use of clarity enhancement techniques (such as the use of flocculant) is recommended [48].

6.3.3. Flocculation

Flocculation is an aggregation process; a polymer with a higher molecular mass, that can carry a strong positive or negative charge, is promoted by flocculant. They increase the aggregations by charge neutralization or bridging. Dispersed particle tends to form the flock, which make it easier to extract solid contaminants and allows cheaper filters to be used [75]. Thus, flocculants can double the effectiveness of filtration in more costly filters such as depth filters. Though, integrating them carefully with the process parameter is very critical. The effectiveness of flocculants is based on the concentration and characteristic of the particle, the nature of the polymers, process conditions such as conductivity and pH, and stream feed properties. Also, the latter stages of the downstream processing such as when charges are used may be influenced by flocculation. For this reason, it is important to take into account the next stages of downstream processing when designing the use of the flocculant [76].

6.3.4. Protein Purification

The main point of the purification is obtaining the extremely purified recombinant proteins from the filtrated extract. Generally, this part of downstream processing comprises with series of successive chromatography steps. But pre-treatment methods such as membrane filtration, precipitations, and aqueous two-phase partitioning may be used alone or for improving chromatography performance. While purification does not require more than 3-processing steps to maintain the maximum efficient protein rate. In most small-scale purification of recombinant proteins, while chromatography is a method of choice. The prospect of large-scale processing of agricultural recombinant proteins in plants raises concerns about scaling up the method of purification. The target is to decrease the step and to exceed the column chromatography capacity. Currently, for production at a large-scale, the non-chromatographic method is becoming increasingly important.

6.3.4.1. Aqueous Two-phase Partitioning

Aqueous two-phase partitioning enables plant phenolic contaminant and alkaloid to be separated from recombinant protein extract. Typically, this approach uses 2-polymers (or 1-salt and 1-polymer) applied to the aqueous solution at some critical concentrations. Dispersed particles settled in the lower phase due to gravity and dissolved particle positioned in the upper phase. In one phase target proteins are separated which depends on their character such as hydrophilic or hydrophobic. Hence, later purification may be further selective [77].

6.3.4.2. Precipitation

For initial sample concentration and clean-up, this approach may be used. But the implementation of the precipitation protocol depends strictly on the characteristic of recombinant proteins and the property of the contaminated protein. This process permits the removal of more than 90% of host cell protein. The extract is applied to the salt, normally $(NH_4)_2SO_4$ (ammonium sulfate), at a concentration that induces protein precipitations. This method's objective is to pick a concentration at which only unwanted protein impurities (not target proteins) precipitate. Before chromatography, such precipitations are often carried out because higher salt concentration increases the hydrophobic reaction between the medium of chromatography and sample component. The target protein may often be precipitated by using heat instead of salt, but this requires a thermostable POI [75]. There are limited heat precipitation approaches, such as a leaf or extract heat precipitation, permitting to scale up and optimize the entire process which depending on condition, protein types, and plant. pH precipitation is beneficial in rare situations, but recombinant proteins must maintain stability at an extreme pH level. Affinity precipitations are a particular precipitation variant in which the affinity ligands are covalently bound to polymers or other molecules, which can bind to structures of higher molecular weight. Purification at large-scale, affinity precipitations is robust, simple, and good [78].

6.3.4.3. Membrane Separation

Membr ane separation, otherwise known as membrane chromatography, is successfully used in several applications such as in the removal of contaminants (anion-exchange membranes with flow *via* mode). As an alternative to column chromatography, it is also tested in the purification of the therapeutic proteins. Also, for larger impurity, the capacity of membrane-binding is very high; thus, using disposable membrane will restrict the problems associated with the large chromatographic column. Membrane chromatography, particularly in processes of production, has the potential to become the preferred non-chromatographic method to remove endotoxin, nucleic acid, and host protein. It offers simple scalability, non-denaturation condition, and lower cost [79].

6.3.4.4. Chromatography

Chromatography leads to the isolation of recombinant proteins that are dissolved in solution. There are several commercially available techniques of chromatography ready to be used. Chromatography has very few different variants as follows: MMC (mixed-mode chromatography), HIC (hydrophobic interaction chromatography), SEC (size exclusion chromatography, also known as gel filtration), IEC (ion-exchange chromatography), AFC (affinity chromatography), AEC (anion exchange chromatography) and CEC (cation exchange chromatography). Most of the ready-to-use, ligand combined with matrices, that is maintaining with GMP [80]. The chromatographic approach can be custom-made, which is based on target proteins parameter and characteristics of host plant proteins. Few chromatography developments allow the removal of contaminants that are bound to the target protein, a method called weak partitioning chromatography. A flow-through mode may be used for leaving recombinant proteins in the liquid phase and bind the impurity to resins of chromatography. Although this is a sensitive and effective small-scale method, this is comparatively costly and thus not appropriate for production on an industrial scale.

6.3.4.4.1. Expanded Bed Adsorption

EBA is a technique that uses the same principle as a hydrophobic reaction, affinity column chromatography, and ion exchange ligand. In this technique, however, adsorption is accompanied by the washing of the fluidized bed. EBA used fluidized molecules to accomplishing the isolation of the product and remove the insoluble ingredients only in one step. Cell debris and particle aggregates do not clog the column, as the bed is fluidized. This method is tolerant to high particle load in the feed stream. It should be noted that it can contribute to the collapse of the fluidized bed by using a molecule that is too large or too heavily charged. An example of good use of the EBA method is the purification of the canola extract. EBA is also used for the purification of patatin from potato and recombinant β-glucuronidase from the canola seeds. This method can lessen the time of the process in contrast to the classical chromatography with less cost of production [81].

6.3.4.4.2. Fusion Tags

Tags are defined as sequences of exogenous amino acids with specific property or affinity for direct ligands. However, they allow the target protein to be purified without prior insights into their possible biochemical properties. Also, they can improve the solubility and stability of target proteins and aid in localization and imaging research. The existence of tags is undesired during the clinical application of recombinant proteins. To overcome this problem, methods such as the use of recombinant end-proteases are developed, which have the ability for cutting specific sequences. Afterward, enzymes can quickly be removed from the solution through the affinity purification method because they have specific tags. Genetic fusion of affinity tag and recombinant protein is intended to promote the purification of the target product and are also widely used tools in DSP [82].

7. RECENT ADVANCES IN BIOPHARMACEUTICAL PRODUCTION

7.1. Cell and Tissue Culture

The cell and tissue cultures include plant suspension culture, hairy root culture, moss culture, and microalga culture. In suspension culture, calli are dispersed in the liquid media and are being propagated under. Simple nutrients are required for this culture. Suspension culture of plant cells is developed for the production of secondary metabolite *e.g.* shikonin, digoxin, artemisinin, paclitaxel, ajmalicine, and ginsenosides with little commercial success. The potential of plant cell culture for the production of heterologous protein has been recognized for the last two decades. Now, it has become a viable substituted platform for the production of plant-based pharmaceutical proteins. The most commonly used cell lines of plants for the production of recombinant biopharmaceuticals are derived from *Nicotiana tabacum*, for example, cultivars BY-2 cells and NT-1 cells. They have some interesting features, such as robust, fast-growing, and capable to transform readily through *Agrobacterium* and cell cycle synchronization. Other commonly used plant cell lines are the edible crop *e.g. Oryza sativa* (rice), *Glycine max* (soybean), *Medicago sativa* (alfalfa), *Daucus carota* (carrot) and *Lycopersicon esculentum* (tomato). These cell lines may be considered more favorable as compared to the tobacco cell line in terms of regulatory compliance and by-product production. The carrot cell line is used for the production of human recombinant proteins such as glucocerebrosidase by Protalix. It was first approved as a plant-derived biopharmaceutical for the market [83].

Hairy root (HR) culture has also been used for the production of heterologous proteins. HR culture offers various benefits *e.g.* genetic stability, sterile condition for growth, fast accumulation of biomass, and a possibility for secreting heterologous protein in a culture medium. The first plant-based-pharmaceutical protein that was expressed in the HR was Anti-Streptococcus mutants MAbs Guy's 13 which was produced successfully in the culture medium. The current example of the MAb expressed in HR is a tumor-targeting antibody (anti-vitronectin mAbs M12). The production of a chimeric gene that encodes human growth hormones first reported in callus tissues of sunflower. This system has numerous advantages such as rapid growth rate, consistency of protein production under controlled environments, has fewer issues regarding pathogen contamination *e.g.* bacterial or viral toxin. It can address environmental and regulatory concerns related to the potential release of GMOs (genetically modified organisms) from the whole plant. Also, extraction and purification of protein from cell culture are more convenient and simpler, and cost-effective in contrast to the whole plant. Cell culture of rice produced rAAT (Recombinant α-1-antitrypsin) at a high rate such as 100-247 mg/L. Induction of recombinant protein secretion into the culture medium is a fine technique for increasing the yield, allow easy harvest and purifications. Moreover, the secretion of recombinant proteins by cell into the medium is a good choice when productions are scaled-up in a bioreactor. Signal peptides of the amylase were used to confirm the secretion of target recombinant proteins by cell. ER signal peptides facilitate the secretion and expression of the intracellular rhEPO from hairy roots culture of *N. tabacum*, which result in up to 66.75 ng/g yield of total soluble proteins (TSP). In several cases, signal peptides and promoter from amylase gene of rice has been used for developing the two-step process. In the first step maintain the cell viability and increase the cell number. In the second step, the recombinant protein is produced which resulted in a high rate of production of secreted protein. Though this system possesses some strength. It increased the cost and risk of contamination when the medium is changing. On the other hand, production has been scaled-up through a more appropriate method is based on air-lift bioreactors. It required no change of the medium. They proved the six-fold increases in yield of the recombinant protein by using a 2-step process. Now, the carrot is the most popular plant species for the manufacture of pharmaceutical, with 10 vaccines such as HIV (human immunodeficiency virus,) measles, Y. pestis, HBV (hepatitis B virus), *Chlamydia trachomatis*, enterotoxigenic *E. coli*, *M. tuberculosis*, *Corynebacterium diphtheria*, *Helicobacter pylori*, and *Clostridium tetani*. Rice is the most famous plant for the establishment of a cell line in suspension cultures. There are following recombinant proteins derived from plants through cell and tissue culture techniques, Table **1** [84].

Table 1. Recombinant therapeutic protein production in plants through the cell and tissue culture.

Recombinant protein	Host	Yield	References
Human erythropoietin	*N. tabacum* cv. BY-2	1 pg/dry weight	[86]
Antibody svFv fragments	*Oryza sativa* cv. Bengal	3.8 µg/callus dry weight	[87]
Human serum albumins	*Nicotiana tabacum*	0.25 mg/l	[88]
Bryodin 1	*N. tabacum* cv. NT-1	30 mg/l	[89]

(Table 1) cont.....

Recombinant protein	Host	Yield	References
Cytotoxic T cell surface antigen	*O. sativa*	76.5 mg/l	[90]
Human α1 antitrypsins	*O. sativa*	200 mg/l	[91]
Human growth hormones	*O. sativa* cv. Donjin	57 mg/l	[92]
Human alpha-l-iduronidase	Tobacco BY-2 cells	10 mg/l	[93]
Human lysosome	*O. sativa* cv. Taipie	3-4% TSP	[94]
Human collagen α1 chains	*Hordeum vulgare*	2-9 mg/l	[95]
Human interleukin-12	Rice	31 mg/l	[96]
Human epidermal growth hormone	Tobacco	2 µg/dry weight	[97]
Human HIV antibodies	Tobacco BY-2 cells	10 mg/l	[98]
Human growth hormones	Tobacco BY-2 cells	35 mg/l	[99]
Tanomatin	Tobacco	2.63 mg/l	[100]
FimA monoclonal antibody	*O. sativa*cv	17.28 mg/l	[101]
Human epidermal growth factor	*N. benthamiana*	6.24% of TSP (Total soluble proteins)	[102]
tumor-r-targeting antibody	*N. benthamiana*	16.2 ± 1.7 µg/g Fresh weight	[103]
Human epidermal growth factor (hEGF)	*N. benthamiana*	15.695 µg/g leaf fresh weight or 0.499% TSP	[104]

Plants are not the only photosynthetic organisms that can be used for recombinant protein production. Moss cells can photosynthesize and are grown in a self-contained system such as a bioreactor. Additionally, the recombinant protein could be secreted into the culture medium. *Physcomitrella patens* (a moss) is usually used for the production of different types of biopharmaceuticals. The aforementioned moss has well been characterized genetically and is taken as an ideal organism for the specific gene targeting through homologous recombination. This is a valuable source to produce glycol-engineered versions of tumor-targeting lewis Y-specifics mAbs MB314. A current achievement made by Greenovation Biopharmaceutical company is the establishment of a moss culture platform. It allows the production of high-quality protein on a large-scale.

Besides, microalgae are also a valuable source for the production of various types of therapeutics and are used as bio-factory of different molecules. The major benefits of algae are a higher rate of growth, rapid biomass accumulation, ease of cultivations in the contained bioreactor, and its ability to conduct post-translational modification of protein including glycosylation. There are several examples of microalgae species such as *Chlamydomonas reinhardtii, Dunaliella salina, Phaeodactylum tricornutum,* and *Schizochytrium* have been used for the production of pharmaceutical proteins *e.g.* vaccines, enzymes, and antibodies. Of the most commonly used green algae is *Chlamydomonas reinhardtii* that had been explored for the cost-effective production of recombinant proteins and ease of scaling up the process of production. Recently, diatom (*P. Tricornutum)* has also been engineered to express mAb, to be used against HBV (Hepatitis B virus) [85].

7.2. Virus-infected Plants – A Valuable Therapeutic Protein Production Source

Plant viruses have been explored as expression vectors to facilitate transgene integration into the plant genome for the expression of vaccines and other proteins of pharmaceutical value. The viruses used for this purpose are predominantly RNA (positive-sense) viruses *e.g.* Cowpea mosaic virus, TMV (tobacco mosaic virus), PVX (potato virus X), CMV (cucumber mosaic virus). Plants virus expression vector has been engineered for better functioning as an inexpensive, rapid, and robust platform for the production of vaccines [105]. Plant viruses have been used for introducing foreign gene in plants since the 1980s and technical advance in plant virology and molecular biology has enabled to generate numerous improved expression system. Virus-infected plants offer several advantages which include rapid, high-level transgene expression and better containment of the transgene.

The plant virus expression system was based on CaMV (cauliflower mosaic virus) of the Caulimoviridae family. It was only a plant virus with a dsDNA genome. Though, the limited capacity of packaging and restricted amount of the viral DNA can be eliminated without affecting the essential function which is a fundamental limitation in the development of viruses as expression vehicles. Fortuitously, technical advances in molecular biology *e.g.* generation of cDNA from RNA template have enabled to expand research and to address the existing bottlenecks.

Gene insertion vector comprises complete functional virus and extra ORF (open reading frame) for the target protein. They are capable of a cell to cell systemic movements and infections. Viruses with both rod and spherical shaped virion have been investigated. Rod-shaped viral particles have better potential because of a limited amount of inserted nucleic acid. The most popular gene insertion vector has been originated from PVX (potato virus X) and TMV (tobacco mosaic virus). Both viruses have a positive ssRNA genome. They used sub-genomic promoter and consequently sub-genomic RNA for expressing their ORF. They have helical virions symmetry which results in the rod-shaped particle. Chimeric TMV constructed with the CAT gene, inserted between MP (movement protein) gene and CP gene. Expressions were regulated by an extra copy of the sub-genomics promoter of the CP gene which as a consequence, was duplicated in hybrid vectors. Vectors can assemble correctly, replicate, to form the sub-genomic RNA, and to produce the activity of the reporter gene. However, duplication of the sub-genomic promoter that sustained the homologous recombination, caused instability and as a result loss of exogenous genes. The problem was solved through sub-genomic CP promoters derived from diverse viruses belonging to the Tobamo virus genus for the prevention of homologous recombination. Similar to TMV, PVX vectors were generated by using duplicated promoters of CP genes. However, in that case, vectors are stable and capable to retain additional coding sequences. Recombination involved in homologous CP promoters' sequence but also other mechanisms mediated by both virus and host plant (Table **2**) [106].

Table 2. Recombinant therapeutic protein production in plants engineered through plant virus vectors.

Recombinant protein	Host plant	Yield	References
Norwalk Virus VLPs	Tobacco, lettuce	0.34 mg/g LFW	[107]
WNV E protein Mab	Tobacco, lettuce	0.23–0.27 mg/g LFW	[108]
SEB	*N. bethamiana*	-	[109]
HBVcAg	*N. benthamiana*	0.8 mg/g LFW	[110]
HPV-1 L1 protein	*N. benthamiana*	-	[111]
HAV VP1	*N. benthamiana*	-	[112]
Ebola Virus GP1Mab	Tobacco, lettuce	0.23–0.27 g/g LFW	[108]
HIV-1 type C p24	*N. benthamiana*	-	[111]
vitronectin	*N. benthamiana*	2.3% TSP	[113]
Influenza vaccine based on M2 protein	*N. benthamiana*	30% of total soluble protein (about 1 mg/g of fresh leaf tissue	[114]
Antifungal protein	*N. benthamiana*	4.3 ± 1.1 mg per gram of leaf fresh weight	[115]
Anticancer Mambalgin-1 peptide	*N. benthamiana*	165 μg/ml (4.5% TSP)	[116]

7.3. Expression of Therapeutic Proteins in Plants Through Agro-infection

Both vacuum and syringe infiltration has been used to produce therapeutic proteins. The simplicity of the syringe infiltration allows quick assessment of the expression level of pharmaceutical candidates under the established condition. Various optima have been worked out by various research groups for the optimization of expression parameters *i.e.* different concentrations of *Agrobacterium* culture was tested on a single leaf for comparing their effects on expression and yield of target proteins. Similarly, the ability of the various expression vectors in driving their expression and accumulation of target protein was also characterized. Other parameters *i.e* growth condition, favorable organelles for the accumulation of target protein, and requirements for the silencing suppressors can be established quickly or recognized by assay with the syringe infiltration. After optimization of these conditions for new protein candidates. Syringe infiltration can be used for infiltrating the full area of the leaf of numerous plants. So that, enough recombinant proteins can be obtained rapidly, for biochemical characterizations, preclinical functional study, and to develop the purification scheme [117]. But poor scalability is a limitation of syringe infiltration. For developing an Agro-infiltration to produce the commercial pharmaceutical proteins, a scalable technology of Agro-infiltration must be established. For this purpose, vacuum infiltration is a prime candidate. In this procedure, the desiccator is connected with a vacuum pump for providing vacuum for the infiltration. Placing the upside-down of a plant on a shelf and submergence of leaves into infiltration media containing *Agrobacterium* results in enhanced protein production. A 100-bar vacuum was applied for one-minute and slowly opening the release valve for releasing the vacuum. Agro-infiltration is achieved by exposure of plants to vacuum. Air is drawn out from the interstitial space of leaves through vacuum and *Agrobacteria* enters in infiltration medium and when the vacuum is released then *Agrobacteria* replaces air from this space. The literature review suggests that a more robust method for infiltration is vacuum infiltration and it significantly reduced the period required to infiltrate each plant. Vacuum infiltration is quite an efficient method of transgene expression and has been explored by many biotechnology companies for the commercial scale production of therapeutic proteins. It has commercially been attempted to process numerous metric tons of *N. benthamiana* plant per-hour through Kentucky Bioprocessing, LLC. Hence, vacuum infiltration has made it possible to produce the gram level of pharmaceuticals product under GMP (Good Manufacture Practice) guidelines which are desirable for both in quantity and quality of therapeutic proteins (Table **3**).

Table 3. Salient examples of production of recombinant proteins in plants through Agro-infection.

Recombinant protein	Host plant	Yield	References
Anti-HCG antibodies	*N. tabaccum*	40 mg/kg FW	[118]
Interferon- α2B	*N. benthamiana*	~10% TSP	[118]
Ag85B antigen	*N. benthamiana*	200-800 mg/kg FW	[120]
Aprotinin	*N. benthamiana*	1 mg/g FW	[121]
Human fibroblast growth factor 8b	*N. tabacum*	4% TSP	[122]
ESAT6 antigen	*N. benthamiana*	2 mg/kg FW	[120]
E. maxima gametocyte antigen (Gam82)	*N. tabacum*	20 mg/kg FW	[123]

(Table 3) cont.....

Recombinant protein	Host plant	Yield	References
CHIKV E1 and	*N. benthamiana*	E28–13 mg/kg of fresh leaf weight	[124]
ZIKV envelope (E) protein	*N. benthamiana*	160 μg/g FW	[125]
HIV Env gp140	*N. benthamiana*	5–6 mg/kg FW	[126]
cD5	*N. benthamiana*	50 μg/g FW	[127]
CHKV mab	*N. benthamiana*	100 μg/g FW	[128]
Human anti-HIV monoclonal antibody 2G12	*N. benthamiana*	110 μg of 2G12 per gram of leaf material	[129]
hemagglutinin H1	*N. benthamiana*	(0.25 M Units/g LFW) (0.05 M Units/g)	[130]
Type 16 E7 protein	*Solanum Lycopersicum*	5.5 μg/g of fresh weight	[131]
Haemagglutinin	*N. benthamiana*	50 mg/kg fresh weight of leaf	[130]
virus-like nanoparticles (VNPs)	*N. benthamiana*	~4 g/kg wet weight of tobacco	[132]
Essential oils	Guava leaves	100 mg/mL	[133]
Flavonoids	*Olea capensi*	0.05g	[134]

8. QUALITY ASSURANCE IN BIOPHARMACEUTICAL PRODUCTION

Biosafety is the fundamental pre-requisite for the commercialization of any product relevant to human health. The safety of patients or consumers is very important before the commercialization of biopharmaceuticals, therapeutics, and nutraceuticals, *etc.* The ability to provide pure, effective, and safe products are the primary factors to determine the success of a product. It must be ensured that biopharmaceuticals (drugs) meet the requirement related to safety, strength, and identity. Downstream-processing is very important in this context as it leads to the development of a purified product, free from any sort of hazardous impurities.

An essential aspect of the quality-assurance is critical monitoring of all of the activities and maintenance of the proper record relevant to their assessment and evaluation Documentations should be adequate for ensuring traceability of every batch production history, including all associated issues *i.e.* raw material, cleaning procedure, labeling, distribution, and packaging. Validation is the action of verifying any materials, processes, procedures, activities, equipment, or mechanisms that are used to achieve the intended and desired results. Therefore, sufficient technically and scientifically sound data should be provided for showing that specification is met and the intended result is attained in a reproducible fashion.

8.1. Biopharmaceutical Production Validation

Authentic validation of the processes could be extended from designing and planning of equipment items to routine inspections within productions, with the entire cycle including numerous elements including (1) DQ (design qualification) includes the user-requirements specification and functional specification for engineering procurement and design. (2) IQ is the abbreviation of installation qualification. It is the process of verifying all keys aspects of the hardware installations. It adheres to apposite code and permitted design intention. (3) OQ (operational qualification). It verifies that the subsystem performs as proposed with the model process material. (4) PQ (performance qualification) of the equipment. Process-change controls, to confirm that product qualities are optimized or maintained after made the changes to the processes. Also, regulatory measures have been developed and defined for the development and commercialization of transgenic plants with valuable proteins

The majority of biotechnological operations are run under aseptic conditions. Aseptic designs have been developed rapidly. But the requirement of hygienic designs, for example, cleaned equipment from the undesired matters, *e.g.* product residues are mostly underestimated. Validation should confirm that the cleaning process is as per guidelines of the equipment and types of contaminations. Hygienic designs of the fermenters are critical for the success of the cleaning procedure. Moreover, the reproducibility of the cleaning procedure can be augmented through developing the equipment having automated CIP (cleaning-i--place) systems. The sterilized equipment should be built on simple criteria. Process validations and GMPs productions can be achieved through checking for those criteria at the early stage of projects. The surface should have resistance against the products and cleaning at all ranges of operating pressure and temperature. The surface must be free from all sorts of cracka and the roughness of surfaces should be less or 0.5 μm. It should be easily reachable for visual inspections and manual cleaning. Equipment should be self-drained and the dead legs positioned correctly for ensuring that the CIP procedure can access all surfaces. The hygienic designs can also be extended to all outer parts, including insulations for avoiding condensation on the external surface of the equipment and such insulations should be sealed with the stainless-steels cladding. Supports of equipment should be sealed to buildings and should be without any gap or pocket and provide appropriate reach for allowing cleaning and inspection. Validations of the CIP procedure have become a major issue. The method has been developed to measure the removability of the model pollutants from the food-processing equipment, to determine in-place cleanability; a validation method is also available to clean fermentation equipment. The maximum permissible carry-over residual concentration for the multipurpose plant can be determined by measuring product harmfulness or maximum product dosage/day, many units of dosage per batch, and amount per batch. Hygienic designs have found broad acceptance for designing fermentation equipment. The purification process should be validated enough to verify that, it is capable of removing the impurity to a satisfactory level. For the production of biopharmaceutical, the emphasis is laid on the components that originate from host cells (DNA and proteins), media component or substance used during the process downstream *e.g.* nutrient, buffer component, stabilizer, chromatography media, and possible external contaminations through an adventitious agent *e.g.* bacteria, virus, and mycoplasma, as well as scrapies-like agent in the cell culture, that must not be existing in all over the processes but it can accidentally contaminate the cultures.

The analytical procedure should have statistical accuracy, precisions, robustness, sensitivity, (capacity of the methods to remain unchanged through small and deliberate variation in the parameters of method), and ruggedness (reproducibility and intermediate precision) tested. The validation included evaluation of the matrix effect, *e.g.* effect of the sample pH or protein content. Final product quality can be evaluated by analytical procedure, have the highest priority for comprehensive and full validation. For this purpose, all the types of analytical equipment are selected properly, installed adequately, and operated by using classical qualification stages such as DQ, OQ, IQ, and PQ.

In case, an automated system is used for the production of pharmaceuticals, it should be fully validated and documented; both software and hardware should be tested. The system should perform within the specified limit such as performance test and with the certain event, *e.g.* erroneous operator input and sensor failure, amongst others. If a regulatory system is not mutually recognized officially, the product can only be marketed with the approval of individuals regulatory bodies, *e.g.* FDA or European Agencies for Evaluation of Medicinal Product with its Committee for Proprietary Medicinal Product (EMEA–CPMP). So, FDA would examine the GMPs acquiescence not only in the USA but also in the foreign company wanted for importing into the USA. As with guidelines of FDA, the European GMP guideline is valid for an entire product marketed in European Community members' state. MRA is based on the negotiation within the International Conference on the Harmonization (ICH) of the Technical Requirement for Registration of the Pharmaceutical for Humans Use. ICH brings together regulatory authorities of Europe, the USA, and Japan, as well as an expert from pharmaceuticals industries. Their objectives are for providing the recommendation, guideline, and requirement for product registrations to eliminate or minimize the requirement to carry out duplicate testing of new registered medicine [135].

9. APPLICATIONS OF BIOPHARMACEUTICAL FOR VETERINARY AND HUMANS

Modern medicine has been taken to another level by protein therapeutics and is currently used for treating various diseases associated with humans namely diabetes, anemia, hepatitis, and cancer. The technique in which the synthesis of a recombinant protein involves the use of plant cells or whole plant is known as molecular farming. Currently, plant-based cultures aided in the production of two vaccines. Among them, one is tobacco cell suspensions that have undergone the development and regulation protocols and hurdles. The second one is a human therapeutic enzyme procured from carrot cell suspension and it is assumed to strike the commercial development shortly. The pledge of the development of products with the aid of plant-based culture is forthcoming [105]. Plant-derived vaccines and other therapeutic agents seem to have a prodigious ability to produce a striking response. These agents produced by plant-based approaches are advantageous as they are cost-effective, don't need the expensive purification procedures, and could be stored for a prolonged period at vast temperature ranges [136].

The idea of veterinary vaccine production in plant-based expression systems has been employed for about two decades; even though only a few have been evaluated in the field. The vaccine against Newcastle disease was the first one to be approved by USDA in 2006 [137, 138]. This was made of a recombinant protein (hemagglutinin-neuraminidase) and expressed in tobacco suspension cells and was in the form of injectable vaccines. Afterward, the same protein was expressed in the leaves of tobacco and potato while in the seeds of maize and rice as reported in other studies [139]. An immunogenic response was observed in the chicken by giving oral dosages of antigens, against Norwalk disease. Other Vaccines against IBV infection in chicken were produced in potato [140]. In another study, maize seeds were used for the expression of a recombinant vaccine for Porcine Transmissible gastroenteritis coronavirus (TGEV) [141]. Later, the hemagglutinin proteins from H5N1 and H1N1 strains of influenza viruses were expresses in *Nicotiana benthaminama* to obtain virus-like proteins [142].

Anthrax is recognized as an extremely powerful source of bioterrorism, which is caused by *Bacillus anthracis*. Weakened strains of *B. anthracis* have been used for vaccine production. Various subunit vaccines have been expressed in transgenic tobacco chloroplasts in vast quantities against the *Variola* virus.

Important protein molecules for the development of vaccines against smallpox are B5R, L1R, A27L, and A33R. These proteins were expressed at high levels in tobacco protoplasts [143]. Enteropathogenic strains of *E. coli* (ETEC) are known to produce a protein called heat-labile toxin (LT). The B subunit of LT has been found highly analogous to cholera toxin protein which is the major reason for diarrhea. Henmce, LT-B protein was expressed in corn seeds and its immunogenicity was verified in mice. Cholera vaccine has been expressed in tobacco leaves and potato tubers. In potato tubers, the cholera toxin B subunit was expressed up to 0.3% of TSP (total soluble protein). It was, then, translocated to and stored in the endoplasmic reticulum (ER). Similarly, the same protein was expressed in tobacco chloroplast yielding up to 4% of TSP [144, 145].

The parasite *S. japonicum* has been identified to contain a protein Sj23 that is used for boosting immunity level in the mice when DNA coding for this protein was injected into mice [146]. Other important bioproducts included monoclonal antibodies that have promising immunity boosting potential against many bacterial and viral diseases [147] and are essential biological compounds, produced on a very large scale since they have a variety of uses in the management, analysis, contagious diseases, and medication of cancer. Antimicrobial agents are used widely since 1950 in the veterinary that has controlled animals' infectious diseases and hence increased the production of animals. There were almost 17 antimicrobial classes permitted to be used in animal food about 15 years ago in the USA [148]. Antibiotics are mainly used in the animal's food to provide treatment to the diseased animals and to avoid the infection spread in other animals and also helps in improving the livestock production rate.

The expression and production of immunoglobulin A have been reported in various edible tissue plants because they can be consumed in raw form with minimal post-harvest processing. It has led to the conclusion that plant tissues are an efficient production platform and safe media for use. For the last few years, another antibody namely 2G12 acting as an HIV neutralizing agent has been expressed and produced in tobacco plants. The process of protein A/G affinity chromatography is routinely employed to purify antibodies from plant tissues. However, the production cost depends upon production scale, transgene expression level, employed expression system, and column life. Using an expression system resulting high level of protein expression may reduce the isolation and purification cost which contributes significantly to reducing the overall production cost of antibodies. Various efforts have been made to substitute chromatographic media suitable for protein purification from plant tissues. The use of fusion products of oleosin and protein has also been found effective and reported in safflower seeds. It was found that fusion product got attached with antibodies and isolated easily through high-speed centrifugation. This helped in removing soluble as well as insoluble impurities in a single step and the antibodies in soluble form were acquired simply by washing and eluting oil bodies.

Recombinant therapeutic molecules and vaccines are always needed in larger amounts, hence various efforts have been made to overexpress them in plant tissue using various transformation methodologies. As a result, a big number of therapeutic proteins are being synthesized in plants on an experimental basis [149] and many of them have been accepted clinically and seem to be offered in the market shortly. Some examples of antibodies currently being produced from plant-based expression systems include MAPP66, 2G12, Antibody M12, BLX-301, Guy's 13 and 4E10 (HIV-neutralizing human antibody) [150], Human serum albumin [149], Ig A/G antibody, mouse IgG1, MGR48, mAb C5-1, HRIG [151].

In 2013, a study was conducted to produce an oral vaccine to immunize them against diarrhea caused which was found to be caused by an enterotoxigenic *E.coli* (ETEC). Hence, the anti-ETEC antibodies were produced in Arabidopsis seeds (Virdi, Coddens *et al.* 2013). Production of vaccines and other therapeutics in plant-based expression systems is considered to be more appropriate as compared to other systems especially microbial and mammalian by apposite protein folding and assembly of IgGs (Hiatt, Caffferkey, *et al.* 1989). A remarkable example is the production of vaccines against ebola outbreak in West Africa. When the antibodies called ZMAPP produced in a plant-based system were assessed in monkeys, it showed 100% efficient results [152].

10. PROSPECTS

The biopharmaceutical industry has been growing at a faster rate compared to the drug or other markets. Analysts proposed that the market would continue to expand quickly. With the sharply increasing epidemics and the emergence of chronic diseases, the need for therapeutics is increasing day by day. Insight into the mechanism underlying infectious disease-causing agents has enabled the identification of specific factors and processes. This has driven ongoing studies into the application of biopharmaceuticals in newly emerging clinical circumstances. The proven effectiveness of these drugs and their acceptability by physicians and patients as a therapeutic solution has led to the increased requests for novel biopharmaceuticals. One realistic benefit of these drugs is the targeted therapy rather than symptomatic therapy [153]. This has made possible the treatment of formerly incurable diseases. Recent scientific interventions have led to the development of therapeutic proteins with superior properties compared to the native proteins hence playing a pivotal role. The highly promising prospect of the biopharmaceuticals market is linked to breakthrough advances such as immunotherapy production, antibody-drug conjugate, and gene therapy. So, the foremost factor responsible for the hindrance of this industry is the higher cost involved in the development of biopharmaceutical [154].

The need for plant-based biopharmaceuticals will continue to rise in the future. More significantly, the economic feasibility of manufacturing effective recombinant protein will determine their viability and wide-spread acceptance at an industrial scale. Transient expression system has been most commonly adopted by various companies and has resulted in the commercialization of environmentally friendly and safe therapeutics. A bioreactor-based cell culture system that can be scaled-up quickly and free from animal pathogens, has been used by conventional fermentation-based companies. Though plants are the ideal pharmaceutical factories for the commercial-scale economical production of pharmaceuticals, certain aspects need further attention and research. The interventions in molecular biology have to resolve the existing bottlenecks in the way of transforming plants into pharmaceutical factories. Further, nanobiotechnology and genome editing are really exciting areas of research that can contribute to the production and delivery of these biopharmaceuticals [83].

CONSENT FOR PUBLICATION

Not applicable.

CONFLICTS OF INTEREST

The author declares no conflict of interest, financial or otherwise.

ACKNOWLEDGMENTS

This work was supported by the Ministry of Science and Technology (MoST), Government of Pakistan to MSK, and by Punjab Agricultural Research Board (PARB), Government of Punjab, Pakistan to MSK.

REFERENCES

[1]   Fischer R, Stoger E, Schillberg S, Christou P, Twyman RM. Plant-based production of biopharmaceuticals. Curr Opin Plant Biol 2004; 7(2): 152-8.
[http://dx.doi.org/10.1016/j.pbi.2004.01.007] [PMID: 15003215]

[2]   Fent K, Weston AA, Caminada D. Ecotoxicology of human pharmaceuticals. Aquat Toxicol 2006; 76(2): 122-59.
[http://dx.doi.org/10.1016/j.aquatox.2005.09.009] [PMID: 16257063]

[3]   Goodman M. Market watch: Sales of biologics to show robust growth through to 2013. Nat Rev Drug Discov 2009; 8(11): 837.
[http://dx.doi.org/10.1038/nrd3040] [PMID: 19876035]

[4]　Hiatt A, Cafferkey R, Bowdish K. Production of antibodies in transgenic plants. Nature 1989; 342(6245): 76-8.
[http://dx.doi.org/10.1038/342076a0] [PMID: 2509938]

[5]　Boothe J, Nykiforuk C, Shen Y, *et al*. Seed-based expression systems for plant molecular farming. Plant Biotechnol J 2010; 8(5): 588-606.
[http://dx.doi.org/10.1111/j.1467-7652.2010.00511.x] [PMID: 20500681]

[6]　Abiri R, Valdiani A, Maziah M, *et al*. A critical review of the concept of transgenic plants: insights into pharmaceutical biotechnology and molecular farming. Curr Issues Mol Biol 2016; 18: 21-42.
[PMID: 25944541]

[7]　Khan MS, Mustafa G, Joyia FA. Enzymes: Plant-based production and their applications. Protein &amp. Peptide Letters 2018; 25(2): 136-47.
[http://dx.doi.org/10.2174/0929866525666180122123722] [PMID: 29359656]

[8]　Kinna AW. Improved production and purification of recombinant proteins from mammalian expression systems. 2017.

[9]　Fischer R, Liao YC, Hoffmann K, Schillberg S, Emans N. Molecular farming of recombinant antibodies in plants. Biol Chem 1999; 380(7-8): 825-39.
[http://dx.doi.org/10.1515/BC.1999.102] [PMID: 10494831]

[10]　Levi G. Vaccine cornucopia. Transgenic vaccines in plants: new hope for global vaccination? EMBO Rep 2000; 1(5): 378-80.
[http://dx.doi.org/10.1093/embo-reports/kvd103] [PMID: 11258471]

[11]　Rybicki EP. Plant-produced vaccines: promise and reality. Drug Discov Today 2009; 14(1-2): 16-24.
[http://dx.doi.org/10.1016/j.drudis.2008.10.002] [PMID: 18983932]

[12]　Mason HS, Lam DM, Arntzen CJ. Expression of hepatitis B surface antigen in transgenic plants. Proc Natl Acad Sci USA 1992; 89(24): 11745-9.
[http://dx.doi.org/10.1073/pnas.89.24.11745] [PMID: 1465391]

[13]　Bucchini L, Goldman LR. Starlink corn: a risk analysis. Environ Health Perspect 2002; 110(1): 5-13.
[http://dx.doi.org/10.1289/ehp.021105] [PMID: 11781159]

[14]　Mihaliak CA, Webb S, Miller T, *et al*. Development of plant cell produced vaccines for animal health applications. Proceedings of the 108th Annual Meeting of the United States Animal Health Association. 158-63.

[15]　Sainsbury F, Lavoie PO, D'Aoust MA, Vézina LP, Lomonossoff GP. Expression of multiple proteins using full-length and deleted versions of cowpea mosaic virus RNA-2. Plant Biotechnol J 2008; 6(1): 82-92.
[PMID: 17986176]

[16]　Khan MS, Joyia FA, Mustafa G. Seeds as economical production platform for recombinant proteins. Protein Pept Lett 2020; 27(2): 89-104.
[http://dx.doi.org/10.2174/0929866526666191014151237] [PMID: 31622192]

[17]　Demain AL, Vaishnav P. Production of recombinant proteins by microbes and higher organisms. Biotechnol Adv 2009; 27(3): 297-306.
[http://dx.doi.org/10.1016/j.biotechadv.2009.01.008] [PMID: 19500547]

[18]　Ferrer-Miralles N, Domingo-Espín J, Corchero JL, Vázquez E, Villaverde A. Microbial factories for recombinant pharmaceuticals. Microb Cell Fact 2009; 8: 17.
[http://dx.doi.org/10.1186/1475-2859-8-17] [PMID: 19317892]

[19]　Svahn KS, Chryssanthou E, Olsen B, Bohlin L, Göransson U. *Penicillium nalgiovense* Laxa isolated from Antarctica is a new source of the antifungal metabolite amphotericin B. Fungal Biol Biotechnol 2015; 2: 1.
[http://dx.doi.org/10.1186/s40694-014-0011-x] [PMID: 28955453]

[20] Wurm FM. Production of recombinant protein therapeutics in cultivated mammalian cells. Nat Biotechnol 2004; 22(11): 1393-8.
[http://dx.doi.org/10.1038/nbt1026] [PMID: 15529164]

[21] Dumont J, Euwart D, Mei B, Estes S, Kshirsagar R. Human cell lines for biopharmaceutical manufacturing: history, status, and future perspectives. Crit Rev Biotechnol 2016; 36(6): 1110-22.
[http://dx.doi.org/10.3109/07388551.2015.1084266] [PMID: 26383226]

[22] Dyck MK, Lacroix D, Pothier F, Sirard MA. Making recombinant proteins in animals--different systems, different applications. Trends Biotechnol 2003; 21(9): 394-9.
[http://dx.doi.org/10.1016/S0167-7799(03)00190-2] [PMID: 12948672]

[23] Bertolini LR, Meade H, Lazzarotto CR, *et al.* The transgenic animal platform for biopharmaceutical production. Transgenic Res 2016; 25(3): 329-43.
[http://dx.doi.org/10.1007/s11248-016-9933-9] [PMID: 26820414]

[24] Harvey AJ, Speksnijder G, Baugh LR, Morris JA, Ivarie R. Expression of exogenous protein in the egg white of transgenic chickens. Nat Biotechnol 2002; 20(4): 396-9.
[http://dx.doi.org/10.1038/nbt0402-396] [PMID: 11923848]

[25] Faye L, Gomord V. Recombinant proteins from plants Methods and Protocols. USA: Humana Press 2009.
[http://dx.doi.org/10.1007/978-1-59745-407-0]

[26] Daniell H, Khan MS, Allison L. Milestones in chloroplast genetic engineering: an environmentally friendly era in biotechnology. Trends Plant Sci 2002; 7(2): 84-91.
[http://dx.doi.org/10.1016/S1360-1385(01)02193-8] [PMID: 11832280]

[27] Reggi S, Marchetti S, Patti T, *et al.* Recombinant human acid beta-glucosidase stored in tobacco seed is stable, active and taken up by human fibroblasts. Plant Mol Biol 2005; 57(1): 101-13.
[http://dx.doi.org/10.1007/s11103-004-6832-x] [PMID: 15821871]

[28] Khan MS, Khan IA. Biopharming: A Biosecurity Measure to Combat Newcastle Disease for Household Food Security. Biosafety (Los Angel) 2015; 4: e156.
[http://dx.doi.org/10.4172/2167-0331.1000e156]

[29] Fischer R, Emans N, Emans N. Molecular farming of pharmaceutical proteins. Transgenic Res 2000; 9(4-5): 279-99. [Commentary].
[http://dx.doi.org/10.1023/A:1008975123362] [PMID: 11131007]

[30] Breedveld FC. Therapeutic monoclonal antibodies. Lancet 2000; 355(9205): 735-40.
[http://dx.doi.org/10.1016/S0140-6736(00)01034-5] [PMID: 10703815]

[31] Ma JK, Hikmat BY, Wycoff K, *et al.* Characterization of a recombinant plant monoclonal secretory antibody and preventive immunotherapy in humans. Nat Med 1998; 4(5): 601-6.
[http://dx.doi.org/10.1038/nm0598-601] [PMID: 9585235]

[32] Humphreys DP, Glover DJ. Therapeutic antibody production technologies: molecules, applications, expression and purification. Curr Opin Drug Discov Devel 2001; 4(2): 172-85.
[PMID: 11378956]

[33] Mahon BP, Moore A, Johnson PA, Mills KHG. Approaches to new vaccines. Crit Rev Biotechnol 1998; 18(4): 257-82.
[http://dx.doi.org/10.1080/0738-859891224167] [PMID: 9887505]

[34] Tacket CO, Mason HS. A review of oral vaccination with transgenic vegetables. Microbes Infect 1999; 1(10): 777-83.
[http://dx.doi.org/10.1016/S1286-4579(99)80080-X] [PMID: 10816083]

[35] Streatfield SJ, Jilka JM, Hood EE, *et al.* Plant-based vaccines: unique advantages. Vaccine 2001; 19(17-19): 2742-8.
[http://dx.doi.org/10.1016/S0264-410X(00)00512-0] [PMID: 11257418]

[36]   Chargelegue D, Obregon P, Drake PMW. Transgenic plants for vaccine production: expectations and limitations. Trends Plant Sci 2001; 6(11): 495-6.
[http://dx.doi.org/10.1016/S1360-1385(01)02123-9] [PMID: 11701351]

[37]   Theisen M. Production of recombinant blood factors in transgenic plants. Chemicals *via* Higher Plant Bioengineering. Springer 1999; pp. 211-20.
[http://dx.doi.org/10.1007/978-1-4615-4729-7_16]

[38]   Miller HI, Longtin D. Down on the Biopharm. Policy Rev 2003; 55.

[39]   Colson G, Huffman WE, Rousu MC. Estimates of the Welfare Impact of Intragenic and Transgenic GM Labeling Policies. 2010.https://ideas.repec.org/p/ags/aaea10/61387.html Agricultural and Applied Economics Association.

[40]   Fernandez-Cornejo J, Caswell MF. The first decade of genetically engineered crops in the United States. USDA-ERS Econ Inf Bull 2006.

[41]   Kapuscinski AR, Goodman RM, Hann SD, *et al.* Making 'safety first' a reality for biotechnology products. Nat Biotechnol 2003; 21(6): 599-601.
[http://dx.doi.org/10.1038/nbt0603-599] [PMID: 12776139]

[42]   Bradford KJ, Van Deynze A, Gutterson N, Parrott W, Strauss SH. Regulating transgenic crops sensibly: lessons from plant breeding, biotechnology and genomics. Nat Biotechnol 2005; 23(4): 439-44.
[http://dx.doi.org/10.1038/nbt1084] [PMID: 15815671]

[43]   Daniell H, Streatfield SJ, Wycoff K. Medical molecular farming: production of antibodies, biopharmaceuticals and edible vaccines in plants. Trends Plant Sci 2001; 6(5): 219-26.
[http://dx.doi.org/10.1016/S1360-1385(01)01922-7] [PMID: 11335175]

[44]   Ma JKC, Drake PMW, Christou P. The production of recombinant pharmaceutical proteins in plants. Nat Rev Genet 2003; 4(10): 794-805.
[http://dx.doi.org/10.1038/nrg1177] [PMID: 14526375]

[45]   Thomas JA. Biotechnology: Safety evaluation of biotherapeutics and agribiotechnology products. Biotechnology and Safety Assessment. Elsevier 2003; pp. 347-84.
[http://dx.doi.org/10.1016/B978-012688721-1/50013-1]

[46]   Goldstein DA, Thomas JA. Biopharmaceuticals derived from genetically modified plants. QJM 2004; 97(11): 705-16.
[http://dx.doi.org/10.1093/qjmed/hch121] [PMID: 15496527]

[47]   Hechavarría-Nuñez Y, Martínez-Muñoz L, Pérez-Massipe RO, *et al.* Metodología para la elaboración de las pautas reguladoras para productos biofarmacéuticos obtenidos de plantas transgénicas en Cuba. Biotecnol Apl 2009; 26: 122-6.

[48]   Fischer R, Schillberg S, Buyel JF, Twyman RM. Commercial aspects of pharmaceutical protein production in plants. Curr Pharm Des 2013; 19(31): 5471-7.
[http://dx.doi.org/10.2174/13816128113199310002] [PMID: 23394566]

[49]   Redkiewicz P, Sirko A, Kamel KA, Góra-Sochacka A. Plant expression systems for production of hemagglutinin as a vaccine against influenza virus. Acta Biochim Pol 2014; 61(3): 551-60.
[http://dx.doi.org/10.18388/abp.2014_1877] [PMID: 25203219]

[50]   Hamamoto H, Sugiyama Y, Nakagawa N, *et al.* A new tobacco mosaic virus vector and its use for the systemic production of angiotensin-I-converting enzyme inhibitor in transgenic tobacco and tomato. Biotechnology (N Y) 1993; 11(8): 930-2.
[PMID: 7763916]

[51]   Mustafa G, Khan MS. Transmission of engineered plastids in sugarcane, a C4 monocotyledonous plant, reveals that sorting of preprogrammed progenitor cells produce heteroplasmy. Plants 2020; 10(1): 26.

[http://dx.doi.org/10.3390/plants10010026] [PMID: 33374390]

[52]   Xu J, Ge X, Dolan MC. Towards high-yield production of pharmaceutical proteins with plant cell suspension cultures. Biotechnol Adv 2011; 29(3): 278-99.
[http://dx.doi.org/10.1016/j.biotechadv.2011.01.002] [PMID: 21236330]

[53]   Tescari S. Production and characterization of therapeutic proteins/peptides: Human Recombinant FSH-beta subunit expressed in Plant Cells and Chemical Synthesis of Human Osteocalcin and Neuritogenic Peptides. PhD Dissertation University of Padova 2017.

[54]   Huang Z, Phoolcharoen W, Lai H, *et al.* High-level rapid production of full-size monoclonal antibodies in plants by a single-vector DNA replicon system. Biotechnol Bioeng 2010; 106(1): 9-17.
[http://dx.doi.org/10.1002/bit.22652] [PMID: 20047189]

[55]   Fujiuchi N, Matoba N, Matsuda R. Environment control to improve recombinant protein yields in plants based on Agrobacterium-mediated transient gene expression. Front Bioeng Biotechnol 2016; 4: 23.
[http://dx.doi.org/10.3389/fbioe.2016.00023] [PMID: 27014686]

[56]   Fujiuchi N, Matsuda R, Matoba N, Fujiwara K. Removal of bacterial suspension water occupying the intercellular space of detached leaves after agroinfiltration improves the yield of recombinant hemagglutinin in a Nicotiana benthamiana transient gene expression system. Biotechnol Bioeng 2016; 113(4): 901-6.
[http://dx.doi.org/10.1002/bit.25854] [PMID: 26461274]

[57]   Twyman RM, Schillberg S, Fischer R. Optimizing the yield of recombinant pharmaceutical proteins in plants. Curr Pharm Des 2013; 19(31): 5486-94.
[http://dx.doi.org/10.2174/1381612811319310004] [PMID: 23394567]

[58]   Arlen PA, Falconer R, Cherukumilli S, *et al.* Field production and functional evaluation of chloroplast-derived interferon-α2b. Plant Biotechnol J 2007; 5(4): 511-25.
[http://dx.doi.org/10.1111/j.1467-7652.2007.00258.x] [PMID: 17490449]

[59]   Lee SJ, Park CI, Park MY, *et al.* Production and characterization of human CTLA4Ig expressed in transgenic rice cell suspension cultures. Protein Expr Purif 2007; 51(2): 293-302.
[http://dx.doi.org/10.1016/j.pep.2006.08.019] [PMID: 17079164]

[60]   Chen L, Yang X, Luo D, Yu W. Efficient production of a bioactive Bevacizumab monoclonal antibody using the 2A self-cleavage peptide in transgenic rice callus. Front Plant Sci 2016; 7: 1156.
[http://dx.doi.org/10.3389/fpls.2016.01156] [PMID: 27555853]

[61]   Kim NS, Yu HY, Chung ND, Shin YJ, Kwon TH, Yang MS. Production of functional recombinant bovine trypsin in transgenic rice cell suspension cultures. Protein Expr Purif 2011; 76(1): 121-6.
[http://dx.doi.org/10.1016/j.pep.2010.10.007] [PMID: 20951807]

[62]   Wilson SA, Roberts SC. Recent advances towards development and commercialization of plant cell culture processes for the synthesis of biomolecules. Plant Biotechnol J 2012; 10(3): 249-68.
[http://dx.doi.org/10.1111/j.1467-7652.2011.00664.x] [PMID: 22059985]

[63]   Santos RB, Abranches R, Fischer R, Sack M, Holland T. Putting the spotlight back on plant suspension cultures. Front Plant Sci 2016; 7: 297.
[http://dx.doi.org/10.3389/fpls.2016.00297] [PMID: 27014320]

[64]   Lee K, Lee SY, Na JG, *et al.* Magnetophoretic harvesting of oleaginous Chlorella sp. by using biocompatible chitosan/magnetic nanoparticle composites. Bioresour Technol 2013; 149: 575-8.
[http://dx.doi.org/10.1016/j.biortech.2013.09.074] [PMID: 24128604]

[65]   Hu YR, Guo C, Wang F, Wang SK, Pan F, Liu CZ. Improvement of microalgae harvesting by magnetic nanocomposites coated with polyethylenimine. Chem Eng J 2014; 242: 341-7.
[http://dx.doi.org/10.1016/j.cej.2013.12.066]

[66]   Lim JK, Chieh DCJ, Jalak SA, *et al.* Rapid magnetophoretic separation of microalgae. Small 2012; 8(11): 1683-92.

[http://dx.doi.org/10.1002/smll.201102400] [PMID: 22438107]

[67]   Wilken LR, Nikolov ZL. Recovery and purification of plant-made recombinant proteins. Biotechnol Adv 2012; 30(2): 419-33.
[http://dx.doi.org/10.1016/j.biotechadv.2011.07.020] [PMID: 21843625]

[68]   Chen Q. Expression and Purification of Pharmaceutical Proteins in Plants. Biol Eng 2008; 1: 291-321.
[http://dx.doi.org/10.13031/2013.26854]

[69]   Gecchele E, Schillberg S, Merlin M, Pezzotti M, Avesani L. A downstream process allowing the efficient isolation of a recombinant amphiphilic protein from tobacco leaves. J Chromatogr B Analyt Technol Biomed Life Sci 2014; 960: 34-42.
[http://dx.doi.org/10.1016/j.jchromb.2014.04.004] [PMID: 24786219]

[70]   Łojewska E, Kowalczyk T, Olejniczak S, Sakowicz T. Extraction and purification methods in downstream processing of plant-based recombinant proteins. Protein Expr Purif 2016; 120: 110-7.
[http://dx.doi.org/10.1016/j.pep.2015.12.018] [PMID: 26742898]

[71]   Azzoni AR, Kusnadi AR, Miranda EA, Nikolov ZL. Recombinant aprotinin produced in transgenic corn seed: extraction and purification studies. Biotechnol Bioeng 2002; 80(3): 268-76.
[http://dx.doi.org/10.1002/bit.10408] [PMID: 12226858]

[72]   Woodard SL, Wilken LR, Barros GOF, White SG, Nikolov ZL. Evaluation of monoclonal antibody and phenolic extraction from transgenic Lemna for purification process development. Biotechnol Bioeng 2009; 104(3): 562-71.
[http://dx.doi.org/10.1002/bit.22428] [PMID: 19575415]

[73]   Gallo M, Vinci G, Graziani G, De Simone C, Ferranti P. The interaction of cocoa polyphenols with milk proteins studied by proteomic techniques. Food Res Int 2013; 54: 406-15.
[http://dx.doi.org/10.1016/j.foodres.2013.07.011]

[74]   Loginov M, Boussetta N, Lebovka N, Vorobiev E. Separation of polyphenols and proteins from flaxseed hull extracts by coagulation and ultrafiltration. J Membr Sci 2013; 442: 177-86.
[http://dx.doi.org/10.1016/j.memsci.2013.04.036]

[75]   Buyel JF, Gruchow HM, Boes A, Fischer R. Rational design of a host cell protein heat precipitation step simplifies the subsequent purification of recombinant proteins from tobacco. Biochem Eng J 2014; 88: 162-70.
[http://dx.doi.org/10.1016/j.bej.2014.04.015]

[76]   Barany S, Szepesszentgyörgyi A. Flocculation of cellular suspensions by polyelectrolytes. Adv Colloid Interface Sci 2004; 111(1-2): 117-29.
[http://dx.doi.org/10.1016/j.cis.2004.07.003] [PMID: 15571665]

[77]   Aguilar O, Rito-Palomares M. Aqueous two-phase systems strategies for the recovery and characterization of biological products from plants. J Sci Food Agric 2010; 90(9): 1385-92.
[http://dx.doi.org/10.1002/jsfa.3956] [PMID: 20549787]

[78]   Mehta A. Downstream processing for biopharmaceutical recovery. In: Arora D, Sharma C, Jaglan S, Licgtfouse E, Eds. Pharmaceuticals from Microbes Environmental Chemsitry for Sustainable World. 163-90.
[http://dx.doi.org/10.1007/978-3-030-01881-8_6]

[79]   Roque ACA, Silva CSO, Taipa MÂ. Affinity-based methodologies and ligands for antibody purification: advances and perspectives. J Chromatogr A 2007; 1160(1-2): 44-55.
[http://dx.doi.org/10.1016/j.chroma.2007.05.109] [PMID: 17618635]

[80]   Nfor BK, Zuluaga DS, Verheijen PJT, Verhaert PDEM, van der Wielen LAM, Ottens M. Model-based rational strategy for chromatographic resin selection. Biotechnol Prog 2011; 27(6): 1629-43.
[http://dx.doi.org/10.1002/btpr.691] [PMID: 22238769]

[81]   Menkhaus TJ, Glatz CE. Antibody capture from corn endosperm extracts by packed bed and expanded bed adsorption. Biotechnol Prog 2005; 21(2): 473-85.

[http://dx.doi.org/10.1021/bp049689s] [PMID: 15801788]

[82]   Conley AJ, Joensuu JJ, Richman A, Menassa R. Protein body-inducing fusions for high-level production and purification of recombinant proteins in plants. Plant Biotechnol J 2011; 9(4): 419-33.
[http://dx.doi.org/10.1111/j.1467-7652.2011.00596.x] [PMID: 21338467]

[83]   Moon KB, Park JS, Park YI, *et al.* Development of systems for the production of plant-derived biopharmaceuticals. Plants 2019; 9(1): E30.
[http://dx.doi.org/10.3390/plants9010030] [PMID: 31878277]

[84]   Xu J, Zhang N. On the way to commercializing plant cell culture platform for biopharmaceuticals: present status and prospect. Pharm Bioprocess 2014; 2(6): 499-518.
[http://dx.doi.org/10.4155/pbp.14.32] [PMID: 25621170]

[85]   Donini M, Marusic C. Current state-of-the-art in plant-based antibody production systems. Biotechnol Lett 2019; 41(3): 335-46.
[http://dx.doi.org/10.1007/s10529-019-02651-z] [PMID: 30684155]

[86]   Matsumoto S, Ikura K, Ueda M, Sasaki R. Characterization of a human glycoprotein (erythropoietin) produced in cultured tobacco cells. Plant Mol Biol 1995; 27(6): 1163-72.
[http://dx.doi.org/10.1007/BF00020889] [PMID: 7766897]

[87]   Torres E, Vaquero C, Nicholson L, *et al.* Rice cell culture as an alternative production system for functional diagnostic and therapeutic antibodies. Transgenic Res 1999; 8(6): 441-9.
[http://dx.doi.org/10.1023/A:1008969031219] [PMID: 10767987]

[88]   Sijmons PC, Dekker BMM, Schrammeijer B, Verwoerd TC, van den Elzen PJ, Hoekema A. Production of correctly processed human serum albumin in transgenic plants. Biotechnology (N Y) 1990; 8(3): 217-21.
[PMID: 1366404]

[89]   Francisco JA, Gawlak SL, Miller M, *et al.* Expression and characterization of bryodin 1 and a bryodin 1-based single-chain immunotoxin from tobacco cell culture. Bioconjug Chem 1997; 8(5): 708-13.
[http://dx.doi.org/10.1021/bc970107k] [PMID: 9327135]

[90]   Park CI, Lee SJ, Kang SH, Jung HS, Kim DI, Lim SM. Fed-batch cultivation of transgenic rice cells for the production of hCTLA4Ig using concentrated amino acids. Process Biochem 2010; 45: 67-74.
[http://dx.doi.org/10.1016/j.procbio.2009.08.004]

[91]   Huang J, Sutliff TD, Wu L, *et al.* Expression and purification of functional human α-1-Antitrypsin from cultured plant cells. Biotechnol Prog 2001; 17(1): 126-33.
[http://dx.doi.org/10.1021/bp0001516] [PMID: 11170490]

[92]   Kim TG, Baek MY, Lee EK, Kwon TH, Yang MS. Expression of human growth hormone in transgenic rice cell suspension culture. Plant Cell Rep 2008; 27(5): 885-91.
[http://dx.doi.org/10.1007/s00299-008-0514-0] [PMID: 18264710]

[93]   Fu LH, Miao Y, Lo SW, *et al.* Production and characterization of soluble human lysosomal enzyme α-iduronidase with high activity from culture media of transgenic tobacco BY-2 cells. Plant Sci 2009; 177: 668-75.
[http://dx.doi.org/10.1016/j.plantsci.2009.08.016]

[94]   Huang J, Nandi S, Wu L, Yalda D, *et al.* Expression of natural antimicrobial human lysozyme in rice grains. Mol Breed 2002; 10: 83-94.
[http://dx.doi.org/10.1023/A:1020355511981]

[95]   Ritala A, Wahlström EH, Holkeri H, *et al.* Production of a recombinant industrial protein using barley cell cultures. Protein Expr Purif 2008; 59(2): 274-81.
[http://dx.doi.org/10.1016/j.pep.2008.02.013] [PMID: 18406168]

[96]   Shin YJ, Lee NJ, Kim J, An XH, Yang MS, Kwon TH. High-level production of bioactive heterodimeric protein human interleukin-12 in rice. Enzyme Microb Technol 2010; 46: 347-51.
[http://dx.doi.org/10.1016/j.enzmictec.2009.12.011]

[97] Parsons J, Wirth S, Dominguez M, Bravo-Almonacid F, Talou AMG, Rodriguez J. Production of Human Epidermal Growth Factor (hEGF) by *in Vitro* Cultures of Nicotiana tabacum. Effect of Tissue Differentiation and Sodium Nitroprusside Addition Int J Biotechnol Biochem 2010; 6: 133-40.

[98] Holland T, Sack M, Rademacher T, *et al.* Optimal nitrogen supply as a key to increased and sustained production of a monoclonal full-size antibody in BY-2 suspension culture. Biotechnol Bioeng 2010; 107(2): 278-89.
[http://dx.doi.org/10.1002/bit.22800] [PMID: 20506104]

[99] Weathers PJ, Towler MJ, Xu J. Bench to batch: advances in plant cell culture for producing useful products. Appl Microbiol Biotechnol 2010; 85(5): 1339-51.
[http://dx.doi.org/10.1007/s00253-009-2354-4] [PMID: 19956945]

[100] Pham NB, Schäfer H, Wink M. Production and secretion of recombinant thaumatin in tobacco hairy root cultures. Biotechnol J 2012; 7(4): 537-45.
[http://dx.doi.org/10.1002/biot.201100430] [PMID: 22125283]

[101] Kim BG, Kim SH, Kim NS, *et al.* Production of monoclonal antibody against FimA protein from Porphyromonas gingivalis in rice cell suspension culture. Plant Cell Tissue Organ Cult 2014; 118: 293-304.
[http://dx.doi.org/10.1007/s11240-014-0481-9]

[102] Thomas DR, Walmsley AM. Improved expression of recombinant plant-made hEGF. Plant Cell Rep 2014; 33(11): 1801-14.
[http://dx.doi.org/10.1007/s00299-014-1658-8] [PMID: 25048022]

[103] Lonoce C, Salem R, Marusic C, *et al.* Production of a tumour-targeting antibody with a human-compatible glycosylation profile in N. benthamiana hairy root cultures. Biotechnol J 2016; 11(9): 1209-20.
[http://dx.doi.org/10.1002/biot.201500628] [PMID: 27313150]

[104] Hanittinan O, Oo Y, Chaotham C, Rattanapisit K, Shanmugaraj B, Phoolcharoen W. Expression optimization, purification and *in vitro* characterization of human epidermal growth factor produced in *Nicotiana benthamiana.* Biotechnol Rep (Amst) 2020; 28: e00524.
[http://dx.doi.org/10.1016/j.btre.2020.e00524] [PMID: 32953470]

[105] Hefferon KL. DNA virus vectors for vaccine production in plants: Spotlight on geminiviruses. Vaccines (Basel) 2014; 2(3): 642-53.
[http://dx.doi.org/10.3390/vaccines2030642] [PMID: 26344750]

[106] Lico C, Chen Q, Santi L. Viral vectors for production of recombinant proteins in plants. J Cell Physiol 2008; 216(2): 366-77.
[http://dx.doi.org/10.1002/jcp.21423] [PMID: 18330886]

[107] Lai H, He J, Engle M, Diamond MS, Chen Q. Robust production of virus-like particles and monoclonal antibodies with geminiviral replicon vectors in lettuce. Plant Biotechnol J 2012; 10(1): 95-104.
[http://dx.doi.org/10.1111/j.1467-7652.2011.00649.x] [PMID: 21883868]

[108] Phoolcharoen W, Bhoo SH, Lai H, *et al.* Expression of an immunogenic Ebola immune complex in Nicotiana benthamiana. Plant Biotechnol J 2011; 9(7): 807-16.
[http://dx.doi.org/10.1111/j.1467-7652.2011.00593.x] [PMID: 21281425]

[109] Shen WH, Hohn B. Vectors based on maize streak virus can replicate to high copy numbers in maize plants. J Gen Virol 1995; 76(Pt 4): 965-9.
[http://dx.doi.org/10.1099/0022-1317-76-4-965] [PMID: 9049343]

[110] Huang Z, Chen Q, Hjelm B, Arntzen C, Mason H. A DNA replicon system for rapid high-level production of virus-like particles in plants. Biotechnol Bioeng 2009; 103(4): 706-14.
[http://dx.doi.org/10.1002/bit.22299] [PMID: 19309755]

[111] Regnard GL, Halley-Stott RP, Tanzer FL, Hitzeroth II, Rybicki EP. High level protein expression in

plants through the use of a novel autonomously replicating geminivirus shuttle vector. Plant Biotechnol J 2010; 8(1): 38-46.
[http://dx.doi.org/10.1111/j.1467-7652.2009.00462.x] [PMID: 19929900]

[112] Chung HY, Lee HH, Kim KI, *et al.* Expression of a recombinant chimeric protein of hepatitis A virus VP1-Fc using a replicating vector based on Beet curly top virus in tobacco leaves and its immunogenicity in mice. Plant Cell Rep 2011; 30(8): 1513-21.
[http://dx.doi.org/10.1007/s00299-011-1062-6] [PMID: 21442402]

[113] Dugdale B, Mortimer CL, Kato M, James TA, Harding RM, Dale JL. Design and construction of an in-plant activation cassette for transgene expression and recombinant protein production in plants. Nat Protoc 2014; 9(5): 1010-27.
[http://dx.doi.org/10.1038/nprot.2014.068] [PMID: 24705598]

[114] Mardanova ES, Kotlyarov RY, Kuprianov VV, *et al.* Rapid high-yield expression of a candidate influenza vaccine based on the ectodomain of M2 protein linked to flagellin in plants using viral vectors. BMC Biotechnol 2015; 15(1): 42.
[http://dx.doi.org/10.1186/s12896-015-0164-6] [PMID: 26022390]

[115] Shi X, Cordero T, Garrigues S, Marcos JF, Daròs JA, Coca M. Efficient production of antifungal proteins in plants using a new transient expression vector derived from tobacco mosaic virus. Plant Biotechnol J 2019; 17(6): 1069-80.
[http://dx.doi.org/10.1111/pbi.13038] [PMID: 30521145]

[116] Khezri G, Baghban Kohneh Rouz B, Ofoghi H, Davarpanah SJ. Heterologous expression of biologically active Mambalgin-1 peptide as a new potential anticancer, using a PVX-based viral vector in *Nicotiana benthamiana.* Plant Cell Tissue Organ Cult 2020; 142(2): 1-11. [PCTOC].
[http://dx.doi.org/10.1007/s11240-020-01838-x] [PMID: 32836586]

[117] Lai H, Engle M, Fuchs A, *et al.* Monoclonal antibody produced in plants efficiently treats West Nile virus infection in mice. Proc Natl Acad Sci USA 2010; 107(6): 2419-24.
[http://dx.doi.org/10.1073/pnas.0914503107] [PMID: 20133644]

[118] Kathuria S, Sriraman R, Nath R, *et al.* Efficacy of plant-produced recombinant antibodies against HCG. Hum Reprod 2002; 17(8): 2054-61.
[http://dx.doi.org/10.1093/humrep/17.8.2054] [PMID: 12151436]

[119] Gils M, Kandzia R, Marillonnet S, Klimyuk V, Gleba Y. High-yield production of authentic human growth hormone using a plant virus-based expression system. Plant Biotechnol J 2005; 3(6): 613-20.
[http://dx.doi.org/10.1111/j.1467-7652.2005.00154.x] [PMID: 17147632]

[120] Dorokhov YL, Sheveleva AA, Frolova OY, *et al.* Superexpression of tuberculosis antigens in plant leaves. Tuberculosis (Edinb) 2007; 87(3): 218-24.
[http://dx.doi.org/10.1016/j.tube.2006.10.001] [PMID: 17182283]

[121] Werner S, Marillonnet S, Hause G, Klimyuk V, Gleba Y. Immunoabsorbent nanoparticles based on a tobamovirus displaying protein A. Proc Natl Acad Sci USA 2006; 103(47): 17678-83.
[http://dx.doi.org/10.1073/pnas.0608869103] [PMID: 17090664]

[122] Potula HHSK, Kathuria SR, Ghosh AK, Maiti TK, Dey S. Transient expression, purification and characterization of bioactive human fibroblast growth factor 8b in tobacco plants. Transgenic Res 2008; 17(1): 19-32.
[http://dx.doi.org/10.1007/s11248-007-9072-4] [PMID: 17265164]

[123] Kota S, Subramanian M, Shanmugaraj BM, Challa H. Subunit vaccine based on plant expressed recombinant eimeria gametocyte antigen gam82 elicit protective immune response against chicken coccidiosis. J vaccines vaccin 2017; 08(06): 6-11.
[http://dx.doi.org/10.4172/2157-7560.1000374]

[124] Iyappan G, Shanmugaraj BM, Inchakalody V, Ma JK-C, Ramalingam S. Potential of plant biologics to tackle the epidemic like situations - case studies involving viral and bacterial candidates. Int J Infect Dis 2018; 73: 363.

[http://dx.doi.org/10.1016/j.ijid.2018.04.4236]

[125] Yang M, Sun H, Lai H, Hurtado J, Chen Q. Plant-produced Zika virus envelope protein elicits neutralizing immune responses that correlate with protective immunity against Zika virus in mice. Plant Biotechnol J 2018; 16(2): 572-80.
[http://dx.doi.org/10.1111/pbi.12796] [PMID: 28710796]

[126] Margolin E, Chapman R, Meyers AE, *et al.* Production and Immunogenicity of Soluble Plant-Produced HIV-1 Subtype C Envelope gp140 Immunogens. Front Plant Sci 2019; 10: 1378.
[http://dx.doi.org/10.3389/fpls.2019.01378] [PMID: 31737007]

[127] Rattanapisit K, Chao Z, Siriwattananon K, Huang Z, Phoolcharoen W. Plant-produced anti-enterovirus 71 (EV71) monoclonal antibody efficiently protects mice against EV71 infection. Plants 2019; 8(12): E560.
[http://dx.doi.org/10.3390/plants8120560] [PMID: 31805650]

[128] Hurtado J, Acharya D, Lai H, *et al. In vitro* and *in vivo* efficacy of anti-chikungunya virus monoclonal antibodies produced in wild-type and glycoengineered Nicotiana benthamiana plants. Plant Biotechnol J 2020; 18(1): 266-73.
[http://dx.doi.org/10.1111/pbi.13194] [PMID: 31207008]

[129] Strasser R, Stadlmann J, Schähs M, *et al.* Generation of glyco-engineered Nicotiana benthamiana for the production of monoclonal antibodies with a homogeneous human-like N-glycan structure. Plant Biotechnol J 2008; 6(4): 392-402.
[http://dx.doi.org/10.1111/j.1467-7652.2008.00330.x] [PMID: 18346095]

[130] Goulet MC, Gaudreau L, Gagné M, *et al.* Production of biopharmaceuticals in Nicotiana benthamiana—axillary stem growth as a key determinant of total protein yield. Front Plant Sci 2019; 10(June): 735.
[http://dx.doi.org/10.3389/fpls.2019.00735] [PMID: 31244869]

[131] Franconi R, Demurtas OC, Massa S. Plant-derived vaccines and other therapeutics produced in contained systems. Expert Rev Vaccines 2010; 9(8): 877-92.
[http://dx.doi.org/10.1586/erv.10.91] [PMID: 20673011]

[132] Rybicki EP. Plant molecular farming of virus-like nanoparticles as vaccines and reagents. Wiley Interdiscip Rev Nanomed Nanobiotechnol 2020; 12(2): e1587.
[http://dx.doi.org/10.1002/wnan.1587] [PMID: 31486296]

[133] Zhang F, Ramachandran G, Mothana RA, *et al.* Anti-bacterial activity of chitosan loaded plant essential oil against multi drug resistant *K. pneumoniae.* Saudi J Biol Sci 2020; 27(12): 3449-55.
[http://dx.doi.org/10.1016/j.sjbs.2020.09.025] [PMID: 33304155]

[134] Ndiege ML, Bo S, Ondijo CO, Jelagat L. Phytochemistry of antiulcer Plant based medicines used by Luhya people of western Kenya. 2020; 9(5): 1-7.

[135] Doblhoff-Dier O, Bliem R. Quality control and assurance from the development to the production of biopharmaceuticals. Trends Biotechnol 1999; 17(7): 266-70.
[http://dx.doi.org/10.1016/S0167-7799(99)01314-1] [PMID: 10370232]

[136] Yusibov V, Shivprasad S, Turpen TH, Dawson W, Koprowski H. Plant viral vectors based on tobamoviruses. Curr Top Microbiol Immunol 1999; 240: 81-94.
[PMID: 10394716]

[137] Rybicki EP, Williamson AL, Meyers A, Hitzeroth II. Vaccine farming in Cape Town. Hum Vaccin 2011; 7(3): 339-48.
[http://dx.doi.org/10.4161/hv.7.3.14263] [PMID: 21358269]

[138] Kolotilin I, Topp E, Cox E, *et al.* Plant-based solutions for veterinary immunotherapeutics and prophylactics. Vet Res 2014; 45(1): 117.
[http://dx.doi.org/10.1186/s13567-014-0117-4] [PMID: 25559098]

[139] Joensuu JJ, Niklander-Teeri V, Brandle JE. Transgenic plants for animal health: plant-made vaccine

antigens for animal infectious disease control. Phytochem Rev 2008; 7(3): 553-77.
[http://dx.doi.org/10.1007/s11101-008-9088-2] [PMID: 32214922]

[140] Zhou J-Y, Cheng L-Q, Zheng X-J, *et al.* Generation of the transgenic potato expressing full-length spike protein of infectious bronchitis virus. J Biotechnol 2004; 111(2): 121-30.
[http://dx.doi.org/10.1016/j.jbiotec.2004.03.012] [PMID: 15219399]

[141] Lamphear BJ, Jilka JM, Kesl L, Welter M, Howard JA, Streatfield SJ. A corn-based delivery system for animal vaccines: an oral transmissible gastroenteritis virus vaccine boosts lactogenic immunity in swine. Vaccine 2004; 22(19): 2420-4.
[http://dx.doi.org/10.1016/j.vaccine.2003.11.066] [PMID: 15193404]

[142] Govea-Alonso DO, Rybickim E, Rosales-Mendoza S. Plant-based vaccines as a global vaccination approach: current perspectives. Genetically engineered plants as a source of vaccines against wide spread diseases. Springer 2014; pp. 265-80.
[http://dx.doi.org/10.1007/978-1-4939-0850-9_13]

[143] Rigano MM, Manna C, Giulini A, Vitale A, Cardi T. Plants as biofactories for the production of subunit vaccines against bio-security-related bacteria and viruses. Vaccine 2009; 27(25-26): 3463-6.
[http://dx.doi.org/10.1016/j.vaccine.2009.01.120] [PMID: 19460602]

[144] Bergquist R, Al-Sherbiny M, Barakat R, Olds R. Blueprint for schistosomiasis vaccine development. Acta Trop 2002; 82(2): 183-92.
[http://dx.doi.org/10.1016/S0001-706X(02)00048-7] [PMID: 12020891]

[145] Pearce EJ. Progress towards a vaccine for schistosomiasis. Acta Trop 2003; 86(2-3): 309-13.
[http://dx.doi.org/10.1016/S0001-706X(03)00062-7] [PMID: 12745147]

[146] Waine GJ, Alarcon JB, Qiu C, McManus DP. Genetic immunization of mice with DNA encoding the 23 kDa transmembrane surface protein of Schistosoma japonicum (Sj23) induces antigen-specific immunoglobulin G antibodies. Parasite Immunol 1999; 21(7): 377-81.
[http://dx.doi.org/10.1046/j.1365-3024.1999.00221.x] [PMID: 10417672]

[147] Ko K, Steplewski Z, Glogowska M, Koprowski H. Inhibition of tumor growth by plant-derived mAb. Proc Natl Acad Sci USA 2005; 102(19): 7026-30.
[http://dx.doi.org/10.1073/pnas.0502533102] [PMID: 15867145]

[148] Anderson AD, Nelson JM, Rossiter S, Angulo FJ. Public health consequences of use of antimicrobial agents in food animals in the United States. Microb Drug Resist 2003; 9(4): 373-9.
[http://dx.doi.org/10.1089/107662903322762815] [PMID: 15000744]

[149] Sijmons PC, Dekker BM, Schrammeijer B, Verwoerd TC, van den Elzen PJ, Hoekema A. Production of correctly processed human serum albumin in transgenic plants. Biotechnology (N Y) 1990; 8(3): 217-21.
[PMID: 1366404]

[150] Häkkinen ST, Raven N, Henquet M, *et al.* Molecular farming in tobacco hairy roots by triggering the secretion of a pharmaceutical antibody. Biotechnol Bioeng 2014; 111(2): 336-46.
[http://dx.doi.org/10.1002/bit.25113] [PMID: 24030771]

[151] Zimmermann J, Saalbach I, Jahn D, *et al.* Antibody expressing pea seeds as fodder for prevention of gastrointestinal parasitic infections in chickens. BMC Biotechnol 2009; 9(1): 79.
[http://dx.doi.org/10.1186/1472-6750-9-79] [PMID: 19747368]

[152] Qiu X, Wong G, Audet J, *et al.* Reversion of advanced Ebola virus disease in nonhuman primates with ZMapp. Nature 2014; 514(7520): 47-53.
[http://dx.doi.org/10.1038/nature13777] [PMID: 25171469]

[153] Minghetti P, Rocco P, Cilurzo F, Vecchio LD, Locatelli F. The regulatory framework of biosimilars in the European Union. Drug Discov Today 2012; 17(1-2): 63-70.
[http://dx.doi.org/10.1016/j.drudis.2011.08.001] [PMID: 21856438]

[154] Kesik-Brodacka M. Progress in biopharmaceutical development. Biotechnol Appl Biochem 2018; 65(3): 306-22.
[http://dx.doi.org/10.1002/bab.1617] [PMID: 28972297]

CHAPTER 7

Green Factories: Plants As A Platform For Cost-effective Production of High-value Targets

Muhammad Omar Khan[1,2]**, Ayesha Siddiqui**[1,2] **and Niaz Ahmad**[1,2,*]

[1] *Agricultural Biotechnology Division, National Institute for Biotechnology & Genetic Engineering (NIBGE), Jhang Road, Faisalabad, 38000, Pakistan*

[2] *Department of Biotechnology, Pakistan Institute of Engineering and Applied Sciences (PIEAS), Nilore, Islamabad, Pakistan*

Abstract: Transgenic plants have been developed since the early 1980s, when researchers were able to transform a piece of foreign DNA into a plant genome. Since then, the technology has expanded enormously, giving rise to many private and public ventures in the field of plant-based recombinant technology. The technology has helped in crop improvement against various biotic and abiotic stresses such as insect resistance and herbicide tolerance, as well as improving their nutritional values, for example, Golden rice. In addition to crop improvement, the technology has enabled plants to be used as green factories for the production of recombinant proteins. Several platforms are available for the heterologous expression of foreign proteins, each of which represents its own set of advantages and limitations. Plants offer many advantages for inexpensive yet large-scale production of high-value targets, making them extremely attractive for commercial applications. In this chapter, we briefly discuss the need for using plants as solar-powered cellular factories to produce recombinant proteins. We provide a snapshot of different expression systems and argue that the plant-based expression system is highly commercially feasible not only for the production of high-value targets but also to help address global challenges like Covid-19.

Keywords: Biopharming, High-Value targets, Green factories, Plants.

1. WHY PLANT-BASED EXPRESSION SYSTEMS?

Heterologous expression of recombinant proteins for different applications has become a focus of intensive research for a while, paving the way for another revolution in the area for the development of new production technologies. The demand for cost-effective yet large-scale production of protein and secondary metabolites for various purposes, such as medical reagents, cosmetic products,

* **Corresponding authors Niaz Ahmad:** Agricultural Biotechnology Division, National Institute for Biotechnology & Genetic Engineering (NIBGE), Jhang Road, Faisalabad, Department of Biotechnology, Pakistan Institute of Engineering and Applied Sciences, (PIEAS), Nilore, Islamabad, Pakistan; E-mail: niazbloch@yahoo.com

and industrial enzymes, in terms of quantity, diversity, and, most importantly, quality has dramatically increased since the past decade [1]. The gap between demand and supply has further increased due to inefficient yet highly expensive production systems [2, 3]. Several systems, including bacteria, yeasts, animal cells, transgenic animals, plant cells, and transgenic plants, are available for the heterologous production of high-value targets [4, 5]. All available expression systems have their pros and cons in terms of cost, time, efficiency, product size, growth conditions, yield, post-translational modification, downstream processing, and regulatory approval [6]. The advantages of plant expression platforms are cited in several earlier reports [7 - 12]. Table **1** shows head-to-head comparisons of all existing platforms. Transgenic plants have become a focus of interest as new generation bioreactors mainly due to: i) reduced up-front production costs, ii) lower risk of endotoxins as well as human pathogen contamination, iii) scal-ability, iv) availability of existing infrastructure for the cultivation of transgenic plants, v) assemble complex protein with eukaryotic-like post-transcriptional modifications. However, plants lack the human-like N-glycosylation mechanism for protein processing that has been overcome by engineering tactics to ensure the authentic quality, homogeneity, and quantity [13]

Table 1. Comparison of different expression system [10].

Parameter	Bacteria	Yeast	Insect cells	Microalgae	Mammalian Cells	Transgenic Plants
Capital cost	Medium	Medium	High	Medium	Very high	Low
Operating cost	Low	Medium	High	Low	Very high	Low
Production scale	Short	Short	Medium	Short	Long	Long
Speed	Fast	Fast	Medium	Fast	Slow	Slow
Multigene engineering	Yes	No	No	Yes	No	Yes
Glycosylation	Absent	Incorrect	Yes	Yes, absent in chloroplast	Yes	Yes, absent in chloroplast
Contamination risk	High	Medium	High	Low	High	Low
Multimeric assembly	No	No	No	Yes	No	Yes
Protein folding	Low	Medium	High	High	High	High
Protein yield	High	Moderate high	Medium	High	Medium	Low-High
Scale up cost	High	High	Very high	Medium	Very high	Very low
Safety	Low	Unknown	Medium		Low	High

(Table 1) cont.....

Parameter	Bacteria	Yeast	Insect cells	Microalgae	Mammalian Cells	Transgenic Plants
Storage	Very cheap	Costly	Expensive	Low	Very expensive	Very cheap
Distribution	Easy	Feasible	Difficult	Easy	Difficult	Easy

Plant molecular farming (PMF) is termed as the technique of producing high-value proteins recombinantly in plants without disturbing their phenotype, metabolism, or performance. The proteins have been produced by this technique for more than 30 years, either in purified form, crude extract, or in planta [3, 14]. The idea of molecular farming based on the genetic transformation of plants was first proposed in the 1980s [15], which has now become a reality and is often termed as the 3rd generation of biotechnology [6]. The first examples of molecular farming using transgenic plants and plant cell suspension cultures involved the production of a human growth hormone, Nopaline synthase [16], and an antibody IgG$_1$ (6D4) [17]. However, the commercial application of this platform came years later when avidin was recombinantly produced in transgenic maize [18]. The breakthrough to commercial success for plant-derived biologics culminated in 2012 when the first plant-made pharmaceutical, Taliglucerase alfa, commercially known as Elelyso®, was developed by Protalix BioTherapeutics, was approved by the US Food and Drug Administration [19]. Elelyso® is a recombinant human glucocerebrocidase used for the treatment of Gaucher's disease (lysosomal storage disorder) [20].

The use of plants for the production of valuable proteins has been refined and improved over the years due to advancements in knowledge and technology. This has led to a major paradigm shift in the pharma sector, as the potential drawbacks associated with the early stages of PMF, including high expression level and efficient downstream processes, have been attained [6]. The product portfolio ranges from pharmaceutical therapeutics to non-pharmaceutical products such as antibodies, vaccine antigens, enzymes, growth factors, research or diagnostic reagents, and cosmetic ingredients. A number of 'proof-of-concept' studies have been performed to evaluate the potential of different plant species as hosts for molecular pharming [21, 22]. The host cells or the plant used for molecular farming purposes, depending upon target protein and its application, range from crop plants (rice, maize, tobacco, alfalfa, safflower, and lettuce) to pondweed, algae, microalgae, and mosses. The *Nicotiana* genus has been widely used for genetic transformation studies as it is easily genetically manipulated and has a fast growth rate. Two species, *Nicotiana benthamiana* and *Nicotiana tabacum* are considered as 'biological warehouses' for the production of many pharma or non-pharma products by the stable and transient expression [21]. Many plant-based proteins, including antibodies, either pharmaceutical or nonpharmaceutical, are in the pipeline of clinical or preclinical trials, and some are in the developmental stage for commercialization (Table **2** and **3**).

Table 2. Selected list of antibodies expressed in plants against various diseases.

Recombinant Protein	Pathogen/Disease	Host Plant	Transformation Method	Expression Level	References
cT84.66	Cancer (tumor marker)	*Nicotiana tabacum*	Transient expression	1 mg/kg FW	[66]
scFvT84.66	Cancer (tumor marker)	*Nicotiana tabacum*	Transient expression	5 mg/kg FW	[66]
scFvT84.66	Cancer (tumor marker)	*Oryza sativa*	Nuclear transformation	3.8 _g/g FW	[67]
scFvT84.66	Cancer (tumor marker)	*Wheat and rice*	Nuclear transformation	30 _g/g FWY	[68]
BR55-2	Human colorectal cancer	*Nicotiana tabacum*	Nuclear transformation	30 mg/kg FW	[69]
2F5	HIV	*Nicotiana benthamiana*	Nuclear transformation	0.01% of TSP	[70]
2G12	HIV	*Nicotiana benthamiana*	Transient expression	0.3 g/kg FW	[71]
2G12	HIV	*Nicotiana benthamiana*	Nuclear transformation	8 mg/L culture medium	[72]
6D8	Ebola virus	*Nicotiana benthamiana*	Transient expression	0.5 mg/g FW	[73]
6D8	Ebola virus	*Lettuce (L. sativa)*	Transient expression	0.23–0.27 mg/g FW	[74]
CO17-1AK	Human colorectal cancer	*Nicotiana tabacum*	Nuclear transformation	0.25 mg/kg FW	[75]
Palivizumab-N	Respiratory syncytial virus	*Nicotiana benthamiana*	Transient expression	180 mg/kg FW	[76]
E559	Rabies	*Nicotiana tabacum*	Nuclear transformation	1.8 mg/kg FW (0.04% of TSP	[77]
pE16	West Nile virus	*Nicotiana benthamiana*	Transient expression	0.74 mg/g FW	[78]
pE16scFv-CH	West Nile virus	*Nicotiana benthamiana*	Transient expression	0.77 mg/g FW	[78]
E60	Dengue virus	*Nicotiana benthamiana*	Transient expression	120 _g/g FW	[79]
2G12	HIV	*Oryza sativa*	Nuclear transformation	46.4 g g DW (seed)	[80]

(Table 2) cont.....

Recombinant Protein	Pathogen/Disease	Host Plant	Transformation Method	Expression Level	References
8B10	Chikungunya virus	*Nicotiana benthamiana*	Transient expression	20–30 mg/kg FW	[81]
5F10	Chikungunya virus	*Nicotiana benthamiana*	Transient expression	20–30 mg/kg FW	[81]
SO57	Rabies virus	*Nicotiana tabacum*	Transient expression	0.014–0.019% of TSP	[82]
cD5	Enterovirus 71	*Nicotiana benthamiana*	Transient expression	50 _g/g FW	[83]
PD1	Cancer	*Nicotiana benthamiana*	Transient expression	140 _g/g FW	[84]
c2A10G6	Zika virus	*Nicotiana benthamiana*	Transient expression	1.47 mg/g FW	[85]
6D8	Ebola	*Nicotiana benthamiana*	Transient expression	1.21 mg/g FW	[85]
HSV8	Herpes simplex virus	*Nicotiana benthamiana*	Transient expression	1.42 mg/g FW	[85]
CHKV mab	Chikungunya virus	*Nicotiana benthamiana*	Transient expression	100 _g/g FW	[86]
2C10	Porcine epidemic diarrhea virus	*Nicotiana benthamiana &* (*L. Sativa*)	Transient expression	NR	[87]
KPF1-Antx	Influenza	*Nicotiana benthamiana*	Transient expression	NR	[88]

Table 3. Plant-derived antibodies in clinical stages of development or on market.

Product	Disease	Plant	Clinical Trial Status	Company	Source
CaroRX	Dental caries	Tobacco	EU approved as medical advice	Planet Biotechnology, USA	www.planetbiotechnology.com
DoxoRX	Side-effects of cancer therapy	Tobacco	Phase I completed	Planet Biotechnology, USA	www.planetbiotechnology.com
RhinoRX	Common cold	Tobacco	Phase I completed	Planet Biotechnology, USA	www.planetbiotechnology.com
Fv antibodies	Non-Hodgkin's lymphoma	Tobacco	Phase I	Large Scale Biology, USA	www.lsbc.coma

(Table 3) cont.....

Product	Disease	Plant	Clinical Trial Status	Company	Source
IgG (ICAM1)	Common cold	Tobacco	Phase I	Planet Biotechnology, USA	www.planetbiotechnology.com
Antibody against hepatitis B	Vaccine purification	Tobacco	On market	CIGB, Cuba Kaiser	www.planetbiotechnology.com/ Kaiser,2008.

The PMF, with more than 120 companies, universities, and research institutes, is involved in realizing the fullest potential of this area. Therefore, it has emerged as a vibrant segment of the biotech industry [10, 14] have critically reviewed the strengths and bottlenecks of the commercial potential of plant-based expression platforms. The pharmaceuticals products which are endorsed from 2014 to mid of 2018 are 7 vaccines, 9 enzymes, 16 coagulating factors, 23 hormones, and 68 monoclonal antibodies [23]. There are several reports to prove the significant efficiency of plant-based pharmaceuticals compared with mammalian cell-based protein, which provides efficacious and less costly strategies to treat emerging infectious diseases. The plant-based expression system can be quickly up-scaled to satisfy the sudden and unexpected arising demands such as Covid-19 pandemic crisis (Covid-19) [24].

This chapter is aimed at describing the principles, current advancements in methodology for plant molecular pharming. We argue that plant molecular pharming has presented itself as a viable as well as a competitive platform to produce recombinant proteins inexpensively at a large scale. Different strategical system for plant transformation and expression is discussed, that have been developed to produce commercially important proteins. The advantages and disadvantages of each system have been well considered. The chapter also reviews the high-value bio-products (pharmaceutical or non-pharmaceutical) that are successfully being produced in the established and emerging plant systems and are in the pipeline of commercialization. The final section focuses on the outlook and perspective of plant molecular pharming as a potential therapeutic intervention against the ongoing human pandemic – COVID19 (SARS-nCoV-2019).

2. DEVELOPMENT OF TRANSGENIC PLANTS

Traditionally, crop plants have been improved through artificial selection and breeding based on phenotypic characteristics such as reduced susceptibility to biotic and abiotic stresses, plant height, grain size, and higher yield [25]. This approach has been quite instrumental in attaining supply-demand equilibrium, but the continuous increase in human population and a decline in arable land area necessitates to devise of new strategies for crop plant improvement. Conventional methods are limited by a narrow gene pool and lengthy procedures of selection [26]; however, transgenic technology offers the ability to develop transgenic plants with novel traits from varied taxonomic groups [27]. In this scenario, several candidate genes conferring novel traits such as insect resistance, salt tolerance, herbicide tolerance, heat tolerance, biofortification, and value addition have been identified and evaluated in model systems for their functioning into higher plants [28 - 63]. In addition, plants can also be developed having improved nutritional status [64] or value-addition such as delayed fruit ripening to improve the shelf life of perishable commodities such as tomatoes, so that they can reach the consumer intact preserving their taste, smell, color, and texture [65]. Transgenic plants can also be developed to produce high-value targets which are cheaper and are in larger quantities, such as recombinant proteins and metabolites of industrial importance, including antibodies (Table **2** and **3**), vaccine antigens (Table **4**), industrial enzymes (Table **5**), non-pharmaceutical recombinant proteins (Table **6**, Table **7**), human therapeutics (Table **8**) and even nutraceuticals (Table **9**). In this section, we will discuss, briefly, that how transgenic plants can be developed, and the advantages and challenges of each methodology used for the development of transgenic plants.

Table 4. Selected vaccine antigens produced in plants.

Expressed Protein	Disease	Gene	Plant	Expression Level	Source
Cholera toxin B subunit	Cholera	*Codon optimized CTB of Vibrio cholerae*	*Nicotiana tabacum*	4.1%	[89]
Bovine group A rotavirus VP6	Rotavirus	*VP6*	*Nicotiana tabacum*	3%	[90]
Canine parvovirus; Cholera toxin B subunit	Parvovirus and Cholera	*CTP-2L21*	*Nicotiana tabacum*	31.3%	[91]
Anthrax protective anti-gen	Anthrax	*pagA*	*Nicotiana tabacum*	14.2%	[92]
Bacterial lipoprotein A	Lyme disease	*OspA*	*Nicotiana tabacum*	10%	[93]
Cysteine rich region of lectin	*Entamoeba histolytica*	*LecA*	*Nicotiana tabacum*	6.3%	[94]
Cholera toxin B subunit–human proinsulin	Cholera	*CTB-Pins*	*Nicotiana tabacum*	16%	[95]

(Table 4) cont.....

Expressed Protein	Disease	Gene	Plant	Expression Level	Source
Cholera toxin B sub-unit–human proinsulin	Cholera	*CTB-Pins*	*Lactuca sativa*	2.5%	[95]
Human papillomavirus L1 protein	Cervical cancer	*L1* HPV-16	*Nicotiana tabacum*	24%	[96]
VP1 of the foot and mouth disease virus	Foot and Mouth Disease	*VP1*	*Stylosanthes guianensis*	0.1-0.5%	[97]
A27L of vaccinia virus	Orthopoxviruses (OPVs)	*A27L*	*Nicotiana tabacum*	18%	[98]
Heat labile toxin B subunit and heat-stable toxin	Enterotoxigenic Escherichia coli (ETEC)	*LTB:ST*	*Nicotiana tabacum*	2.3%	[99]
LTB–HN-neutralizing epitope	New Castle Disease Virus (NDV)	*LTB-HNE*	*Nicotiana tabacum*	0.5%	[100]
Cholera toxin-B subunit fused with apical membrane antigen-1	*CTB:AMA1*	Cholera and Malaria	*Nicotiana tabacum*	13.17%	[101]
Cholera toxin-B subunit fused with merozoite surface protein-1	*CTB:MSP1*	Cholera and Malaria	*Nicotiana tabacum*	10.11%	[101]
Cholera toxin-B subunit (CTB) fused with apical membrane antigen-1	*CTB:AMA1*	Cholera and Malaria	*Lactuca sativa*	7.3%	[101]
Cholera toxin-B subunit (CTB) fused with mero-zoite surface protein-1	*CTB:MSP1*	Cholera and Malaria	*Lactuca sativa*	6.1%	[101]
human b-site APP cle-aving enzyme	*BACE*	Alzheimer disease	*Nicotiana tabacum*	2%	[102]
GRA4 antigen	*GRA4*	*Toxoplasma gondii*	*Nicotiana tabacum*	0.2%	[103]
Hemagglutinin (HA) pro-teins	*HA*	Influenza virus	*Nicotiana benthamiana*	1300mg/kg	[104]
Mtb72F fused with cho-lera toxin B-subunit	*CTB-Mtb72F*	*Mycobacterium tuberculosis*	*Nicotiana tabacum*	1.2%	[105]
ESAT-6 fused with cho-lera toxin B-subunit	*CTB-ESAT6*	*Mycobacterium tuberculosis*	*Nicotiana tabacum*	7.5%	[105]
ESAT-6 fused with cholera toxin B-subunit	*CTB-ESAT6*	*Mycobacterium tuberculosis*	*Lactuca sativa*	0.75%	[105]
Tetra-epitope antigen	*cE-DI/IIp*	Dengue virus	*Lactuca sativa*	-	[106]

(Table 4) cont.....

Expressed Protein	Disease	Gene	Plant	Expression Level	Source
Hepatitis C virus E1E2 heterodimer	*HCVE1E2*	Hepatitis C virus (HCV)	*Lactuca sativa*	1.6 ug/mL	[107]
M2e Peptide fused with Ricin Toxin B Chain	*M2e*	Avian Influenza Virus	*Wolffia globosa*	0.01%	[108]
E2 protein of classical swine fever virus	*E2*	classical swine fever virus (CSFV)	*Arabidopsis thaliana*	0.7%	[109]
Human Papillomavirus (HPV) type 16 E7 protein	*E7*	Human Papillomavirus (HPV)	*Solanum lycopersicum*	35.5 µg/g	[110]
Enterotoxin B subunit	*LTB-Syn*	Parkinson's disease	*Nicotiana tabacum*	0.27ug/g	[111]
A region of PAc protein and cholera toxin B subunit	*PAcA-ctxB*	Streptococcus mutans	*Solanum lycopersicum*	-	[112]
VP1 capsid protein of FMDV serotype O	*VP1*	Foot and Mouth Disease (FMDV)	*Nicotiana tabacum*	0.72%	[113]
Non-toxic carboxylterminal domain of a-toxin (PlcC) and attenuated mutant of NetB (NetB-W262A)	*PlcC-NetB*	Clostridium perfringens	*Nicotiana benthamiana*	20%	[114]
LamB outer membrane protein of Vibrio bacteria	*LamB*	*Vibrio alginolyticus*	*Wolffia globosa*	-	[115]
Four African horse sickness (AHS) capsid proteins	*AHSV-1 VP2, VP5, VP3 and VP7*	African horse sickness virus (AHSV)	*Nicotiana benthamiana*	-	[116]

Table 5. Industrial enzymes and biomaterials obtained via chloroplast production in tobacco.

Gene/s	Product	Host Plant	Expression Level	References
bgl-1	β-Glucosidase	Tobacco Chloroplast	44.4 U/g FW	[117]
bgl1, celA, celB	β-Glucosidase, Cellulases	Tobacco Chloroplast	9.9–58.2 U/mg of TSP	[118]
endo, celB, xyn	Cellulases, Xylanase	Tobacco Chloroplast	0.38–75.6% TSP	[119]
bgl1C, cel6B, cel9A, xeg74	Cell wall-degrading enzyme	Tobacco Chloroplast	5–40% TSP	[120]
manI	β-Mannanase	Tobacco Chloroplast	25 U/g FW	[121]
xynA, xyn10A, xyn11B	Xylanase	Tobacco Chloroplast	0.2–6% TSP	[122]
UbiC	p-Hydroxybenzoic acid	Tobacco Chloroplast	25% DW	[123]

(Table 5) cont.....

Gene/s	Product	Host Plant	Expression Level	References
PHB pathway genes	Polyhydroxybutyrate	Tobacco Chloroplast	18.8% TSP	[124]
xynA	xylanase	Tobacco Chloroplast	6% TSP	[125]
xyn2	xylanase	Tobacco Chloroplast	421 U/mg TSP	[126]
xyl10B	xylanase	Tobacco Chloroplast	13% TSP, 61.9 U/mg D	[127]
E1	Endo- β -1,4-glucanase	Tobacco Chloroplast	1.35% TSP	[128 - 130]
E1	Endo- β -1,4-glucanase	Potato chloroplast	2.6% TSP	[131]
EG	Endo- β -1,4-glucanase	Sugarcane chloroplast	Avg 223.8 ng/mg,	[132]
CBH I, CBH II	Cellobiohydrolase	Sugarcane chloroplast	-	[132]
Cel6A,Cel6B	Endo- β -1,4-glucanase	Tobacco chloroplast	2-4% TSP	[133]
BglB	Betaglucosidase	Tobacco chloroplast	5.8% TSP	[134]
BglB	Betaglucosidase	Tobacco chloroplast	9.3% TSP	[135]
XylII	Endo-1,4-β –xylanase	Arabidopsis Chloroplast	3-4.8% TSP	[136]
XynII	Endo-1,4-β –xylanase	Arabidopsis Chloroplast	1.4-3.2% TSP	[137]
Chitinase	Chitinase	Tobacco Chloroplast	0.8-1% TSP	[138]
Chitinase	Glucanase	Tobacco Chloroplast	0.3% TSP	[139]

Table 6. Selected list of various non-pharmaceutical proteins produced in plants.

Recombinant Protein	Host Plant	Transformation Method	Expression Level	References
Human serum albumin	*Solanum tuberosum*	Nuclear transformation	0.25 g/mg (0.02% of TSP)	[140]
Erythropoietin	*Nicotiana tabacum*	Nuclear transformation	4.6–5.7 mg/g dry cell	[141]
1-antitrypsin	*Oriza sativa japonica*	Nuclear transformation	4.6–5.7 mg/g dry cell	[142]
Aprotinin	*Zea mays*	Nuclear transformation	0.069% of total seed protein	[143]
Human-secreted alkaline phosphatase	*Nicotiana tabacum*	Nuclear transformation	1.1 _g/g FW (3% of TSPs	[144]
Collagen	*Nicotiana tabacum*	Nuclear transformation	0.03 g/kg powdered plants	[145]
Human somatotropin	*Nicotiana tabacum*	Chloroplast transformation	>7% of TSP	[146]

(Table 6) cont.....

Recombinant Protein	Host Plant	Transformation Method	Expression Level	References
Bacillus thuringiensis (Bt) cry2Aa2	*Nicotiana tabacum*	Chloroplast transformation	5 mg/g FW (45.3–46.1% of TSPs	[147]
Human serum albumin	*Nicotiana tabacum*	Chloroplast transformation	11.1% of TPs	[148]
Human epidermal growth factor	*Nicotiana tabacum*	Nuclear transformation	34.2 _g/g FW	[149]
Human basic fibroblast growth factor	*Glycine max*	Nuclear transformation	2.3% of TSP	[150]
Type I interferon (IFN_2b)	*Nicotiana tabacum*	Chloroplast transformation	3 mg/g FW (20% of TSP	[151]
Human growth hormone	*Oryza sativa*	Nuclear transformation	57 mg/L culture medium	[152]
PlyGBS lysin	*Nicotiana tabacum*	Chloroplast transformation	>70% of TSP	[153]
Human growth hormone	*Tobacco BY-2 cells*	Nuclear transformation	35 mg/L or 2-4% of TSP	[154]
Human basic fibroblast growth factor	*Oryza sativa*	Nuclear transformation	185.66 mg/kg	[155]
Lumbrokinase	*Helianthus annuus*	Nuclear transformation	5.1 g/kg seeds	[156]
Human acidic fibroblast growth factor 1	*Salvia miltiorrhiza*	Nuclear transformation	272 ng/g FW	[157]
Glucocerebrosidase	*Nicotiana benthamiana*	Nuclear transformation	68 _g/g FW (1.45% of TSP	[158]
Human acid alpha glucosidase	*Nicotiana tabacum*	Chloroplast transformation	6.38 _g/g FW	[159]
Human basic fibroblast growth factor	*Nicotiana tabacum*	Chloroplast transformation	0.1% of TSP	[160]
Endo-_-1,4-xylanase	*Nicotiana tabacum*	Chloroplast transformation	35.7% of TSP	[119]
Glucosidase	*Nicotiana tabacum*	Chloroplast transformation	>75% of TSP	[119]
Osteopontin	*Nicotiana benthamiana*	Transient expression	100 ng/g FW	[87]
Dentin matrix protein-1	*Nicotiana benthamiana*	Transient expression	0.3 _g/g FW	[161]

Table 7. Commercial development of nonpharmaceutical proteins produced in plants.

Product	Application	Plant Species	Processing Degree	Advantage	Development Stage	Company	Source
Trypsin	Technical reagents	Maize seeds	Purified	Cost, animal-free	Commercialized	ProdiGene/ Sigma-Aldrich, United States	www.sigmaaldrich.com
Avidin	Technical reagents	Maize seeds	Purified	Cost, animal-free	Commercialized	ProdiGene/ Sigma-Aldrich, United States	www.sigmaaldrich.com
Endo-1,4-β-D-glucanase	Technical reagents	Maize seeds	Purified	Cost, animal-free	Commercialized	ProdiGene/ Sigma-Aldrich, United States	www.sigmaaldrich.com
Cellobiohydrolase I	Technical reagent	Maize seeds	Purified	Cost, integrated production	Commercialized	Infinite Enzymes/ Sigma Aldrich, United States	www.sigmaaldrich.com
Growth factors	Research reagents	Tobacco leaves, transient	Purified	Cost, animal-free	Commercialized	Agrenvec, Spain	www.agrenvec.com
Cytokines	Research reagents	Tobacco leaves, transient	Purified	Cost, animal-free	Commercialized	Agrenvec, Spain	www.agrenvec.com
Thioredoxin	Research reagents	Tobacco leaves, transient	Purified	Cost, animal-free	Commercialized	Agrenvec, Spain	www.agrenvec.com
TIMP-2	Research reagents	Tobacco leaves, transient's	Purified	Cost, animal-free	Commercialized	Agrenvec, Spain	www.agrenvec.com
Growth factors	Research reagent	Barley seeds	Purified	Cost, animal-free	Commercialized	ORF Genetics, Iceland	www.orfgenetics.com
Cytokines	Research reagent	Barley seeds	Purified	Cost, animal-free	Commercialized	ORF Genetics, Iceland	www.orfgenetics.com
Epithelial growth factor	Cosmetics	Barley seeds	Purified	Cost, animal-free	Commercialized	Sif Cosmetics, Iceland	www.sifcosmetics.com
Albumin	Research reagents	Rice seeds	Purified	Cost, animal-free	Commercialized	Ventria Bioscience/ InVitria, United States	www.invitria.com
Lactoferrin	Research reagents	Rice seeds	Purified	Cost, animal-free	Commercialized	Ventria Bioscience/ InVitria, United States	www.invitria.com www.invitria.com
Lysozyme	Research reagents	Rice seeds	Purified	Cost, animal-free	Commercialized	Ventria Bioscience/ InVitria, United States	www.invitria.com
Transferrin	Research reagents	Rice seeds	Purified	Cost, animal-free	Commercialized	Ventria Bioscience/ InVitria, United States	www.invitria.com

(Table 7) cont.....

Product	Application	Plant Species	Processing Degree	Advantage	Development Stage	Company	Source
Insulin	Research reagents	Rice seeds	Purified	Cost, animal-free	Commercialized	Ventria Bioscience/ InVitria, United States	www.invitria.com
Aprotinin	Research reagent	Tobacco leaves, transient	Purified	Cost	Commercialized	Kentucky Bioprocessing, United States	www.kbpllc.com
Collagen	Research reagent, tissue culture, health applications	Transgenic tobacco	Purified	Cost, animal-free	Commercialized	CollPlant, Israel	www.collplant.com
Trypsin	Research reagents, cosmetic ingredients	Rice cell suspension	Purified	Cost, animal-free	Commercialized	Natural BioMaterials, South Korea	www.nbms.co.kr
Enterokinase	Research reagents, cosmetic ingredients	Rice cell suspension	Purified	Cost, animal-free	Commercialized	Natural BioMaterials, South Korea	www.nbms.co.kr
Growth factors	Research reagents, cosmetic ingredients	Rice cell suspension	Purified	Cost, animal-free	Commercialized	Natural BioMaterials, South Korea	www.nbms.co.kr
Cytokines	Research reagents, cosmetic ingredients	Rice cell suspension	Purified	Cost, animal-free	Commercialized	Natural BioMaterials, South Korea	www.nbms.co.kr
Antibody	Purification of a hepatitis B vaccine	Transgenic tobacco	Purified	Cost	Commercial application	Center for Genetic Engineering and Biotechnology, Cuba	gndp.cigb. edu.cu
A-Amylase	Bioethanol production	Maize seeds	Biomass extract	Cost, integrated production	Commercialized	Syngenta, United States	www.syngenta.com
Phytase	Feed	Maize seeds	Delivered in biomass	Increased mineral availability, integrated production	Commercialization pending	Origin Agritech, China	www.originseed.com. cn
Growth factors	Tissue culture reagent	Tobacco leaves, transient	Purified	Cost, animal-free	Commercialized	NexGen, South Korea	www.exgen.com

Table 8. Plant-derived therapeutic human protein in clinical stages of development or on market.

Product	Disease	Plant	Clinical Trial Statuts	Company	Source
Gastric lipase, Merispase®	Cystic fibrosis	Maize	On market	Meristem Therapeutics France	www.meristem-therapeutics.com
α-Galactosidase	Fabry disease	Tobacco	Phase I	Planet Biotechnology, USA	www.planetbiotechnology.com
Lactoferon™ (α-interferon)	Hepatitis B and C	Duckweed	Phase II	Biolex, USA	www.biolex.com
Fibrinolytic drug (thrombolytic drug)	Blood clot	Duckweed	Phase I	Biolex, USA	www.biolex.com
Human glucocerebrosidase	Gaucher 's disease	Carrot suspension cells	Awaiting USDA's approval	Protalix Biotherapeutics, Israel	www.protalix.com
Insulin	Diabetes	Safflower	Phase III	SemBioSys, Canada	www.sembiosys.com
Apolipoprotein	Cardiovascular	Safflower	Phase I	SemBioSys, Canada	www.sembiosys.com

Table 9. Plant-derived nutraceuticals in advanced clinical stages of development or on market.

Product	Disease	Plant	Clinical Trial Statuts	Company	Source
ISOkine™	Human growth factor	Barley	On market	ORF Genetics	www.orfgenetics.com
DERMOkine™	Human growth factor	Barley	On market	ORF Genetics	www.orfgenetics.com
Human intrinsic factor	Vitamin B12 deficiency	Arabidopsis	On market	Cobento Biotech AS	www.cobento.dk
Coban	Vitamin B12 deficiency	Arabidopsis	On market	Cobento Biotech AS	www.cobento.dk
Human lactoferrin	Anti-infection, anti-inflammatory	Rice	Advanced stage	Ventria, USA	www.ventriabio.com
Human lysozyme	Anti-infection, anti-inflammatory	Rice	Advanced stage	Ventria, USA	www.ventria.com
Immunosphere™	Food additive for shrimps	Safflower	On market	SemBioSys, Canada	www.sembiosys.com

Genetic engineering of plants comprises of several steps: identifying and isolating the gene of interest, choice of the promoter, construction of expression and selection cassette, selection of a suitable cloning and expression vector, an appropriate method for stable or transient DNA introduction into the plant genome, tissue culture system that allows the regeneration of whole plants, selection pressure for the distinction of transgenic plants from non-transformants followed by biochemical methods to detect the expression of foreign genes.

The plant genome is compartmentalized into the nucleus, mitochondria, and plastids; each of which possesses its genome and genetic machinery. The development of transgenic plants depends on recombinant DNA technology for the stable or transient expression of foreign gene(s) into any of the plant genomes. Stable transformation refers to the permanent integration of exogenous gene(s) into the plant genome while transient expression refers to a temporary high-level transgene expression, generally but not limited to validate the constructs. Transformation of the nuclear genome is now a routine in tailoring agronomical traits [31, 54, 62, 162] while that of organellar genomes, mitochondrial or plastids, is also emerging as an alternative target for the transformation process. Mitochondrial genome transformation is relatively a new concept and very limited success has been achieved so far, however, the transformation of the plastid genome has become an established platform for the production of commercially important compounds due to high expression and confinement into bona fide structure [153]. Plastids are a group of semi-autonomous organelles found in green plants, algae, and cyanobacteria that possess a great capacity for differentiation, de-differentiation, and re-differentiation. Their major roles include photosynthesis, storage of various products, and synthesis of key molecules which maintain the basic architecture and operation of cells. These organelles vary in size, shape, content, and function. Pro-plastids are the precursors of all plastids and are present in the meristematic regions of the plant. The plastid DNA (ptDNA) is present in the form of DNA-protein complexes known as plastid nucleoids and are attached to the inner membrane of the plastids. Each plastid nucleoid possesses 10-100 copies of ptDNA. The size of chloroplast genomes is ~140 kbp in higher plants while <200 kbp in unicellular eukaryotes. The copy number of the chloroplast genome is variable and ranges between 1,000-10,000 per plant cell. Transformation of both the chloroplast genome as well as the nuclear genome offers its own set of features which are discussed in Table **10**.

Table 10. Comparison of nuclear and chloroplast expression approaches.

Feature	Nuclear Transformation	Chloroplast Transformation
Copy number	Low copy number Specific chromosome number in each species	High copy number 2 Inverted repeats (IR)/ptDNA 10-100 ptDNA/plastid 10-100 plastid/cell ~20,000 transgene copies/cell
Level of gene expression	Low 1-2% of TSP	High 5-25% of TSP
Integration	Random May interrupt expression of other genes	Site specific No effect on the expression of other genes due to homologous recombination
Gene silencing	Yes Natural phenomenon of RNAi in eukaryotes	No Prokaryotic in nature
Gene containment	Transgene escape Non-maternal mode of inheritance	No gene pollution Maternal mode of inheritance
Formation of disulphides	Yes	Yes
Toxic proteins	No Severe pleiotropic effects	Yes Plastid expressed proteins are contained within the plastid
Gene expression	Monocistronic	Polycistronic (Operon)

There are various methods to introduce transgene into the plant genome, however, in this chapter, we will discuss the commonly used system such as *Agrobacterium*-mediated transformation, biolistic, and viral (transient expression).

Agrobacterium is a gram-negative soil bacterium that causes crown gall or hairy roots in most of the dicotyledonous plants to produce amino acids and sugar-phosphate compounds known as opines for its energy source. The cells of *Agrobacterium tumefaciens* and *Agrobacterium rhizogenes* contain Ti (tumor-inducing) plasmid and Ri (roots-inducing) plasmids, respectively. Commonly *Agrobacterium tumefaciens* is used for plant transformation by replacing the T-DNA region with the gene of interest and selection cassette. Injured plant tissues release a phenolic compound (acetosyringone) that triggers *Agrobacterium* recognition in which VirA protein acts as an antenna and autophosphorylates itself consequently phosphorylation VirG [163]. This triggers a cascade of chemical reactions wherein type IV secretion system is formed consisting of VirD4 and 11 VirB proteins [164]. T-DNA is replicated to produce T-DNA/VirD2 complex; Vir protein complex (VirB2, VirB5, VirB7) work together to allow VirD2-DNA complex (immature T-DNA complex) to enter the cytoplasm of the recipient cell, VirE2 is combined with T-strand to form mature T-DNA complex [165], which passes through the cytoplasm to reach the nucleus; the T-DNA complex enters the nucleus of the recipient cell through the nuclear pore targeting the integration site. The T-complex removes the guard protein, and the T-DNA is integrated into the nuclear genome. The gene expression regulatory sequence in the T-DNA region is like that of eukaryotes, so it can be expressed in plant cells. *Agrobacterium*-mediated transformation is a method of choice for transformation of the nuclear genome of dicotyledonous plants due to simple operation, low cost, high success rate [166]. However naturally the bacterium offers limited host range and organellar genomes cannot be transformed. Different steps involved in the *Agrobacterium*-mediated transformation procedure are shown in Fig. (**1**).

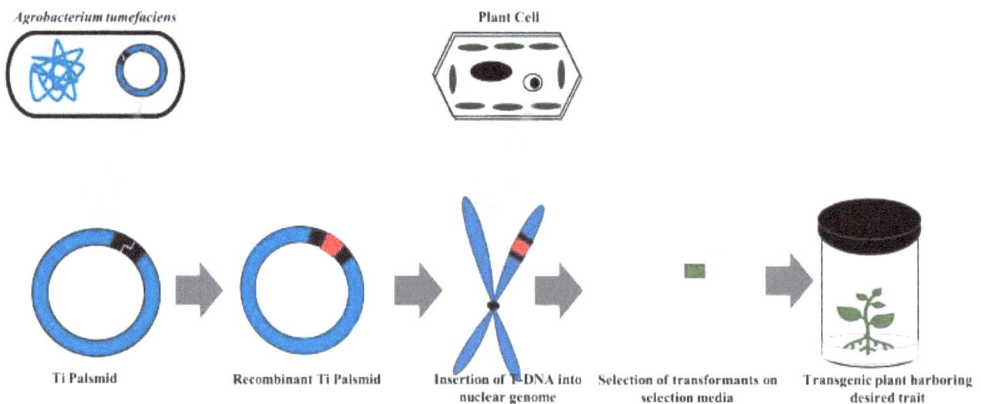

Fig. (1). Illustration of different steps involved in the development of transgenic plants using nuclear transformation.

As illustrated in Fig. (**2**), the microprojectile bombardment method employs physical means of transforming cells; also known as biolistics or gene-gun method. It accelerates the metal particles (tungsten, gold) coated with foreign DNA so that they can penetrate the cell wall [167]. It can be used to introduce foreign genes into plant cells, tissues, and organs. PDS-1000/He uses rupture discs of different thicknesses to regulate the helium pressure. When the helium pressure reaches the capacity of rupture disc it bursts to generate a strong acceleration of macrocarrier to carry the microcarriers at high speed. When it encounters the rigid blocking mesh, the macrocarrier is blocked while microcarriers use inertia to continue to move forward at high speed, bombarding target cells or tissues, thereby carrying foreign genes into the cells. The gene bombardment method is not restricted by the recipient's genotype, a wide range of explants, rapid and simple operation, and can effectively transform plastids, thus turning a new page in the genetic transformation of plants. However, this technology is expensive, and large DNA fragments can be easily sheared during the transformation process, multiple insertions are also a limitation.

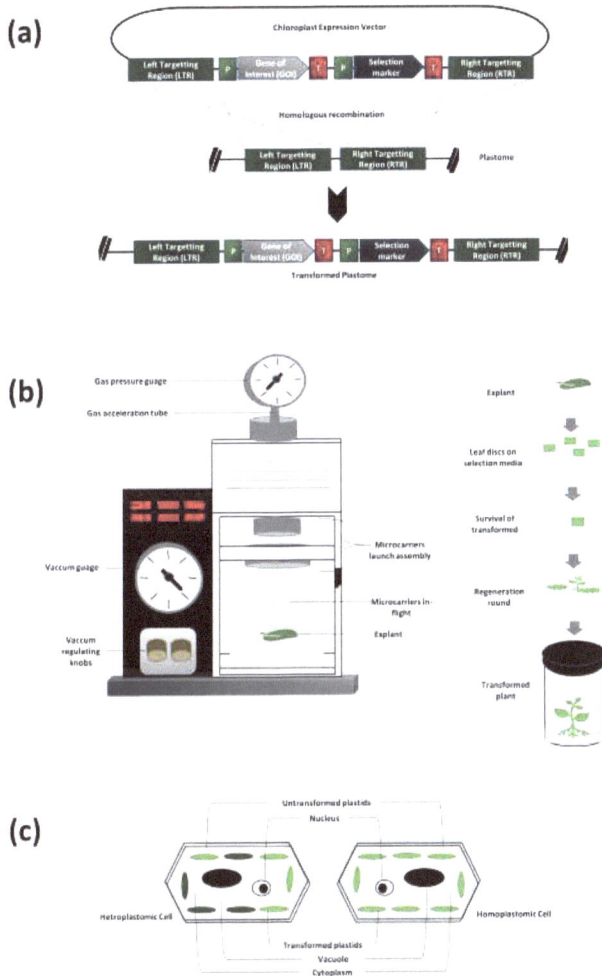

Fig. (2). Development of transgenic plants via transformation of the chloroplast genome. **a)** Displays a typical vector construction scheme and the homologous recombination for the integration of transgene into chloroplast genome at the chosen location. **b)** Shows the delivery of transgene into chloroplast genome via particle bombardment system using a gene gun. **c)** Demonstrates the progression of homoplasmy (right; in which all chloroplasts are transformed) a heteroplasmic cell (left; in which untransformed chloroplasts are also present).

Plant viruses such as Tobacco Mosaic Virus (TMV), and Potato Virus X (PVX) can also be used for plant transformation exploiting their natural capability [168].

Viruses can spread rapidly and produce a large number of foreign protein(s) due to their efficient self-replication and expression ability [169]. The introduction and expression of new genes can be achieved in most cells of a mature plant without the need to undergo a long-term transformation process from explants to regenerated plants. This method is used for transient expression since the transgene is not integrated into plant genome and therefore, cannot be transferred to next generation [168]. It will not affect the expression of other functional genes in the recipient plant. Two types of viral expression systems are available: i) independent viral vectors-harboring the transgene in place of capsid protein [170] or alongside all required viral proteins. However, this system suffered from transgene size limitation, ii) minimal viral vectors-harboring the transgene with minimal viral proteins capable of accommodating larger transgene at the cost of systemic infection. This was compensated by the transformation of agrobacterium with minimal viral vector and infiltrating the inoculum into the host plant leaf using vacuum. It helps to quickly validate the cassette for its successful expression in plants. However, there are several limitations associated with virus-mediated plant transformation such as viral vectors cannot integrate foreign genes into chromosomes, so they cannot be passed to the offspring according to Mendelian laws, and they have no advantage in long-term expression of foreign proteins. The frequency of mutations in the genome is relatively high. The instability of the viral vector itself can easily cause the loss of foreign genes. In view of this, viral vector-mediated genetic transformation is mainly used in two fields: the application of virus-induced gene silencing (VIGS) and high efficiency transient expression of foreign protein(s).

3. SUMMARY AND OUTLOOK

It has now been established that plants are capable of producing recombinant proteins of industrial importance at commercially feasible levels. Plants offer several advantages compared to conventional systems. Although, the lengthy procedures to produce transgenic plants are considered a big hurdle at present, several tissue-culture independent methods such as the floral dip method to generate transgenic plants have been devised. However, they are currently limited to few plant species such as Arabidopsis. The success of such methods depends on the use of an efficient selectable marker system that could effectively suppress the growth of non-transformed plants at the seedling stage. Such methods have been attempted for other species as well however, they have not been successful yet. Once, the transgenic plants have been developed, then they offer significant advantages compared to other systems particularly in terms of scaling up. With minimum efforts and existing infrastructure deployed, transgenic plants can be grown on several thousand hectares in one planting season.

In situations like Covid-19 pandemic calling for sudden, huge, and cheap production of vaccines, the plant-based production system of recombinant vaccines also holds great promise. Covid-19, a novel coronavirus with the potential of lethality has created an alarming situation at a global scale. The virus was first identified in late December 2019 in Wuhan, China, and has been declared a serious global health concern by WHO [171]. The virus is responsible for acute respiratory (pneumonia-like) infection characterized by different symptoms. Critical cases can cause respiratory failure, septic shock, or organ failure which then requires intensive care support. The rapid outbreak of this deadly virus through human transmission has provoked governments across the world to ensure and address the emergency control and containment measures, treating patients with quarantine facilities and vaccine development. As the virus has emerged suddenly with no available vaccination and other treatments, prevention of infection is the current priority to control the pandemic. This outbreak with massive mortalities and newly reported cases has created an urgent demand for vaccine development. Although the traditional expression system is flexible for biopharmaceutical production the transient expression in plants has carved a niche in the biopharmaceutical sector for producing biopharma products. Genetic engineering of plant has evolved smarter for the transient expression with profound benefits and have been substantially fruitful in achieving its worth in biopharmaceutical sector. Therefore, in such emergencies, vaccine antigens can be produced transiently without stably integrating the transgenes into the plant genome [172]. The amazing speed of this system was very recently demonstrated by Medicago Inc., a pioneer of plant-based transient expression and manufacturing, by producing VLPs (Virus-Like Particles) in just 20 days after having access to SARS-nCoV-2's spike (S) protein sequence [173].

There are other concerns of transgene spread to weedy relatives through pollen. However, the chance of spreading transgenes become meager if the transgenes-coding for a fitness-enhancing trait under certain circumstances such as herbicide tolerance is not used. Any transgene offering no selective advantage under certain circumstances to the host plant would remain neutral and pose no threat to the population. Research shows that transgenes offering no selection advantage to the host would rather become a source of extra metabolic burden at the cellular level. Nevertheless, researchers have developed alternative approaches such as the manipulation of non-nuclear genomes to address such pollen-mediated gene transfer to weedy relatives from transgenic plants when cultivated in open fields. Non-nuclear genomes such as plastids and mitochondria are often inherited maternally, and therefore, provide a sort of natural gene compartment for the transgene(s). Manipulation of the plastid genome has emerged as a successful alternative to nuclear transformation while the efforts to engineer the mitochondrial genome are underway. Another significant advantage of using plants as a host for the production of recombinant proteins lies in their suitability for the development of edible vaccines. Vaccine antigen production in edible parts of plants would offer a significant advance not only in the cost-effective production but also downstream application including purification, distribution as well as administration of the vaccines-the three major phases incurring high input costs. The seeds of plants expressing edible vaccine/recombinant protein can be stored at room temperature for quite long periods and can be distributed to remote distances without requiring special cold storage as well as special transportation facilities. Despite these advantages, plant-made pharmaceuticals have not yet reached the market or captured the market share. However, the situation is changing now. Many funding agencies are funding proposals revolving the recombinant protein productions using plant-based expression systems.

CONSENT FOR PUBLICATION

Not applicable.

CONFLICT OF INTEREST

The author declares no conflict of interest, financial or otherwise.

ACKNOWLEDGMENTS

Work in the authors' lab is supported by Higher Education Commission (HEC), Pakistan, and ICGEB, Italy. We apologize to those colleagues whose work could not be cited due to space limitations.

REFERENCES

[1]   Insights FB. Insights, F.B. Pharmaceuticals Market to Reach USD 1,310.0 Billion in 2020; Eruption of the COVID-19 Pandemic to Accelerate the Demand for Effective Treatments and Drugs Worldwide 2020.https://www.globenewswire.com/news-release/2020/04/27/2022157/0/en/Pharmaceuticals- Market-to-Reach-USD-1-310-0-Billion-in-2020-Eruption-of-the-COVID-19-Pandemic-to-Accelerate- the-Demand-for-Effective-Treatments-and-Drugs-Worldwide-Fortune-Business.html

[2]   Knäblein J. Plant-based expression of biopharmaceuticals. 2nd Ed. Encyclopedia of Molecular Cell Biology and Molecular Medicine. 2006; 10. Wiley-VCH Verlag GmbH and Co. KGaA.

[3]   Obembe OO, Popoola JO, Leelavathi S, Reddy SV. Advances in plant molecular farming. Biotechnol Adv 2011; 29(2): 210-22.
        [http://dx.doi.org/10.1016/j.biotechadv.2010.11.004] [PMID: 21115109]

[4]   Siddiqui A, Wei Z, Boehm M, Ahmad N. Engineering microalgae through chloroplast transformation to produce high-value industrial products. Biotechnol Appl Biochem 2020; 67(1): 30-40.
        [http://dx.doi.org/10.1002/bab.1823] [PMID: 31538673]

[5]   Mahmoud K. Recombinant protein production: strategic technology and a vital research tool. Research Journal of Cell and Molecular Biology 2007; 1(1): 9-22.

[6]   Shanmugaraj B, I Bulaon CJ, Phoolcharoen W. Plant molecular farming: a viable platform for recombinant biopharmaceutical production. Plants 2020; 9(7): 842.

[http://dx.doi.org/10.3390/plants9070842] [PMID: 32635427]

[7]   Rybicki EP. Wiley Interdisciplinary Reviews: Nanomedicine 2020; 12(2): e1587.

[8]   Daniell H, Singh ND, Mason H, Streatfield SJ. Plant-made vaccine antigens and biopharmaceuticals. Trends Plant Sci 2009; 14(12): 669-79.
[http://dx.doi.org/10.1016/j.tplants.2009.09.009] [PMID: 19836291]

[9]   Ma JKC, Christou P, Chikwamba R, *et al.* Realising the value of plant molecular pharming to benefit the poor in developing countries and emerging economies. Plant Biotechnol J 2013; 11(9): 1029-33.
[http://dx.doi.org/10.1111/pbi.12127] [PMID: 24119183]

[10]   Schillberg S, Raven N, Spiegel H, Rasche S, Buntru M. Critical analysis of the commercial potential of plants for the production of recombinant proteins. Front Plant Sci 2019; 10: 720.
[http://dx.doi.org/10.3389/fpls.2019.00720] [PMID: 31244868]

[11]   Siddiqui A, Wei Z, Boehm M, Ahmad N. Engineering microalgae through chloroplast transformation to produce high-value industrial products. Biotechnol Appl Biochem 2020; 67(1): 30-40.
[http://dx.doi.org/10.1002/bab.1823] [PMID: 31538673]

[12]   Ahmad N, Mukhtar Z. Green factories: plastids for the production of foreign proteins at high levels. Gene Ther Mol Biol 2013; 15: 14-29.

[13]   Montero-Morales L, Steinkellner H. Advanced plant-based glycan engineering. Front Bioeng Biotechnol 2018; 6: 81.
[http://dx.doi.org/10.3389/fbioe.2018.00081] [PMID: 29963553]

[14]   Christou P. Plant genetic engineering and agricultural biotechnology 1983-2013. Trends Biotechnol 2013; 31(3): 125-7.
[http://dx.doi.org/10.1016/j.tibtech.2013.01.006] [PMID: 23375945]

[15]   Bevan MW, Flavell RB, Chilton M-D. A chimaeric antibiotic resistance gene as a selectable marker for plant cell transformation. Nature 1983; 304(5922): 184-7.
[http://dx.doi.org/10.1038/304184a0]

[16]   Barta A, Sommergruber K, Thompson D, Hartmuth K, Matzke MA, Matzke AJ. The expression of a nopaline synthase - human growth hormone chimaeric gene in transformed tobacco and sunflower callus tissue. Plant Mol Biol 1986; 6(5): 347-57.
[http://dx.doi.org/10.1007/BF00034942] [PMID: 24307385]

[17]   Hiatt A, Cafferkey R, Bowdish K. Production of antibodies in transgenic plants. Nature 1989; 342(6245): 76-8.
[http://dx.doi.org/10.1038/342076a0] [PMID: 2509938]

[18]   Hood EE, *et al.* Commercial production of avidin from transgenic maize: characterization of transformant, production, processing, extraction and purification. Mol Breed 1997; 3(4): 291-306.
[http://dx.doi.org/10.1023/A:1009676322162]

[19]   Fox JL. First plant-made biologic approved. Nature 2012.

[20]   Kaiser J. Is the drought over for pharming? Science 2008; 320(5875): 473-5.
[http://dx.doi.org/10.1126/science.320.5875.473] [PMID: 18436771]

[21]   Burnett MJ, Burnett AC. Therapeutic recombinant protein production in plants: Challenges and opportunities. Plants, People. Planet 2020; 2(2): 121-32.

[22]   Siddiqui A, Wei Z, Boehm M, Ahmad N. Engineering microalgae through chloroplast transformation to produce high-value industrial products. Biotechnol Appl Biochem 2020; 67(1): 30-40.
[http://dx.doi.org/10.1002/bab.1823] [PMID: 31538673]

[23]   Walsh G. Biopharmaceutical benchmarks 2018. Nat Biotechnol 2018; 36(12): 1136-45.
[http://dx.doi.org/10.1038/nbt.4305] [PMID: 30520869]

[24]   Capell T, Twyman RM, Armario-Najera V, Ma JK, Schillberg S, Christou P. Potential applications of

plant biotechnology against SARS-CoV-2. Trends Plant Sci 2020; 25(7): 635-43.
[http://dx.doi.org/10.1016/j.tplants.2020.04.009] [PMID: 32371057]

[25]   Council NR. Safety of Genetically Engineered Foods: Approaches to Assessing Unintended Health Effects. 2004.

[26]   Ahmad N, Mukhtar Z. Genetic manipulations in crops: Challenges and opportunities. Genomics 2017; 109(5-6): 494-505.
[http://dx.doi.org/10.1016/j.ygeno.2017.07.007] [PMID: 28778540]

[27]   Varshney RK, Bansal KC, Aggarwal PK, Datta SK, Craufurd PQ. Agricultural biotechnology for crop improvement in a variable climate: hope or hype? Trends Plant Sci 2011; 16(7): 363-71.
[http://dx.doi.org/10.1016/j.tplants.2011.03.004] [PMID: 21497543]

[28]   Kumari P, Kant S, Zaman S, Mahapatro GK, Banerjee N, Sarin NB. A novel insecticidal GroEL protein from Xenorhabdus nematophila confers insect resistance in tobacco. Transgenic Res 2014; 23(1): 99-107.
[http://dx.doi.org/10.1007/s11248-013-9734-3] [PMID: 23888329]

[29]   Liu Y, Wu H, Chen H, *et al.* A gene cluster encoding lectin receptor kinases confers broad-spectrum and durable insect resistance in rice. Nat Biotechnol 2015; 33(3): 301-5.
[http://dx.doi.org/10.1038/nbt.3069] [PMID: 25485617]

[30]   Misra P, Pandey A, Tiwari M, *et al.* Modulation of transcriptome and metabolome of tobacco by Arabidopsis transcription factor, AtMYB12, leads to insect resistance. Plant Physiol 2010; 152(4): 2258-68.
[http://dx.doi.org/10.1104/pp.109.150979] [PMID: 20190095]

[31]   Naqvi RZ, Asif M, Saeed M, *et al.* Development of a triple gene *cry1ac-cry2ab-epsps* construct and its expression in *nicotiana benthamiana* for insect resistance and herbicide tolerance in plants. Front plant sci 2017; 8(55): 55.
[http://dx.doi.org/10.3389/fpls.2017.00055] [PMID: 28174591]

[32]   Mamta , Reddy KR, Rajam MV. Targeting chitinase gene of Helicoverpa armigera by host-induced RNA interference confers insect resistance in tobacco and tomato. Plant Mol Biol 2016; 90(3): 281-92.
[http://dx.doi.org/10.1007/s11103-015-0414-y] [PMID: 26659592]

[33]   Senthilkumar R, Cheng C-P, Yeh K-W. Genetically pyramiding protease-inhibitor genes for dual broad-spectrum resistance against insect and phytopathogens in transgenic tobacco. Plant Biotechnol J 2010; 8(1): 65-75.
[http://dx.doi.org/10.1111/j.1467-7652.2009.00466.x] [PMID: 20055959]

[34]   Sun H, Lang Z, Zhu L, Huang D. Acquiring transgenic tobacco plants with insect resistance and glyphosate tolerance by fusion gene transformation. Plant Cell Rep 2012; 31(10): 1877-87.
[http://dx.doi.org/10.1007/s00299-012-1301-5] [PMID: 22777591]

[35]   Tahir HAS, Gu Q, Wu H, Niu Y, Huo R, Gao X. Bacillus volatiles adversely affect the physiology and ultra-structure of Ralstonia solanacearum and induce systemic resistance in tobacco against bacterial wilt. Sci Rep 2017; 7(1): 40481.
[http://dx.doi.org/10.1038/srep40481] [PMID: 28091587]

[36]   Thakur N, Upadhyay SK, Verma PC, Chandrashekar K, Tuli R, Singh PK. Enhanced whitefly resistance in transgenic tobacco plants expressing double stranded RNA of v-ATPase A gene. PLoS One 2014; 9(3): e87235.
[http://dx.doi.org/10.1371/journal.pone.0087235] [PMID: 24595215]

[37]   Zhu J-Q, Liu S, Ma Y, *et al.* Improvement of pest resistance in transgenic tobacco plants expressing dsRNA of an insect-associated gene EcR. PLoS One 2012; 7(6): e38572.
[http://dx.doi.org/10.1371/journal.pone.0038572] [PMID: 22685585]

[38]   Gao Z, He X, Zhao B, *et al.* Overexpressing a putative aquaporin gene from wheat, TaNIP, enhances salt tolerance in transgenic Arabidopsis. Plant Cell Physiol 2010; 51(5): 767-75.

[http://dx.doi.org/10.1093/pcp/pcq036] [PMID: 20360019]

[39]   Kotula L, *et al.* Improving crop salt tolerance using transgenic approaches: An update and physiological analysis. Plant Cell Environ 2020; 43(12): 2932-56.
[http://dx.doi.org/10.1111/pce.13865]

[40]   Pang Q, Chen S, Dai S, Chen Y, Wang Y, Yan X. Comparative proteomics of salt tolerance in Arabidopsis thaliana and Thellungiella halophila. J Proteome Res 2010; 9(5): 2584-99.
[http://dx.doi.org/10.1021/pr100034f] [PMID: 20377188]

[41]   Pasapula V, Shen G, Kuppu S, *et al.* Expression of an Arabidopsis vacuolar H^+-pyrophosphatase gene (AVP1) in cotton improves drought- and salt tolerance and increases fibre yield in the field conditions. Plant Biotechnol J 2011; 9(1): 88-99.
[http://dx.doi.org/10.1111/j.1467-7652.2010.00535.x] [PMID: 20492547]

[42]   Alvarez Viveros MF, Inostroza-Blancheteau C, Timmermann T, González M, Arce-Johnson P. Overexpression of GlyI and GlyII genes in transgenic tomato (Solanum lycopersicum Mill.) plants confers salt tolerance by decreasing oxidative stress. Mol Biol Rep 2013; 40(4): 3281-90.
[http://dx.doi.org/10.1007/s11033-012-2403-4] [PMID: 23283739]

[43]   Wu H-J, Zhang Z, Wang JY, *et al.* Insights into salt tolerance from the genome of Thellungiella salsuginea. Proc Natl Acad Sci USA 2012; 109(30): 12219-24.
[http://dx.doi.org/10.1073/pnas.1209954109] [PMID: 22778405]

[44]   Xianjun P, Xingyong M, Weihong F, *et al.* Improved drought and salt tolerance of Arabidopsis thaliana by transgenic expression of a novel DREB gene from Leymus chinensis. Plant Cell Rep 2011; 30(8): 1493-502.
[http://dx.doi.org/10.1007/s00299-011-1058-2] [PMID: 21509473]

[45]   Yue Y, Zhang M, Zhang J, Duan L, Li Z. SOS1 gene overexpression increased salt tolerance in transgenic tobacco by maintaining a higher K^+/Na^+ ratio. J Plant Physiol 2012; 169(3): 255-61.
[http://dx.doi.org/10.1016/j.jplph.2011.10.007] [PMID: 22115741]

[46]   Zhou G-A, Chang R-Z, Qiu L-J. Overexpression of soybean ubiquitin-conjugating enzyme gene GmUBC2 confers enhanced drought and salt tolerance through modulating abiotic stress-responsive gene expression in Arabidopsis. Plant Mol Biol 2010; 72(4-5): 357-67.
[http://dx.doi.org/10.1007/s11103-009-9575-x] [PMID: 19941154]

[47]   Zhou J, Wang J, Bi Y, *et al.* Overexpression of PtSOS2 enhances salt tolerance in transgenic poplars. Plant Mol Biol Report 2014; 32(1): 185-97.
[http://dx.doi.org/10.1007/s11105-013-0640-x] [PMID: 24465084]

[48]   Liang C, Sun B, Meng Z, *et al.* Co-expression of GR79 EPSPS and GAT yields herbicide-resistant cotton with low glyphosate residues. Plant Biotechnol J 2017; 15(12): 1622-9.
[http://dx.doi.org/10.1111/pbi.12744] [PMID: 28418615]

[49]   Ahn Y-K, Yoon M-K, Jeon J-S. Development of an efficient Agrobacterium-mediated transformation system and production of herbicide-resistant transgenic plants in garlic (Allium sativum L.). Mol Cells 2013; 36(2): 158-62.
[http://dx.doi.org/10.1007/s10059-013-0142-6] [PMID: 23832764]

[50]   Tian S, Jiang L, Cui X, *et al.* Engineering herbicide-resistant watermelon variety through CRISPR/Cas9-mediated base-editing. Plant Cell Rep 2018; 37(9): 1353-6.
[http://dx.doi.org/10.1007/s00299-018-2299-0] [PMID: 29797048]

[51]   Imran M, Asad S, Barboza AL, Galeano E, Carrer H, Mukhtar Z. Genetically transformed tobacco plants expressing synthetic EPSPS gene confer tolerance against glyphosate herbicide. Physiol Mol Biol Plants 2017; 23(2): 453-60.
[http://dx.doi.org/10.1007/s12298-017-0424-0] [PMID: 28461732]

[52]   Latif A, Rao AQ, Khan MA, *et al.* Herbicide-resistant cotton (Gossypium hirsutum) plants: an alternative way of manual weed removal. BMC Res Notes 2015; 8(1): 453.

[http://dx.doi.org/10.1186/s13104-015-1397-0] [PMID: 26383095]

[53]   Benekos K, Kissoudis C, Nianiou-Obeidat I, *et al.* Overexpression of a specific soybean GmGSTU4 isoenzyme improves diphenyl ether and chloroacetanilide herbicide tolerance of transgenic tobacco plants. J Biotechnol 2010; 150(1): 195-201.
[http://dx.doi.org/10.1016/j.jbiotec.2010.07.011] [PMID: 20638428]

[54]   Zhao H, *et al.* Production of EPSPS and bar gene double-herbicide resistant castor (Ricinus communis L.). Biotechnol Biotechnol Equip 2020; 34(1): 825-40.
[http://dx.doi.org/10.1080/13102818.2020.1804450]

[55]   Liu Y, Yang SX, Cheng Y, *et al.* Production of herbicide-resistant medicinal plant Salvia miltiorrhiza transformed with the bar gene. Appl Biochem Biotechnol 2015; 177(7): 1456-65.
[http://dx.doi.org/10.1007/s12010-015-1826-5] [PMID: 26364310]

[56]   Wu JX, *et al.* Scarabaeid Larvae- and Herbicide-Resistant Transgenic Perennial Ryegrass (Lolium perenne L.) Obtained by Agrobacterium tumefaciens-Mediated Transformation of cry8Ca2, cry8Ga and bar Genes. J Integr Agric 2012; 11(1): 53-61.
[http://dx.doi.org/10.1016/S1671-2927(12)60782-2]

[57]   Chhapekar S, Raghavendrarao S, Pavan G, *et al.* Transgenic rice expressing a codon-modified synthetic CP4-EPSPS confers tolerance to broad-spectrum herbicide, glyphosate. Plant Cell Rep 2015; 34(5): 721-31.
[http://dx.doi.org/10.1007/s00299-014-1732-2] [PMID: 25537885]

[58]   El-Esawi MA, Alayafi AA. Overexpression of rice Rab7 gene improves drought and heat tolerance and increases grain yield in rice (*Oryza sativa* L.). Genes (Basel) 2019; 10(1): 56.
[http://dx.doi.org/10.3390/genes10010056] [PMID: 30658457]

[59]   Wang X, Chen J, Liu C, *et al.* Over-expression of a protein disulfide isomerase gene from Methanothermobacter thermautotrophicus, enhances heat stress tolerance in rice. Gene 2019; 684: 124-30.
[http://dx.doi.org/10.1016/j.gene.2018.10.064] [PMID: 30367983]

[60]   Shah Z, Shah SH, Ali GS, *et al.* Introduction of Arabidopsis's heat shock factor HsfA1d mitigates adverse effects of heat stress on potato (*Solanum tuberosum* L.) plant. Cell Stress Chaperones 2020; 25(1): 57-63.
[http://dx.doi.org/10.1007/s12192-019-01043-6] [PMID: 31898287]

[61]   Biswas S, Islam MN, Sarker S, Tuteja N, Seraj ZI. Overexpression of heterotrimeric G protein beta subunit gene (OsRGB1) confers both heat and salinity stress tolerance in rice. Plant Physiol Biochem 2019; 144: 334-44.
[http://dx.doi.org/10.1016/j.plaphy.2019.10.005] [PMID: 31622936]

[62]   Wang J, Gao X, Dong J, *et al.* Over-Expression of the Heat-Responsive Wheat Gene *TaHSP23.9* in Transgenic *Arabidopsis* Conferred Tolerance to Heat and Salt Stress. Front Plant Sci 2020; 11(243): 243.
[http://dx.doi.org/10.3389/fpls.2020.00243] [PMID: 32211001]

[63]   Xue Y, *et al.* Over-expression of heat shock protein gene hsp26 in Arabidopsis thaliana enhances heat tolerance. Biol Plant 2010; 54(1): 105-11.
[http://dx.doi.org/10.1007/s10535-010-0015-1]

[64]   Paine JA, Shipton CA, Chaggar S, *et al.* Improving the nutritional value of Golden Rice through increased pro-vitamin A content. Nat Biotechnol 2005; 23(4): 482-7.
[http://dx.doi.org/10.1038/nbt1082] [PMID: 15793573]

[65]   Diretto G, Frusciante S, Fabbri C, *et al.* Manipulation of β-carotene levels in tomato fruits results in increased ABA content and extended shelf life. Plant Biotechnol J 2020; 18(5): 1185-99.
[http://dx.doi.org/10.1111/pbi.13283] [PMID: 31646753]

[66]   Vaquero C, Sack M, Chandler J, *et al.* Transient expression of a tumor-specific single-chain fragment

and a chimeric antibody in tobacco leaves. Proc Natl Acad Sci USA 1999; 96(20): 11128-33.
[http://dx.doi.org/10.1073/pnas.96.20.11128] [PMID: 10500141]

[67]   Torres E, Vaquero C, Nicholson L, *et al.* Rice cell culture as an alternative production system for functional diagnostic and therapeutic antibodies. Transgenic Res 1999; 8(6): 441-9.
[http://dx.doi.org/10.1023/A:1008969031219] [PMID: 10767987]

[68]   Stöger E, Vaquero C, Torres E, *et al.* Cereal crops as viable production and storage systems for pharmaceutical scFv antibodies. Plant Mol Biol 2000; 42(4): 583-90.
[http://dx.doi.org/10.1023/A:1006301519427] [PMID: 10809004]

[69]   Brodzik R, Glogowska M, Bandurska K, *et al.* Plant-derived anti-Lewis Y mAb exhibits biological activities for efficient immunotherapy against human cancer cells. Proc Natl Acad Sci USA 2006; 103(23): 8804-9.
[http://dx.doi.org/10.1073/pnas.0603043103] [PMID: 16720700]

[70]   Floss DM, Sack M, Stadlmann J, *et al.* Biochemical and functional characterization of anti-HIV antibody-ELP fusion proteins from transgenic plants. Plant Biotechnol J 2008; 6(4): 379-91.
[http://dx.doi.org/10.1111/j.1467-7652.2008.00326.x] [PMID: 18312505]

[71]   Sainsbury F, Lomonossoff GP. Extremely high-level and rapid transient protein production in plants without the use of viral replication. Plant Physiol 2008; 148(3): 1212-8.
[http://dx.doi.org/10.1104/pp.108.126284] [PMID: 18775971]

[72]   Holland T, Sack M, Rademacher T, *et al.* Optimal nitrogen supply as a key to increased and sustained production of a monoclonal full-size antibody in BY-2 suspension culture. Biotechnol Bioeng 2010; 107(2): 278-89.
[http://dx.doi.org/10.1002/bit.22800] [PMID: 20506104]

[73]   Huang Z, Phoolcharoen W, Lai H, *et al.* High-level rapid production of full-size monoclonal antibodies in plants by a single-vector DNA replicon system. Biotechnol Bioeng 2010; 106(1): 9-17.
[http://dx.doi.org/10.1002/bit.22652] [PMID: 20047189]

[74]   Lai H, He J, Engle M, Diamond MS, Chen Q. Robust production of virus-like particles and monoclonal antibodies with geminiviral replicon vectors in lettuce. Plant Biotechnol J 2012; 10(1): 95-104.
[http://dx.doi.org/10.1111/j.1467-7652.2011.00649.x] [PMID: 21883868]

[75]   So Y, *et al.* Glycomodification and characterization of anti-colorectal cancer immunotherapeutic monoclonal antibodies in transgenic tobacco. Plant Cell Tissue Organ Cult 2013; 113(1): 41-9.
[PCTOC].
[http://dx.doi.org/10.1007/s11240-012-0249-z]

[76]   Zeitlin L, *et al.* Prophylactic and therapeutic testing of Nicotiana-derived RSV-neutralizing human monoclonal antibodies in the cotton rat model.MAbs. Taylor & Francis 2013.
[http://dx.doi.org/10.4161/mabs.23281]

[77]   van Dolleweerd CJ, Teh AY, Banyard AC, *et al.* Engineering, expression in transgenic plants and characterisation of E559, a rabies virus-neutralising monoclonal antibody. J Infect Dis 2014; 210(2): 200-8.
[http://dx.doi.org/10.1093/infdis/jiu085] [PMID: 24511101]

[78]   Lai H, He J, Hurtado J, *et al.* Structural and functional characterization of an anti-West Nile virus monoclonal antibody and its single-chain variant produced in glycoengineered plants. Plant Biotechnol J 2014; 12(8): 1098-107.
[http://dx.doi.org/10.1111/pbi.12217] [PMID: 24975464]

[79]   Dent M, Hurtado J, Paul AM, *et al.* Plant-produced anti-dengue virus monoclonal antibodies exhibit reduced antibody-dependent enhancement of infection activity. J Gen Virol 2016; 97(12): 3280-90.
[http://dx.doi.org/10.1099/jgv.0.000635] [PMID: 27902333]

[80]   Vamvaka E, Twyman RM, Murad AM, *et al.* Rice endosperm produces an underglycosylated and

potent form of the HIV-neutralizing monoclonal antibody 2G12. Plant Biotechnol J 2016; 14(1): 97-108.
[http://dx.doi.org/10.1111/pbi.12360] [PMID: 25845722]

[81]   Iyappan G, *et al.* Potential of plant biologics to tackle the epidemic like situations-case studies involving viral and bacterial candidates. Int J Infect Dis 2018; 73: 363.
[http://dx.doi.org/10.1016/j.ijid.2018.04.4236]

[82]   Shafaghi M, Maktoobian S, Rasouli R, Howaizi N, Ofoghi H, Ehsani P. Transient expression of biologically active anti-rabies virus monoclonal antibody in tobacco leaves. Iran J Biotechnol 2018; 16(1): e1774.
[http://dx.doi.org/10.21859/ijb.1774] [PMID: 30555840]

[83]   Rattanapisit K, Chao Z, Siriwattananon K, Huang Z, Phoolcharoen W. Plant-produced anti-Enterovirus 71 (EV71) monoclonal antibody efficiently protects mice against EV71 infection. Plants 2019; 8(12): 560.
[http://dx.doi.org/10.3390/plants8120560] [PMID: 31805650]

[84]   Rattanapisit K, Phakham T, Buranapraditkun S, *et al.* Structural and in vitro functional analyses of novel plant-produced anti-human PD1 antibody. Sci Rep 2019; 9(1): 15205.
[http://dx.doi.org/10.1038/s41598-019-51656-1] [PMID: 31645587]

[85]   Diamos AG, Hunter JGL, Pardhe MD, *et al.* High level production of monoclonal antibodies using an optimized plant expression system. Front Bioeng Biotechnol 2020; 7: 472.
[http://dx.doi.org/10.3389/fbioe.2019.00472] [PMID: 32010680]

[86]   Hurtado J, Acharya D, Lai H, *et al.* In vitro and in vivo efficacy of anti-chikungunya virus monoclonal antibodies produced in wild-type and glycoengineered Nicotiana benthamiana plants. Plant Biotechnol J 2020; 18(1): 266-73.
[http://dx.doi.org/10.1111/pbi.13194] [PMID: 31207008]

[87]   Rattanapisit K, Abdulheem S, Chaikeawkaew D, *et al.* Recombinant human osteopontin expressed in Nicotiana benthamiana stimulates osteogenesis related genes in human periodontal ligament cells. Sci Rep 2017; 7(1): 17358.
[http://dx.doi.org/10.1038/s41598-017-17666-7] [PMID: 29229947]

[88]   Park J-G, Ye C, Piepenbrink MS, *et al.* A Broad and potent H1-specific human monoclonal antibody produced in plants prevents Influenza virus infection and transmission in Guinea Pigs. Viruses 2020; 12(2): 167.
[http://dx.doi.org/10.3390/v12020167] [PMID: 32024281]

[89]   Daniell H, Lee SB, Panchal T, Wiebe PO. Expression of the native cholera toxin B subunit gene and assembly as functional oligomers in transgenic tobacco chloroplasts. J Mol Biol 2001; 311(5): 1001-9.
[http://dx.doi.org/10.1006/jmbi.2001.4921] [PMID: 11531335]

[90]   Birch-Machin I, Newell CA, Hibberd JM, Gray JC. Accumulation of rotavirus VP6 protein in chloroplasts of transplastomic tobacco is limited by protein stability. Plant Biotechnol J 2004; 2(3): 261-70.
[http://dx.doi.org/10.1111/j.1467-7652.2004.00072.x] [PMID: 17147617]

[91]   Molina A, Hervás-Stubbs S, Daniell H, Mingo-Castel AM, Veramendi J. High-yield expression of a viral peptide animal vaccine in transgenic tobacco chloroplasts. Plant Biotechnol J 2004; 2(2): 141-53.
[http://dx.doi.org/10.1046/j.1467-7652.2004.00057.x] [PMID: 17147606]

[92]   Koya V, Moayeri M, Leppla SH, Daniell H. Plant-based vaccine: mice immunized with chloroplast-derived anthrax protective antigen survive anthrax lethal toxin challenge. Infect Immun 2005; 73(12): 8266-74.
[http://dx.doi.org/10.1128/IAI.73.12.8266-8274.2005] [PMID: 16299323]

[93]   Glenz K, Bouchon B, Stehle T, Wallich R, Simon MM, Warzecha H. Production of a recombinant bacterial lipoprotein in higher plant chloroplasts. Nat Biotechnol 2006; 24(1): 76-7.
[http://dx.doi.org/10.1038/nbt1170] [PMID: 16327810]

[94]   Chebolu S, Daniell H. Stable expression of Gal/GalNAc lectin of Entamoeba histolytica in transgenic chloroplasts and immunogenicity in mice towards vaccine development for amoebiasis. Plant Biotechnol J 2007; 5(2): 230-9.
[http://dx.doi.org/10.1111/j.1467-7652.2006.00234.x] [PMID: 17309678]

[95]   Ruhlman T, Ahangari R, Devine A, Samsam M, Daniell H. Expression of cholera toxin B-proinsulin fusion protein in lettuce and tobacco chloroplasts--oral administration protects against development of insulitis in non-obese diabetic mice. Plant Biotechnol J 2007; 5(4): 495-510.
[http://dx.doi.org/10.1111/j.1467-7652.2007.00259.x] [PMID: 17490448]

[96]   Fernández-San Millán A, Ortigosa SM, Hervás-Stubbs S, *et al.* Human papillomavirus L1 protein expressed in tobacco chloroplasts self-assembles into virus-like particles that are highly immunogenic. Plant Biotechnol J 2008; 6(5): 427-41.
[http://dx.doi.org/10.1111/j.1467-7652.2008.00338.x] [PMID: 18422886]

[97]   Wang DM, Zhu JB, Peng M, Zhou P. Induction of a protective antibody response to FMDV in mice following oral immunization with transgenic Stylosanthes spp. as a feedstuff additive. Transgenic Res 2008; 17(6): 1163-70.
[http://dx.doi.org/10.1007/s11248-008-9188-1] [PMID: 18651235]

[98]   Rigano MM, Manna C, Giulini A, *et al.* Transgenic chloroplasts are efficient sites for high-yield production of the vaccinia virus envelope protein A27L in plant cellsdagger. Plant Biotechnol J 2009; 7(6): 577-91.
[http://dx.doi.org/10.1111/j.1467-7652.2009.00425.x] [PMID: 19508274]

[99]   Rosales-Mendoza S, Alpuche-Solís AG, Soria-Guerra RE, *et al.* Expression of an *Escherichia coli* antigenic fusion protein comprising the heat labile toxin B subunit and the heat stable toxin, and its assembly as a functional oligomer in transplastomic tobacco plants. Plant J 2009; 57(1): 45-54.
[http://dx.doi.org/10.1111/j.1365-313X.2008.03666.x] [PMID: 18764920]

[100]   Sim J-S, *et al.* Expression and characterization of synthetic heat-labile enterotoxin B subunit and hemagglutinin–neuraminidase-neutralizing epitope fusion protein in *Escherichia coli* and tobacco chloroplasts. Plant Mol Biol Report 2009; 27(3): 388-99.
[http://dx.doi.org/10.1007/s11105-009-0114-3]

[101]   Davoodi-Semiromi A, Schreiber M, Nalapalli S, *et al.* Chloroplast-derived vaccine antigens confer dual immunity against cholera and malaria by oral or injectable delivery. Plant Biotechnol J 2010; 8(2): 223-42.
[http://dx.doi.org/10.1111/j.1467-7652.2009.00479.x] [PMID: 20051036]

[102]   Youm JW, Jeon JH, Kim H, *et al.* High-level expression of a human β-site APP cleaving enzyme in transgenic tobacco chloroplasts and its immunogenicity in mice. Transgenic Res 2010; 19(6): 1099-108.
[http://dx.doi.org/10.1007/s11248-010-9383-8] [PMID: 20229285]

[103]   Del L Yácono M, Farran I, Becher ML, *et al.* A chloroplast-derived Toxoplasma gondii GRA4 antigen used as an oral vaccine protects against toxoplasmosis in mice. Plant Biotechnol J 2012; 10(9): 1136-44.
[http://dx.doi.org/10.1111/pbi.12001] [PMID: 23020088]

[104]   Karimi F, *et al.* Immunogenicity of EIT chimeric protein expressed in transplastomic tobacco plants towards development of an oral vaccine against *Escherichia coli* O157: H7. Plant Biotechnol Rep 2013; 7(4): 535-46.
[http://dx.doi.org/10.1007/s11816-013-0296-x]

[105]   Lakshmi PS, Verma D, Yang X, Lloyd B, Daniell H. Low cost tuberculosis vaccine antigens in capsules: expression in chloroplasts, bio-encapsulation, stability and functional evaluation in vitro. PLoS One 2013; 8(1): e54708.
[http://dx.doi.org/10.1371/journal.pone.0054708] [PMID: 23355891]

[106]   Maldaner FR, Aragão FJ, dos Santos FB, *et al.* Dengue virus tetra-epitope peptide expressed in lettuce

chloroplasts for potential use in dengue diagnosis. Appl Microbiol Biotechnol 2013; 97(13): 5721-9.
[http://dx.doi.org/10.1007/s00253-013-4918-6] [PMID: 23615743]

[107] Clarke JL, Paruch L, Dobrica MO, *et al.* Lettuce-produced hepatitis C virus E1E2 heterodimer triggers immune responses in mice and antibody production after oral vaccination. Plant Biotechnol J 2017; 15(12): 1611-21.
[http://dx.doi.org/10.1111/pbi.12743] [PMID: 28419665]

[108] Firsov A, Tarasenko I, Mitiouchkina T, *et al.* Expression and Immunogenicity of M2e Peptide of Avian Influenza Virus H5N1 Fused to Ricin Toxin B Chain Produced in Duckweed Plants. Front Chem 2018; 6(22): 22.
[http://dx.doi.org/10.3389/fchem.2018.00022] [PMID: 29487846]

[109] Sohn E-J, *et al.* Development of plant-produced E2 protein for use as a green vaccine against classical swine fever virus. J Plant Biol 2018; 61(4): 241-52.
[http://dx.doi.org/10.1007/s12374-018-0133-4]

[110] Ilyas S, *et al.* Genetic transformation of carrot (*Daucas carota* L.) to express Hepatitis B surface antigen gene for edible vaccine of HBV. J Anim Plant Sci 2020; 30(5): 1263-72.

[111] Arevalo-Villalobos JI, Govea-Alonso DO, Monreal-Escalante E, Zarazúa S, Rosales-Mendoza S. LTB-Syn: a recombinant immunogen for the development of plant-made vaccines against synucleinopathies. Planta 2017; 245(6): 1231-9.
[http://dx.doi.org/10.1007/s00425-017-2675-y] [PMID: 28315001]

[112] Bai G, Tian Y, Wu J, *et al.* Construction of a fusion anti-caries DNA vaccine in transgenic tomato plants for PAcA gene and cholera toxin B subunit. Biotechnol Appl Biochem 2019; 66(6): 924-9.
[http://dx.doi.org/10.1002/bab.1806] [PMID: 31434162]

[113] Habibi M, *et al.* Expression of an epitope-based recombinant vaccine against Foot and Mouth Disease (FMDV) in tobacco plant *(Nicotiana tabacum).* J Plant Mol Breed 2019; 7(1): 1-9.

[114] Hunter JGL, Wilde S, Tafoya AM, *et al.* Evaluation of a toxoid fusion protein vaccine produced in plants to protect poultry against necrotic enteritis. PeerJ 2019; 7: e6600.
[http://dx.doi.org/10.7717/peerj.6600] [PMID: 30944775]

[115] Heenatigala PPM, Sun Z, Yang J, Zhao X, Hou H. Expression of lamb vaccine antigen in *wolffia globosa* (duck weed) against fish vibriosis. Front immunol 2020; 11: 1857-7.
[http://dx.doi.org/10.3389/fimmu.2020.01857] [PMID: 32973766]

[116] Rutkowska DA, Mokoena NB, Tsekoa TL, Dibakwane VS, O'Kennedy MM. Plant-produced chimeric virus-like particles - a new generation vaccine against African horse sickness. BMC Vet Res 2019; 15(1): 432.
[http://dx.doi.org/10.1186/s12917-019-2184-2] [PMID: 31796116]

[117] Jin S, Kanagaraj A, Verma D, Lange T, Daniell H. Release of hormones from conjugates: chloroplast expression of β-glucosidase results in elevated phytohormone levels associated with significant increase in biomass and protection from aphids or whiteflies conferred by sucrose esters. Plant Physiol 2011; 155(1): 222-35.
[http://dx.doi.org/10.1104/pp.110.160754] [PMID: 21068365]

[118] Espinoza-Sánchez EA, *et al.* Production and characterization of fungal β-glucosidase and bacterial cellulases by tobacco chloroplast transformation. Plant Biotechnol Rep 2016; 10(2): 61-73.
[http://dx.doi.org/10.1007/s11816-016-0386-7]

[119] Castiglia D, Sannino L, Marcolongo L, *et al.* High-level expression of thermostable cellulolytic enzymes in tobacco transplastomic plants and their use in hydrolysis of an industrially pretreated Arundo donax L. biomass. Biotechnol Biofuels 2016; 9(1): 154.
[http://dx.doi.org/10.1186/s13068-016-0569-z] [PMID: 27453729]

[120] Petersen K, Bock R. High-level expression of a suite of thermostable cell wall-degrading enzymes from the chloroplast genome. Plant Mol Biol 2011; 76(3-5): 311-21.

[http://dx.doi.org/10.1007/s11103-011-9742-8] [PMID: 21298465]

[121] Agrawal P, Verma D, Daniell H. Expression of Trichoderma reesei β-mannanase in tobacco chloroplasts and its utilization in lignocellulosic woody biomass hydrolysis. PLoS One 2011; 6(12): e29302.
[http://dx.doi.org/10.1371/journal.pone.0029302] [PMID: 22216240]

[122] Kolotilin I, Kaldis A, Pereira EO, Laberge S, Menassa R. Optimization of transplastomic production of hemicellulases in tobacco: effects of expression cassette configuration and tobacco cultivar used as production platform on recombinant protein yields. Biotechnol Biofuels 2013; 6(1): 65.
[http://dx.doi.org/10.1186/1754-6834-6-65] [PMID: 23642171]

[123] Viitanen PV, Devine AL, Khan MS, Deuel DL, Van Dyk DE, Daniell H. Metabolic engineering of the chloroplast genome using the Echerichia coli ubiC gene reveals that chorismate is a readily abundant plant precursor for p-hydroxybenzoic acid biosynthesis. Plant Physiol 2004; 136(4): 4048-60.
[http://dx.doi.org/10.1104/pp.104.050054] [PMID: 15563620]

[124] Bohmert-Tatarev K, McAvoy S, Daughtry S, Peoples OP, Snell KD. High levels of bioplastic are produced in fertile transplastomic tobacco plants engineered with a synthetic operon for the production of polyhydroxybutyrate. Plant Physiol 2011; 155(4): 1690-708.
[http://dx.doi.org/10.1104/pp.110.169581] [PMID: 21325565]

[125] Leelavathi S, *et al.* Overproduction of an alkali-and thermo-stable xylanase in tobacco chloroplasts and efficient recovery of the enzyme. Mol Breed 2003; 11(1): 59-67.
[http://dx.doi.org/10.1023/A:1022168321380]

[126] Verma D, Kanagaraj A, Jin S, Singh ND, Kolattukudy PE, Daniell H. Chloroplast-derived enzyme cocktails hydrolyse lignocellulosic biomass and release fermentable sugars. Plant Biotechnol J 2010; 8(3): 332-50.
[http://dx.doi.org/10.1111/j.1467-7652.2009.00486.x] [PMID: 20070870]

[127] Kim JY, Kavas M, Fouad WM, Nong G, Preston JF, Altpeter F. Production of hyperthermostable GH10 xylanase Xyl10B from Thermotoga maritima in transplastomic plants enables complete hydrolysis of methylglucuronoxylan to fermentable sugars for biofuel production. Plant Mol Biol 2011; 76(3-5): 357-69.
[http://dx.doi.org/10.1007/s11103-010-9712-6] [PMID: 21080212]

[128] Dai Z, Hooker BS, Anderson DB, Thomas SR. Expression of Acidothermus cellulolyticus endoglucanase E1 in transgenic tobacco: biochemical characteristics and physiological effects. Transgenic Res 2000; 9(1): 43-54.
[http://dx.doi.org/10.1023/A:1008922404834] [PMID: 10853268]

[129] Jin R, Richter S, Zhong R, Lamppa GK. Expression and import of an active cellulase from a thermophilic bacterium into the chloroplast both *in vitro* and *in vivo*. Plant Mol Biol 2003; 51(4): 493-507.
[http://dx.doi.org/10.1023/A:1022354124741] [PMID: 12650616]

[130] Ziegelhoffer T, Raasch JA, Austin-Phillips S. Dramatic effects of truncation and sub-cellular targeting on the accumulation of recombinant microbial cellulase in tobacco. Mol Breed 2001; 8(2): 147-58.
[http://dx.doi.org/10.1023/A:1013338312948]

[131] Dai Z, *et al.* Improved plant-based production of E1 endoglucanase using potato: expression optimization and tissue targeting. Mol Breed 2000; 6(3): 277-85.
[http://dx.doi.org/10.1023/A:1009653011948]

[132] Harrison MD, Geijskes J, Coleman HD, *et al.* Accumulation of recombinant cellobiohydrolase and endoglucanase in the leaves of mature transgenic sugar cane. Plant Biotechnol J 2011; 9(8): 884-96.
[http://dx.doi.org/10.1111/j.1467-7652.2011.00597.x] [PMID: 21356003]

[133] Yu L-X, Gray BN, Rutzke CJ, Walker LP, Wilson DB, Hanson MR. Expression of thermostable microbial cellulases in the chloroplasts of nicotine-free tobacco. J Biotechnol 2007; 131(3): 362-9.
[http://dx.doi.org/10.1016/j.jbiotec.2007.07.942] [PMID: 17765995]

[134] Jung S, Kim S, Bae H, Lim HS, Bae HJ. Expression of thermostable bacterial β-glucosidase (BglB) in transgenic tobacco plants. Bioresour Technol 2010; 101(18): 7155-61.
[http://dx.doi.org/10.1016/j.biortech.2010.03.140] [PMID: 20427180]

[135] Jung S, Lee DS, Kim YO, Joshi CP, Bae HJ. Improved recombinant cellulase expression in chloroplast of tobacco through promoter engineering and 5′ amplification promoting sequence. Plant Mol Biol 2013; 83(4-5): 317-28.
[http://dx.doi.org/10.1007/s11103-013-0088-2] [PMID: 23771581]

[136] Hyunjong B, Lee D-S, Hwang I. Dual targeting of xylanase to chloroplasts and peroxisomes as a means to increase protein accumulation in plant cells. J Exp Bot 2006; 57(1): 161-9.
[http://dx.doi.org/10.1093/jxb/erj019] [PMID: 16317036]

[137] Bae H-J, Kim HJ, Kim YS. Production of a recombinant xylanase in plants and its potential for pulp biobleaching applications. Bioresour Technol 2008; 99(9): 3513-9.
[http://dx.doi.org/10.1016/j.biortech.2007.07.064] [PMID: 17889523]

[138] Chen PJ, Senthilkumar R, Jane WN, He Y, Tian Z, Yeh KW. Transplastomic Nicotiana benthamiana plants expressing multiple defence genes encoding protease inhibitors and chitinase display broad-spectrum resistance against insects, pathogens and abiotic stresses. Plant Biotechnol J 2014; 12(4): 503-15.
[http://dx.doi.org/10.1111/pbi.12157] [PMID: 24479648]

[139] Lebel EG, *et al.* Transgenic plants expressing a cellulase. Google Patents 2008.

[140] Sijmons PC, Dekker BM, Schrammeijer B, Verwoerd TC, van den Elzen PJ, Hoekema A. Production of correctly processed human serum albumin in transgenic plants. Biotechnology (N Y) 1990; 8(3): 217-21.
[PMID: 1366404]

[141] Matsumoto S, Ikura K, Ueda M, Sasaki R. Characterization of a human glycoprotein (erythropoietin) produced in cultured tobacco cells. Plant Mol Biol 1995; 27(6): 1163-72.
[http://dx.doi.org/10.1007/BF00020889] [PMID: 7766897]

[142] Terashima M, Murai Y, Kawamura M, *et al.* Production of functional human α 1-antitrypsin by plant cell culture. Appl Microbiol Biotechnol 1999; 52(4): 516-23.
[http://dx.doi.org/10.1007/s002530051554] [PMID: 10570799]

[143] Zhong G-Y, D Peterson, DE Delaney, *et al.* Commercial production of aprotinin in transgenic maize seeds. Mol Breed 1999; 5(4): 345-56.
[http://dx.doi.org/10.1023/A:1009677809492]

[144] Komarnytsky S, Borisjuk NV, Borisjuk LG, Alam MZ, Raskin I. Production of recombinant proteins in tobacco guttation fluid. Plant Physiol 2000; 124(3): 927-34.
[http://dx.doi.org/10.1104/pp.124.3.927] [PMID: 11080270]

[145] Ruggiero F, Exposito JY, Bournat P, *et al.* Triple helix assembly and processing of human collagen produced in transgenic tobacco plants. FEBS Lett 2000; 469(1): 132-6.
[http://dx.doi.org/10.1016/S0014-5793(00)01259-X] [PMID: 10708770]

[146] Staub JM, Garcia B, Graves J, *et al.* High-yield production of a human therapeutic protein in tobacco chloroplasts. Nat Biotechnol 2000; 18(3): 333-8.
[http://dx.doi.org/10.1038/73796] [PMID: 10700152]

[147] De Cosa B, Moar W, Lee SB, Miller M, Daniell H. Overexpression of the Bt cry2Aa2 operon in chloroplasts leads to formation of insecticidal crystals. Nat Biotechnol 2001; 19(1): 71-4.
[http://dx.doi.org/10.1038/83559] [PMID: 11135556]

[148] Fernández-San Millán A, Mingo-Castel A, Miller M, Daniell H. A chloroplast transgenic approach to hyper-express and purify Human Serum Albumin, a protein highly susceptible to proteolytic degradation. Plant Biotechnol J 2003; 1(2): 71-9.
[http://dx.doi.org/10.1046/j.1467-7652.2003.00008.x] [PMID: 17147744]

[149] Wirth S, *et al.* Expression of active human epidermal growth factor (hEGF) in tobacco plants by integrative and non-integrative systems. Mol Breed 2004; 13(1): 23-35.
[http://dx.doi.org/10.1023/B:MOLB.0000012329.74067.ca]

[150] Ding S-H, Huang LY, Wang YD, Sun HC, Xiang ZH. High-level expression of basic fibroblast growth factor in transgenic soybean seeds and characterization of its biological activity. Biotechnol Lett 2006; 28(12): 869-75.
[http://dx.doi.org/10.1007/s10529-006-9018-6] [PMID: 16786271]

[151] Arlen PA, Falconer R, Cherukumilli S, *et al.* Field production and functional evaluation of chloroplast-derived interferon-α2b. Plant Biotechnol J 2007; 5(4): 511-25.
[http://dx.doi.org/10.1111/j.1467-7652.2007.00258.x] [PMID: 17490449]

[152] Kim T-G, Baek MY, Lee EK, Kwon TH, Yang MS. Expression of human growth hormone in transgenic rice cell suspension culture. Plant Cell Rep 2008; 27(5): 885-91.
[http://dx.doi.org/10.1007/s00299-008-0514-0] [PMID: 18264710]

[153] Oey M, Lohse M, Kreikemeyer B, Bock R. Exhaustion of the chloroplast protein synthesis capacity by massive expression of a highly stable protein antibiotic. Plant J 2009; 57(3): 436-45.
[http://dx.doi.org/10.1111/j.1365-313X.2008.03702.x] [PMID: 18939966]

[154] Xu J, Okada S, Tan L, Goodrum KJ, Kopchick JJ, Kieliszewski MJ. Human growth hormone expressed in tobacco cells as an arabinogalactan-protein fusion glycoprotein has a prolonged serum life. Transgenic Res 2010; 19(5): 849-67.
[http://dx.doi.org/10.1007/s11248-010-9367-8] [PMID: 20135224]

[155] An N, Ou J, Jiang D, *et al.* Expression of a functional recombinant human basic fibroblast growth factor from transgenic rice seeds. Int J Mol Sci 2013; 14(2): 3556-67.
[http://dx.doi.org/10.3390/ijms14023556] [PMID: 23434658]

[156] Guan C, *et al.* Expression of biologically active anti-thrombosis protein lumbrokinase in edible sunflower seed kernel. J Plant Biochem Biotechnol 2014; 23(3): 257-65.
[http://dx.doi.org/10.1007/s13562-013-0209-7]

[157] Tan Y, Wang KY, Wang N, Li G, Liu D. Ectopic expression of human acidic fibroblast growth factor 1 in the medicinal plant, Salvia miltiorrhiza, accelerates the healing of burn wounds. BMC Biotechnol 2014; 14(1): 74.
[http://dx.doi.org/10.1186/1472-6750-14-74] [PMID: 25106436]

[158] Limkul J, Misaki R, Kato K, Fujiyama K. The combination of plant translational enhancers and terminator increase the expression of human glucocerebrosidase in Nicotiana benthamiana plants. Plant Sci 2015; 240: 41-9.
[http://dx.doi.org/10.1016/j.plantsci.2015.08.018] [PMID: 26475186]

[159] Su J, Sherman A, Doerfler PA, Byrne BJ, Herzog RW, Daniell H. Oral delivery of Acid Alpha Glucosidase epitopes expressed in plant chloroplasts suppresses antibody formation in treatment of Pompe mice. Plant Biotechnol J 2015; 13(8): 1023-32.
[http://dx.doi.org/10.1111/pbi.12413] [PMID: 26053072]

[160] Wang Y-P, Wei ZY, Zhong XF, *et al.* Stable expression of basic fibroblast growth factor in chloroplasts of tobacco. Int J Mol Sci 2015; 17(1): 19.
[http://dx.doi.org/10.3390/ijms17010019] [PMID: 26703590]

[161] Ahmad AR, Kaewpungsup P, Khorattanakulchai N, Rattanapisit K, Pavasant P, Phoolcharoen W. Recombinant Human Dentin Matrix Protein 1 (hDMP1) Expressed in *Nicotiana benthamiana* Potentially Induces Osteogenic Differentiation. Plants 2019; 8(12): 566.
[http://dx.doi.org/10.3390/plants8120566] [PMID: 31816999]

[162] Hong Y, Zhang H, Huang L, Li D, Song F. Overexpression of a Stress-Responsive NAC Transcription Factor Gene ONAC022 Improves Drought and Salt Tolerance in Rice. Front Plant Sci 2016; 7(4): 4.
[http://dx.doi.org/10.3389/fpls.2016.00004] [PMID: 26834774]

[163] Jin SG, Prusti RK, Roitsch T, Ankenbauer RG, Nester EW. Phosphorylation of the VirG protein of Agrobacterium tumefaciens by the autophosphorylated VirA protein: essential role in biological activity of VirG. J Bacteriol 1990; 172(9): 4945-50.
[http://dx.doi.org/10.1128/jb.172.9.4945-4950.1990] [PMID: 2394678]

[164] Vergunst AC, Schrammeijer B, den Dulk-Ras A, de Vlaam CM, Regensburg-Tuïnk TJ, Hooykaas PJ. VirB/D4-dependent protein translocation from Agrobacterium into plant cells. Science 2000; 290(5493): 979-82.
[http://dx.doi.org/10.1126/science.290.5493.979] [PMID: 11062129]

[165] Howard E, Citovsky V. The emerging structure of the Agrobacterium T-DNA transfer complex. BioEssays 1990; 12(3): 103-8.
[http://dx.doi.org/10.1002/bies.950120302]

[166] Hwang H-H, Yu M, Lai E-M. Agrobacterium-mediated plant transformation: biology and applications. Arabidopsis Book 2017; 15: e0186.
[http://dx.doi.org/10.1199/tab.0186] [PMID: 31068763]

[167] Sanford JC. Biolistic plant transformation. Physiol Plant 1990; 79(1): 206-9.
[http://dx.doi.org/10.1111/j.1399-3054.1990.tb05888.x]

[168] Pogue GP, Holzberg S. Transient virus expression systems for recombinant protein expression in dicot-and monocotyledonous plants. Plant Sci 2012; 191-216.

[169] Meyers AJ, Grohs BM, Hall JC. Antibody Production in planta, in Comprehensive Biotechnology (Second Edition), M. Moo-Young, Editor. Academic Press: Burlington 2011; pp. 287-300.

[170] French R, Janda M, Ahlquist P. Bacterial gene inserted in an engineered RNA virus: efficient expression in monocotyledonous plant cells. Science 1986; 231(4743): 1294-7.
[http://dx.doi.org/10.1126/science.231.4743.1294] [PMID: 17839568]

[171] Shanmugaraj B, Malla A, Phoolcharoen W. Emergence of novel coronavirus 2019-nCoV: need for rapid vaccine and biologics development. Pathogens 2020; 9(2): 148.
[http://dx.doi.org/10.3390/pathogens9020148] [PMID: 32098302]

[172] LeBlanc Z, Waterhouse P, Bally J. Plant-based vaccines: the way ahead? Viruses 2020; 13(1): 5.
[http://dx.doi.org/10.3390/v13010005] [PMID: 33375155]

[173] COVID-19. Medicago's development program. Available from: https://www.medicago.com/en/covid-19-programs/

CHAPTER 8

Analysis of Cross-Reactivity, Specificity and the Use of Optimised ELISA for Rapid Detection of *Fusarium* Spp.

Phetole Mangena[1,*] and **Phumzile Mkhize**[2]

[1] *Department of Biodiversity, School of Molecular and Life Sciences, Faculty of Science and Agriculture, University of Limpopo, Private Bag X1106, Sovenga 0727, Republic of South Africa*

[2] *Department of Microbiology, Biochemistry and Biotechnology, School of Molecular and Life Sciences, Faculty of Science and Agriculture, University of Limpopo, Private Bag X1106, Sovenga 0727, Republic of South Africa*

Abstract: Many strides have been made in the development of antibody-based detection systems for rapid and sensitive analysis of *Fusarium* pathogens and their toxins. Antibody cross-reactivity, specificity, and binding affinity with antigenic molecules affect the efficacy in which these molecules serve their own functions. Researchers are, therefore, directed in investigating the principles that govern cross-reactivity, specificity, and the relationship between them, using various tools such as optimised ELISA. This is important because the ability of *Fusarium* spp. to infect and produce mycotoxins in agronomic crops passes these toxins to animals and humans after contact or ingestion. Antibodies that recognise and bind particular antigens with great affinity and specificity, especially for the effective relief of unwanted *Fusarium* pathogenic materials in humans and animals, are thus required. Furthermore, the demand for fungal contaminants free agriculture, emerging antifungal drug resistance, and the fatal health effects of fungal infections in immunocompromised humans and animals drive the need for the development of a rapid, sensitive, reliable, and accurate relief system for these pathogens. Therefore, this chapter provides a succinct review on the role of antibody cross-reactivity and specificity, with reference to basic principles, challenges, and detection for rapid and reliable assessment in *Fusarium* pathogens.

Keywords: Antibody specificity, Antigens, Cross-reactivity, ELISA, *Fusarium*, Immunoglobulins, Mycotoxins, Somatic hypermutation.

* **Corresponding author Phetole Mangena:** Department of Biodiversity, School of Molecular and Life Sciences, Faculty of Science and Agriculture, University of Limpopo, Private Bag X1106, Sovenga 0727, Republic of South Africa; Tel: 015-268-4715; E-mail: Phetole.Mangena@ul.ac.za

Muhammad Sarwar Khan (Ed)

1. INTRODUCTION

The genus *Fusarium* contains a group of filamentous fungi commonly found in soils, in which its species are linked with animal and plant pathogenesis. *Fusarium* species are toxigenic, and the mycotoxins (*i.e.*, compounds with deleterious effects on susceptible host organisms) produced by these fungi are often associated with animal and human infections, including some of the diseases affecting seedlings and mature plants. According to Moretti [1], these organisms remain among some of the widely occurring plant-pathogenic species, causing diseases in several agriculturally important crops, particularly cereals, lumber, and pulses. Many of *Fusarium* spp. produce a wide range of biologically active secondary metabolites (Table **1**), with tremendous accompanying chemical diversity. Some of the bioactive metabolites serve as products of primary and secondary metabolic value, characterised by a distinct and unusual chemical structure with varying molecular weights.

Table 1. Brief outline of mycotoxins and valuable compounds produced by *Fusarium* spp., including the early and recently discovered examples of bioactive compounds.

Fungal spp.	Mycotoxins/ SMs	Brief Description	References
F. sambucinum TE-6L	Indole alkaloids	Amoenamide C and Sclereotiamide B, from angularly prenylated indole alkaloids with pyrano [2,3-g] indole moieties *Application:* Antimicrobial and insecticidal activity	Zhang *et al.* [4]
Unclassified *Fusarium* spp.	Terpenes	Orcyl aldehyde units (condensed with farnesyl side chain terminally cyclised to cyclohexanone ring in γ, ε, δ, ζ, α and β) designated LL-Z1272 *Application:* anti-*Tetrahymena pyriformis* activity	Ellestad *et al.* [13]
F. fujikuroi	Polyketides	Yellow and/ orange pigmented fusarins, encoded by a *fusA* gene *Application:* Antimicrobial activity	Diaz-Sanchez *et al.* [14]
F. graminearum	Non-ribosomal peptides	Fasaoctaxin A, product of ectopic expression of *fg3_54* gene cluster *Role:* A as a virulence factor required for cell-to-cell invasion of wheat by *F. graminearum*	Jia *et al.* [15]
F. incarnatum-equiseti	PKs + NRPs derived	Polyketides (PKs) and Nonribosomal peptides (NRPs) secondary metabolites synthesized by NRP synthetase and type-1 PK synthase *e.g.* Carotenoid, fusarubin pigment, enniatin, fusarin and zearalenone mycotoxins	Villani *et al.* [16]

(Table 1) cont.....

Fungal spp.	Mycotoxins/ SMs	Brief Description	References
F. crookwellense, F. culmorum and *F. graminearum*	Trichothecenes	Nivalenol (NIV), deoxynivalenol (DON) or scirpentriol (STO) chemotypes	Lauren *et al.* [17]
F. oxysporum, F. proliferatum, F. moniliforme and *F. nygami*	Fumonisins	Fumonisins A (N-acetyl analogs, FAs), B (sphingosine N-acyltransferase, ceramide synthase), C (dimethyl analogs, FCs) and P (N-3-hydroxypiridinium analogs, FPs)	Sewran *et al.* [8], Tamura *et al.* [9] and Rheeder *et al.* [10]
F. graminearum, F. culmorum and *F. crookwellense*	Zearalenone	Zearalenone (ZAN), α-zearalenol (α-ZOL), β-zearalenol (β-ZOL), α-zearalanol (α-ZAL) and β-zearalanol (β-ZAL)	Tian *et al.* [11]
F. proliferatum, F. solani and *F. pseudonygamai*	Fusaproliferin	Fusaproliferin, a sesterterpene mycotoxin C_{20}-core carbon skeleton possessing an extra C_5-head/C_5-tail unit	Fotso *et al.* [18] and Liuzzi *et al.* [19]
F. concentricum	Beauvericin	Structurally related beauvericin (BEA) and enniatins B (ENN B) synthesised by multifunctional enzyme enniatin synthetase containing both peptide synthetase and S-adenosyl-L-methionine-dependent N-methyltransferase activities	Liuzzi *et al.* [19]
F. Orthoceras var. *enniatimum*	Enniatins	Structurally related beauvericin (BEA) and enniatins B (ENN B) synthesised by multifunctional enzyme enniatin synthetase containing both peptide synthetase and S-adenosyl-L-methionine-dependent N-methyltransferase activities	Liuzzi *et al.* [19]
F. proliferatum, F. phyllophilum and *F. subglutinans*	Moniliformin	Small ionic carboxylate potassium/ sodium salt (1-hydroxycyclobut-1-ene-3,4 dione) with a highly toxic one water crystallisation	Fotso *et al.* [18]
Note: SMs- Secondary metabolites Spp. – species			

Not only are mycotoxins produced by species in this genus, but some secondary metabolites potentially serve as a source of many useful novel compounds with enzymatic capability, antibacterial activity, antiviral, anti-parasitic, and growth-promoting effects, as well as other properties can be obtained [2, 3]. *Fusarium* could, furthermore, be used in industrial processing to directly or indirectly provide enzymes used to catalyse the generation of various other novel substances such as pigments, cosmetic and food compounds that may be used to substitute synthetic compounds [2, 4]. These compounds indicated in Table **1** provide the basis for developing new promising and prolific sources of bioactive secondary metabolites with prominent biotechnological applications, as well as the detection of *Fusarium* spp. using specific mycotoxin profiles. However, most studies indicated that the diversity of mycotoxins or the use of distinctive characters of their shapes and sizes of macro-and microconidia, among others, for the identification of species within the genera remains the most critical issue. The morphological characterisation used for species identification also includes colony appearance, pigmentation, growth rate, and the absence/presence of chlamydospores [1, 5].

Other studies have identified and characterised *Fusarium* spp. using these characters in combination with several other molecular techniques (for example, polymerase chain reaction, amplified fragment length polymorphism, and DNA markers) to assess the genetic diversity of species responsible for inducing host's symptoms [1, 5, 6]. In addition, various molecular techniques are still continuously tested to develop straightforward detection protocols for *Fusarium* species, especially those that may be rapid, efficient, and reliable, in making appropriate and timely diagnoses enabling efficacious disease management. Although most of the *Fusarium* spp. infections are asymptomatic or cause mild symptoms widespread in plants, there is evidence that these species also cause significant physiological and neurological complications after infecting their animal and human hosts. *Fusarium* infections such as keratitis, onychomycosis, skin, and disseminated multiorgan infections were discussed by Nelson *et al.* [7] as highly neurological. The irreversible association of carcinogenic fumonisins (Table **1**) produced by clustered gene expression in *Fusarium moniliforme* and other humans infecting *Fusarium* spp. were also reported [7–12]. Some of these infections have emerged significantly fatal in immunocompromised hosts, both in humans and animals [13–18].

This is particularly based on the observations that the treatment of *Fusarium* infection in such hosts do not produce desired results due to the limited susceptibility of these pathogens to antifungal agents that are currently available. As reported by Tortorano *et al.* [20], most species have also shown resistance to antifungal treatments, such as azoles, polyenes, and echinocandins. The experienced resistance is often exacerbated by the lack of highly efficient detection technologies and clinical sensitivity in achieving favourable outcomes based on pathological processes. Although, various immunofluorescence, immunoblotting, enzyme-linked immunosorbent assay (EIA) and chromato-graphic assays have been successfully developed. A major challenge in determining immunoglobulin-antigen cross-reactivity and specificity for fungal pathogens is the occurrence of false-positive results largely related to cross-reactivity with other pathogens.

For example, an epitope binding by EB-A2 monoclonal antibody used for EIA analysis for the detection of *Aspergillus* species was found not exclusively expressed in this genera. But the epitope was also contained in antigens of *Geotrichum*, *Myceliophthora* (such as *Blastomyces dermatitidis* and *Histoplasma capsulatum*), *Paecilomyces*, *Penicillium*, and *Trichothecium* fungi, as well as yeasts [20]. Therefore, variations in molecular structures attributed to both the hosts and pathogens that determine immune recognition pose several challenges in the development of efficient detection assays and subsequent preventions and management of pathogenic populations.

The analysis of information regarding antigen-immunoglobulin interactions in terms of specificity and cross-reactivity will provide an understanding of the failure of antibody binding, providing the basis for the development of more sensitive and efficient antigen detection tools. Furthermore, such analysis will provide the molecular understanding of host-pathogen recognition that may also lead to the establishment of insights that clearly determine forces of antigen variations and, thus, prevention of any obstructions to antigen detection and specific antibody expression for treatment. This chapter, therefore, reviews the affinity of antigen-antibody complex and the possible effects that low affinity may have on populations of antibodies that bind to epitopes on other antigens in *Fusarium* species. It discusses how this cross-reactivity may result in over- or under-estimation of *Fusarium* antigen concentration and the negative role that this may have on immunoassays' efficiency. The chapter will finally discuss the benefits of developing optimised ELISA for sensitive, accurate, and rapid detection of highly specific antibody binding to antigens, especially in the presence of elevated levels of contaminating molecules.

2. FUNGAL DISEASES AND HOST RANGE OF *FUSARIUM* SPP. INFECTING PLANTS AND ANIMALS

Fungi occupy a broad range of environmental conditions, particularly the inhabitable ecological niche on earth. However, species that derive their nutrients from living organisms, which are primarily parasitic fungi, completely live on their animal and plant hosts. Most of these fungal species were reported to exhibit the competency to expand their host range and cause opportunistic infections [21, 22]. Other ubiquitous fungal isolates indicate single-host-specificity and while others serve as broad-host-range pathogens that can cause diseases in a large number of different host organisms [21]. Among *Fusarium* species, *F. oxysporum f. sp. Chrysanthemi* was identified as the causal agent of wilt and rot disease in more than 100 plant species [22]. Additionally, emerging pathogens of clinical importance in these genera that include members of the *F. solani* species complex

(FSSC), were also found to be responsible for the majority of systemic fusarial infections in immunocompromised humans and animals [23].

The host range of these fungi still emerges to be very diverse, causing symptoms such as tissue necrosis and abscess in all host plants and animals. In humans, compromised immune systems, particularly lung transplantation, hematologic malignancy, burns, and skin conditions, were reported as the most common infection sites for invasive fusariosis. A report by Muhammed *et al.* [23] indicated that FSSC was identified as the most frequent species, respectively, causing 50, 40 and 37.5% mortality rates in patients with disseminated, skin, and pulmonary fusariosis. Phylogenetic analysis of *Fusarium* species of Hypocreales and Necriaceae, that infects both humans and plants elucidated the distribution of genes required to synthesise enzymes and secondary metabolites that cause health complications in the hosts. According to O'Donnell *et al.* [24], diversification that occurred since the middle Cretaceous (91.3 million years ago) is responsible for the production of important secondary metabolites that cause cereal diseases such as *Fusarium* head blight. Fungal species like *F. graminearum* were strongly linked to the cereal head blight pathogenic isolates that radiated in the Pleistocene, often with the origin of concomitant toxic secondary metabolites such as trichothecenes, fumonisins, and furasin mycotoxins illustrated in Fig. (**1**). These secondary metabolites are well-known as mycotoxins and are capable of causing mycotoxicosis (diseases and deaths) in plants and animals.

The contamination of surface areas, air, and foodstuffs by fungi is predominantly accompanied by the presence of these fungal mycotoxins, which enter living organisms through dermal, ingestions and inhalation [25]. Therefore, the existence of these harmful metabolites in our midst implies that fungal pathogens, together with a myriad of challenges that impede the development of the most accurate measures used to identify, manage and treat diseases caused by fungal infections, should never be overlooked by researchers. Some of the paramount obstacles faced in dealing with these fungal pathogens include the following: (i) the diverse host-range that exists both in the species complex and within a single species leading to misidentifications/false positive diagnostic results, (ii) lack of reliably accurate evidence-based molecular diagnostic tools and better biomarkers, (iii) asymptomatic hosts indicating large quantities of pathogen DNA, (iv) many countries with no stringent import regulations on mycotoxin contamination, and (v) the lack of affordable, practical techniques used for mycotoxin removals following the exposure to fungal contaminations [21, 25, 26].

In agriculture, it was reported by Liew and Mohd-Redzwan [25] that an estimated 25% of the world's crops (*e.g.*, barley, corn, rice, sorghum, wheat, and nuts) are frequently contaminated by mold and fungal growth, as indicated by FAO (Food and Agriculture Organisation of the United Nations). Intrinsically, comprehensive research knowledge on fungal infection, mycotoxin type, physiological and metabolic effects of the pathogens must be gathered in order to reduce health problems and potential market losses of the exported agricultural products. Furthermore, such insights will potentially complement the development of novel and innovative strategies for rapid treatment and prevention of mycotoxin contamination and mycotoxicosis already suffered by plants and animals, as indicated by FAO [25].

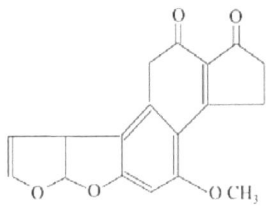

Aflatoxin B1

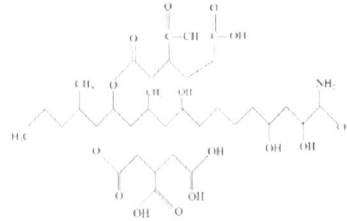

Fumonisin B1

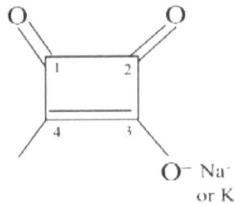

Moniliformin

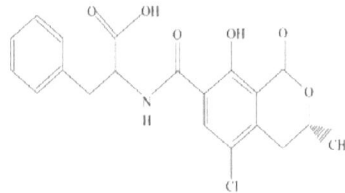

Ochratoxin A

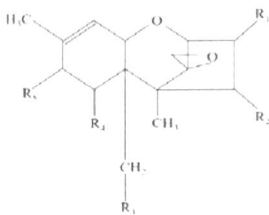

Group A Trichothecenes

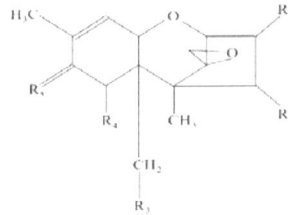

Group B Trichothecenes

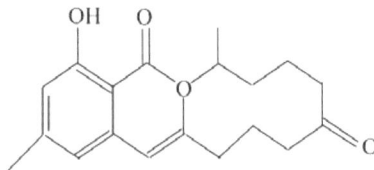

Fig. (1). Chemical structures of secondary metabolites serving as key mycotoxins produced by *Fusarium* species.

3. DETECTION SYSTEMS FOR *FUSARIUM* PATHOGENS

Like other microbial pathogens, *Fusarium* spp. also cause systemic illnesses to their hosts. The conditions are presented in the form of certain clinical signs and symptoms, while other hosts remain asymptomatic. More information on the detection and increasing host range of these pathogens will be valuable for determining the appropriate disease management strategies [27]. However, due to pathogenicity and host range, *Fusarium* continues to cause severe diseases across all agriculturally important crops, humans and animals. Cases of human fungal keratitis were reported by Murphy [28], showing severe infection with *F. solani* (Fig. **2a**). Colonisation of internal cavities of potato tubers creating dry rot infections also remain wide-spread in most potato cultivating areas (Fig. **2b**). Furthermore, *F. verticillioides*, *F. sporotrichiella var. sporotrichioides* and *F. oxysporum* were also implicated in causing ear rot in corn, tongue lesion mycotoxicosis in broiler chickens and deaths of zebrafish embryos as demonstrated in Fig. (2 **c**, **d** and **e**), respectively [29 - 31].

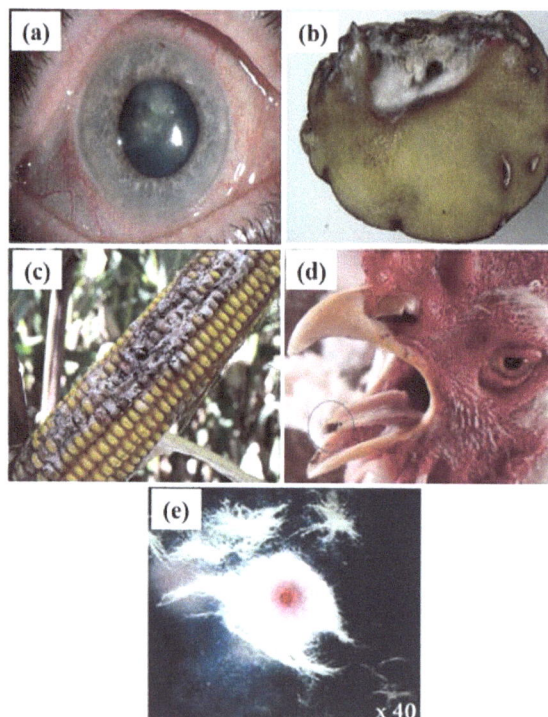

Fig. (2). Disease symptoms caused by fungal pathogens on plants and animals. **(a)** Fungal keratitis eye infection caused by *F. solani*, **(b)***Fusarium* dry rot of potato associated with *F. sambucinum*, *F. solani*, *F. culmorum* and *F. avenaceum* (Department of Primary Industries and Regional- Western Australia), **(c)***Fusarium* ear rot disease primarily caused by *F. virticillioides* (Crop Protection Network), **(d)** tongue lesion mycotoxicosis induced by *F. sporotrichiella*, and **(e)** fungal mycelium growth on infected zebrafish (dead) embryos.

The management of diseases mentioned above require not only the positive identification of destructive fungal pathogens, but also need a highly efficacious, reliable and rapid diagnostic system. Thus, eradication and prevention of these pathogens demand accurate detection systems suitable for both laboratory and field assessments to monitor the disease and optimise fungicide application. Moreover, Lin *et al.* [32] reported threshold pathogen density, changes in pathogen distribution and the interaction/synergy between pathogen, environmental factors and hosts as the additional factors that also need thorough consideration before a robustly efficient detection system can be developed. Some of the detection systems that are commonly and currently used are discussed below.

3.1. Traditional and Current Methods of *Fusarium* Detection

Fusarium species are some of the most widely distributed fungal pathogens in nature serving as common contaminants of agricultural commodities, foods, beverages and feeds. According to Gourama and Bullerman [33] the contaminated products serve as important medium for fungal growth because they constitute rich habitats for microorganisms since they contain sufficient water and nutrients such as carbohydrates, proteins, lipids, amino acids and nucleic acids. However, the toxic nature of mycotoxins produced by these organisms previously warranted the development of several detection techniques. Traditional detection protocols were based on bacteriological methods such as plate count techniques and use of selective media, including chemical assays as well as various biochemical tools [33]. Malachite green agar 2.5 ppm is one of the developed potent selective medium for the isolation and enumeration of *Fusarium* spp. In addition, potato dextrose agar (PDA), corn meal agar, malt agar and carnation leaf agar served as some of the most widely used selective media [34]. These selective media are still currently used by many plant pathologists because they can provide key ingredients supporting the growth of specific and isolated species.

As previously indicated, *Fusarium* species are very diverse in their morphology, mycotoxin profile and host associations. This implies that visibly identifying and quantifying one species from another based on an array of molecular, metabolic and morphological data is critical given the need for proper controls and precautionary measures [34]. These measures are deemed necessary to ensure high yields, high quality products and good value of agricultural produce, as well as improved wellbeing of people and animals. Currently, *Fusarium* species are detected using polymerase chain reaction (PCR), which is one of the most reliable, sensitive and accurate methods. For example, a specific primer pair CIRC1A-CIRC4A, which amplifies a 360-bp DNA fragment in the intergenic spacer region of the nuclear ribosomal operon during this technique has been used to successfully detect *F. circinatum* [35]. PCR has proved useful for detecting fungal pathogens in asymptomatic hosts and in distinguishing the pathogens up to the species level. Furthermore, antibody based detection system that use enzyme linked immunosorbent assay (ELISA) have also been used for *Fusarium* spp. A dipstick enzyme immunoassay for the rapid detection of *Fusarium* T-2 toxin, one of the trichothecene (shown in Fig. **1**), was also developed for detection in wheat and maize [36].

3.2. Limitations in the Traditional and Current *Fusarium* Detection Systems

Literature indicate great efforts in the past decades towards the development of highly specific and sensitive diagnostic methods, such as the use of antibody-based detections, biosensors and nucleotides-based evaluations to assess *Fusarium* pathogenicity and identification. During this period, antibody-based detection systems were widely reported, together with the nucleotide-based assays. However, increased reports were observed in the detection of *Fusarium* species using PCR assays than the ELISAs, which nearly doubled by 2010, and little focus placed on other techniques like the use of biosensors [37]. Reduction in the utilisation of other detection techniques may be attributed to the lack of high sensitivity, specificity and reproducibility relating to those applications. Furthermore, Saeger and Peteghem [36] reported the requirement of sophisticated equipment, cost issues and qualified personnel with specific expertise, as well as restrictions of the applications mainly to laboratories, as reasons behind these critical setbacks. The accuracy of results, consistency and reliability of detection systems are among some of the most paramount factors that need proper considerations for the optimisation and adoption of new detection tools.

Other issues include multiplexing, involving the screening of a large number of samples simultaneously or in a short period of time, and the lack of self-contained testing systems. However, all of the abovementioned challenges strongly impedes the opportunity for any developed detection system to meet market needs, and provide results rapidly with little or no instrumentation. In the use of enzyme immunoassays for example, these approaches often use passive absorption of proteins on hydrophobic surfaces that may cause undesirable changes in the detected proteins. These techniques rely on passive absorption for the immobilisation of antibodies or antigens on the microtiter plates. Although, ELISAs are very sensitive and require little sample preparation which makes them suitable for screening of a large number of fungal sample. These tools, including PCR are still marked by numerous inaccuracies or false positive identifications, including the inability to identify multiple infections on a single host caused by different pathogens [37].

4. ANTIBODY REACTIVITY AND SPECIFICITY INVOLVING *FUSARIUM* SPP.

Antibody reactivity and specificity has been widely studied with the aim of obtaining species specific antibodies that are important for the accurate and early detection of different *Fusarium* pathogens. Reports showed that one of the main limitations to raising antibodies with uniform reactivity against more complex organisms such as fungi is the fact that these organisms may have different morphological features and express different structural proteins at each stage of their life cycle. Moreover, some of the fungi, especially those found in the same genera demonstrate serological similarities, with some of the species expressing reactive antigens [38]. Such similarities then induce dissimilar antibody reactivities from different samples, especially when polyclonal antibodies are being used. Similarly, monoclonal antibodies that are vulnerable to slight changes of epitope, which may be any changes in the protein conformation may lead to drastic reduction in the antibody binding capacity. These observations imply that, raising antibodies that are specific to species level is also limited by the similarities in the cellular biochemistry of *Fusarium* species. These may include glycoprotein scaffolds, carbohydrate epitopes and cell wall materials of the fungus, especially chitin as described by Wycoff *et al.* [39], De-Bernardis *et al.* [40] and Hitchcock *et al.* [41].

Recent developments in genome sequencing, protein mass spectrometry and the use of two-dimensional electrophoresis (2D- electrophoresis) [isoelectric focusing (IEF) sodium-dodecyl polyacrylamide gel electrophoresis (SDS-PAGE)] have made it possible to identify unique and specific protein with the potential to produce highly specific antibodies [42, 43]. But the diversity highlights the complex nature of developing specific antibody-based detection system for different *Fusarium* spp. This directs research in the development of an ELISA protocol that uses a mixture of antibodies containing different specificity for the detection of different *Fusarium* spp. In summary, Table **2** gives examples of reports showing different antibody reactivity and specificity for the detection of different *Fusarium* species, which were found to cause diseases in both plants and animals. Additionally, it also highlights protein sources used for antigens preparation. However, during attempts to improve specificity and success in triggering efficient immune response, detection tools must be optimized followed by applications such as antifungal fusion with the antibodies for pathogen eradication. Antibody fusion holds an immense potential as an effective tool for the prevention of mycotoxin contamination affecting plants and animals, and the role of antibody-specificity in this resistance cannot be over-emphasised [44].

Table 2. Reports elucidating antibody reactivity and specificity in the detection of different *Fusarium* species.

Species for antisera production	Antigen Source /Preparation	Specificity-cross reactivity levels	References
F. oxysporum f. sp. cubense (Foc)	Heat-killed conidia	Species specific	Wong *et al.* [48]
F. solani f.sp. pisi	Purified cutinase	Genus specific and cross-reacted with *Botrytis cinerea* and *Puccinia* sp.	Coleman *et al.* [45]
F. moniliforme	Extracellular polysaccharide	Genus specific	Kwak *et al.* [49], Kwon *et al.* [50]
F. oxysporum	Homogenised mycelium	Genus specific	Notermans and Heuvekman [51],
F. verticilliodes	Exo-antigens and gel-purified 67 KDa protein	Genus specific and cross-reacted with *Aspergillus* sp., *Penicillium* sp.,	Omori *et al.* [52]
F. verticilliodes	Exo-antigens to spore suspension	Genus specific and cross-reacted with *Aspergillus* sp., *Penicillium* sp., *Cladosporium, Mucor, Alternaria* sp., *Cercospora* sp. and *Colletotrichum* sp.	Biazon *et al.* [53]
F. oxysporum f. sp. cucumerinum	Cell wall fragments	Genus specific	Kitagawa *et al.* [54]
F. culmorum	Soluble protein fractions	Genus specific and cross-reacted with *Pseudo cercosporella spp, Microdochium nivale, Drechslera sorokiniana* and *Ceratobasidium cereale*	Beyer *et al.* [55]
F. oxysporum	Whole mycelium	Genus specific	Arie *et al.* [56]
F. graminearum, F moniliforme	Mycelium soluble proteins	Genus specific and cross-reacted with *Monascus spp* and *Phoma exigua*	Iyer and Cousin [57]

Current data indicate that any approach targeting the commonly expressed proteins across different fungal isolates, particularly those directly involved in pathogenesis and other developmental stages could be used to overcome non-uniform antibody reactivity and specificity issues against *Fusarium* species [45]. Cloning of antibody binding domain as pure antibodies or fusion proteins indicated above has been used to increase binding specificity [46]. This procedure has managed to generate single-chain antibodies showing specificity and affinity identical to monoclonal antibodies. According to Hu *et al.* [44], the differences in antigen-binding specificities between antibodies which lies entirely within variable regions that are directly involved in the antibody interactions with antigen forms the basis of this approach. Any further groundbreaking advances in the production of recombinant antibodies will play a significant role in enhancing immune recognition and thus, reduce the negative impacts of *Fusarium* mycotoxin-producing pathogens in human, plants, and animals. Liu *et al.* [47] additionally, reported the attainment of 15-fold, 11-fold and 7-fold higher affinity recombinant single-chain antibodies subjected to direct evolution by error-prone PCR and DNA shuffling, generating a mutated library to confer resistance against *Fusarium graminearum* infections.

5. NATURE OF ANTIGEN VARIATION AND DISTRIBUTION

Fungi, like viruses, bacteria and protozoan organisms continue to respond proactively to the host's ever-evolving and multifaceted immune response through a number of phenotypical and genetic mechanisms. Animal and plant hosts respond to the ever-present threats of infections by fungi and other pathogenic microorganisms by evolving an elaborate system that prevents any physiological or clinical destruction by pathogens. Similarly, pathogenic microbes respond by evolving an equally elaborate system, through a process known as coevolution that results in the development of complex genetic systems that underlie antigenic variations [58]. Earlier reports dating back to 1910 elucidated how pathogens like trypanosomes use this coevolution of antigenic variation to evade host immunity and clearance [59]. This phenomenon of antigenic variation is focussed on the host-pathogen interface and at cell surfaces of the infecting organisms. Molecules that include proteins, polysaccharides and glycoproteins found on the pathogen's cell wall surfaces often mediate adhesion within specific niches and serve as virulence determinants [58, 60].

The expression of surface molecules responsible for adhesion or virulence involves a gene system that is intrinsically activated or silenced in order to achieve the sort of antigenic variation required by the evading pathogen. In fungi specifically, the variation process involves a couple of fungal growth characteristics which include phenotypic variations, which has so far proved to be critical for the organism's life cycle. *Fusarium* phenotypic variability is also classified into a number of these characteristics. This genera comprise among others the environmentally induced morphological transitions involving the whole fungal population and reversible phenotypic switching which appears restricted to a small fraction of the specific species within the *Fusarium* population [58]. However, these variations are all associated with the antigenic variation of surface antigens that occur in individual *Fusarium* species. Both the phenotypic and antigenic variations allows for a rapid adaptation of these pathogens to the constantly changing environment, facilitate evasion of an evolving host immune recognition and improve host's susceptibility to mycotoxigenesis.

The effectiveness of antigenic variations is species specific and remain unevenly distributed among certain individuals of fungal species belonging to a particular population. Generally, this mechanism is distributed in all microbial pathogens, assisting the pathogen to overcome innate and adaptive response to ensure successful establishment of infections [60, 61]. But, despite this and the diversity of pathogens and their ability to alter recognition by established immune responses, the mechanism often appears to be infrequent in some fungal species. In line with these observations, Mbofung *et al.* [62] reported low antigenic variation in the detection and aggressiveness of fungal isolates examined in soybean plants infected with *F. virguliforme*. Disease severity and plant growth varied significantly among *Fusarium* isolates, but with less correlation between antigenic variation and genetic variation of the fungal isolates. Such considerable variations in antigenic variation, species specificity and hosts' immune response among the isolates of the same species were also reported by Roy *et al.* [63] and Malvick and Bussey [64] in their studies involving *Fusarium* population under greenhouse and field conditions.

6. ANTIBODY RECOGNITION

All vertebral cells and their immune recognitions are responsible for the production of defense signalling molecules, detoxification enzymes, resistance and antifungal proteins as illustrated on Fig. (**3**) [65]. These living systems produce antibodies/immunoglobulins that use fragment antigen binding (Fab) variable region to recognise a pathogen's antigen. The six hypervariable loops found within the antibody's variable domain known as the complementarity determining regions (CDRs) are assumed to be responsible for antigen recognition. Antibodies use this regions to perform functions involving CDRs recognition, specific binding to antigens and use of constant domains to mediate effectors or activation of other components of the immune system to fight the pathogen [66]. Antibodies serve a critical role as a defense molecule synthesised in all vertebrate. In contrast, plants do not naturally produce antibodies, but plantibodies are in turn produced by plants that have been genetically engineered with animal DNA encoding a specific human antibody. Transgenic plants are used as bioreactors to synthesise plantibodies that are also used in neutralising mycotoxins.

As described by Oluwayela and Adebiyi [67], the plant's antibody produced by transgenic plants with antibody-coding genes of animal origin have shown to function in the same manner as mammalian antibodies. This report has also indicated that antibodies production in plants substantially costs lesser than animal antibody production. However, production takes longer duration, yields is relatively low, and it precludes the transfer of pathogens to the end product, which may serve as a setback when highly specific antibodies are required [68]. Plantibodies production was pioneered by Hiatt *et al.* [69] and Düring *et al.* [70], demonstrating that plants can also express and assemble functionally active antibodies with identical specificity and affinity to monoclonal antibodies produced by vertebrates [67]. Since then, numerous reports, including that of Hu *et al.* [44] and Edque *et al.* [71] provided further evidence which clearly illustrated that antibody DNA sequence or antibody fused with antifungal proteins can be used or produced in plants to eradicate *Fusarium* mycotoxigenesis.

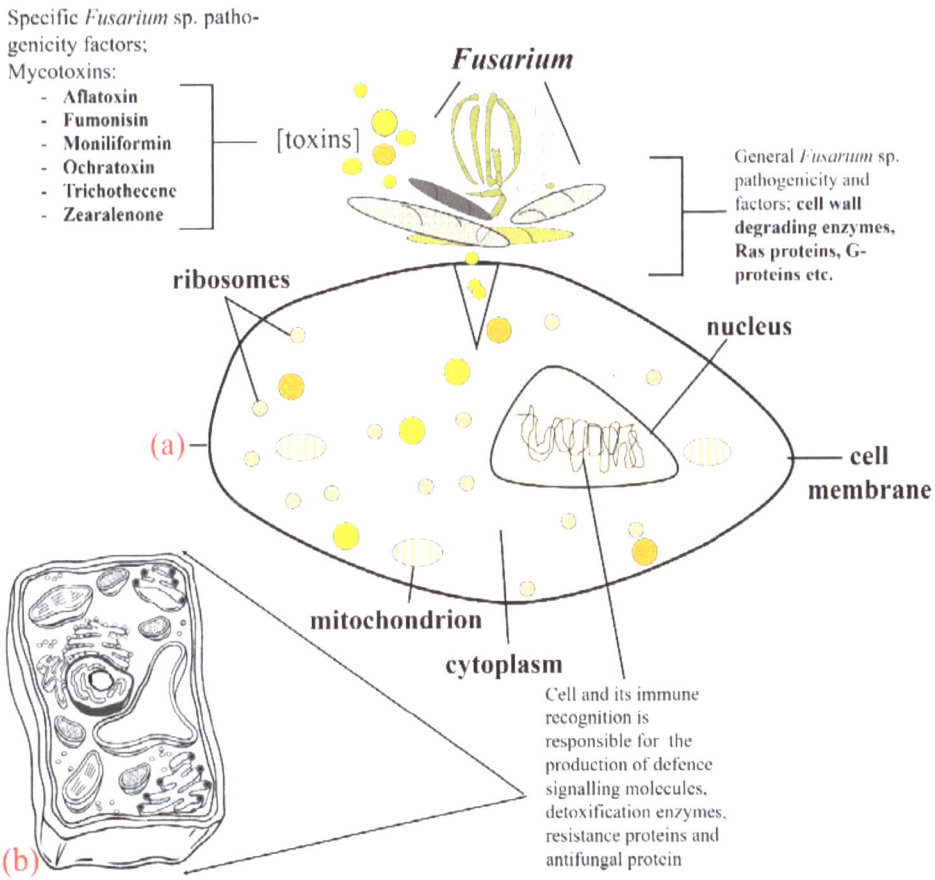

Fig. (3).   Illustration of *Fusarium* pathogenicity and host defense mechanism in animal and plant host. Mechanism of mycotoxin secretion and immune response between *Fusarium* and animal **(a)** plant **(b)** cells are as described by Ma *et al.* [65].

Antibodies that recognise and bind particular antigens with great affinity and specificity, especially for effective relief of unwanted pathogenic materials from humans or agriculturally important crops are required. With *Fusarium* like *F. oxysporum f.* sp. *lycopersici* (fol) being able to manipulate plant hosts by employing effector proteins Avr2 to retain its full virulence [72, 73]. Both the antibody fusion and plantibody engineering needs to be optimised and intensified for the prevention of mycotoxin contaminations, particularly those that are produced by fungal agents. The need to reduce pathogen infection in agricultural fields by expressing endogenous resistance genes or developing resistant germplasm is a key step in modern biotechnology. This is important because the ability of *Fusarium* spp. to infect and produce mycotoxins in agronomic crops pass these toxins to animal and human after ingestion, subsequently leading to a spread of opportunistic fungi complexes of *F. fujikuroi*, *F. oxysporum* and *F. solani* [65].

6.1. Epitope Structure and Functionality

The different molecular components of specific pathogens trigger the antibody-antigen protein interactions involving immune recognition receptors. It is the host's innate immune system function and critical role to use combinatorial receptor recognition patterns to decode the nature of mycotoxigenic materials and tailor the ensuing early innate response, followed by antigen specific response to combat mycotoxigenesis [74]. The innate immune system constitutes the first line of defense against invading pathogen and uses diverse pattern recognition receptors for the detection of pathogenic antigens (see Fig. **3** for a brief overview). Antigens contain epitopes whose analysis indicates that, it serves as a structural topographical area of the protein surface contacted by antibodies or what is referred to as antibody-antigen interface [75]. Epitopes serve as antigen determinants for immune system recognition. The antibody region which recognises the epitope in order to precisely fit and bind to it is the paratope [66, 75]. Both paratope and epitope should be thoroughly analysed to inform the process of eliciting functional and therapeutic antibodies. Steve and Lindpaintner [76] found distribution of antibody contact residue in three prominent contact residue regions, particularly on light (L) and heavy (H) chains that overlaps, but without coinciding antibody CDRs. These insights, together with other findings in this study gave valuable information on the size of a complete structural epitope (with a range of 20 to 400 amino acids) or paratope that would be required for understanding the role of epitope structure and function, even early during antibody-antigen conformations. Such information is necessary for developing functional and therapeutic antibodies against microbial pathogens.

6.2. Paratope Binding

Upon invasion by potential harmful pathogens, pattern recognition receptors must alert the body to signal for an immune response. A paratope, serving as an antigen-binding site is an integral part of the antibody which recognises this and binds to the antigen that is signalling the immune response in dealing with the threat. Paratope are produced by the CDRs of L and H chains, estimated to be around 10^4 L and 10^8 H amino acid chains long, respectively [76]. According to Spillner *et al.* [77], paratopes can also be used for the selection of antibodies in the detection of unknown protein antigens using a peptide library. Such use of paratope can serve as a useful alternative for the identification of proteins of interest when protein levels are extremely low or when there is instability under the conditions of purification. Generally, when the antibody paratope matches the antigen epitope, a reaction to suppress the antigen is then initiated. Furthermore, a cell uses somatic hypermutation if the antibody paratope do not match the antigen epitope [78].

Somatic hypermutation introduces nucleotide alteration at the variable region of both the L and H chains to generate high-affinity antibodies for effective immune response [79]. Evaluation with ELISA, furthermore, indicated that monoclonal antibodies paratopes composed of fluctuation-regulated affinity proteins (FLAPs) from CDR loops and a protein scaffold was developed to improve paratope binding for targeted protein binding [80]. The small FLAPs proteins can be expressed in host organism as fusion proteins and show high specificity binding affinity to certain disease causing antigens [81].

6.3. Maturation of Antibody Specificity

Since millions of unique epitopes are potentially generated by pathogens, vertebrates use somatic hypermutation to code for every possible antigenic-binding site. Lymphocytes, that constitutes a major part of evolved humoral immune response in vertebrates [82], expresses a number of unique genes responsible for mutating these antigen-binding sites, causing rapid responses to the induced antigen. Only antigen-binding sites that recognise the antigen with high affinity produce antibodies, meanwhile those with low affinity get eliminated. According to Manivel *et al.* [83] this maturation process of antibody response reflects a procedure in which paratope exhibit conformational flexibility in order to effectively facilitate the antigen-binding to a variety of antigens, whose relative rigidity accounts for the high affinity and fidelity of binding. Furthermore, it was reported by Shehata *et al.* [84] that, although, somatic hypermutation is associated with increased antibody specificity, the process also diminishes conformational stability. This outcome consequently implies that the numerous coding of antigen-binding sites compromises the efforts for antibody discovery and engineering. However, the spontaneous hypermutation associated DNA breaks and error-prone repair are provided by DNA polymerases to stabilise the mutation. Several studies providing evidence with regard to the role of these DNA polymerases in generating the required mutations were also reported by Zeng *et al.* [85] and Rogozin *et al.* [86].

6.4. Contrast Binding Affinity and Specificity

Factors leading to high binding affinity and specificity are responsible for the strong antibody reaction against the targeted molecules. Understanding the functional characteristics of naturally acquired antibodies against any mycotoxigenesis like bacterial and fungal antigens, is critical in determining the protective function of antibodies and their subsequent engineering. However, the product of high affinity antibodies is said to be an indication of a successful priming by an antigen or an indication that immune cells have undergone affinity maturation [87]. Priming is part of the filtering mechanisms, which enrich the immune cells receptors for favourable biophysical properties. Apart from biophysical properties of host tissues involved and the general antibody pharmacology, antibody-antigen interactions affecting binding, specificity and efficacy of antigen-binding sites serve as the major contributing factors to antibody binding affinity and specificity [88]. The affinity, and the overall strength of binding between an antibody and an antigen known as avidity, of the binding interactions plays a key role in this relationship. Reverberi and Reverberi [89] reported other several factors that include low ionic strength, highest serum/cell ratio, amount of immune cells and cell homozygosity which influence antibody affinity and specificity. Affinity of antibodies also depends on the genetic system, and immune system response maturity, in addition to antigen-antibody concentration. This study also reported that the weak bonds interaction with steric complementarity between molecules are required for stable binding.

7. DIVERSITY OF *FUSARIUM* ANTIBODIES AND THEIR SPECIFICITY

The specificity of antibodies is essentially important for rapid and accurate detection of microbial pathogens such as *Fusarium* species. But the development of the best antibodies that specifically and effectively binds to the targeted antigens can be very challenging. Experimental monoclonal and polyclonal antibodies derived from various clonal populations' presents varying specificities, particularly due to epitope structures and functions [90]. Also, their reactivity varies according to species, and are subject to diversity and character change during the production period. However, monoclonal antibody diversity is limited as these antibodies are derived from a single epitope. Therefore, careful consideration should be given to antigen design or isolation and preparation when raising antibodies in order to improve efficiency in detection systems. Various antigen isolation and preparation procedures for *Fusarium* pathogens have been used both in agricultural and food sector to improve the specificity of antibodies produced for immunological assays.

The assays involve the use of partially purified cell walls, exo-proteins, glycoproteins and specific proteins produced by fungal pathogens during the infection process, such as cutinase and gel-purified proteins [52, 56, 91]. The isolation of antigenic molecules (*e.g.* mycelium-soluble antigens) play a critical role in determining the diversity and specificity of antibodies used in immunological detection systems. Furthermore, the antibodies developed using these mycelium-soluble antigens like those obtained from *F. culmorum* (Wm. G. Sm.) Sacc, *F. graminearum* (Schwein.) Petch., *Fusarium oxysporum* Schlecht and *F. avenaceum* (Fr.) Sacc often show specificity up to a genus level [54 - 56]. Brill *et al.* [38] also reported that polyclonal antibodies raised to culture filtrate were more specific, but with less cross-reactivity than antibodies raised against mycelia extracts. Abovementioned studies indicate a requirement for proper antigen isolation and analysis of antigenic proteins for proper detection of microbial pathogens, including *Fusarium* species.

8. CROSS-REACTIVITY IN MONOCLONAL AND POLYCLONAL ANTIBODIES AGAINST *FUSARIUM* ANTIGENS

Ward *et al.* [92] identified polyclonal and monoclonal as two routes for the production of antibodies. Polyclonal antibodies are a mixture of heterogeneous clones which are usually produced by different B cell lines in the body. They can recognize and bind to many epitopes of a single antigen. Whereas monoclonal antibodies are identical B cells which are clones from a single parent cell. Therefore, monoclonal antibodies have a monogeneous affinity and only bind the same epitope of an antigen. Experimental animals are injected with a specific antigen to elicit primary immune response, and the secondary as well as tertiary injections are administered only to produce higher titers of antibodies [93]. Researchers use purified antibodies for antigen detection through a number of immunological techniques including immunoblotting, ELISA, immunohisto-chemistry and immunoprecipitation. Such detections, especially for *Fusarium* species using either monoclonal or polyclonal antibodies and their specificity are outlined in Table **3**.

Table 3. The use of monoclonal and polyclonal antibodies for detection of *Fusarium* species and their specificities.

Antibody Used	Immunological Technique	Comments on Specificity	References
Monoclonal anti-*F. oxysporum*	Dot-immunobinding	Specific to the genus	Arie *et al.* [56]
Polyclonal anti-*F. solani*	Immunohistochemistry	Increased reactivity with *F. solani* with minimal cross-reactivity to *Aspergillus* spp.	Kaufman *et al.* [100]
Polyclonal anti-*F. oxysporum*	Indirect ELISA and western blotting	At high antigen dilution species specific reaction obtained	Ray *et al.* [101]
Polyclonal anti-*F. oxysporum*	ELISA	Species specific antibodies but could not differentiate individual strains of *F. oxysporum*	Eparvier and Alabouvette [102]
Polyclonal anti- *F. oxysporum f.sp. narcissi*	ELISA	Specific and detected all nine isolates of *F. oxysporum f.sp. narcissi* with minimal cross-reactivity with other *Fusarium* spp.	Olal *et al.* [94], Linfield [103]
Polyconal anti-*F. verticillioides*	Indirect competitive ELISA	Genus specific with 4.5 – 14.7% cross-reactivity with *Aspergillus* spp., *Penicillium* spp. and *Colletotrichum* spp.	Olal *et al.* [94], Meirelles *et al.* [104]
Monoclonal anti-T-2 toxin from *F. sporotrichioides*	Indirect ELISA	Genus specific	Nagayama *et al.* [105]
Fusarium specific monoclonal antibody ED7	ELISA	Detected the targeted *F. solani* spp. and *F. oxysporum*	Al-Maqtoofi and Thornton [95]
Anti-zearalenone monoclonal antibody	ELISA	Only detected zearalenone producing *Fusarium* spp. (*F. graminearum*)	Ward *et al.* [92], Pie *et al.* [106]
Anti-fumonisins monoclonal antibody	ELISA	Specific for fumonisins producing *Fusarium* spp. (*F. moniliforme*)	Azcona-Olivera *et al.* [107]
Anti-endopolygalacturonase monoclonal antibody	Western blotting	Only detected endopolygalacturonase producing *Fusarium* spp. (*F. moniliforme* strains)	Al-Maqtoofi and Thornton [95], Daroda *et al.* [108]

Numerous studies investigating pathogenic organisms (viruses, bacteria, spirochete, protozoans and fungi) showed the role of antibody cross-reactivity among others in triggering tissue-specific autoimmunity via antibody amino acid sequence homology [93]. Moreover, cross-reactivity also intensified the difficulty in finding a suitable and specific antigen-antibody immune reactivity. Furthermore, with affinity mutations compromising conformational binding stability during antigen-antibody interactions. In contrast, Olal *et al.* [94] reported the use of polyclonal antibodies packaged into an immunodiagnostic procedure integrated with the common detection techniques for improved detection of *F. xylarioides* Steyaert. This fungal species remain one of the most virulent species that is highly destructive in Robusta coffee, especially in the planting fields of Uganda. This study indicated that polyclonal antibodies raised against *F. xylarioides* endoantigens of 24kDa was promising for the rapid, sensitive and accurate detection of the pathogen since there was no cross-reactivity that was observed.

9. DEVELOPMENT OF OPTIMISED ELISA FOR RAPID DETECTION OF *FUSARIUM* SPECIES

ELISAs are ideal methods that have been used for preliminary screening, detection, and quantification of *Fusaium* in foods and different crops. *Fusarium* pathogens currently detected using this tool include *F. poae, F. graminearum, F. sporotrichides, F. oxysporum, F. culmorum, F. graminearum, F. solani* and *F. avenaceum* [49, 50, 54 - 56, 95]. Optimised ELISA formats such as indirect ELISA or double antibody sandwich (DAS) ELISA can allow quantification of fungi in plant tissue and complex mixtures such as soil and plant extracts. Specific antibodies in the assay preferentially bind to the target antigens of interest in the mixture, and unbound materials are subsequently washed off. This method presents a better advantage over PCR as it allows for the detection of multiple samples in a single test simultaneously, rapidly, reliably and economically [54]. In the indirect ELISA, antigen is first bound to a solid support and the subsequent addition of a substrate solution gives a colour change that can be measured to quantify the amount of antigen present in a sample. Meanwhile, in DAS-ELISA (double-antibody-sandwich ELISA), antibody (capture antibody) is immobilized on a polystyrene microtitre plate solid support [54, 96].

Antigen is then added which complexes with the antibody and a detector antibody (secondary antibody) bound to biotin, followed by avidin complexed with an enzyme [horseradish peroxidase or alkaline phosphatise (HRPO) or alkaline phosphatase (AP)]. Substrate solution is added which then reacts with the enzyme bound to the detector antibody or avidin. This produces a coloured product that indicates the amount of antigen present in the samples [96]. Brill *et al.* [97] showed that between the indirect and DAS-ELISA the latter provides increased levels of specificity in detection of *Fusarium* species. More than a 100-fold increase in sensitivity was observed with sandwich ELISA and consistent results between different polyclonal antibodies were also obtained [54]. Other factors to consider when optimising ELISA include varying incubation periods for antibodies, changing the antibody concentration, improving antigen isolation method and the correct preparation and use of buffers [98, 99]. Important to note, optimized ELISA should be selected based on the availability of materials required for a quicker detection in the control of *Fusarium* and other pathogens spread in crop farms, humans and animals.

FINAL CONSIDERATIONS OR CONCLUSIONS

A demand for fungal contaminants free agricultural marker, emerging antifungal drug resistance and the fatal health effects of fungal infections in humans and animals drive the need for the development of rapid, sensitive, reliable and accurate detection systems. Marker pressure, rather than regulatory oversight will effect changes on demand for improved infection assessments, enhanced decontamination tools and the production and validation of novel antibodies for application in various industries. Researchers should continue bolstering efforts made on antibody selection, use application-specific methods and detection in prevention of mycotoxigenesis spreads [108]. However, extreme specificity of potential antibodies for the intended target molecules is critically important. Specificity is crucial for antigen-binding site, pathogen elimination from the hosts' system and subsequently in applications for drug discovery as well as for toxicity profiles [109].

Antibody cross-reactivity, specificity and binding affinity of interactions with antigenic molecules affect how well molecules on both sides serve their own functions. Researchers are therefore, led to investigate the principles that governs cross-reactivity and specificity, and the relationship between them using various tools such as optimised ELISA. Advances also indicate that other methods have been improved on the basic principles of ELISA, allowing on-site detections rather than carrying samples to the laboratory. Such methods can be used by farmers in the greenhouses and in the field, nurses in the health centres as well as staff in the food storage facilities. Examples include detection systems based on the dip-stick principle for the detection of human and some plant diseases that are currently well established [110]. ELISA dip-stick require little training and are based on an easy to interpret colour development without the need for equipment such as spectrophotometers and micro-titre plates. Efficacy of this method has been reported in the detection of *Fusarium* mycotoxins (T-2 toxin) and *Phytophthora cinnamomi* Rands. zoospores in soils [96, 111]. Increasing developments of *Fusarium* mycotoxigenesis and various other pathogens create the need to enhance antibody specificity, homogeneity and the production of valuable antibodies, including the demand to develop optimised DAS-ELISA for quick screening of large quantities of samples as alluded by Olal *et al.* [94].

ABBREVIATIONS

CDR Complementarity determining region

DAS-ELISA Double antibody sandwich enzyme-linked immunosorbent assay

DNA Deoxyribonucleic acid

EIA Enzyme-immunosorbent assay

ELISA Enzyme-linked immunosorbent assay

FAO Food and Agriculture Organisation of the United Nations

FLAPs Fluctuation-regulated affinity proteins

FSSC*Fusarium solani* species complex

H-Chains Heavy-chains

IEF Isoelectric focussing

L-Chains Light-chains

PAGE Polyacrylamide gel electrophoresis

PCR Polymerase chain reaction

PDA Potato dextrose agar

SDS Sodium dodecyl sulphate

2-D-SDS-PAGE 2-dimensional-sodium dodecyl sulphate polyacrylamide gel

electrophoresis

spp. Species (plural)

CONSENT FOR PUBLICATION

Not applicable.

CONFLICT OF INTEREST

The author declares no conflict of interest, financial or otherwise.

ACKNOWLEDGEMENTS

Authors would like to thank the Department of Biodiversity, Department of Microbiology, Biochemistry and Biotechnology, and the Department of Research Development and Administration at the University of Limpopo for their continued support.

REFERENCES

[1]   Morretti AN. Taxonomy of *Fusarium* genus: A continuous fight between lumbers and splitters. Prog Nat Sci 2009; 117: 7-13.
[http://dx.doi.org/10.2298/ZMSPN0917007M]

[2]   Pessôa MG, Paulino BN, Mano MCR, Neri-Numa IA, Molina G, Pastore GM. *Fusarium* species-a promising tool box for industrial biotechnology. Appl Microbiol Biotechnol 2017; 101(9): 3493-511.
[http://dx.doi.org/10.1007/s00253-017-8255-z] [PMID: 28343243]

[3]   Solecka J, Zajko J, Postek M, Rajnisz A. Biologically active secondary metabolites from Actinomycetes. Cent Eur J Biol 2012; 7(3): 373-90.
[http://dx.doi.org/10.2478/s11535-012-0036-1]

[4]   Zhang P, Yuan XL, Du YM, *et al.* Angularly prenylated indole alkaloids with antimicrobial and insecticidal activities from an endophytic fungus *Fusarium* sambucinum TE-6L. J Agric Food Chem 2019; 67(43): 11994-2001.
[http://dx.doi.org/10.1021/acs.jafc.9b05827] [PMID: 31618578]

[5]   Hafizi R, Salleh B, Latiffah Z. Morphological and molecular characterization of Fusarium. solani and F. oxysporum associated with crown disease of oil palm. Braz J Microbiol 2014; 44(3): 959-68.
[http://dx.doi.org/10.1590/S1517-83822013000300047] [PMID: 24516465]

[6]   Lievens B, Rep M, Thomma BPJ. Recent developments in the molecular discrimination of formae speciales of *Fusarium oxysporum.* Pest Manag Sci 2008; 64(8): 781-8.
[http://dx.doi.org/10.1002/ps.1564] [PMID: 18335459]

[7]   Nelson PE, Dignani MC, Anaissie EJ. Taxonomy, biology, and clinical aspects of *Fusarium* species. Clin Microbiol Rev 1994; 7(4): 479-504.
[http://dx.doi.org/10.1128/CMR.7.4.479] [PMID: 7834602]

[8]   Sewram V, Mshicileli N, Shephard GS, *et al.* Production of fumonisin B and C analogues by several *fusarium* species. J Agric Food Chem 2005; 53(12): 4861-6.
[http://dx.doi.org/10.1021/jf050307n] [PMID: 15941327]

[9]   Tamura M, Mochizuki N, Nagatomi Y, Toriba A, Hayakawa K. Characterization of fumonisin A-series by high-resolution liquid chromatography-orbitrap mass spectrometry. Toxins (Basel) 2014; 6(8): 2580-93.
[http://dx.doi.org/10.3390/toxins6082580] [PMID: 25153258]

[10]   Rheeder JP, Marasas WF, Vismer HF. Production of fumonisin analogs by *Fusarium* species. Appl Environ Microbiol 2002; 68(5): 2101-5.
[http://dx.doi.org/10.1128/AEM.68.5.2101-2105.2002] [PMID: 11976077]

[11]   Tian Y, Tan Y, Yan Z, *et al.* Antagonistic and detoxification potentials of Trichoderma isolates for control of zearalenone (ZEN) producing *Fusarium graminearum.* Front Microbiol 2018; 8: 2710.
[http://dx.doi.org/10.3389/fmicb.2017.02710]

[12]   Sumikova T, Remesova J, Leisova L, Kucera L, Chrpova J, Sip V. AFLP genotyping improves the level of discrimination between the *Fusarium* species responsible for head blight in wheat. Cereal Res Commun 2010; 38(4): 533-40.
[http://dx.doi.org/10.1556/CRC.38.2010.4.10]

[13]   Ellestad GA, Evans RH Jr, Kunstmann MP. Some new terpenoid metabolites from an unidentified *fusarium* species. Tetrahedron 1969; 25(6): 1323-34.
[http://dx.doi.org/10.1016/S0040-4020(01)82703-4] [PMID: 5346204]

[14]   Díaz-Sánchez V, Avalos J, Limón MC. Identification and regulation of fusA, the polyketide synthase gene responsible for fusarin production in *Fusarium fujikuroi.* Appl Environ Microbiol 2012; 78(20): 7258-66.
[http://dx.doi.org/10.1128/AEM.01552-12] [PMID: 22865073]

[15]   Jia LJ, Tang HY, Wang WQ, *et al.* A linear nonribosomal octapeptide from *Fusarium graminearum* facilitates cell-to-cell invasion of wheat. Nat Commun 2019; 10(1): 922.
[http://dx.doi.org/10.1038/s41467-019-08726-9] [PMID: 30804501]

[16]   Villani A, Proctor RH, Kim HS, *et al.* Variation in secondary metabolite production potential in the *Fusarium* incarnatum-equiseti species complex revealed by comparative analysis of 13 genomes. BMC Genomics 2019; 20(1): 314.
[http://dx.doi.org/10.1186/s12864-019-5567-7] [PMID: 31014248]

[17]   Lauren DR, Sayer ST, di Menna ME. Trichothecene production by *Fusarium* species isolated from grain and pasture throughout New Zealand. Mycopathologia 1992; 120: 167-76.
[http://dx.doi.org/10.1007/BF00436395]

[18]   Fotso J, Leslie JF, Smith JS. Production of beauvericin, moniliformin, fusaproliferin, and fumonisins b(1), b(2), and b(3) by fifteen ex-type strains of *fusarium* species. Appl Environ Microbiol 2002; 68(10): 5195-7.
[http://dx.doi.org/10.1128/AEM.68.10.5195-5197.2002] [PMID: 12324376]

[19]   Liuzzi VC, Mirabelli V, Cimmarusti MT, *et al.* Enniatin and beauvericin biosynthesis in *Fusarium* species: Production profiles and structural determinant prediction. Toxins (Basel) 2017; 9(2): 45-62.
[http://dx.doi.org/10.3390/toxins9020045] [PMID: 28125067]

[20]   Tortorano AM, Esposto MC, Prigitano A, *et al.* Cross-reactivity of *Fusarium* spp. in the *Aspergillus* Galactomannan enzyme-linked immunosorbent assay. J Clin Microbiol 2012; 50(3): 1051-3.
[http://dx.doi.org/10.1128/JCM.05946-11] [PMID: 22205818]

[21]   Sharon A, Shlezinger N. Fungi infecting plants and animals: killers, non-killers, and cell death. PLoS Pathog 2013; 9(8): e1003517.
[http://dx.doi.org/10.1371/journal.ppat.1003517] [PMID: 24009499]

[22]   Matic S, Gilardi G, Gullino ML, Garibaldi A. Evidence of an expanded host range of *Fusarium oxysporum f. sp. chysanthemi.* J Plant Pathol 2018; 100: 97-104.
[http://dx.doi.org/10.1007/s42161-018-0021-2]

[23]   Muhammed M, Anagnostou T, Desalermos A, *et al.* Fusarium infection: report of 26 cases and review of 97 cases from the literature. Medicine (Baltimore) 2013; 92(6): 305-16.
[http://dx.doi.org/10.1097/MD.0000000000000008] [PMID: 24145697]

[24]   O'Donnell K, Rooney AP, Proctor RH, *et al.* Phylogenetic analyses of RPB1 and RPB2 support a middle Cretaceous origin for a clade comprising all agriculturally and medically important *fusaria.* Fungal Genet Biol 2013; 52: 20-31.
[http://dx.doi.org/10.1016/j.fgb.2012.12.004] [PMID: 23357352]

[25]   Liew WPP, Mohd-Redzwan S. Mycotoxin: Its impact on gut health and microbiota. Front Cell Infect Microbiol 2018; 8: 60.
[http://dx.doi.org/10.3389/fcimb.2018.00060] [PMID: 29535978]

[26]   Coleman JJ. The *Fusarium solani* species complex: ubiquitous pathogens of agricultural importance. Mol Plant Pathol 2016; 17(2): 146-58.
[http://dx.doi.org/10.1111/mpp.12289] [PMID: 26531837]

[27]   Safarieskandari S, Chatterton S, Hall LM. Pathogenicity and host range of *Fusarium* species associated with pea root rot in Alberta, Canada. Can J Plant Pathol 2020; 43(1)
[http://dx.doi.org/10.1080/07060661.2020.1730442]

[28]   Murphy J. Fusarium outbreak continues to mystify. Rev Optom 2006; 145: 5.

[29]   Wise K, Allen T, Chilvers M, *et al.* Corn disease management- Ear rots. Crop Prot Net 2016; pp. 1-12.

[30]   Hoerr FJ. Mycotoxicoses in poultry- MSD and the MSD veterinary manual. USA: Merck and Co Inc 2019.

[31]   Hoerr FJ, Carlton WW, Tuite J, Vesonder RF, Rohwedder WK, Szigeti G. Experimental trichothecene

mycotoxicosis produced in broiler chickens by *Fusarium sporotrichiella var. sporotrichioides.* Avian Pathol 1982; 11(3): 385-405.
[http://dx.doi.org/10.1080/03079458208436113] [PMID: 18770204]

[32]   Lin Z, Xu S, Que Y, *et al.* Species-specific detection and identification of *fusarium* species complex, the causal agent of sugarcane pokkah boeng in China. PLoS One 2014; 9(8): e104195.
[http://dx.doi.org/10.1371/journal.pone.0104195] [PMID: 25141192]

[33]   Gourama H, Bullerman LB. Detection of molds in food and feeds: Potential rapid and selective methods. J Food Prot 1995; 58(12): 1389-94.
[http://dx.doi.org/10.4315/0362-028X-58.12.1389] [PMID: 31159040]

[34]   Thompson RS, Aveling TAS, Prieto RB. A new semi-selective medium for *Fusarium graminearum, F. proliferatum, F. subglutinans* and *F. verticillioides* in maize seed. S Afr J Bot 2013; 84: 94-101.
[http://dx.doi.org/10.1016/j.sajb.2012.10.003]

[35]   Schweigkofler W, O'Donnell K, Garbelotto M. Detection and quantification of airborne conidia of *Fusarium circinatum*, the causal agent of pine pitch canker, from two California sites by using a real-time PCR approach combined with a simple spore trapping method. Appl Environ Microbiol 2004; 70(6): 3512-20.
[http://dx.doi.org/10.1128/AEM.70.6.3512-3520.2004] [PMID: 15184151]

[36]   Saeger SD, Peteghem C. Dipstick enzyme immunoassay to detect Fusarium T-2 toxin in wheat. App Environ Microbiol 1996; 62(6).

[37]   Skottrup PD, Nicolaisen M, Justesen AF. Towards on-site pathogen detection using antibody-based sensors. Biosens Bioelectron 2008; 24(3): 339-48.
[http://dx.doi.org/10.1016/j.bios.2008.06.045] [PMID: 18675543]

[38]   de Biazio GR, Leite GGS, Tessmann DJ, Barbosa-Tessmann IP. A new PCR approach for the identification of *Fusarium* graminearum. Braz J Microbiol 2008; 39(3): 554-60.
[http://dx.doi.org/10.1590/S1517-83822008000300028] [PMID: 24031265]

[39]   Wycoff KL, Jellison J, Ayers AR. Monoclonal antibodies to glycoprotein antigens of a fungal plant pathogen, *Phytophthora megasperma f.sp. glycinea.* Plant Physiol 1987; 85(2): 508-15.
[http://dx.doi.org/10.1104/pp.85.2.508] [PMID: 16665728]

[40]   De Bernardis F, Molinari A, Boccanera M, *et al.* Modulation of cell surface-associated mannoprotein antigen expression in experimental *candidal vaginitis.* Infect Immun 1994; 62(2): 509-19.
[http://dx.doi.org/10.1128/iai.62.2.509-519.1994] [PMID: 7507895]

[41]   Hitchcock P, Gray TRG, Frankland JC. Production of a monoclonal antibody specific to *Mycena galopus* mycelium. Mycol Res 1997; 101: 1051-9.
[http://dx.doi.org/10.1017/S0953756297003602]

[42]   Arruda MP, Brown P, Brown-Guedira G, *et al.* Genome-wide association mapping of *Fusarium* head blight resistance in wheat using genotyping-by-sequencing. Plant Genome 2016; 9(1): 22-33.
[http://dx.doi.org/10.3835/plantgenome2015.04.0028] [PMID: 27898754]

[43]   Manikandan R, Harish S, Karthikeyan G, Raguchander T. Differentially expressed proteins responsible for virulence on tomato plants. Front Microbiol 2018; 9: 420.
[http://dx.doi.org/10.3389/fmicb.2018.00420] [PMID: 29559969]

[44]   Hu ZQ, Li HP, Zhang JB, Glinka E, Liao YC. Antibody-mediated prevention of *Fusarium* mycotoxins in the field. Int J Mol Sci 2008; 9(10): 1915-26.
[http://dx.doi.org/10.3390/ijms9101915] [PMID: 19325726]

[45]   Coleman JOD, Hiscock SJ, Dewey FM. Monoclonal antibodies to purified cutinase from *Fusarium solani f.sp. pisi.* Physiol Mol Plant Pathol 1993; 43: 391-401.
[http://dx.doi.org/10.1006/pmpp.1993.1067]

[46]   Liao YC, Li HP, Zhao CS, Yao MJ, Zhang JB, Liu JL. Plantibodies: a novel strategy to create pathogen-resistant plants. Biotechnol Genet Eng Rev 2006; 23: 253-71.

[http://dx.doi.org/10.1080/02648725.2006.10648087] [PMID: 22530511]

[47]   Liu JL, Hu ZQ, Xing S, *et al.* Attainment of 15-fold higher affinity of a *Fusarium*-specific single-chain antibody by directed molecular evolution coupled to phage display. Mol Biotechnol 2012; 52(2): 111-22.
[http://dx.doi.org/10.1007/s12033-011-9478-3] [PMID: 22161226]

[48]   Wong WC, White M, Wright IG. Production of monoclonal antibodies to *Fusarium f.sp. cubense* race 4. Lett Appl Microbiol 1988; 6: 14-20.
[http://dx.doi.org/10.1111/j.1472-765X.1988.tb01210.x]

[49]   Kwak B, Kwon B, Kweon C, Shon D. Detection of *Fusarium* species by Enzyme-Linked Immunosorbent Assay using monoclonal antibody. J Microbiol Biotechnol 2003; 13: 794-9.

[50]   Kwon KBB, Kweon C, Shon D. Detection of *Aspergillus, Penicillium*, and *Fusarium* species by sandwich Enzyme-Linked Immunosorbent Assay using mixed monoclonal antibodies. J Microbiol Biotechnol 2004; 14: 385-9.

[51]   Notermans S, Heuvelman CJ. Immunological detection of moulds in food by using the enzyme-linked immunosorbent assay (ELISA); preparation of antigens. Int J Food Microbiol 1985; 2: 247-58.
[http://dx.doi.org/10.1016/0168-1605(85)90015-7]

[52]   Omori AM, Ono EYS, Hirozawa MT, *et al.* Development of enzyme-linked immunosorbent assay to detect *Fusarium vericillioides* in poultry feed samples. Toxins (Basel) 2019; 11: 1-13.
[http://dx.doi.org/10.3390/toxins11010048]

[53]   Biazon L, Meirelle PG, Ono MA, *et al.* Development of polyclonal antibodies against *Fusarium verticillioides* exo-antigens. Food Agric Immunol 2006; 17: 69-77.
[http://dx.doi.org/10.1080/09540100600621458]

[54]   Kitagawa T, Sakamoto Y, Furumi K, Ogura H. Novel enzyme immunoassays for specific detection of *Fusarium oxysporum f. sp. cucumerinum* and for general detection of various *Fusarium* species. Phytopathology 1989; 79: 162-5.
[http://dx.doi.org/10.1094/Phyto-79-162]

[55]   Beyer W, Hoxter H, Medaner T, Sander E, Geiger HH. 1993.https://www.jstor.org/stable/43386175

[56]   Arie T, Hayashi Y, Nagatani KYA, Furuya M, Yamagichi I. Detection of *Fusarium spp.* in plants with monoclonal antibody. Ann Phytopathol Soc Jpn 1995; 61: 311-7.
[http://dx.doi.org/10.3186/jjphytopath.61.311]

[57]   Iyer MS, Cousin MA. Immunological detection of Fusarium species in cornmeal. J Food Prot 2003; 66(3): 451-6.
[http://dx.doi.org/10.4315/0362-028X-66.3.451] [PMID: 12636300]

[58]   Deitsch KW, Lukehart SA, Stringer JR. Common strategies for antigenic variation by bacterial, fungal and protozoan pathogens. Nat Rev Microbiol 2009; 7(7): 493-503.
[http://dx.doi.org/10.1038/nrmicro2145] [PMID: 19503065]

[59]   Palmer GH, Bankhead T, Seifert HS. Antigenic variation in bacterial pathogens. Microbiol Spectr 2016; 4(1): 10.
[http://dx.doi.org/10.1128/microbiolspec.VMBF-0005-2015] [PMID: 26999387]

[60]   Stow JL, Condon ND. The cell surface environment for pathogen recognition and entry. Clin Transl Immunology 2016; 5(4): e71.
[http://dx.doi.org/10.1038/cti.2016.15] [PMID: 27195114]

[61]   Ernst JD. Antigenic variation and immune escape in the MTBC. Adv Exp Med Biol 2017; 1019: 171-90.
[http://dx.doi.org/10.1007/978-3-319-64371-7_9] [PMID: 29116635]

[62]   Mbofung GYC, Harrington TC, Steimel JT, Nari SS, Yang XB, Leandro LF. Genetic structure and variation in aggressiveness in *Fusarium virguliforme* in the Midwest United States. Can J Plant Pathol

2012; 34(1): 83-97.
[http://dx.doi.org/10.1080/07060661.2012.664564]

[63]   Roy KW, Hershman DE, Rupe JC, Abney TS. Sudden death syndrome of soybean. Plant Dis 1997; 81(10): 1100-11.
[http://dx.doi.org/10.1094/PDIS.1997.81.10.1100] [PMID: 30861702]

[64]   Malvick DK, Bussey KE. Comparative analysis and characterisation of the soybean sudden death syndrome pathogen Fusarium virguliforme in the northern United States. Can J Plant Pathol 2008; 30: 467-76.
[http://dx.doi.org/10.1080/07060660809507544]

[65]   Ma LJ, Geiser DM, Proctor RH, *et al. Fusarium* pathogenomics. Annu Rev Microbiol 2013; 67: 399-416.
[http://dx.doi.org/10.1146/annurev-micro-092412-155650] [PMID: 24024636]

[66]   Sela-Culang I, Kunik V, Ofran Y. The structural basis of antibody-antigen recognition. Front Immunol 2013; 4: 302.
[http://dx.doi.org/10.3389/fimmu.2013.00302] [PMID: 24115948]

[67]   Oluwayelu DO, Adebiyi AI. Plantibodies in human and animal health: a review. Afr Health Sci 2016; 16(2): 640-5.
[http://dx.doi.org/10.4314/ahs.v16i2.35] [PMID: 27605982]

[68]   Jain P, Pandey P, Jain D, Dwivedi P. Plantibody: An overview. Asian J Pharm Life Sci 2011; 1(1): 87-94.

[69]   Hiatt A, Cafferkey R, Bowdish K. Production of antibodies in transgenic plants. Nature 1989; 342(6245): 76-8.
[http://dx.doi.org/10.1038/342076a0] [PMID: 2509938]

[70]   Düring K, Hippe S, Kreuzaler F, Schell J. Synthesis and self-assembly of a functional monoclonal antibody in transgenic *Nicotiana tabacum.* Plant Mol Biol 1990; 15(2): 281-93.
[http://dx.doi.org/10.1007/BF00036914] [PMID: 2129424]

[71]   Edgue G, Twyman RM, Beiss V, Fischer R, Sack M. Antibodies from plants for bionanomaterials. Wiley Interdiscip Rev Nanomed Nanobiotechnol 2017; 9(6): e1462.
[http://dx.doi.org/10.1002/wnan.1462] [PMID: 28345261]

[72]   Liu S, Anderson JA. Marker-assisted evaluation of *Fusarium* head blight resistant wheat germplasm. Crop Sci 2003; 43: 760-6.
[http://dx.doi.org/10.2135/cropsci2003.7600]

[73]   Di X, Cao L, Hughes RK, Tintor N, Banfield MJ, Takken FLW. Structure-function analysis of the *Fusarium oxysporum* Avr2 effector allows uncoupling of its immune-suppressing activity from recognition. New Phytol 2017; 216(3): 897-914.
[http://dx.doi.org/10.1111/nph.14733] [PMID: 28857169]

[74]   Mogensen TH. Pathogen recognition and inflammatory signaling in innate immune defenses. Clin Microbiol Rev 2009; 22(2): 240-73.
[http://dx.doi.org/10.1128/CMR.00046-08] [PMID: 19366914]

[75]   Smith-Gill SJ. Protein epitopes: functional *vs.* structural definitions. Res Immunol 1994; 145(1): 67-70.
[http://dx.doi.org/10.1016/S0923-2494(94)80047-2] [PMID: 7516566]

[76]   Stave JW, Lindpaintner K. Antibody and antigen contact residues define epitope and paratope size and structure. J Immunol 2013; 191(3): 1428-35.
[http://dx.doi.org/10.4049/jimmunol.1203198] [PMID: 23797669]

[77]   Spillner E, Deckers S, Grunwald T, Bredehorst R. Paratope-based protein identification by antibody and peptide phage display. Anal Biochem 2003; 321(1): 96-104.
[http://dx.doi.org/10.1016/S0003-2697(03)00439-1] [PMID: 12963060]

[78]   Chingthan TS, Sahoo G, Ghore MK. An artificial immune system model for multi agents resources sharing in distributed environments. Int J Comput Sci Eng 2010; 2(5): 1813-8.

[79]   Papavasiliou FN, Schatz DG. Somatic hypermutation of immunoglobulin genes: merging mechanisms for genetic diversity. Cell 2002; 109 (Suppl.): S35-44.
[http://dx.doi.org/10.1016/S0092-8674(02)00706-7] [PMID: 11983151]

[80]   See K, Kadonosono T, Ota Y, Miyamoto K, Yimchuen W, Kizaka-Kondoh S. Reconstitution of an anti-HR2 antibody paratope by grafting dual CDR-derived peptides onto a small protein scaffold. Biotechnol J 2020; 15(12): e2000078.
[http://dx.doi.org/10.1002/biot.202000078] [PMID: 32975036]

[81]   Kadonosono T, Kizaka-Kondoh S. [Semi-rational Design of Target-binding Small Proteins for Cancer Treatment]. Yakugaku Zasshi 2020; 140(2): 159-62.
[http://dx.doi.org/10.1248/yakushi.19-00187-4] [PMID: 32009038]

[82]   Parra D, Takizawa F, Sunyer JO. Evolution of B cell immunity. Annu Rev Anim Biosci 2013; 1: 65-97.
[http://dx.doi.org/10.1146/annurev-animal-031412-103651] [PMID: 25340015]

[83]   Manivel V, Sahoo NC, Salunke DM, Rao KVS. Maturation of an antibody response is governed by modulations in flexibility of the antigen-combining site. Immunity 2000; 13(5): 611-20.
[http://dx.doi.org/10.1016/S1074-7613(00)00061-3] [PMID: 11114374]

[84]   Shehata L, Maurer DP, Wec AZ, *et al.* Affinity maturation enhances antibody specificity but compromises conformational stability. Cell Rep 2019; 28(13): 3300-3308.e4.
[http://dx.doi.org/10.1016/j.celrep.2019.08.056] [PMID: 31553901]

[85]   Zeng X, Winter DB, Kasmer C, Kraemer KH, Lehmann AR, Gearhart PJ. DNA polymerase eta is an A-T mutator in somatic hypermutation of immunoglobulin variable genes. Nat Immunol 2001; 2(6): 537-41.
[http://dx.doi.org/10.1038/88740] [PMID: 11376341]

[86]   Rogozin IB, Pavlov YI, Bebenek K, Matsuda T, Kunkel TA. Somatic mutation hotspots correlate with DNA polymerase eta error spectrum. Nat Immunol 2001; 2(6): 530-6.
[http://dx.doi.org/10.1038/88732] [PMID: 11376340]

[87]   Tijani MK, Reddy SB, Langer C, *et al.* Factors influencing the induction of high affinity antibodies to *Plasmodium falciparum* merozoite antigens and how affinity changes over time. Sci Rep 2018; 8(1): 9026.
[http://dx.doi.org/10.1038/s41598-018-27361-w] [PMID: 29899351]

[88]   Rudnick SI, Adams GP. Affinity and avidity in antibody-based tumor targeting. Cancer Biother Radiopharm 2009; 24(2): 155-61.
[http://dx.doi.org/10.1089/cbr.2009.0627] [PMID: 19409036]

[89]   Reverberi R, Reverberi L. Factors affecting the antigen-antibody reaction. Blood Transfus 2007; 5(4): 227-40.
[http://dx.doi.org/10.2950/2007.0047-07] [PMID: 19204779]

[90]   Newcombe C, Newcombe AR. Antibody production: polyclonal-derived biotherapeutics. J Chromatogr B Analyt Technol Biomed Life Sci 2007; 848(1): 2-7.
[http://dx.doi.org/10.1016/j.jchromb.2006.07.004] [PMID: 16893686]

[91]   Glancy H, Palfreyman JW, Button D, Bruce A, King B. An immunological method for the detection of *Lentinus lepideus* in distribution poles. J Inst Wood Sci 1990; 12: 59-64.
[http://dx.doi.org/10.1007/978-94-009-1363-9_91]

[92]   Ward E, Foster SJ, Fraaije BA, McCartney HA. Plant pathogen diagnostics: immunological and nucleic acid-based approaches. Ann Appl Biol 2004; 145: 1-16.
[http://dx.doi.org/10.1111/j.1744-7348.2004.tb00354.x]

[93] Kharrazian D, Herbert M, Vojdani A. Immunological reactivity using monoclonal and polyclonal antibodies of autoimmune thyroid target sites with dietary proteins. J Thyroid Res 2017; 2017: 4354723.
[http://dx.doi.org/10.1155/2017/4354723] [PMID: 28894619]

[94] Olal S, Atuhaire KD, Ochwo S, *et al.* Immunodiagnostic potential of a 27kDa protein of Fusarium xylarioides, the cause of coffee wilt disease in Robusta coffee in Uganda. Afr J Biotechnol 2014; 13(29): 2922-9.
[http://dx.doi.org/10.5897/AJB2013.13451]

[95] Al-Maqtoofi M, Thornton CR. Detection of human pathogenic *Fusarium* species in hospital and communal sink biofilms by using a highly specific monoclonal antibody. Environ Microbiol 2016; 18(11): 3620-34.
[http://dx.doi.org/10.1111/1462-2920.13233] [PMID: 26914362]

[96] Ali-Shtayeh MS, MacDonald JD, Kabashima J. A method for using commercial ELISA tests to detect zoospores of *Phytophthora* and *Pythium* species in irrigation water. Plant Dis 1991; 75: 305-11.
[http://dx.doi.org/10.1094/PD-75-0305]

[97] Brill LM, McClary RD, Sinclair JB. Analysis of two ELISA formats and antigen preparations using polyclonal antibodies against *Phomopsis longicolla.* Phytopathology 1994; 84: 178-9.
[http://dx.doi.org/10.1094/Phyto-84-173]

[98] Banks JN, Cox SJ. The solid phase attachment of fungal hyphae in an ELISA to screen for antifungal antibodies. Mycopathologia 1992; 120(2): 79-85.
[http://dx.doi.org/10.1007/BF00578292] [PMID: 1480211]

[99] Rohde S, Rabenstein F. Standardization of an indirect PTA-ELISA for detection of *Fusarium* spp. in infected grains. Mycotoxin Res 2005; 21(2): 100-4.
[http://dx.doi.org/10.1007/BF02954429] [PMID: 23605267]

[100] Kaufman L, Standard PG, Jalbert M, Kraft DE. Immunohistologic identification of *Aspergillus* spp. and other hyaline fungi by using polyclonal fluorescent antibodies. J Clin Microbiol 1997; 35(9): 2206-9.
[http://dx.doi.org/10.1128/jcm.35.9.2206-2209.1997] [PMID: 9276388]

[101] Ray M, Dash S, Achary KG, Nayak S, Singh S. Development and evaluation of polyclonal antibodies for detection of *Pythium aphanidermatum* and *Fusarium oxysporum* in ginger. Food Agric Immunol 2017; 29: 204-15.
[http://dx.doi.org/10.1080/09540105.2017.1365820]

[102] Eparvier A, Alabouvette C. Use of ELISA and GUS-transformed strains to study completion between pathogenic and non-pathogenic *Fusarium oxysporum* for root colonization. Biocontrol Sci Technol 1993; 4: 35-47.
[http://dx.doi.org/10.1080/09583159409355310]

[103] Linfield CA. A rapid serological test detecting *F. oxysporum f.sp. narcissi* in Narcissus. Ann Appl Biol 1993; 123: 29-33.
[http://dx.doi.org/10.1111/j.1744-7348.1993.tb04938.x]

[104] Meirelles PG, Ono MA, Ohe MCT, Maroneze DM, Itano EN, Garcia GT. Detection of *Fusarium* sp. contamination in corn using enzyme-linked immunosorbent assay. Food Agric Immunol 2006; 17: 79-89.
[http://dx.doi.org/10.1080/09540100600688754]

[105] Nagayama S, Kawamura O, Ohtani K, *et al.* Application of an enzyme-linked immunosorbent assay for screening of T-2 toxin-producing *Fusarium* spp. Appl Environ Microbiol 1988; 54(5): 1302-3.
[http://dx.doi.org/10.1128/aem.54.5.1302-1303.1988] [PMID: 3389821]

[106] Pei SC, Zhen YP, Gao JW, *et al.* Screening and monitoring zearalenone-producing *Fusarium* species by PCR and zearalenone by monoclonal antibodies in feed from China. Food Addit Contam Part B

Surveill 2014; 7(4): 282-7.
[http://dx.doi.org/10.1080/19393210.2014.925981] [PMID: 24867386]

[107] Azcona-Olivera JI, Abouzied MM, Plattner RD, Pestka JJ. Production of monoclonal antibodies to the mycotoxins fumonisins B$_1$, B$_2$ and B$_3$. J Agric Food Chem 1992; 40: 531-4.
[http://dx.doi.org/10.1021/jf00015a034]

[108] Daroda L, Hahn K, Pashkoulov D, Benvenuto E. Molecular characterization and *in planta* detection of *Fusarium moniliforme* endopolygalacturonase isoforms. J Physiol Mol Plant Pathol 2001; 59: 317-25.
[http://dx.doi.org/10.1006/pmpp.2001.0370]

[109] Ascoli CA, Aggeler B. Overlooked benefits of using polyclonal antibodies. Biotechniques 2018; 65(3): 127-36.
[http://dx.doi.org/10.2144/btn-2018-0065] [PMID: 30089399]

[110] Dewey FM, Meyer U. Rapid, quantitative tube immunoassays for on-site detection of *Botrytis, Aspergillus* and *Penicillium* antigens in grape juice. Anal Chim Acta 2004; 513: 11-9.
[http://dx.doi.org/10.1016/j.aca.2003.11.088]

[111] Cahill DM, Hardham AR. Exploitation of zoospore taxis in the development of a novel dipstick immunoassay for the specific detection of *Phytophthora cinnamomi*. Phytopathology 1994; 84: 193-200.
[http://dx.doi.org/10.1094/Phyto-84-193]

CHAPTER 9

Plant Molecular Pharming For Human Diseases

Kiran Saba[1], Muhammad Suleman Malik[1], Sara Latif[1], Fatima Ijaz[1], Muhammad Sameeullah[2] and Mohammad Tahir Waheed[1,*]

[1] *Department of Biochemistry, Faculty of Biological Sciences, Quaid-i-Azam University, 45320, Islamabad, Pakistan*

[2] *Innovative Food Technologies Development Application and Research Center, Faculty of Engineering, Bolu Abant Izzet Baysal University, 14030, Bolu, Turkey*

Abstract: Infectious diseases pose an increasing threat to global health. The world has experienced many outbreaks due to Emerging Infectious Diseases (EIDs) in the 21st century. Vaccination proves to be the most successful public health intervention to counter such outbreaks. Vaccines against many diseases are available. Most of these vaccines either consist of live or attenuated strains, thus posing health risks. There is a need for new and safe vaccines to prevent and mitigate the impact of outbreaks due to emerging and endemic infectious diseases. The requisition of plant-based medicine is increasing day by day because of their non-toxic nature with no to very few side effects and readily available at a reasonable cost. In the present chapter, we will discuss the importance of plant molecular pharming (PMP) with its perspective on human diseases. Several advantages of PMP in relation to the United Nations' sustainable development goals (SDGs) will also be deliberated.

Keywords: COVID-19, Human diseases, Plant molecular pharming, Sustainable development goals, World health organization.

1. INTRODUCTION

Kingdom Plantae is a huge source of biologically active molecules and compounds and it is the biggest known source of medicines since pre-historic times. According to the oldest available record (5000-3000 BCE by Sumerians on clay tablets), humans understood diseases and the use of medicines to cure their ailments [1]. Independent of each other, all big ancient civilizations such as China, Greece, and the Arab developed their medical systems, but all of them

Correspondence author Mohammad Tahir Waheed: Department of Biochemistry, Faculty of Biological Sciences, Quaid-i-Azam University, 45320, Islamabad, Pakistan; E-mail: tahirwaheed@qau.edu.pk

Muhammad Sarwar Khan (Ed)

primarily depended on plants and plant-derived components [2]. To fight against natural selection pressure, human beings always turned to the plants, either to make them a source of food or to fight against pandemics. Along with the bigger civilizations, the smaller communities in Asia, Africa, and Latin America have a known history of reliance on traditional medicaments that are largely based on plants for prompt access to a rather safe, cost-effective, competent, and culturally acceptable solution to the primary health care [3].

Since ancient times, people around the world have relied primarily on plants to fulfill all their medicinal needs, for alleviating ailments, and discovering a cure to relieve pain and discomfort. The early man was encouraged to explore his immediate natural environment and try many plant and animal products, minerals, and a range of therapeutic agents [4]. Plants play an appealing role in the modulation of human and non-human diseases [5]. More than 75% of the medicines to control infectious diseases are of plant origin, whereas about 61% of the drugs approved by the FDA are either isolated completely from plants or are the derivative of plant-based active compounds. Not only in primitive times, but plants are also still the best source of food and medicine [6]. According to the World Health organization (WHO), about 80% of the population of the developed and developing countries believes in traditional medicines or plant-derived drugs for their primary health needs [7]. The era of pharmaceutical sciences and industries is well developed now and growing rapidly, and the market is being introduced to a variety of synthetic drugs. The modern pharmacopeia contains at least 25% of drugs that are plant derivatives or semi-synthetic, made up of prototype compounds derived from plants [8]. The requisition of plant-based medicine is increasing day by day because of the flourishing cognizance of natural products, being nontoxic, with few side effects. These are readily available at a reasonable cost and sometimes the only available source of health care available to the poor or low-income communities. Hence, plant-based medical practices hold an imperative spot in the socio-cultural and economic values of both developing and developed countries traditionally [9]. There is an undisputed belief that 'green medicine' is better than synthetic drugs. Since the last decade, an upsurge in the utilization of herbal medicines has been seen even in industrialized countries [10]. It is reported that over 90% of medicinal plants utilized by the pharmaceutical industries are collected in their raw form from natural sources. In about 800 species utilized by industries, less than 20 species of plants are cultivated commercially [11]. Hence, the plant collection for pharmaceutical purposes involves destructive harvesting. This massive harvesting is a big threat to the reservoirs and natural diversity of medicinal plant resources and ultimately to the economy of the country if the biodiversity is not sustainably used. The commercial cultivation of medicinal plants is more appropriate for use in the production of drugs [12].

Molecular pharming presents the best alternative to produce economic modern medicines and ensure their large-scale availability around the globe. Molecular pharming is defined as the synthesis of recombinant pharmaceutical products using plants. The field of plant molecular pharming emerged in the early 1990s as a subdivision of plant biotechnology, and the prime purpose was widely scalable production of recombinant therapeutic proteins at an affordable cost [13]. One of the main characteristics of plant molecular pharming is that it includes diverse platforms and technologies developed with recognizable and overlapping interests, linked only by the utilization of plants as bio-factory. Similarly, it includes plant cells or only the part of plants growing in small containers on the synthetic nutrient media or the plant grown on the soil at a large scale, containing a stably integrated transgene for the expression of recombinant protein [14]. Plant molecular pharming has existed since the successful transformation of the first higher plant in 1983 [15]. The expression of the first human antibodies in the plants was done by During [16] and was broaden to express secretory antibodies by Hiatt *et al.* [17]. The first protein (Avidin) expressed in plants for the commercialization purpose was done by Hood *et al.* [18], followed by Aprotinin (used as an anti-inflammatory and wound healing agent), a recombinant pharmaceutical drug [19].

For the last two decades, plant molecular pharming has been resuscitated by the commercialization of biopharmaceutical products. A significant breakthrough was achieved when the first-ever molecular pharming product enzyme taliglucerase alfa was approved for human use in 2012. This was the recombinant form of human glucocerebrosidase designed and developed by Protalix Biotherapeutics® for the treatment of Gaucher's disease, a lysosomal storage disorder [20]. Two more plant-derived pharmaceutical products approved by the European Union were insulin expressed in safflower developed by SemBioSys Genetics and HIV-neutralizing monoclonal antibody expressed in tobacco by a publicly funded consortium (Pharma-Planta; http://www.pharma-planta.net). In the production of both recombinant products, the protocols adapted were following the advanced GMP (pharmaceutical good manufacturing practice) [21].

A plant-based system has several advantages, mainly safety and high expression of recombinant proteins [22, 23]. During the Ebola outbreak in 2014, Zmapp (a plant-made monoclonal antibody cocktail) has shown the ability to fight against this disease [24]. The most promising candidates are the influenza vaccines developed by Medicago Inc. that rely on using a non-replicative vector carrying viral regulatory sequences to mediate the transient expression of Hemagglutinin (HA) in *N. benthamiana*, which has led to injectable vaccine candidates [25, 26]. Recently, functional single-chain monoclonal antibodies are also produced in plant systems, providing simpler molecules for viral neutralization for rabies [27].

Plant-made antigens and antibodies can also be suitable tools for diagnosis, providing low-cost proteins with preserved antigenic determinants and specificity [28, 29].

The recent COVID-19 outbreak has caused 85.9 million cases with more than 1.87 million deaths [30]. Many control measures are proposed and practiced for treating the disease. However, the outcome of the treatment strategies is variable and some drugs used for the treatment are also linked to side effects. But to control such pandemic, an effective vaccine for immunization of a large number of people is urgently needed. This vaccine cannot be produced in a bulk amount and with fewer or no side effects with the conventional methods. So plants can be used for cost-effective vaccine production with very high yield [31].

The infection has spread quickly, and millions of individuals have been influenced across 6 continents, representing a steady danger to worldwide wellbeing. Given the speed advantages and proven viability of the plant production platform, the transient expression system, in particular, could be employed to produce recombinant proteins at high levels to meet the sudden demand for production of viral antigens or antiviral proteins that could be used as research reagents, emergency vaccines (SARS-CoV-2 subunit and virus-like particle vaccines), or other biopharmaceuticals to fight against COVID-19 [32, 33]. The neutralizing monoclonal antibodies against SARS-CoV-2 could also be produced in plants with minimal investment, which could be used for passive immunotherapy [34].

Medicago Inc. (www.medicago.com/en/pipeline/) has announced to express the virus-like particles (VLPs) in transient expression systems to develop plant-based vaccines and antibodies against the SARS-CoV-2 virus. This company has estimated to hold a production capacity of 10 million doses a month of VLPs-based vaccines, which is an attractive production capacity. iBio Inc. (www.ibioinc.com/ pipeline) has also announced development of an epitope-based vaccine against SARS-CoV-2 to reduce the risk of this disease. Some other companies are also interested in the development of plant-made vaccines against COVID-19, including: Nomad (www.nomadbioscience.com/), Ventria (ventria. com/), Greenovation Biopharmaceuticals (www. greenovation.com/), protalix (ww.protalix.com) and Kentucky Bioprocessing (www.kentucky bioprocess-ing.com) [33].

2. MILLENNIUM DEVELOPMENT GOALS (MDGS)

In September 2000, at the Millennium Summit being the largest get-together in history, the world leaders accepted the United Nations Millennium Declaration.

The leaders committed their nations to a novel global partnership aimed at eradicating extreme poverty and set out a chain of targets to be accomplished in a time frame of 15 years, *i.e.*, by 2015. These targets were known as the Millennium Development Goals (MDGs) shown in Table **1**. The MDGs are the world's quantified and time-bound targets that address poverty in its various magnitudes, such as hunger, absence of enough shelter, income poverty, diseases, and exclusion, along with the promotion of education, equality of gender, and sustainability of the environment. Basic human rights are also there, such as the rights of people on earth to education, shelter, health, and security [35, 36].

Table 1. The united nations millennium development goals.

1	Goal 1: Eradicate extreme poverty and hunger
2	Goal 2: Achieve Universal Primary Education
3	Goal 3: Promote Gender Equality and Empower Women
4	Goal 4: Reduce Child Mortality
5	Goal 5: Improve Maternal Health
6	Goal 6: Combat HIV/AIDS, Malaria, and other diseases
7	Goal 7: Ensure Environmental Sustainability
8	Goal 8: Develop a Global Partnership for Development

The world has made remarkable progress in attaining many of its goals. Approximately, 21% increase has been seen in overall incomes from 1990 till 2002. There was a decline of an estimated 130 million people living below the poverty line. Mortality rates of children reduced to 88 deaths per 1000 live births from 103 deaths a year. Life expectancy raised to almost 65 years which was previously 63 years or even lower. An additional 8% of the people from the developing world gained access to water while an additional 15% got access to better sanitation facilities. But this progress was not uniform across the world or the set goals. There were a lot of discrepancies across and within countries [37].

3. SUSTAINABLE DEVELOPMENT GOALS (SDGS)

At the 70th session of the United Nations General Assembly held on September 25, 2015 at New York, the post-2015 development agenda was set in the form of 17 Sustainable Development Goals with 169 related targets. These newly adopted goals go further than the Millennium Development Goals. The idea of sustainable development was suggested by the seminal Brundtland Report [38] as being a directorial norm bridging environmental and human development concerns [39, 40].

This universal set comprising of 17 goals (shown in Table **2**), 169 targets, and 304 indicators was accepted by all 193 UN member states in the UN General Assembly. It is meant to be executed for the next fifteen years starting January 2016 to end hunger and poverty, to protect plants, and to make sure that every human on earth is enjoying prosperous life [41]. The SDGs are the extensions of the MDGs. These Sustainable development Goals are being stated as "The Global Goals" in policy circles [42]. In general, it is hoped that by 2030, poverty and hunger would be eliminated, quality of life will be greatly improved, all forms of capital will be intact and functioning under ideal climatic situations, and peace and prosperity will be shared by all. Sustainable Development Goals "assist as guideposts for a challenging transition to sustainable development" [43, 44].

Table 2. The united nations sustainable development goals summary.

	Sustainable Development Goals	
	Goals	**Outline Description**
	No poverty	End poverty in all its forms everywhere
	Zero Hunger	End hunger, achieve food security, improve nutrition and promote sustainable agriculture
	Good health and well-being	Ensure healthy lives and promote well-being for all at all ages

(Table 2) cont.....

	Quality education	Ensure inclusive and equitable quality education and promote lifelong learning opportunities for all
	Gender equality	Achieve gender equality and empower all women and girls
	Clean water and sanitation	Ensure availability and sustainable management of water and sanitation for all
	Affordable and clean energy	Ensure access to affordable, reliable, sustainable, and modern energy for all
	Decent work and economic growth	Promote sustained, inclusive, and sustainable economic growth, full and productive employment, and decent work for all
	Industry, innovation, and infrastructure	Build resilient infrastructure, promote inclusive and sustainable industrialization, and foster innovation
	Reduced inequalities	Reduce income inequality within and among countries
	Sustainable cities and communities	Make cities and human settlements inclusive, safe, resilient, and sustainable
	Responsible consumption and production	Ensure sustainable consumption and production patterns
	Climate action	Take urgent action to combat climate change and its impacts by regulating emissions and promoting developments in renewable energy
	Life below water	Conserve and sustainably use the oceans, seas, and marine resources for sustainable development
	Life on land	Protect, restore and promote sustainable use of terrestrial ecosystems, sustainably manage forests, combat desertification, halt and reverse land degradation and halt biodiversity loss

(Table 2) cont.....

	Peace, justice, and strong institutions	Promote peaceful and inclusive societies for sustainable development, provide access to justice for all, and build effective, accountable, inclusive institutions at all levels
	Partnerships for the goals	Strengthen the means of implementation and revitalize the global partnership for sustainable development

3.1. Role of Plant Molecular Pharming in Achieving Sustainable Development Goals

Plant molecular pharming can play a significant role in the better living of humans and achieving several SDGs. Some of the sustainable development goals can be addressed directly or indirectly through plant molecular pharming such as No Poverty, Zero Hunger, Good Health and well-being, Decent work and economic growth, Industry innovation & infrastructure, and Responsible consumption and production.

One of the total 17 SDGs is devoted precisely to health, and has been outlined in deliberately wide terms relevant to all populations and all countries: "Ensure healthy lives and promote well-being for all at all ages". Plant molecular pharming can be used for improved access to modern medicines, and improve the health of the poor in developing countries and emerging economies. Low- and middle-income countries (LMICs) face health problems at several stages. Good healthcare comes at a cost, but poorer countries face hurdles including very high costs of pharmaceutical products. Through plant pharming, growing plants facilitate a significant decline in manufacturing costs. Production of pharmaceuticals through this platform has significant financial benefits associated with cheap infrastructure investment costs at the early stages of product development. The costs of building a genetically modified plant's compliant facility for growing plants are much lower than for a cell fermentation facility having equivalent production capacity. Lower up-front costs for early-stage assessment of drug candidates would allow testing of many more potential products. Good health and well-being of people attained through plant molecular pharming will indirectly help to eliminate poverty and overcome food shortage [45]. When people have access to better and cheap health facilities, they can work better and earn a living for themselves. Better health will lead to reducing the number of people living below the poverty line. Similarly, people associated with the agriculture and food industry can contribute to remove hunger. Production scalability is another significant feature of plant pharming. The majority of the pharmaceutical targets for different diseases and ailments are needed in very large quantities. The need for such products often prevails over the present production capacity, which is one of the main reasons for the lack of availability for medicines in developing countries. Whereas the production of pharmaceutical products and other recombinant proteins (such as monoclonal antibodies) at a larger scale could be achieved through plant pharming, thus opening new and innovative approaches for disease management [46].

Plant molecular pharming offers opportunities to develop innovative manufacturing technologies, which not only can help pharmaceutical industry to flourish locally but also provide a 'push' for molecular pharming products to both regional and international markets, thus overall contributing towards the industrial uplift. In addition to cost and scalability as the key benefits for using plants, other emerging advantages of plant molecular pharming include the fastest route for producing vaccine using transient expression systems [47], assembly of complex proteins [48], and expression of compounds which are toxic in conventionally used expression systems [49]. The ability of edible plants to be used for oral delivery as formulated products addresses the number of problems related to the delivery of medicines such as cool chain maintenance, costs associated with injectable drugs, and contamination [50]. Additionally, plant molecular pharming can contribute to economic growth through decent work in the pharmaceutical industry.

4. CURRENT STATUS OF MOLECULAR PHARMING IN THE LAST DECADE FOR EMERGING INFECTIOUS DISEASES

According to the World health organization (WHO), infectious diseases are one of the top 10 threats to global health [51]. Since the beginning of the 21st century, many outbreaks of Emerging Infectious Diseases (EIDs) have been experienced by the world with substantial concerns about public health till the present day. Severe Acute Respiratory Syndrome-related *Coronavirus* (SARS-CoV) outbreak in 2003-2004, H1N1 "swine flu" in 2009, Middle East Respiratory Syndrome *Coronavirus* (MERS-CoV) since 2012, Ebola virus in 2013-2016 and 2018 onward, Zika virus in 2015-2016, and COVID-19 since December 2019 are some examples of 21st century outbreaks [52 - 58]. The WHO has developed a list under the WHO Research and Development (WHO R&D) blueprint priority diseases title for EIDs and pathogens to be prioritized. The viruses included in the 2018 WHO R&D list was Ebola, Lassa, Marburg, MERS-CoV, Nipah, and Rift Valley fever (RVF). In the same year alone, the epidemic was caused by 6 of the 10 priority diseases listed in the WHO R&D blueprint [57]. WHO R&D blueprint list of 2020 includes COVID-19, Crimean-Congo hemorrhagic fever, Ebola virus disease, Marburg virus disease, Lassa fever, MERS-CoV, SARS-CoV, Nipah, and *Henipaviral* diseases, RVF disease, Zika virus disease, and "Disease X". Disease-

X represents the fact that a severe international outbreak could be triggered by a pathogen currently unknown to cause human disease [59]. Fig. (**1**) represents the outbreaks due to emerging and prioritized pathogens in the last decade.

Fig. (1). The outbreaks due to different pathogens worldwide in the last decade.

Vaccination has been called the most successful public health intervention to date. Since the modern era of vaccinations, a significant reduction in disease, disability, and deaths from several infectious diseases around the world has been observed [60]. Despite this success, there is still a great need for new vaccines to prevent and mitigate the impact of outbreaks due to emerging and endemic infectious diseases [61]. To develop a vaccine against EIDs is challenging because identifying the causative agent and the time and place of the next outbreak cannot be predicted accurately using current means [62, 63]. The foundation of Coalition for Epidemic Preparedness Innovations (CEPI) was established in 2017 during World Economic Forum's annual meeting, held at Davos in 2016, after the devastating West African Ebola epidemic in 2013-2016 [64, 65]. This epidemic claimed the lives of about 11,000 individuals alone with the estimated economic and social burden of more than US$ 53 billion [64]. CEPI is investing US$ 458.4 million as of November 2019, for the development of vaccines against a set of highly prioritized pathogens causing infectious diseases, which currently includes Chikungunya virus, Influenza virus, Lassa fever virus, Marburg virus, MERS-CoV, Nipah virus, Rabies virus, RVF virus and Respiratory syncytial virus, *via* technology platforms ranging from recombinant viral vectors, recombinant proteins, nucleic acids, and attenuated virus [57]. To accelerate nine programs to develop a vaccine for COVID-19 CEPI has provided up to US$ 895 million; eight of such vaccines have entered clinical trials [66]. Generally, the development of a vaccine is time consuming, risky, and expensive endeavor. The development of vaccines against EIDs faces several obstacles. Testing of such vaccines is problematic, and the market of the vaccine against these infectious diseases is limited [67, 68]. Plants serve as a convenient platform to produce recombinant proteins by reducing the cost for the development of vaccine exponentially as compared to traditional microbial fermentation or cell expression systems (mammalian/insect) [69].

Nowadays, the pressure on natural resources and their use for human benefit is increasing exponentially. For natural resources a question arises related to their ability to meet human needs [70]. In addition to food security, malnutrition and food deficiency are serious problems worldwide. At the same time, in the Asian region, a significant segment of the population is facing health challenges and growth anomalies [71]. The nexus between human health and disease is highlighting the need to develop high-efficiency and low-cost medicine using technology to improve human life [72].

Plant biotechnology offers new strategies for the growth of high yielding, stress-tolerant, disease-resistant transgenic crops. Also, aspects such as medicine, nutrition, and biofuel preparation are focused on human benefits [73 - 77]. High-efficacy vaccines to fight disease are expected for humans. However, due to the high cost of classical vaccines, people in many developing countries cannot afford them. Due to which many deaths occur from various diseases every year [78]. For the production of important pharmaceuticals transgenic plants have been widely adopted [79]. A plant expression system employs the expression of a specific antigenic peptide for the preparation of the edible vaccine. So far, different types of plants have been transformed to make edible vaccines [72]. A summarized list of peptides expressed in plants in the last decade is presented in Table **3**. Some most important diseases that are caused by different emerging and prioritized pathogens, against which immune response generating peptides have been expressed in plants are discussed below.

Table 3. Published studies on peptides expressed in plants for the last decade.

Disease/Disorder	Protein/Peptide/Antibodies Expressed	Pathogen	Transgenic Plant	References
Bacterial Diseases				
Anthrax	Protective antigen (PA)	*Bacillus anthracis*	*Nicotiana benthamiana, Brassica juncea*	[153]
	pp-PA83 and PNGase F	*Bacillus anthracis*	*Nicotiana benthamiana*	[154]
Brucellosis	U-Omp19	*Brucella abortis*	*Nicotiana benthamiana*	[155]

(Table 3) cont.....

Disease/Disorder	Protein/Peptide/Antibodies Expressed	Pathogen	Transgenic Plant	References
Botulism	BoHC	*Clostridium botulinum*	*Oryza sativa*	[156]
Cholera	CTB	*Vibrio cholerae*	*Oryza sativa*	[161]
Diarrhoea/ Gastroenteritis	rLT-B protein	Enterotoxigenic *Escherichia coli*	*Solanum lycopersicum*	[171]
	VP4-VP7 fusion protein	*Salmonella typhimurium*	*Nicotiana benthamiana*	[176]
	OmpA	*Salmonella typhimurium*	*Medicago sativa*	[177]
Tuberculosis	Ag85B, MPT83, MPT64, ESAT6	*Mycobacterium tuberculosis*	*Solanum tuberosum*	[238]
	CFP10-ESAT6-dIFN	*Mycobacterium tuberculosis*	*Daucus carota*	[239]
	CTB-ESAT6 fusion protein, CTB-Mtb72F fusion protein, LipY Protein	*Mycobacterium tuberculosis*	*Nicotiana benthamiana*	[240]
	CTB-ESAT6 fusion protein, LipY Protein	*Mycobacterium tuberculosis*	*Lactuca sativa*	[241]
	ESAT6 and CFP10	*Mycobacterium tuberculosis*	*Daucus carota*	[241]
	GroES TB antigen	*Mycobacterium tuberculosis*	*Nicotiana benthamiana, Solanum tuberosum*	[242]
Cancers				
Cancer	PD1		*Nicotiana benthamiana*	[157]
Colorectal cancer	GA733-Fc		*Nicotiana benthamiana*	[162]
	GA733-FcK		*Nicotiana benthamiana*	[163]
	GA733K, GA733-FcK, and GA733-Fc		*Nicotiana benthamiana*	[164]
	CO17-1AK		*Nicotiana tabacum*	[165]
Follicular lymphoma	Idiotypic Ig (tumor-specific antigen)		*Nicotiana benthamiana*	[180]
Genetic Disorders				
Haemophilia	Factor VIII (FVIII)		*Nicotiana benthamiana*	[181]

(Table 3) cont.....

Disease/Disorder	Protein/Peptide/Antibodies Expressed	Pathogen	Transgenic Plant	References
Parasitic Diseases				
Cysticercosis	HP6/TSOL18 antigen	*Taenia solium*	*Daucus carota*	[167]
Leishmaniasis	promastigote surface antigen (PSA)	*Leishmania infantum*	*Nicotiana benthamiana*	[217]
Malaria	Multi-domain antigen	*Plasmodium falciparum*	*Nicotiana benthamiana*	[218]
	Multi-domain antigen (PfCSP, PfTRAP, PfCelTOS)	*Plasmodium falciparum*	*Nicotiana benthamiana*	[219]
	Pf38	*Plasmodium falciparum*	*Nicotiana benthamiana*	[220]
	PfCP-2.9	*Plasmodium falciparum*	*Solanum lycopersicum*	[221]
	pfGAP50	*Plasmodium falciparum*	*Nicotiana benthamiana*	[222]
	Pfs25	*Plasmodium falciparum*	*Nicotiana benthamiana*	[223]
	Pfs230	*Plasmodium falciparum*	*Nicotiana benthamiana*	[224]
	Pfs25, Pfs230_C0	*Plasmodium falciparum*	*Nicotiana benthamiana*	[225]
	PyMSP1$_{19}$	*Plasmodium yoelii*	*Nicotiana benthamiana*	[226]
Plague	F1–V fusion protein	*Yersinia pestis*	*Daucus carota*	[227]
Psittacosis	MOMP gene	*Chlamydia psittaci*	*Oryza sativa*	[229]
Toxoplasmosis	GRA4 antigen	*Toxoplasma gondii*	*Nicotiana benthamiana*	[236]
	LiHsp83□SAG1 fusion protein	*Toxoplasma gondii*	*Nicotiana benthamiana*	[237]
Viral Diseases				

(Table 3) cont.....

Disease/Disorder	Protein/Peptide/Antibodies Expressed	Pathogen	Transgenic Plant	References
Acquired Immunodeficiency syndrome	HIV Env gp140	Human immunodeficiency virus	*Nicotiana benthamiana*	[148]
	P24	Human immunodeficiency virus	*Arabidopsis thaliana*, Daucus carota	[149]
	Multi-HIV (gp120 and gp41)	Human immunodeficiency virus	*Nicotiana benthamiana*	[150]
	Gp120 and gp41	Human immunodeficiency virus	*Nicotiana benthamiana*	[151]
	2G12	Human immunodeficiency virus	*Oryza sativa*	[152]
Cervical cancer	HPV16	Human papillomavirus	*Nicotiana benthamiana*	[158]
	HPV16 L1 (fused with LTB)	Human papillomavirus	*Nicotiana benthamiana*	[159]
	HPV16 L1 protein	Human papillomavirus	*Nicotiana benthamiana*	[160]
	HPV 16 L2 peptides (SAC 108–120)	Human papillomavirus	*Nicotiana benthamiana*	[69]
Chikungunya	CHIKV E1 and E2	Chikungunya virus	*Nicotiana benthamiana*	[166]
	8B10 and 5F10	Chikungunya virus	*Nicotiana benthamiana*	[166]
	CHKV mab	Chikungunya virus	*Nicotiana benthamiana*	[80]
Crimean-Congo Haemorrhagic fever	G1 & G2	Crimean-Congo Haemorrhagic fever virus	*Nicotiana benthamiana*	[86]
Dengue	Consensus domain III glycoprotein (cEDIII)	Dengue virus E	*Nicotiana benthamiana*	[168]
	Envelop protein domain III (EDIII)	Dengue virus E	*Nicotiana tabacum*	[169]
	E60	Dengue virus E	*Nicotiana tabacum*	[170]

(Table 3) cont.....

Disease/Disorder	Protein/Peptide/Antibodies Expressed	Pathogen	Transgenic Plant	References
Diarrhoea/ Gastroenteritis	C486 BRV VP8 protein	Rotavirus	*Nicotiana tabacum*	[172]
	MucoRice-ARP1	Rotavirus	*Oryza sativa*	[173]
	Narita 104 virus virus-like particles	Narita 104 virus	*Nicotiana benthamiana*	[174]
	NVCP	Norwalk virus	*Nicotiana benthamiana*	[175]
	VP4-VP7 fusion protein	Rotavirus	*Nicotiana benthamiana*	[176]
Ebola	A mix of trio chimeral monoclonal Ebola antibodies (drug ZMapp)	Ebola virus	*Nicotiana benthamiana*	[58]
	6D8	Ebola virus	*Nicotiana benthamiana*	[178]
	6D8	Ebola virus	*Lactuca sativa*	[179]
Hand-foot-mouth disease	cD5	Enterovirus 71	*Nicotiana benthamiana*	[157]
Hepatitis	HAV VP1-Fc	Hepatitis A Virus	*Nicotiana benthamiana*	[182]
	HAV VP1-Fc	Hepatitis A Virus	*Solanum lycopersicum*	[183]
	HA78 mAbs	Hepatitis A Virus	*Arabidopsis thaliana*	[184]
	HBcAg	Hepatitis B Virus	*Chlamydomonas reinhardtii*	[185]
	S-HBsAg	Hepatitis B Virus	*Nicotiana benthamiana*	[186]
	S-HBsAg	Hepatitis B Virus	*Solanum lycopersicum*	[187]

(Table 3) cont.....

Disease/Disorder	Protein/Peptide/Antibodies Expressed	Pathogen	Transgenic Plant	References
Hepatitis	S-HBsAg	Hepatitis B Virus	*Zea mays*	[188]
	M-HBsAg	Hepatitis B Virus	*Nicotiana benthamiana*	[186]
	M-HBsAg	Hepatitis B Virus	*Nicotiana benthamiana, Solanum lycopersicum, Lactuca sativa*	[189]
	L-HBsAg	Hepatitis B Virus	*Nicotiana benthamiana, Lactuca sativa*	[189]
	Hepatitis B core antigen	Hepatitis B Virus	*Nicotiana benthamiana*	[190]
	HCVpc–HBsAg	Hepatitis C Virus	*Nicotiana benthamiana*	[191]
	E1E2	Hepatitis C Virus	*Lactuca sativa*	[192]
	HCVcp	Hepatitis C Virus	*Nicotiana benthamiana*	[193]
	HCVcp	Hepatitis C Virus	*Brassica napus*	[194]
	HEV capsid protein	Hepatitis E virus	*Nicotiana benthamiana*	[195]
Herpes	HSV8	Herpes simplex virus	*Nicotiana benthamiana*	[178]
Human respiratory syncytial disease	RSV-F protein	Respiratory syncytial virus	*Malus domestica*	[97]
	Palivizumab-N	Respiratory syncytial virus	*Nicotiana benthamiana*	[196]
	RhoA peptide	Respiratory syncytial virus	*Nicotiana benthamiana*	[98]

(Table 3) cont.....

Disease/Disorder	Protein/Peptide/Antibodies Expressed	Pathogen	Transgenic Plant	References
Influenza	ELPylated Hemagglutinin (HA)	Avian influenza (H5N1)	*Nicotiana benthamiana*	[197]
	H1 HA	Influenza A virus (H1N1)	*Nicotiana benthamiana*	[198]
	H1 & H5 HA	Influenza A virus (H1N1)	*Nicotiana benthamiana*	[199]
	H1 & H5 HA	Influenza A virus (H5N1), Influenza A virus (H1N1)	*Nicotiana benthamiana*	[200]
	H5 HA (H5 Hemagglutinin)	Influenza A virus (H5N1)	*Nicotiana benthamiana*	[201]
	H5 HA	Avian influenza A virus (H5N1)	*Nicotiana benthamiana*	[202]
	H7 HA	Avian influenza A virus (H7N9)	*Nicotiana benthamiana*	[203]
	H6 subtype haemagglutinin (HA)	Influenza A virus (H6N2)	*Nicotiana benthamiana*	[204]
	HAC1	Influenza A virus (H1N1)	*Nicotiana benthamiana*	[205]
	Hemagglutinin (HA)	Avian influenza (H5N1)	*Arabidopsis thaliana*	[206]
	Hemagglutinin (HA)	Avian influenza A virus (H7N7)	*Nicotiana benthamiana*	[207]

(Table 3) cont.....

Disease/Disorder	Protein/Peptide/Antibodies Expressed	Pathogen	Transgenic Plant	References
Influenza	Hemagglutinin (HA)	Influenza virus	*Nicotiana benthamiana*	[208]
	Hemagglutinin (HA)	Influenza A virus (H5N1), Influenza A virus (H1N1)	*Nicotiana benthamiana*	[209]
	Matrix protein 2 ectodomain (M2e)	Avian influenza (H5N1)	*Lemna minor*	[210]
	Matrix protein 2 ectodomain (M2e)	Avian influenza (H5N1)	*Nicotiana benthamiana*	[211]
	M2eHBc	Avian influenza (H5N1)	*Nicotiana benthamiana*	[212]
	Matrix protein 2 Extracellular domain (M2e)	Influenza A virus	*Nicotiana benthamiana*	[213]
	NA protein	Swine influenza virus (H1N1)	*Lactuca sativa*	[214]
	rNP	Swine influenza virus (H3N2)	*Zea mays*	[215]
Japanese Encephalitis	E Domain III (EDIII) fused to BaMV coat protein (CP)	Japanese encephalitis virus	*Chenopodium quinoa*	[216]
Marburg virus disease	MR 191-N	Marburg virus	*Nicotiana benthamiana*	[120]
Nipah virus disease	NiV-G protein	Nipah virus	*Nicotiana benthamiana*	[134]
Polio	PV3 VLPs	Poliovirus	*Nicotiana benthamiana*	[228]
Rabies	E559	Rabies virus	*Nicotiana tabacum*	[230]
	G-protein	Rabies virus	*Zea mays*	[231]
	RGP–RTB fusion protein	Rabies virus	*Solanum lycopersicum*	[232]
	SO57	Rabies virus	*Nicotiana tabacum*	[233]
Rift Valley fever	N protein	Rift Valley fever virus	*Arabidopsis thaliana*	[234]
	N protein	Rift Valley fever virus	*Nicotiana benthamiana*	[235]
West Nile virus disease	pE16	**West Nile virus**	*Nicotiana benthamiana*	[243]
	pE16scFv-CH	**West Nile virus**	*Nicotiana benthamiana*	[243]
	VLPs	West Nile virus	*Lactuca sativa*	[179]

(Table 3) cont.....

Disease/Disorder	Protein/Peptide/Antibodies Expressed	Pathogen	Transgenic Plant	References
Zika virus disease	ZIKV envelope (E) protein	Zika virus	*Nicotiana benthamiana*	[144]
	c2A10G6	Zika virus	*Nicotiana benthamiana*	[177]

4.1. Chikungunya

Chikungunya is caused by the Chikungunya virus (CHIKV), which is a member of the alphavirus family. CHIKV has a positive-sense RNA virus that is transmitted to humans *via* mosquitos [80]. In humans, CHIKV can cause devastating polyarthralgia that lasts from months to years and affects joints of ankles, fingers, knees, and wrists [81]. In five continents, the re-emergence of CHIKV since 2004 has caused severe and often chronic arthralgia in millions of individuals [82]. Therapeutic advances are needed to counter this vulnerable virus because of its continuous outbreaks and its threat of spreading to new areas. For human use, currently no licensed vaccines or therapeutics are available against CHIKV [80]. Previous vaccine studies have suggested that the potential utility of antibodies can be efficacious therapy against CHIKV. In various mouse models protective effect of monoclonal antibodies produced against CHIKV envelope E1 and E2 proteins have been reported against infection due to CHIKV [83]. CHIKV E1, E2, 8B10, and 5F10 are some recombinant proteins expressed in tobacco *via* *Agrobacterium*-mediated transient transformation, which showed *in vivo* efficacy in animal models [59].

4.2. Crimean-Congo Hemorrhagic Fever

Crimean-Congo hemorrhagic fever (CCHF) is a tick-borne virulent disease of humans with a 10% to 80% mortality rate. The causative agent of CCHF is a negative sense tri-segmented RNA virus named Crimean-Congo hemorrhagic fever virus (CCHFV). CCHFV is a member of the family *Nairoviridae* classified within the genus *Orthonairovirus* [84 - 86]. In Bulgaria, Central Asia, and the Soviet Union several epidemics related to CCHF have been reported after the first outbreak of CCHF in the Crimean Peninsula in 1944 [84]. The constant presence of CCHFV has been reported in Africa, Asia, Eastern Europe, and the Middle East [87, 88]. CCHFV is an emerging and re-emerging virus with an uninterrupted enzootic cycle involving ticks and many craniate animals with human beings considered as dead-end hosts. CCHFV can infect humans *via* the bite of an infected tick, contact with CCHF patients, and with tissue or blood from diseased animals [84, 59]. The WHO has listed CCHFV as a priority pathogen.

The emergence of diseases for the first time in Spain (2016) and its re-emergence in other parts of the world after dormancy periods emphasize the need for approved therapies or vaccines to prevent the emergence of CCHFV in currently non-endemic regions [89, 90]. Targeted vaccination of infected groups is proposed because of the sporadic nature of CCHF cases as compared to mass vaccination [91]. The development of the CCHF vaccine has accelerated since the discovery of CCHF animal models [90]. Under experimental conditions, the plant-based antigens have proved to be protective against the onset of disease in animal models and their efficacy and safety in human clinical trials [86].

4.3. Ebola Virus Disease

Ebola virus disease (EVD) is a severe and often lethal disease caused by the Ebola virus (EBOV) [92]. So far, 12 different filoviruses have been identified. The seven filoviruses that cause infection in humans is related to either genus *Marburgvirus* (Marburg virus) and Ravn virus (RAVV) or genus *Ebolavirus* (Bundibugyo virus (BDBV), Ebola virus (EBOV), Reston virus (RESTV), Sudan virus (SUDV), and Tai Forest virus (TAFV). The tenth known outbreak of EVD in the Democratic Republic of Congo claimed the lives of more than 11,000 individuals in August 2018 [93]. In the absence of therapeutics and licensed vaccines, controlling outbreaks of such magnitudes due to EBOV is challenging. For affected patients who are outside of supportive care, very little can be done, which includes administration of antivirals, replenishments of fluids, and secondary symptoms management [94 - 96]. During the outbreak of Ebola in 2014, the plant transient expression system has shown its incredible ability to produce swift responding proteins, biologics, or emergency vaccines. To treat humans during an outbreak of Ebola in 2014; Mapp Biopharmaceutical Inc., USA has expressed an anti-Ebola antibody (drug ZMapp), a mix of trio chimeral monoclonal antibodies, through transient expression in *Nicotiana benthamiana* [58].

4.5. Human Respiratory Syncytial Virus

The human respiratory syncytial virus (RSV) is a single-stranded negative-sense RNA virus, belonging to the family *Paramyxoviridae*. RSV causes infection in the lower respiratory tract in infants and young children worldwide [97, 98]. RSV is responsible for 64 million low respiratory tract infections and claimed 19.9 million lives annually around the world. RSV infects individuals with the compromised or immature immune system; therefore, its victims include infants, elderly people, and patients with immunodeficiencies, pulmonary, and cardiac diseases [97, 99]. To prevent infants from infections, the recommended treatment is the passive vaccination containing monoclonal RSV F-specific antibodies. This is an expensive treatment and mass treatments are not possible at the time of need [100]. The best solution against RSV infectious disorder is a licensed vaccine that is currently not available. Researchers are trying to develop a vaccine that can mimic immunogenicity without any side effects for more than half a century [99]. The need for new platforms for the development of vaccines is increasing with the hope to avoid limitations such as low efficacy, safety issues, and high cost [101]. Plants have been used as expressing systems to produce immunogenic proteins against RSV. The immunogenic response was observed through oral vaccination achieved by feeding transgenic tomato expressing RSV-F protein in mice [97].

4.6. Influenza

Influenza virus causes a highly contagious and acute feverish respiratory disorder known as "influenza" [102]. The influenza virus, a negative-sense single-stranded RNA virus, is a member of the family *Orthomyxoviridae* [103]. Based on the composition of two out of eight-segmented genetic material *i.e.*, Matrix protein (M) and nucleoprotein (NP), the virus is divided into four groups A, B, C, and D. Based on two surface glycoproteins (Hemaglutinin (HA) and Neuraminidase (NA)) antigenicity, its subtypes are determined [104 - 108]. Out of four groups (A, B, C, and D), the influenza A virus (IAV) is the most contagious and infects humans, horses, pigs, sea mammals, and a variety of bird species [108]. Because of the high rate of mutation in the influenza virus, the world population remains at risk each year to an influenza pandemic, which occurs multiple times with approximately 40 years of inter-pandemic intervals. The last influenza outbreak was in 2009, which claimed 4500 lives worldwide in one year of the pandemic [109, 110]. The influenza virus has an RNA genome that is more susceptible to mutation during replication because it lacks a proof-reading mechanism and due to genetic drifts and genetic shifts. These mutations enable the virus to cause epidemics in certain seasons and serve as a bottleneck for the development of antiviral drugs or vaccines [111, 112].

A vaccine against the influenza epidemic, according to WHO guidelines, must include characteristics such as rapid scalability and the lesser amount required to induce immunogenicity [113]. Currently plan to counter the pandemics worldwide is to ensure the availability of the massive quantity of immunogenic antigen. The technologies like an egg- and cell culture, to produce the low-dose vaccine, cannot help to immunize world population when the first wave of infection reaches its peak during the pandemic outbreak [114]. To meet the anticipated surge requirement, one of the few production systems is a plant-based transient expression system [115].

4.7. Marburg Virus Disease

Marburg virus is highly virulent with a high mortality rate, and belongs to the family *Filoviridae*. Human beings after getting an infection from Marburg virus-infected animals further transmit the virus among human beings *via* direct or indirect contact with blood or other body fluids of a human being either infected or died from Marburg virus disease (MVD) [116]. Sporadic outbreaks and cases have been reported in Angola, the Democratic Republic of the Congo, Kenya, the Netherland, South Africa, Uganda, and the United States. A single fatal case of Marburg virus from an unknown source was reported in Uganda (September 2014) with no secondary cases [117]. In 2017, seven cases of MVD have been reported with two confirmed deaths in Uganda [118]. MVD because of its high case-mortality rate reaching 90% remains of great public health importance [117]. Currently, no treatment is available against MVD [118]. The Mapp Biopharmaceutical Inc., USA has expressed human monoclonal antibody through transient expression in *Nicotiana benthamiana*. The post-infection immunogenicity was tested in *Rhesus macaques* with 100% protection but thedetails of test results on humans have not been made public [119, 120].

4.8. Rift Valley Fever

Rift Valley fever (RVF) is a mosquito-borne emerging viral infectious disease affecting both livestock and humans. The causative agent is the Rift Valley Fever Virus (RVFV). RVFV is a negative-sense RNA virus with the tripartite genome, belonging to the family *Bunyaviridae* [121]. RVFV outbreaks result in human morbidity and fatality and cause significant economic losses due to death and abortion among the infected livestock [122, 123]. The disease is predominant in sub-Saharan Africa and many outbreaks have occurred in the African continent, Saudi Arabia, and Yemen. Humans are highly susceptible to RVFV and are infected with the virus in many ways such as: from the bite of infected mosquitoes, blood contact, contact with other body fluids or tissues of infected animals, and consumption of uncooked meat and raw milk from infected animals. People working in slaughter facilities, laboratories or hospitals are at increased risk of acquiring infections [124]. There is a need to develop a vaccine to counteract the catastrophic effects on the economy and health by RVFV [125]. Attenuated virus vaccine, inactivated virus vaccine, and live attenuated mutant vaccine are available for use in animals. However, these vaccines have major safety concerns [124]. No vaccine against RVFV is licensed for human use. Oral delivery of vaccines can overcome safety concerns and can reduce the cost of production and logistics [126].

4.9. Severe Acute Respiratory Syndrome

Severe acute respiratory syndrome (SARS) is caused by a positive-sense single-stranded RNA virus [127]. SARS is a hazardous viral infection with the potential to become pandemic. Its causative agent is named SARS Coronavirus (SARS-CoV) which has shown high morbidity and death rates [128]. In 2002, SARS-CoV by crossing the species barrier infected humans resulting in its first outbreak in China which spread to 29 countries later with a 9.6% mortality rate [129]. The virus is transmitted from human-to-human *via* sneezing or coughing [130]. Centers for Disease Control and Prevention (CDC) in 2012 declared SARS-CoV as a potential biological weapon due to its highly contagious nature [131]. Currently, no licensed vaccine is available for SARS-CoV. Hence, it is necessary to develop an effective therapeutic tool against SARS-CoV to counter its use as a biological weapon or any other outbreak [129]. Research has shown that SARS-CoV spike S protein plays important role in inducing neutralizing antibodies and provide immunogenicity during SARS-CoV infection [132]. The S protein of SARS-CoV has been expressed in a few plants to produce plant-based oral, economical, effective, and safe vaccine against SARS-CoV [133].

4.10. Nipah Virus Disease

Nipah virus (NiV) is a pathogenic paramyxovirus belonging to the genus *Henipavirus* [134]. Pteropid fruit bats (flying foxes) are known natural reservoirs of NiV [135]. NiV with high mortality can infect and cause Nipah virus disease (NVD) in several species [136, 137]. It is the virulent causative agent in both swine and humans [134]. Since 2001, annual outbreaks of fatal NiV have been reported in humans especially in Bangladesh and India. The estimated case-mortality rate due to poor surveillances in Bangladesh rural areas during endemic was 67% [138, 134]. Behavioral modification is the current prevention method in countries under the NiV outbreak. Such preventive modifications to avoid spreading include decreasing the exposure of livestock to bats, to prevent contamination of date palm sap by use of baboo skirts and by pasteurization. Such practices are proven to be effective but cultural norms are limiting factors [139]. The recent outbreak of NiV in May 2018 claimed the lives of 17 patients out of 19 infected individuals with an 89% fatality rate in Kerala, India [140]. To counter the NiV posed threats, effective therapy or vaccine is needed. NiV outer membrane proteins, the glycoprotein (G) and fusion (F) protein, are reported to induce immunogenicity in different animal models when challenged to post-vaccination infection with NiV. To date, only one plant has been used to transiently express to produce NiV G protein as an alternative platform to develop a vaccine [134].

4.11. Zika Virus Disease

Zika virus (ZIKV) was known to cause only febrile illness until 2013 in humans but later outbreaks of ZIKV have linked this obscure pathogen infection to other serious and lethal neurological disorders namely microcephaly and Guillain-Barre' syndrome in infants and adults, respectively [140, 141]. ZIKV belongs to the genus *Flavivirus* [143]. ZIKV is closely related to dengue virus, tick-borne encephalitis virus, West Nile virus, and yellow fever virus. Currently, no licensed vaccine or treatment for the humans is available to counter the threat to global health due to ZIKV. Thus, it is important to develop affordable, safe, and effective vaccines against ZIKV [144]. Published data has shown that for the development of the ZIKV vaccine four different techniques are being used; adenovirus-vectored DNA, naked plasmid DNA, lipid-nanoparticle-encapsulated nucleoside-modified mRNA (mRNA-LNP), and inactivated virus [145 - 147]. Vaccine candidates for ZIKV under development through these techniques include immunogenic pre membrane (prM) and pre membrane E protein (prM-E) of ZIKV. Problems associated with their safety and cost before they become licensed vaccines still need to be addressed [144]. In response to these problems, ZIKV envelope (E) protein and codon-optimized region of murine antibody were transiently expressed in tobacco (*Nicotiana benthamiana*) which shows significant increase in titers of specific and neutralizing antibodies in test animal models. These results have shown that plant-produced vaccine candidates can be cost-effective and safe vaccines against ZIKV [144 - 147].

CONCLUDING REMARKS

Plant-based pharmaceuticals have strong go-to-market potential for their important advantages. The advancements of several plant-expressed products in clinical trials have shown their safety and efficacy. Specifically, in post-COVID-19 era, plants as expression platforms for valuable and important pharmaceuticals such as vaccines are gaining significant attention.

CONSENT FOR PUBLICATION

Not applicable.

CONFLICT OF INTEREST

The author declares no conflict of interest, financial or otherwise.

ACKNOWLEDGEMENTS

Declared none.

REFERENCES

[1]   Inoue M, Craker LE. Medicinal and aromatic plants- uses and functions. Horticulture: Plants for people and places. 2nd vol.Dordrecht: Springer 2014; pp. 645-69.
[http://dx.doi.org/10.1007/978-94-017-8581-5_3]

[2]   Jamshidi-Kia F, Lorigooini Z, Amini-Khoei H. Medicinal plants: Past history and future perspective. J Herbmed Pharmacol 2018; 7(1): 1-7.
[http://dx.doi.org/10.15171/jhp.2018.01]

[3]   Kelly K. The history of medicine. Early civilizations, prehistoric times to 500 C.E. New York: Facts on file 2009; 29-50.

[4]   Petrovska BB. Historical review of medicinal plants' usage. Pharmacogn Rev 2012; 6(11): 1-5.
[http://dx.doi.org/10.4103/0973-7847.95849] [PMID: 22654398]

[5]   Newman DJ, Cragg GM. Natural products as sources of new drugs over the 30 years from 1981 to 2010. J Nat Prod 2012; 75(3): 311-35.
[http://dx.doi.org/10.1021/np200906s] [PMID: 22316239]

[6]   Dong J. The relationship between traditional Chinese medicine and modern medicine. Evid Based Complementary Altern Med 2013.
[http://dx.doi.org/10.1155/2013/153148]

[7]   Palhares RM, Drummond MG, BSAF Brasil. Medicinal plants recommended by the world health organization: DNA barcode identification associated with chemical analyses guarantees their quality. PLoS One 2015; 10(5): e0127866.
[http://dx.doi.org/10.1371/journal.pone.0127866] [PMID: 25978064]

[8]   Taylor L. Plant based drugs and medicines. Rain tree Nutrition Inc Carson City: 2000; 1-5.

[9]   Bukar BB, Dayom DW, Uguru MO. The growing economic importance of medicinal plants and the need for developing countries to harness from it: A mini review. IOSR J Pharm 2016; 6: 2250-3013.

[10]   Grabley S, Sattler I. Natural products for lead identification: nature is a valuable resource for providing tools. Modern methods of drug discovery. Birkhäuser Basel 2003; pp. 87-107.
[http://dx.doi.org/10.1007/978-3-0348-7997-2_5]

[11]   Schippman U, Danna J, Cunningham AB. Impact of cultivation and gathering of medicinal plants and biodiversity: Global trends and issues. FAO 2002.

[12]   Aslam MS, Ahmad MS. Worldwide importance of medicinal plants: current and historical perspectives. Recent Adv Biol Med 2016; 2: 909.
[http://dx.doi.org/10.18639/RABM.2016.02.338811]

[13]   Tschofen M, Knopp D, Hood E, Stooger E. Plant Molecular Farming: Much More than Medicines. Annu Rev Anal Chem 2016; (9): 11.1-11.24.

[14]   Paul MJ, Teh AY, Twyman RM, Ma JK. Target product selection - where can Molecular Pharming make the difference? Curr Pharm Des 2013; 19(31): 5478-85.
[http://dx.doi.org/10.2174/1381612811319310003] [PMID: 23394563]

[15]   Fraley RT, Rogers SG, Horsch RB, *et al.* Expression of bacterial genes in plant cells. Proc Natl Acad Sci USA 1983; 80(15): 4803-7.
[http://dx.doi.org/10.1073/pnas.80.15.4803] [PMID: 6308651]

[16]   During K. Wound-inducible expression and secretion of T4 lysozyme and monoclonal antibodies in Nicotiana tabacum. 1988.

[17]   Hiatt A, Cafferkey R, Bowdish K. Production of antibodies in transgenic plants. Nature 1989; 342(6245): 76-8.
[http://dx.doi.org/10.1038/342076a0] [PMID: 2509938]

[18]   Hood EE, Witcher DR, Maddock S, *et al.* Commercial production of avidin from transgenic maize:

characterization of transformant, production, processing, extraction and purification. Mol Breed 1997; 3(4): 291-306.
[http://dx.doi.org/10.1023/A:1009676322162]

[19] Zhong GY, Peterson D, Delaney DE, *et al.* Commercial production of aprotinin in transgenic maize seeds. Mol Breed 1999; 5(4): 345-56.
[http://dx.doi.org/10.1023/A:1009677809492]

[20] Ratner M. Pfizer stakes a claim in plant cell-made biopharmaceuticals. Nat Biotechnol 2010; 28(2): 107-8.
[http://dx.doi.org/10.1038/nbt0210-107] [PMID: 20139928]

[21] Moustafa K, Makhzoum A, Trémouillaux-Guiller J. Molecular farming on rescue of pharma industry for next generations. Crit Rev Biotechnol 2016; 36(5): 840-50.
[http://dx.doi.org/10.3109/07388551.2015.1049934] [PMID: 26042351]

[22] Lössl AG, Waheed MT. Chloroplast-derived vaccines against human diseases: achievements, challenges and scopes. Plant Biotechnol J 2011; 9(5): 527-39.
[http://dx.doi.org/10.1111/j.1467-7652.2011.00615.x] [PMID: 21447052]

[23] Saba K, Sameeullah M, Asghar A, *et al.* Expression of ESAT-6 antigen from *Mycobacterium* tuberculosis in broccoli: An edible plant. Biotechnol Appl Biochem 2020; 67(1): 148-57.
[http://dx.doi.org/10.1002/bab.1867] [PMID: 31898361]

[24] Qiu X, Wong G, Audet J, *et al.* Reversion of advanced Ebola virus disease in nonhuman primates with ZMapp. Nature 2014; 514(7520): 47-53.
[http://dx.doi.org/10.1038/nature13777] [PMID: 25171469]

[25] McNulty MJ, Gleba Y, Tusé D, *et al.* Techno-economic analysis of a plant-based platform for manufacturing antimicrobial proteins for food safety. Biotechnol Prog 2020; 36(1): e2896.
[http://dx.doi.org/10.1002/btpr.2896] [PMID: 31443134]

[26] Peyret H, Brown JKM, Lomonossoff GP. Improving plant transient expression through the rational design of synthetic 5′ and 3′ untranslated regions. Plant Methods 2019; 15(1): 108.
[http://dx.doi.org/10.1186/s13007-019-0494-9] [PMID: 31548848]

[27] Phoolcharoen W, Banyard AC, Prehaud C, *et al. In vitro* and *in vivo* evaluation of a single chain antibody fragment generated in planta with potent rabies neutralisation activity. Vaccine 2019; 37(33): 4673-80.
[http://dx.doi.org/10.1016/j.vaccine.2018.02.057] [PMID: 29523449]

[28] Marques LÉC, Silva BB, Dutra RF, Florean EOPT, Menassa R, Guedes MIF. Transient expression of dengue virus NS1 antigen in *Nicotiana benthamiana* for use as a diagnostic antigen. Front Plant Sci 2020; 10: 1674.
[http://dx.doi.org/10.3389/fpls.2019.01674] [PMID: 32010161]

[29] Rybicki EP. Plant molecular farming of virus-like nanoparticles as vaccines and reagents. Wiley Interdiscip Rev Nanomed Nanobiotechnol 2020; 12(2): e1587.
[http://dx.doi.org/10.1002/wnan.1587] [PMID: 31486296]

[30] World health organization, WHO. WHO Coronavirus disease (COVID-19) dashboard. 2021.
https://covid19.who.int/?gclid=CjwKCAiA_9r_BRBZEiwAHZ_v19M0hlzycTqZ2DiPK3qtk4NKoT4 PvxIXiZNieHHvtaeUQP4Skzxw3hoC1D0QAvD_BwE

[31] Rosales-Mendoza S, Márquez-Escobar VA, González-Ortega O, Nieto-Gómez R, Arévalo-Villalobos JI. What does plant-based vaccine technology offer to the fight against COVID-19? Vaccines (Basel) 2020; 8(2): 183.
[http://dx.doi.org/10.3390/vaccines8020183] [PMID: 32295153]

[32] Shanmugaraj B, Malla A, Phoolcharoen W. Emergence of novel coronavirus 2019-nCoV: need for rapid vaccine and biologics development. Pathogens 2020; 9(2): 148.

[http://dx.doi.org/10.3390/pathogens9020148] [PMID: 32098302]

[33]   Rosales-Mendoza S. Will plant-made biopharmaceuticals play a role in the fight against COVID-19? Expert Opin Biol Ther 2020; 20(6): 545-8.
[http://dx.doi.org/10.1080/14712598.2020.1752177] [PMID: 32250170]

[34]   Shanmugaraj B, Siriwattananon K, Wangkanont K, Phoolcharoen W. Perspectives on monoclonal antibody therapy as potential therapeutic intervention for Coronavirus disease-19 (COVID-19). Asian Pac J Allergy Immunol 2020; 38(1): 10-8.
[PMID: 32134278]

[35]   Malik K. Human development report 2014: Sustaining human progress: Reducing vulnerabilities and building resilience. New York: UNDP 2014.

[36]   Schmidt-Traub G, De la Mothe Karoubi E, Espey J. Indicators and a Monitoring Framework for the Sustainable Development Goals: Launching a Data Revolution for the SDGs. SDSN 2015.

[37]   Servaes J. Introduction: From MDGs to SDGs InSustainable Development Goals in the Asian Context. Singapore: Springer 2017; pp. 1-21.

[38]   Session SW. World commission on environment and development Our Common Future. Oxford, UK: Oxford University Press 1987.

[39]   Bebbington J, Larrinaga C. Accounting and sustainable development: An exploration. Account Organ Soc 2014; 39(6): 395-413.
[http://dx.doi.org/10.1016/j.aos.2014.01.003]

[40]   Bebbington J, Unerman J, O'Dwyer B. Sustainability Accounting and Accountability. 2nd ed., London: Routledge 2014.

[41]   Desa UN. Transforming our world: The 2030 agenda for sustainable development. New York, NY: United Nations 2012.

[42]   Bebbington J, Unerman J. Achieving the United Nations Sustainable Development Goals. Account Audit Account J 2018; 31(1): 2-24.
[http://dx.doi.org/10.1108/AAAJ-05-2017-2929]

[43]   Le Blanc D. Towards integration at last? The sustainable development goals as a network of targets. Sustain Dev (Bradford) 2015; 23(3): 176-87.
[http://dx.doi.org/10.1002/sd.1582]

[44]   The Sustainable Development Goals Report. United Nations Publications 2016.

[45]   Maliga P, Bock R. Plastid biotechnology: food, fuel, and medicine for the 21st century. Plant Physiol 2011; 155(4): 1501-10.
[http://dx.doi.org/10.1104/pp.110.170969] [PMID: 21239622]

[46]   Stoger E, Fischer R, Moloney M, Ma JK. Plant molecular pharming for the treatment of chronic and infectious diseases. Annu Rev Plant Biol 2014; 65: 743-68.
[http://dx.doi.org/10.1146/annurev-arplant-050213-035850] [PMID: 24579993]

[47]   Landry N, Ward BJ, Trépanier S, *et al.* Preclinical and clinical development of plant-made virus-like particle vaccine against avian H5N1 influenza. PLoS One 2010; 5(12): e15559.
[http://dx.doi.org/10.1371/journal.pone.0015559] [PMID: 21203523]

[48]   Ma JK, Hiatt A, Hein M, *et al.* Generation and assembly of secretory antibodies in plants. Science 1995; 268(5211): 716-9.
[http://dx.doi.org/10.1126/science.7732380] [PMID: 7732380]

[49]   Sehnke PC, Pedrosa L, Paul AL, Frankel AE, Ferl RJ. Expression of active, processed ricin in transgenic tobacco. J Biol Chem 1994; 269(36): 22473-6.
[http://dx.doi.org/10.1016/S0021-9258(17)31668-X] [PMID: 8077191]

[50]   Arntzen CJ, Ahoney RT. Plant-derived vaccines: a new approach to international public health. J food

sci 2004; 69(1): CRH8-CRH10.
[http://dx.doi.org/10.1111/j.1365-2621.2004.tb17838.x]

[51]   World Health Organizatioin, WHO. Ten threats to global health in 2019. . 2019.https://www.who.int/emergencies/ten-threats-to-global-health-in-2019

[52]   Bell DM. World Health Organization Working Group on prevention of international and community transmission of SARS. Public health interventions and SARS spread 2003; 1900-6.

[53]   Dandagi GL, Byahatti SM. An insight into the swine-influenza A (H1N1) virus infection in humans. Lung India 2011; 28(1): 34-8.
[http://dx.doi.org/10.4103/0970-2113.76299] [PMID: 21654984]

[54]   Kaner J, Schaack S. Understanding Ebola: the 2014 epidemic. Global Health 2016; 12(1): 53.
[http://dx.doi.org/10.1186/s12992-016-0194-4] [PMID: 27624088]

[55]   Aleanizy FS, Mohmed N, Alqahtani FY, El Hadi Mohamed RA. Outbreak of Middle East respiratory syndrome *coronavirus* in Saudi Arabia: a retrospective study. BMC Infect Dis 2017; 17(1): 23.
[http://dx.doi.org/10.1186/s12879-016-2137-3] [PMID: 28056850]

[56]   Pereira AM, Monteiro DLM, Werner H, *et al.* Zika virus and pregnancy in Brazil: What happened? J Turk Ger Gynecol Assoc 2018; 19(1): 39-47.
[http://dx.doi.org/10.4274/jtgga.2017.0072] [PMID: 29503261]

[57]   Bernasconi V, Kristiansen PA, Whelan M, *et al.* Developing vaccines against epidemic-prone emerging infectious diseases. Bundesgesundheitsblatt Gesundheitsforschung Gesundheitsschutz 2020; 63(1): 65-73.
[http://dx.doi.org/10.1007/s00103-019-03061-2] [PMID: 31776599]

[58]   Shanmugaraj B, I Bulaon CJ, Phoolcharoen W. I Bulaon CJ, Phoolcharoen W. Plant molecular farming: a viable platform for recombinant biopharmaceutical production. Plants 2020; 9(7): 842.
[http://dx.doi.org/10.3390/plants9070842] [PMID: 32635427]

[59]   2020.https://www.who.int/activities/prioritizing-diseases-for-research-and-developm-nt-in-emergency-contexts

[60]   Orenstein WA, Ahmed R. Simply put: Vaccination saves lives. Proc Natl Acad Sci USA 2017; 114(16): 4031-3.
[http://dx.doi.org/10.1073/pnas.1704507114] [PMID: 28396427]

[61]   Røttingen JA, Gouglas D, Feinberg M, *et al.* New vaccines against epidemic infectious diseases. N Engl J Med 2017; 376(7): 610-3.
[http://dx.doi.org/10.1056/NEJMp1613577] [PMID: 28099066]

[62]   Oyston P, Robinson K. The current challenges for vaccine development. J Med Microbiol 2012; 61(Pt 7): 889-94.
[http://dx.doi.org/10.1099/jmm.0.039180-0] [PMID: 22322337]

[63]   Mukherjee S. Emerging infectious diseases: epidemiological perspective. Indian J Dermatol 2017; 62(5): 459-67.
[PMID: 28979007]

[64]   Huber C, Finelli L, Stevens W. The economic and social burden of the 2014 Ebola outbreak in West Africa. J Infect Dis 2018; 218 (Suppl. 5): S698-704.
[http://dx.doi.org/10.1093/infdis/jiy213] [PMID: 30321368]

[65]   Anonymous . Coalition for Epidemic Preparedness Innovation CEPI 2019.http://www.cepi.net

[66]   Jodie R. Private sector companies provide financial support for CEPI's COVID-19 vaccine programmes. CEPI 2020.https://cepi.net/news_cepi/private-sector-companies-provide-finan-ial-support-for- cepis-covid-19-vaccine-programmes/

[67]   Plotkin S, Robinson JM, Cunningham G, Iqbal R, Larsen S. The complexity and cost of vaccine manufacturing - An overview. Vaccine 2017; 35(33): 4064-71.

[http://dx.doi.org/10.1016/j.vaccine.2017.06.003] [PMID: 28647170]

[68] Drury G, Jolliffe S, Mukhopadhyay TK. Process mapping of vaccines: Understanding the limitations in current response to emerging epidemic threats. Vaccine 2019; 37(17): 2415-21.
[http://dx.doi.org/10.1016/j.vaccine.2019.01.050] [PMID: 30910404]

[69] Chabeda A, van Zyl AR, Rybicki EP, Hitzeroth II. Substitution of human papillomavirus type 16 l2 neutralizing epitopes into l1 surface loops: the effect on virus-like particle assembly and immunogenicity. Front plant sci 2019; 10: 779.
[http://dx.doi.org/10.3389/fpls.2019.00779] [PMID: 31281327]

[70] Ahmad P, Ashraf M, Younis M, *et al.* Role of transgenic plants in agriculture and biopharming. Biotechnol Adv 2012; 30(3): 524-40.
[http://dx.doi.org/10.1016/j.biotechadv.2011.09.006] [PMID: 21959304]

[71] Batabyal AA, Higano Y, Nijkamp P, Eds. Disease, Human Health, and Regional Growth and Development in Asia. Springer 2019.38
[http://dx.doi.org/10.1007/978-981-13-6268-2]

[72] Aqeel M, Noman A, Sanaullah T, *et al.* Characterization of genetically modified plants producing bioactive compounds for human health: a systemic review. Int j agric biol 2019; 22(6): 1293-304.

[73] Noman A, Aqeel M, Deng J, Khalid N, Sanaullah T, Shuilin H. Biotechnological advancements for improving floral attributes in ornamental plants. Front Plant Sci 2017; 8: 530.
[http://dx.doi.org/10.3389/fpls.2017.00530] [PMID: 28473834]

[74] Hussain A, Li X, Weng Y, *et al.* CaWRKY22 acts as a positive regulator in pepper response to Ralstonia solanacearum by constituting networks with CaWRKY6, CaWRKY27, CaWRKY40, and CaWRKY58. Int J Mol Sci 2018; 19(5): 1426.
[http://dx.doi.org/10.3390/ijms19051426] [PMID: 29747470]

[75] Ifnan Khan M, Zhang Y, Liu Z, *et al.* CaWRKY40b in pepper acts as a negative regulator in response to Ralstonia solanacearum by directly modulating defense genes including CaWRKY40. Int J Mol Sci 2018; 19(5): 1403.
[http://dx.doi.org/10.3390/ijms19051403] [PMID: 29738468]

[76] Islam W, Noman A, Qasim M, Wang L. Plant responses to pathogen attack: small RNAs in focus. Int J Mol Sci 2018; 19(2): 515.
[http://dx.doi.org/10.3390/ijms19020515] [PMID: 29419801]

[77] Maughan RJ, Burke LM, Dvorak J, *et al.* IOC consensus statement: dietary supplements and the high-performance athlete. INT J SPORT NUTR EXE 2018; 28(2): 104-25.
[http://dx.doi.org/10.1123/ijsnem.2018-0020] [PMID: 29589768]

[78] Sym D, Patel PN, El-Chaar GM. Seasonal, avian, and novel H1N1 influenza: prevention and treatment modalities. Ann Pharmacother 2009; 43(12): 2001-11.
[http://dx.doi.org/10.1345/aph.1M557] [PMID: 19920156]

[79] Rybicki E. History and Promise of plant-made vaccines for animals. Prospects of plant-based vaccines in veterinary medicine. Cham: Springer 2018; pp. 1-22.
[http://dx.doi.org/10.1007/978-3-319-90137-4_1]

[80] Hurtado J, Acharya D, Lai H, *et al. In vitro* and *in vivo* efficacy of anti-chikungunya virus monoclonal antibodies produced in wild-type and glycoengineered *Nicotiana benthamiana* plants. Plant Biotechnol J 2020; 18(1): 266-73.
[http://dx.doi.org/10.1111/pbi.13194] [PMID: 31207008]

[81] Schilte C, Staikowsky F, Couderc T, *et al.* Chikungunya virus-associated long-term arthralgia: a 36-month prospective longitudinal study. PLoS Negl Trop Dis 2013; 7(3): e2137.
[http://dx.doi.org/10.1371/journal.pntd.0002137] [PMID: 23556021]

[82] Weaver SC, Forrester NL. Chikungunya: Evolutionary history and recent epidemic spread. Antiviral Res 2015; 120: 32-9.

[http://dx.doi.org/10.1016/j.antiviral.2015.04.016] [PMID: 25979669]

[83] Pal P, Dowd KA, Brien JD, *et al.* Development of a highly protective combination monoclonal antibody therapy against Chikungunya virus. PLoS Pathog 2013; 9(4): e1003312.
[http://dx.doi.org/10.1371/journal.ppat.1003312] [PMID: 23637602]

[84] Hoogstraal H. The epidemiology of tick-borne Crimean-Congo hemorrhagic fever in Asia, Europe, and Africa. J Med Entomol 1979; 15(4): 307-417.
[http://dx.doi.org/10.1093/jmedent/15.4.307] [PMID: 113533]

[85] Whitehouse CA. Crimean-Congo hemorrhagic fever. Antiviral Res 2004; 64(3): 145-60.
[http://dx.doi.org/10.1016/j.antiviral.2004.08.001] [PMID: 15550268]

[86] Ghiasi SM, Salmanian AH, Chinikar S, Zakeri S. Mice orally immunized with a transgenic plant expressing the glycoprotein of Crimean-Congo hemorrhagic fever virus. Clin Vaccine Immunol 2011; 18(12): 2031-7.
[http://dx.doi.org/10.1128/CVI.05352-11] [PMID: 22012978]

[87] Wilson ML, Gonzalez JP, LeGuenno B, *et al.* Epidemiology of crimean-congo hemorrhagic fever in senegal: temporal and spatial patterns.Hemorrhagic fever with renal syndrome, tick-and mosquito-borne viruses. Vienna: Springer 1990; pp. 323-40.
[http://dx.doi.org/10.1007/978-3-7091-9091-3_35]

[88] Kautman M, Tiar G, Papa A, Široký P. AP92-like Crimean-Congo hemorrhagic fever virus in Hyalomma aegyptium ticks, Algeria. Emerg Infect Dis 2016; 22(2): 354-6.
[http://dx.doi.org/10.3201/eid2202.151528] [PMID: 26812469]

[89] Negredo A, de la Calle-Prieto F, Palencia-Herrejón E, *et al.* Autochthonous Crimean-Congo Hemorrhagic Fever in Spain. N Engl J Med 2017; 377(2): 154-61.
[http://dx.doi.org/10.1056/NEJMoa1615162] [PMID: 28700843]

[90] Tipih T, Burt FJ. Crimean congo hemorrhagic fever virus: advances in vaccine development. Bioresearch open access 2020; 9(1): 137-50.

[91] Whitehouse CA. Risk groups and control measures for Crimean-Congo hemorrhagic fever.Crimean-Congo Hemorrhagic Fever. Dordrecht, Netherlands: Springer 2007; pp. 273-80.
[http://dx.doi.org/10.1007/978-1-4020-6106-6_20]

[92] Jacob ST, Crozier I, Fischer WA II, *et al.* Ebola virus disease. Nat Rev Dis Primers 2020; 6(1): 13.
[http://dx.doi.org/10.1038/s41572-020-0147-3] [PMID: 32080199]

[93] Kuhn JH, Amarasinghe GK, Basler CF, *et al.* ICTV virus taxonomy profile: Filoviridae. J Gen Virol 2019; 100(6): 911-2.
[http://dx.doi.org/10.1099/jgv.0.001252] [PMID: 31021739]

[94] Kuhn JH, Adachi T, Adhikari NKJ, *et al.* New filovirus disease classification and nomenclature. Nat Rev Microbiol 2019; 17(5): 261-3.
[http://dx.doi.org/10.1038/s41579-019-0187-4] [PMID: 30926957]

[95] Mulangu S, Dodd LE, Davey RT Jr, *et al.* A randomized, controlled trial of Ebola virus disease therapeutics. N Engl J Med 2019; 381(24): 2293-303.
[http://dx.doi.org/10.1056/NEJMoa1910993] [PMID: 31774950]

[96] Clark DV, Jahrling PB, Lawler JV. Clinical management of filovirus-infected patients. Viruses 2012; 4(9): 1668-86.
[http://dx.doi.org/10.3390/v4091668] [PMID: 23170178]

[97] Lau JM, Korban SS. Transgenic apple expressing an antigenic protein of the human respiratory syncytial virus. J Plant Physiol 2010; 167(11): 920-7.
[http://dx.doi.org/10.1016/j.jplph.2010.02.003] [PMID: 20307914]

[98] Ortega-Berlanga B, Musiychuk K, Shoji Y, *et al.* Engineering and expression of a RhoA peptide against respiratory syncytial virus infection in plants. Planta 2016; 243(2): 451-8.

[http://dx.doi.org/10.1007/s00425-015-2416-z] [PMID: 26474991]

[99]   Rudraraju R, Jones BG, Sealy R, Surman SL, Hurwitz JL. Respiratory syncytial virus: current progress in vaccine development. Viruses 2013; 5(2): 577-94.
[http://dx.doi.org/10.3390/v5020577] [PMID: 23385470]

[100]   Fernández P, Trenholme A, Abarca K, *et al.* A phase 2, randomized, double-blind safety and pharmacokinetic assessment of respiratory syncytial virus (RSV) prophylaxis with motavizumab and palivizumab administered in the same season. BMC Pediatr 2010; 10(1): 38.
[http://dx.doi.org/10.1186/1471-2431-10-38] [PMID: 20525274]

[101]   Márquez-Escobar VA, Rosales-Mendoza S, Beltrán-López JI, González-Ortega O. Plant-based vaccines against respiratory diseases: current status and future prospects. Expert Rev Vaccines 2017; 16(2): 137-49.
[http://dx.doi.org/10.1080/14760584.2017.1232167] [PMID: 27599605]

[102]   X. JJ. Traditional Chinese Medicine Treatment of Viral Infectious Diseases. Beijing: Science and Technology Press 2010.

[103]   H Brody. Influenza. Nature 2019; 573(7774): S49.
[http://dx.doi.org/10.1038/d41586-019-02750-x] [PMID: 31534258]

[104]   Webster RG, Bean WJ, Gorman OT, Chambers TM, Kawaoka Y. Evolution and ecology of influenza A viruses. Microbiol Rev 1992; 56(1): 152-79.
[http://dx.doi.org/10.1128/mr.56.1.152-179.1992] [PMID: 1579108]

[105]   Kawaoka Y. Influenza Virology: Current Topics. England: Caister Academic Press 2006.

[106]   Huang IC, Li W, Sui J, Marasco W, Choe H, Farzan M. Influenza A virus neuraminidase limits viral superinfection. J Virol 2008; 82(10): 4834-43.
[http://dx.doi.org/10.1128/JVI.00079-08] [PMID: 18321971]

[107]   Lu ZYZN. Internal Medicine. Beijing: People's Medical Publishing House 2008.

[108]   Kumar B, Asha K, Khanna M, Ronsard L, Meseko CA, Sanicas M. The emerging influenza virus threat: status and new prospects for its therapy and control. Arch Virol 2018; 163(4): 831-44.
[http://dx.doi.org/10.1007/s00705-018-3708-y] [PMID: 29322273]

[109]   Taubenberger JK, Morens DM. Influenza: the once and future pandemic. Public Health Rep 2010; 125(3) (Suppl. 3): 16-26.
[PMID: 20568566]

[110]   Patel M, Dennis A, Flutter C, Khan Z. Pandemic (H1N1) 2009 influenza. Br J Anaesth 2010; 104(2): 128-42.
[http://dx.doi.org/10.1093/bja/aep375] [PMID: 20053625]

[111]   Treanor J. Influenza vaccine--outmaneuvering antigenic shift and drift. N Engl J Med 2004; 350(3): 218-20.
[http://dx.doi.org/10.1056/NEJMp038238] [PMID: 14724300]

[112]   Xiong Y, Li NX, Duan N, *et al.* Traditional Chinese Medicine in Treating Influenza: From Basic Science to Clinical Applications. Front Pharmacol 2020; 11: 575803.
[http://dx.doi.org/10.3389/fphar.2020.575803] [PMID: 33041821]

[113]   2006.http://www.who.int/csr/resources/publications/influenza/StregPlanEPR_GIP_2006_2.pdf

[114]   D'Aoust MA, Lavoie PO, Couture MM, *et al.* Influenza virus-like particles produced by transient expression in *Nicotiana benthamiana* induce a protective immune response against a lethal viral challenge in mice. Plant Biotechnol J 2008; 6(9): 930-40.
[http://dx.doi.org/10.1111/j.1467-7652.2008.00384.x] [PMID: 19076615]

[115]   Giritch A, Marillonnet S, Engler C, *et al.* Rapid high-yield expression of full-size IgG antibodies in plants coinfected with noncompeting viral vectors. Proc Natl Acad Sci USA 2006; 103(40): 14701-6.
[http://dx.doi.org/10.1073/pnas.0606631103] [PMID: 16973752]

[116] 2016.https://www.who.int/csr/disease/ebola/situation-reports/archive/en/

[117] Nyakarahuka L, Ojwang J, Tumusiime A, *et al.* Isolated case of Marburg virus disease, Kampala, Uganda, 2014. Emerg Infect Dis 2017; 23(6): 1001-4.
[http://dx.doi.org/10.3201/eid2306.170047] [PMID: 28518032]

[118] Selvaraj SA, Lee KE, Harrell M, Ivanov I, Allegranzi B. Infection rates and risk factors for infection among health workers during Ebola and Marburg virus outbreaks: a systematic review. J Infect Dis 2018; 218 (Suppl. 5): S679-89.
[http://dx.doi.org/10.1093/infdis/jiy435] [PMID: 30202878]

[119] Mire CE, Geisbert JB, Borisevich V, *et al.* Therapeutic treatment of Marburg and Ravn virus infection in nonhuman primates with a human monoclonal antibody. Sci Transl Med 2017; 9(384): eaai8711.
[http://dx.doi.org/10.1126/scitranslmed.aai8711] [PMID: 28381540]

[120] Kortepeter MG, Dierberg K, Shenoy ES, Cieslak TJ. Marburg virus disease: A summary for clinicians. Int J Infect Dis 2020; 99: 233-42.
[http://dx.doi.org/10.1016/j.ijid.2020.07.042] [PMID: 32758690]

[121] Pepin M, Bouloy M, Bird BH, Kemp A, Paweska J. Rift Valley fever virus(Bunyaviridae: Phlebovirus): an update on pathogenesis, molecular epidemiology, vectors, diagnostics and prevention. Vet Res 2010; 41(6): 61.
[http://dx.doi.org/10.1051/vetres/2010033] [PMID: 21188836]

[122] Rich KM, Wanyoike F. An assessment of the regional and national socio-economic impacts of the 2007 Rift Valley fever outbreak in Kenya. Am J Trop Med 2010; 83(2) (Suppl.): 52-7.
[http://dx.doi.org/10.4269/ajtmh.2010.09-0291] [PMID: 20682906]

[123] Hassan OA, Ahlm C, Sang R, Evander M. The 2007 Rift Valley fever outbreak in Sudan. PLoS Negl Trop Dis 2011; 5(9): e1229.
[http://dx.doi.org/10.1371/journal.pntd.0001229] [PMID: 21980543]

[124] Dungu B, Donadeu M, Bouloy M. Vaccination for the control of Rift Valley fever in enzootic and epizootic situations.Vaccines and Diagnostics for Transboundary Animal Diseases. Karger Publishers 2013; 135: pp. 61-72.
[http://dx.doi.org/10.1159/000157178]

[125] Bird BH, Nichol ST. Breaking the chain: Rift Valley fever virus control *via* livestock vaccination. Curr Opin Virol 2012; 2(3): 315-23.
[http://dx.doi.org/10.1016/j.coviro.2012.02.017] [PMID: 22463980]

[126] Rosales-Mendoza S, Govea-Alonso DO, Monreal-Escalante E, Fragoso G, Sciutto E. Developing plant-based vaccines against neglected tropical diseases: where are we? Vaccine 2012; 31(1): 40-8.
[http://dx.doi.org/10.1016/j.vaccine.2012.10.094] [PMID: 23142588]

[127] Zhong NS, Zheng BJ, Li YM, *et al.* Epidemiology and cause of severe acute respiratory syndrome (SARS) in Guangdong, People's Republic of China, in February, 2003. Lancet 2003; 362(9393): 1353-8.
[http://dx.doi.org/10.1016/S0140-6736(03)14630-2] [PMID: 14585636]

[128] Demurtas OC, Massa S, Illiano E, *et al.* Antigen production in plant to tackle infectious diseases flare up: the case of SARS. Front Plant Sci 2016; 7: 54.
[http://dx.doi.org/10.3389/fpls.2016.00054] [PMID: 26904039]

[129] Bhatnagar PK, Das D, Suresh MR. Molecular targets for diagnostics and therapeutics of severe acute respiratory syndrome (SARS-CoV). Journal of pharmacy and pharmaceutical sciences: a publication of the Canadian Society for Pharmaceutical Sciences, Societe canadienne des sciences pharmaceutiques 2008; 11(2)

[130] Stockman LJ, Bellamy R, Garner P. SARS: systematic review of treatment effects. PLoS Med 2006; 3(9): e343.
[http://dx.doi.org/10.1371/journal.pmed.0030343] [PMID: 16968120]

[131] Department of Health, and Human Services (HHS): possession, use, and transfer of select agents and toxins; biennial review. Final Rule Fed Regist 2012; 77(194): 61083-115.
[PMID: 23038847]

[132] Du L, He Y, Zhou Y, Liu S, Zheng BJ, Jiang S. The spike protein of SARS-CoV--a target for vaccine and therapeutic development. Nat Rev Microbiol 2009; 7(3): 226-36.
[http://dx.doi.org/10.1038/nrmicro2090] [PMID: 19198616]

[133] Pogrebnyak N, Golovkin M, Andrianov V, *et al.* Severe acute respiratory syndrome (SARS) S protein production in plants: development of recombinant vaccine. Proc Natl Acad Sci USA 2005; 102(25): 9062-7.
[http://dx.doi.org/10.1073/pnas.0503760102] [PMID: 15956182]

[134] Swan GK, Seng TC, Othman RY, Harikrishna JA. Transient expression of an immunogenic envelope attachment glycoprotein of Nipah virus in *Nicotiana benthamiana.* Sains Malays 2018; 47(3): 499-509.
[http://dx.doi.org/10.17576/jsm-2018-4703-09]

[135] Halpin K, Hyatt AD, Fogarty R, *et al.* Pteropid bats are confirmed as the reservoir hosts of henipaviruses: a comprehensive experimental study of virus transmission. Am J Trop 2011; 85(5): 946-51.
[http://dx.doi.org/10.4269/ajtmh.2011.10-0567] [PMID: 22049055]

[136] Wong KT, Ong KC. Pathology of acute henipavirus infection in humans and animals. Pathol Int 2011.
[http://dx.doi.org/10.4061/2011/567248]

[137] Geisbert TW, Feldmann H, Broder CC. Animal challenge models of henipavirus infection and pathogenesis.Henipavirus Berlin. Heidelberg: Springer 2012; pp. 153-77.
[http://dx.doi.org/10.1007/82_2012_208]

[138] Broder CC. Henipavirus outbreaks to antivirals: the current status of potential therapeutics. Curr Opin Virol 2012; 2(2): 176-87.
[http://dx.doi.org/10.1016/j.coviro.2012.02.016] [PMID: 22482714]

[139] Satterfield BA, Dawes BE, Milligan GN. Status of vaccine research and development of vaccines for Nipah virus. Vaccine 2016; 34(26): 2971-5.
[http://dx.doi.org/10.1016/j.vaccine.2015.12.075] [PMID: 26973068]

[140] Kumar AA, Kumar AA. Deadly Nipah outbreak in Kerala: Lessons learned for the future. Indian J Crit Care Med: peer-reviewed, official publication of Indian Society of Critical Care Medicine 2018; 22(7): 475.

[141] Attar N. Zika virus circulates in new regions. Nat Rev Microbiol 2016; 14(2): 62-2.
[http://dx.doi.org/10.1038/nrmicro.2015.28] [PMID: 26751511]

[142] Cao-Lormeau VM, Blake A, Mons S, *et al.* Guillain-Barré Syndrome outbreak associated with Zika virus infection in French Polynesia: a case-control study. Lancet 2016; 387(10027): 1531-9.
[http://dx.doi.org/10.1016/S0140-6736(16)00562-6] [PMID: 26948433]

[143] Gubler DJ, Kuno G, Markoff L. Flaviviruses.Knipe DM, Howley PM, Griffi n DE, Lamb RA, Martin MA, et al. 5th ed. Philadelphia, PA: Lippincott Williams & Wilkins Publishers 2007; pp. 1155-227.

[144] Yang M, Sun H, Lai H, Hurtado J, Chen Q. Plant-produced Zika virus envelope protein elicits neutralizing immune responses that correlate with protective immunity against Zika virus in mice. Plant Biotechnol J 2018; 16(2): 572-80.
[http://dx.doi.org/10.1111/pbi.12796] [PMID: 28710796]

[145] Abbink P, Larocca RA, De La Barrera RA, *et al.* Protective efficacy of multiple vaccine platforms against Zika virus challenge in rhesus monkeys. Science 2016; 353(6304): 1129-32.
[http://dx.doi.org/10.1126/science.aah6157] [PMID: 27492477]

[146] Larocca RA, Abbink P, Peron JP, *et al.* Vaccine protection against Zika virus from Brazil. Nature

2016; 536(7617): 474-8.
[http://dx.doi.org/10.1038/nature18952] [PMID: 27355570]

[147] Pardi N, Weissman D. Nucleoside modified mRNA vaccines for infectious diseases.New York: Human Press 2017; pp. 109-21.
[http://dx.doi.org/10.1007/978-1-4939-6481-9_6]

[148] Margolin E, Chapman R, Meyers AE, *et al.* Production and immunogenicity of soluble plant-produced HIV-1 subtype C envelope gp140 immunogens. Front Plant Sci 2019; 10: 1378.
[http://dx.doi.org/10.3389/fpls.2019.01378] [PMID: 31737007]

[149] Lindh I, Bråve A, Hallengärd D, *et al.* Oral delivery of plant-derived HIV-1 p24 antigen in low doses shows a superior priming effect in mice compared to high doses. Vaccine 2014; 32(20): 2288-93.
[http://dx.doi.org/10.1016/j.vaccine.2014.02.073] [PMID: 24631072]

[150] Rosales-Mendoza S, Rubio-Infante N, Monreal-Escalante E, *et al.* Chloroplast expression of an HIV envelop-derived multiepitope protein: towards a multivalent plant-based vaccine. Plant Cell Tissue Organ Cult 2014; 116(1): 111-23.
[http://dx.doi.org/10.1007/s11240-013-0387-y]

[151] Rubio-Infante N, Govea-Alonso DO, Romero-Maldonado A, *et al.* A Plant-Derived Multi-HIV Antigen Induces Broad Immune Responses in Orally Immunized Mice. Mol Biotechnol 2015; 57(7): 662-74.
[http://dx.doi.org/10.1007/s12033-015-9856-3] [PMID: 25779638]

[152] Vamvaka E, Twyman RM, Murad AM, *et al.* Rice endosperm produces an underglycosylated and potent form of the HIV-neutralizing monoclonal antibody 2G12. Plant Biotechnol J 2016; 14(1): 97-108.
[http://dx.doi.org/10.1111/pbi.12360] [PMID: 25845722]

[153] Gorantala J, Grover S, Rahi A, *et al.* Generation of protective immune response against anthrax by oral immunization with protective antigen plant-based vaccine. J Biotechnol 2014; 176: 1-10.
[http://dx.doi.org/10.1016/j.jbiotec.2014.01.033] [PMID: 24548460]

[154] Mamedov T, Chichester JA, Jones RM, *et al.* Production of functionally active and immunogenic non-glycosylated protective antigen from Bacillus anthracis in Nicotiana benthamiana by co-expression with peptide-N-glycosidase F (PNGase F) of Flavobacterium meningosepticum. PLoS One 2016; 11(4): e0153956.
[http://dx.doi.org/10.1371/journal.pone.0153956] [PMID: 27101370]

[155] Pasquevich KA, Ibañez AE, Coria LM, *et al.* An oral vaccine based on U-Omp19 induces protection against B. abortus mucosal challenge by inducing an adaptive IL-17 immune response in mice. PLoS One 2011; 6(1): e16203.
[http://dx.doi.org/10.1371/journal.pone.0016203] [PMID: 21264260]

[156] Yuki Y, Mejima M, Kurokawa S, *et al.* RNAi suppression of rice endogenous storage proteins enhances the production of rice-based Botulinum neutrotoxin type A vaccine. Vaccine 2012; 30(28): 4160-6.
[http://dx.doi.org/10.1016/j.vaccine.2012.04.064] [PMID: 22554467]

[157] Rattanapisit K, Phakham T, Buranapraditkun S, *et al.* Structural and *in vitro* functional analyses of novel plant-produced anti-human PD1 antibody. Sci Rep 2019; 9(1): 15205.
[http://dx.doi.org/10.1038/s41598-019-51656-1] [PMID: 31645587]

[158] Buyel JF, Bautista JA, Fischer R, Yusibov VM. Extraction, purification and characterization of the plant-produced HPV16 subunit vaccine candidate E7 GGG. J Chromatogr B Analyt Technol Biomed Life Sci 2012; 880(1): 19-26.
[http://dx.doi.org/10.1016/j.jchromb.2011.11.010] [PMID: 22134037]

[159] Waheed MT, Thönes N, Müller M, *et al.* Plastid expression of a double-pentameric vaccine candidate containing human papillomavirus-16 L1 antigen fused with LTB as adjuvant: transplastomic plants show pleiotropic phenotypes. Plant Biotechnol J 2011; 9(6): 651-60.

[http://dx.doi.org/10.1111/j.1467-7652.2011.00612.x] [PMID: 21447051]

[160] Zahin M, Joh J, Khanal S, *et al.* Scalable production of HPV16 L1 protein and VLPs from tobacco leaves. PLoS One 2016; 11(8): e0160995.
[http://dx.doi.org/10.1371/journal.pone.0160995] [PMID: 27518899]

[161] Yuki Y, Mejima M, Kurokawa S, *et al.* Induction of toxin-specific neutralizing immunity by molecularly uniform rice-based oral cholera toxin B subunit vaccine without plant-associated sugar modification. Plant Biotechnol J 2013; 11(7): 799-808.
[http://dx.doi.org/10.1111/pbi.12071] [PMID: 23601492]

[162] Park SR, Lim CY, Kim DS, Ko K. Optimization of ammonium sulfate concentration for purification of colorectal cancer vaccine candidate recombinant protein GA733-Fck isolated from plants. Front Plant Sci 2015; 6(NOVEMBER): 1040.
[http://dx.doi.org/10.3389/fpls.2015.01040] [PMID: 26640471]

[163] Ahn J, Lee KJ, Ko K. Optimization of ELISA conditions to quantify colorectal cancer antigen-antibody complex protein (GA733-FcK) expressed in transgenic plant. Monoclon Antib Immunodiagn Immunother 2014; 33(1): 1-7.

[164] Lu Z, Lee KJ, Shao Y, *et al.* Expression of GA733-Fc fusion protein as a vaccine candidate for colorectal cancer in transgenic plants. J Biomed Biotechnol 2012.

[165] So Y, Lee KJ, Kim DS, *et al.* Glycomodification and characterization of anti-colorectal cancer immunotherapeutic monoclonal antibodies in transgenic tobacco. Plant Cell Tissue Organ Cult 2013; 113(1): 41-9.
[http://dx.doi.org/10.1007/s11240-012-0249-z]

[166] Iyappan G, Shanmugaraj BM, Inchakalody V, Ma J-C, Ramalingam S. Potential of plant biologics to tackle the epidemic like situations-case studies involving viral and bacterial candidates. Int J Infect Dis 2018; 73: 363.
[http://dx.doi.org/10.1016/j.ijid.2018.04.4236]

[167] Monreal-Escalante E, Govea-Alonso DO, Hernández M, *et al.* Towards the development of an oral vaccine against porcine cysticercosis: expression of the protective HP6/TSOL18 antigen in transgenic carrots cells. Planta 2016; 243(3): 675-85.
[http://dx.doi.org/10.1007/s00425-015-2431-0] [PMID: 26613600]

[168] Kim M-Y, Jang Y-S, Yang M-S, Kim T-G. High expression of consensus dengue virus envelope glycoprotein domain III using a viral expression system in tobacco. Plant Cell Tissue Organ Cult 2015; 122(2): 445-51.
[http://dx.doi.org/10.1007/s11240-015-0781-8]

[169] Gottschamel J, Lössl A, Ruf S, *et al.* Production of dengue virus envelope protein domain III-based antigens in tobacco chloroplasts using inducible and constitutive expression systems. Plant Mol Biol 2016; 91(4-5): 497-512.
[http://dx.doi.org/10.1007/s11103-016-0484-5] [PMID: 27116001]

[170] Dent M, Hurtado J, Paul AM, *et al.* Plant-produced anti-dengue virus monoclonal antibodies exhibit reduced antibody-dependent enhancement of infection activity. J Gen Virol 2016; 97(12): 3280-90.
[http://dx.doi.org/10.1099/jgv.0.000635] [PMID: 27902333]

[171] Loc NH, Long DT, Kim T-G, Yang M-S. Expression of Escherichia coli heat-labile enterotoxin B subunit in transgenic tomato (*Solanum lycopersicum* L.) fruit. Czech J Genet Plant Breed 2014; 50(1): 26-31.
[http://dx.doi.org/10.17221/77/2013-CJGPB]

[172] Lentz EM, Mozgovoj MV, Bellido D, Dus Santos MJ, Wigdorovitz A, Bravo-Almonacid FF. VP8* antigen produced in tobacco transplastomic plants confers protection against bovine rotavirus infection in a suckling mouse model. J Biotechnol 2011; 156(2): 100-7.
[http://dx.doi.org/10.1016/j.jbiotec.2011.08.023] [PMID: 21893114]

[173] Tokuhara D, Álvarez B, Mejima M, *et al.* Rice-based oral antibody fragment prophylaxis and therapy against rotavirus infection. J Clin Invest 2013; 123(9): 3829-38.
[http://dx.doi.org/10.1172/JCI70266] [PMID: 23925294]

[174] Mathew LG, Herbst-Kralovetz MM, Mason HS. Norovirus Narita 104 virus-like particles expressed in Nicotiana benthamiana induce serum and mucosal immune responses. Biomed Res Int 2014.

[175] Souza AC, Vasques RM, Inoue-Nagata AK, *et al.* Expression and assembly of Norwalk virus-like particles in plants using a viral RNA silencing suppressor gene. Appl Microbiol Biotechnol 2013; 97(20): 9021-7.
[http://dx.doi.org/10.1007/s00253-013-5077-5] [PMID: 23925532]

[176] Bergeron-Sandoval LP, Girard A, Ouellet F, Archambault D, Sarhan F. Production of human rotavirus and Salmonella antigens in plants and elicitation of fljB-specific humoral responses in mice. Mol Biotechnol 2011; 47(2): 157-68.
[http://dx.doi.org/10.1007/s12033-010-9324-z] [PMID: 20725806]

[177] Dadmehr M, Korouzhdehi B, Rahbarizadeh F, Piri I. Isolation of OmpA gene from Salmonella typhimurium and transformation into alfalfa in order to develop an edible plant based vaccine. Afr J Biotechnol 2011; 10(5): 854-9.

[178] Diamos AG, Hunter JGL, Pardhe MD, *et al.* High level production of monoclonal antibodies using an optimized plant expression system. Front Bioeng Biotechnol 2020; 7: 472.
[http://dx.doi.org/10.3389/fbioe.2019.00472] [PMID: 32010680]

[179] Lai H, He J, Engle M, Diamond MS, Chen Q. Robust production of virus-like particles and monoclonal antibodies with geminiviral replicon vectors in lettuce. Plant Biotechnol J 2012; 10(1): 95-104.
[http://dx.doi.org/10.1111/j.1467-7652.2011.00649.x] [PMID: 21883868]

[180] Tusé D, Ku N, Bendandi M, *et al.* Clinical safety and Immunogenicity of tumor-targeted, plant-made id-klh conjugate vaccines for follicular lymphoma. Biomed Res Int 2015.
[http://dx.doi.org/10.1155/2015/648143]

[181] Sherman A, Su J, Lin S, Wang X, Herzog RW, Daniell H. Suppression of inhibitor formation against FVIII in a murine model of hemophilia A by oral delivery of antigens bioencapsulated in plant cells. Blood 2014; 124(10): 1659-68.
[http://dx.doi.org/10.1182/blood-2013-10-528737] [PMID: 24825864]

[182] Chung HY, Lee HH, Kim KI, *et al.* Expression of a recombinant chimeric protein of hepatitis A virus VP1-Fc using a replicating vector based on Beet curly top virus in tobacco leaves and its immunogenicity in mice. Plant Cell Rep 2011; 30(8): 1513-21.
[http://dx.doi.org/10.1007/s00299-011-1062-6] [PMID: 21442402]

[183] Chung HY, Park JH, Lee HH, *et al.* Expression of a functional recombinant chimeric protein of human hepatitis A virus VP1 and an Fc antibody fragment in transgenic tomato plants. Plant Biotechnol Rep 2014; 8(3): 243-9.
[http://dx.doi.org/10.1007/s11816-014-0318-3]

[184] Loos A, Van Droogenbroeck B, Hillmer S, *et al.* Production of monoclonal antibodies with a controlled N-glycosylation pattern in seeds of Arabidopsis thaliana. Plant Biotechnol J 2011; 9(2): 179-92.
[http://dx.doi.org/10.1111/j.1467-7652.2010.00540.x] [PMID: 20561245]

[185] Soria-Guerra RE, Ramírez-Alonso JI, Ibáñez-Salazar A, *et al.* Expression of an HBcAg-based antigen carrying angiotensin II in Chlamydomonas reinhardtii as a candidate hypertension vaccine. Plant Cell Tissue Organ Cult 2014; 116(2): 133-9.
[http://dx.doi.org/10.1007/s11240-013-0388-x]

[186] Fedorowicz-Strońska O, Kapusta J, Czyż M, Kaczmarek M, Pniewski T. Immunogenicity of parenterally delivered plant-derived small and medium surface antigens of hepatitis B virus. Plant Cell

Rep 2016; 35(5): 1209-12.
[http://dx.doi.org/10.1007/s00299-016-1944-8] [PMID: 26905723]

[187] Rukavtsova EB, Rudenko NV, Puchko EN, Zakharchenko NS, Buryanov YI. Study of the immunogenicity of hepatitis B surface antigen synthesized in transgenic potato plants with increased biosafety. J Biotechnol 2015; 203: 84-8.
[http://dx.doi.org/10.1016/j.jbiotec.2015.03.019] [PMID: 25840367]

[188] Hayden CA, Egelkrout EM, Moscoso AM, *et al.* Production of highly concentrated, heat-stable hepatitis B surface antigen in maize. Plant Biotechnol J 2012; 10(8): 979-84.
[http://dx.doi.org/10.1111/j.1467-7652.2012.00727.x] [PMID: 22816734]

[189] Pniewski T, Kapusta J, Bociąg P, *et al.* Plant expression, lyophilisation and storage of HBV medium and large surface antigens for a prototype oral vaccine formulation. Plant Cell Rep 2012; 31(3): 585-95.
[http://dx.doi.org/10.1007/s00299-011-1223-7] [PMID: 22246107]

[190] Peyret H, Gehin A, Thuenemann EC, *et al.* Tandem fusion of hepatitis B core antigen allows assembly of virus-like particles in bacteria and plants with enhanced capacity to accommodate foreign proteins. PLoS One 2015; 10(4): e0120751.
[http://dx.doi.org/10.1371/journal.pone.0120751] [PMID: 25830365]

[191] Mohammadzadeh S, Roohvand F, Memarnejadian A, *et al.* Co-expression of hepatitis C virus polytope-HBsAg and p19-silencing suppressor protein in tobacco leaves. Pharm Biol 2016; 54(3): 465-73.
[http://dx.doi.org/10.3109/13880209.2015.1048371] [PMID: 25990925]

[192] Clarke JL, Paruch L, Dobrica MO, *et al.* Lettuce-produced hepatitis C virus E1E2 heterodimer triggers immune responses in mice and antibody production after oral vaccination. Plant Biotechnol J 2017; 15(12): 1611-21.
[http://dx.doi.org/10.1111/pbi.12743] [PMID: 28419665]

[193] Mohammadzadeh S, Khabiri A, Roohvand F, *et al.* Enhanced-transient expression of hepatitis C virus core protein in nicotiana tabacum, a protein with potential clinical applications. Hepat Mon 2014; 14(11): e20524.
[http://dx.doi.org/10.5812/hepatmon.20524] [PMID: 25598788]

[194] Mohammadzadeh S, Roohvand F, Ajdary S, Ehsani P, Hatef Salmanian A. Heterologous expression of hepatitis C virus core protein in oil seeds of *Brassica napus* L. Jundishapur J Microbiol 2015; 8(11): e25462.
[http://dx.doi.org/10.5812/jjm.25462] [PMID: 26855744]

[195] Mardanova ES, Takova KH, Toneva VT, Zahmanova GG, Tsybalova LM, Ravin NV. A plant-based transient expression system for the rapid production of highly immunogenic Hepatitis E virus-like particles. Biotechnol Lett 2020; 42(11): 2441-6.
[http://dx.doi.org/10.1007/s10529-020-02995-x] [PMID: 32875477]

[196] Zeitlin L, Bohorov O, Bohorova N, *et al.* . mAbs Prophylactic and therapeutic testing of Nicotiana-derived RSV-neutralizing human monoclonal antibodies in the cotton rat mode. In mAbs Taylor Fr 2013; 5(2): 263-9.

[197] Phan HT, Pohl J, Floss DM, *et al.* ELPylated haemagglutinins produced in tobacco plants induce potentially neutralizing antibodies against H5N1 viruses in mice. Plant Biotechnol J 2013; 11(5): 582-93.
[http://dx.doi.org/10.1111/pbi.12049] [PMID: 23398695]

[198] Iyer V, Liyanage R, Shoji Y, *et al.* Human Vaccines & Immunotherapeutics Formulation development of a plant-derived h1n1 influenza vaccine containing purified recombinant hemagglutinin antigen View supplementary material. Vaccines Immunother 2012; 8(4): 453-64.
[http://dx.doi.org/10.4161/hv.19106]

[199] Cummings JF, Guerrero ML, Moon JE, *et al.* Safety and immunogenicity of a plant-produced

recombinant monomer hemagglutinin-based influenza vaccine derived from influenza A (H1N1)pdm09 virus: a Phase 1 dose-escalation study in healthy adults. Vaccine 2014; 32(19): 2251-9.
[http://dx.doi.org/10.1016/j.vaccine.2013.10.017] [PMID: 24126211]

[200] Le Mauff F, Mercier G, Chan P, *et al.* Biochemical composition of haemagglutinin-based influenza virus-like particle vaccine produced by transient expression in tobacco plants. Plant Biotechnol J 2015; 13(5): 717-25.
[http://dx.doi.org/10.1111/pbi.12301] [PMID: 25523794]

[201] Major D, Chichester JA, Pathirana RD, *et al.* Intranasal vaccination with a plant-derived H5 HA vaccine protects mice and ferrets against highly pathogenic avian influenza virus challenge. Hum Vaccin Immunother 2015; 11(5): 1235-43.
[http://dx.doi.org/10.4161/21645515.2014.988554] [PMID: 25714901]

[202] Mortimer E, Maclean JM, Mbewana S, *et al.* Setting up a platform for plant-based influenza virus vaccine production in South Africa. BMC Biotechnol 2012; 12: 14.
[http://dx.doi.org/10.1186/1472-6750-12-14] [PMID: 22536810]

[203] Pillet S, Racine T, Nfon C, *et al.* Plant-derived H7 VLP vaccine elicits protective immune response against H7N9 influenza virus in mice and ferrets. Vaccine 2015; 33(46): 6282-9.
[http://dx.doi.org/10.1016/j.vaccine.2015.09.065] [PMID: 26432915]

[204] Smith T, O'Kennedy MM, Wandrag DBR, Adeyemi M, Abolnik C. Efficacy of a plant-produced virus-like particle vaccine in chickens challenged with Influenza A H6N2 virus. Plant Biotechnol J 2020; 18(2): 502-12.
[http://dx.doi.org/10.1111/pbi.13219] [PMID: 31350931]

[205] Jul-Larsen Å, Madhun AS, Brokstad KA, Montomoli E, Yusibov V, Cox RJ. Human Vaccines & Immunotherapeutics The human potential of a recombinant pandemic influenza vaccine produced in tobacco plants The human potential of a recombinant pandemic influenza vaccine produced in tobacco plants. Taylor Fr 2012; 8(5): 653-61.

[206] Lee G, Na YJ, Yang BG, *et al.* Oral immunization of haemaggulutinin H5 expressed in plant endoplasmic reticulum with adjuvant saponin protects mice against highly pathogenic avian influenza A virus infection. Plant Biotechnol J 2015; 13(1): 62-72.
[http://dx.doi.org/10.1111/pbi.12235] [PMID: 25065685]

[207] Kanagarajan S, Tolf C, Lundgren A, Waldenström J, Brodelius PE. Transient expression of hemagglutinin antigen from low pathogenic avian influenza A (H7N7) in Nicotiana benthamiana. PLoS One 2012; 7(3): e33010.
[http://dx.doi.org/10.1371/journal.pone.0033010] [PMID: 22442675]

[208] Shoji Y, Farrance CE, Bautista J, *et al.* A plant-based system for rapid production of influenza vaccine antigens. Influenza Other Respir Viruses 2012; 6(3): 204-10.
[http://dx.doi.org/10.1111/j.1750-2659.2011.00295.x] [PMID: 21974811]

[209] Ward BJ, Landry N, Trépanier S, *et al.* Human antibody response to N-glycans present on plant-made influenza virus-like particle (VLP) vaccines. Vaccine 2014; 32(46): 6098-106.
[http://dx.doi.org/10.1016/j.vaccine.2014.08.079] [PMID: 25240757]

[210] Firsov A, Tarasenko I, Mitiouchkina T, *et al.* High-Yield Expression of M2e Peptide of Avian Influenza Virus H5N1 in Transgenic Duckweed Plants. Mol Biotechnol 2015; 57(7): 653-61.
[http://dx.doi.org/10.1007/s12033-015-9855-4] [PMID: 25740321]

[211] Mbewana S, Mortimer E, Pêra FFPG, Hitzeroth II, Rybicki EP. Production of H5N1 influenza virus matrix protein 2 ectodomain protein bodies in tobacco plants and in insect cells as a candidate universal influenza vaccine. Front Bioeng Biotechnol 2015; 3: 197.
[http://dx.doi.org/10.3389/fbioe.2015.00197] [PMID: 26697423]

[212] Ravin NV, Kotlyarov RY, Mardanova ES, *et al.* Plant-produced recombinant influenza vaccine based on virus-like HBc particles carrying an extracellular domain of M2 protein. Biochemistry (Mosc) 2012; 77(1): 33-40.

[http://dx.doi.org/10.1134/S000629791201004X] [PMID: 22339631]

[213] Mardanova ES, Kotlyarov RY, Kuprianov VV, *et al.* Rapid high-yield expression of a candidate influenza vaccine based on the ectodomain of M2 protein linked to flagellin in plants using viral vectors. BMC Biotechnol 2015; 15(1): 42.
[http://dx.doi.org/10.1186/s12896-015-0164-6] [PMID: 26022390]

[214] Liu CW, Chen JJW, Kang CC, Wu CH, Yiu JC. Transgenic lettuce (*Lactuca sativa* L.) expressing H1N1 influenza surface antigen (neuraminidase). Sci Hortic (Amsterdam) 2012; 139: 8-13.
[http://dx.doi.org/10.1016/j.scienta.2012.02.037]

[215] Nahampun HN, Bosworth B, Cunnick J, Mogler M, Wang K. Expression of H3N2 nucleoprotein in maize seeds and immunogenicity in mice. Plant Cell Rep 2015; 34(6): 969-80.
[http://dx.doi.org/10.1007/s00299-015-1758-0] [PMID: 25677970]

[216] Chen TH, Hu CC, Liao JT, *et al.* Production of Japanese encephalitis virus antigens in plants using bamboo mosaic virus-based vector. Front Microbiol 2017; 8(MAY): 788.
[http://dx.doi.org/10.3389/fmicb.2017.00788] [PMID: 28515719]

[217] Lacombe S, Bangratz M, Brizard JP, *et al.* Optimized transitory ectopic expression of promastigote surface antigen protein in Nicotiana benthamiana, a potential anti-leishmaniasis vaccine candidate. J Biosci Bioeng 2018; 125(1): 116-23.
[http://dx.doi.org/10.1016/j.jbiosc.2017.07.008] [PMID: 28803053]

[218] Spiegel H, Boes A, Voepel N, *et al.* Application of a scalable plant transient gene expression platform for malaria vaccine development. Front Plant Sci 2015; 6(DEC): 1169.
[http://dx.doi.org/10.3389/fpls.2015.01169] [PMID: 26779197]

[219] Voepel N, Boes A, Edgue G, *et al.* Malaria vaccine candidate antigen targeting the pre-erythrocytic stage of Plasmodium falciparum produced at high level in plants. Biotechnol J 2014; 9(11): 1435-45.
[http://dx.doi.org/10.1002/biot.201400350] [PMID: 25200253]

[220] Feller T, Thom P, Koch N, *et al.* Plant-based production of recombinant Plasmodium surface protein pf38 and evaluation of its potential as a vaccine candidate. PLoS One 2013; 8(11): e79920.
[http://dx.doi.org/10.1371/journal.pone.0079920] [PMID: 24278216]

[221] Kantor M, Sestras R, Chowdhury K. Transgenic tomato plants expressing the antigen gene PfCP-2.9 of Plasmodium falciparum. Pesq agropec bras 2013; 48(1): 73-9.

[222] Beiss V, Spiegel H, Boes A, *et al.* Plant expression and characterization of the transmission-blocking vaccine candidate PfGAP50. BMC Biotechnol 2015; 15(1): 108.
[http://dx.doi.org/10.1186/s12896-015-0225-x] [PMID: 26625934]

[223] Jones RM, Chichester JA, Mett V, *et al.* A plant-produced Pfs25 VLP malaria vaccine candidate induces persistent transmission blocking antibodies against Plasmodium falciparum in immunized mice. PLoS One 2013; 8(11): e79538.
[http://dx.doi.org/10.1371/journal.pone.0079538] [PMID: 24260245]

[224] Farrance CE, Rhee A, Jones RM, *et al.* A plant-produced Pfs230 vaccine candidate blocks transmission of Plasmodium falciparum. Clin Vaccine Immunol 2011; 18(8): 1351-7.
[http://dx.doi.org/10.1128/CVI.05105-11] [PMID: 21715576]

[225] Beiss V, Spiegel H, Boes A, *et al.* Heat-precipitation allows the efficient purification of a functional plant-derived malaria transmission-blocking vaccine candidate fusion protein. Biotechnol Bioeng 2015; 112(7): 1297-305.
[http://dx.doi.org/10.1002/bit.25548] [PMID: 25615702]

[226] Ma C, Wang L, Webster DE, Campbell AE, Coppel RL. Production, characterisation and immunogenicity of a plant-made Plasmodium antigen--the 19 kDa C-terminal fragment of Plasmodium yoelii merozoite surface protein 1. Appl Microbiol Biotechnol 2012; 94(1): 151-61.
[http://dx.doi.org/10.1007/s00253-011-3772-7] [PMID: 22170105]

[227] Rosales-Mendoza S, Soria-Guerra RE, Moreno-Fierros L, Han Y, Alpuche-Solís ÁG, Korban SS.

Transgenic carrot tap roots expressing an immunogenic F1-V fusion protein from Yersinia pestis are immunogenic in mice. J Plant Physiol 2011; 168(2): 174-80.
[http://dx.doi.org/10.1016/j.jplph.2010.06.012] [PMID: 20655621]

[228] Marsian J, Fox H, Bahar MW, *et al.* Plant-made polio type 3 stabilized VLPs-a candidate synthetic polio vaccine. Nat Commun 2017; 8(1): 245.
[http://dx.doi.org/10.1038/s41467-017-00090-w] [PMID: 28811473]

[229] Zhang XX, Yu H, Wang XH, *et al.* Protective efficacy against Chlamydophila psittaci by oral immunization based on transgenic rice expressing MOMP in mice. Vaccine 2013; 31(4): 698-703.
[http://dx.doi.org/10.1016/j.vaccine.2012.11.039] [PMID: 23196208]

[230] van Dolleweerd CJ, Teh AYH, Banyard AC, *et al.* Engineering, expression in transgenic plants and characterisation of E559, a rabies virus-neutralising monoclonal antibody. J Infect Dis 2014; 210(2): 200-8.
[http://dx.doi.org/10.1093/infdis/jiu085] [PMID: 24511101]

[231] Loza-Rubio E, Rojas-Anaya E, López J, Olivera-Flores MT, Gómez-Lim M, Tapia-Pérez G. Induction of a protective immune response to rabies virus in sheep after oral immunization with transgenic maize, expressing the rabies virus glycoprotein. Vaccine 2012; 30(37): 5551-6.
[http://dx.doi.org/10.1016/j.vaccine.2012.06.039] [PMID: 22749836]

[232] Singh A, Srivastava S, Chouksey A, *et al.* Expression of rabies glycoprotein and ricin toxin B chain (RGP-RTB) fusion protein in tomato hairy roots: a step towards oral vaccination for rabies. Mol Biotechnol 2015; 57(4): 359-70.
[http://dx.doi.org/10.1007/s12033-014-9829-y] [PMID: 25519901]

[233] Shafaghi M, Maktoobian S, Rasouli R, Howaizi N, Ofoghi H, Ehsani P. Transient expression of biologically active anti-rabies virus monoclonal antibody in tobacco leaves. Iran J Biotechnol 2018; 16(1): e1774.
[http://dx.doi.org/10.21859/ijb.1774] [PMID: 30555840]

[234] Kalbina I, Lagerqvist N, Moiane B, *et al.* Arabidopsis thaliana plants expressing Rift Valley fever virus antigens: Mice exhibit systemic immune responses as the result of oral administration of the transgenic plants. Protein Expr Purif 2016; 127: 61-7.
[http://dx.doi.org/10.1016/j.pep.2016.07.003] [PMID: 27402440]

[235] Mbewana S, Meyers AE, Weber B, *et al.* Expression of Rift Valley fever virus N-protein in Nicotiana benthamiana for use as a diagnostic antigen. BMC Biotechnol 2018; 18(1): 77.
[http://dx.doi.org/10.1186/s12896-018-0489-z] [PMID: 30537953]

[236] Del L Yácono M, Farran I, Becher ML, *et al.* A chloroplast-derived Toxoplasma gondii GRA4 antigen used as an oral vaccine protects against toxoplasmosis in mice. Plant Biotechnol J 2012; 10(9): 1136-44.
[http://dx.doi.org/10.1111/pbi.12001] [PMID: 23020088]

[237] Albarracín RM, Becher ML, Farran I, *et al.* The fusion of Toxoplasma gondii SAG1 vaccine candidate to Leishmania infantum heat shock protein 83-kDa improves expression levels in tobacco chloroplasts. Biotechnol J 2015; 10(5): 748-59.
[http://dx.doi.org/10.1002/biot.201400742] [PMID: 25823559]

[238] Zhang Y, Chen S, Li J, Liu Y, Hu Y, Cai H. Oral immunogenicity of potato-derived antigens to Mycobacterium tuberculosis in mice. Acta Biochim Biophys Sin (Shanghai) 2012; 44(10): 823-30.
[http://dx.doi.org/10.1093/abbs/gms068] [PMID: 22917938]

[239] Permyakova N V, Zagorskaya AA, Belavin PA, *et al.* Transgenic carrot expressing fusion protein comprising M. tuberculosis antigens induces immune response in mice. Biomed Res Int 2015.

[240] Lakshmi PS, Verma D, Yang X, Lloyd B, Daniell H. Low cost tuberculosis vaccine antigens in capsules: expression in chloroplasts, bio-encapsulation, stability and functional evaluation in vitro. PLoS One 2013; 8(1): e54708.
[http://dx.doi.org/10.1371/journal.pone.0054708] [PMID: 23355891]

[241]   Uvarova EA, Belavin PA, Permyakova NV, *et al.* Oral immunogenicity of plant-made mycobacterium tuberculosis ESAT6 and CFP10. Biomed Res Int 2013.

[242]   Jose S, Ignacimuthu S, Ramakrishnan M, *et al.* Expression of GroES TB antigen in tobacco and potato. Plant Cell Tissue Organ Cult 2014; 119(1): 157-69.
[http://dx.doi.org/10.1007/s11240-014-0522-4]

[243]   Lai H, He J, Hurtado J, *et al.* Structural and functional characterization of an anti-West Nile virus monoclonal antibody and its single-chain variant produced in glycoengineered plants. Plant Biotechnol J 2014; 12(8): 1098-107.
[http://dx.doi.org/10.1111/pbi.12217] [PMID: 24975464]

Frontiers in Protein and Peptide Sciences, 2021, *Vol. 2*, 267-297

Plant Molecular Farming for Human Therapeutics: Recent Advances and Future Prospects

Amna Ramzan[1], Zainab Y. Sandhu[2], Saba Altaf[1], Aisha Tarar[1], Iqra Arshad[1], Sumera Rashid[1], Huma Shakoor[1], Rabia Abbas[1] and Bushra Rashid[1,*]

[1] *Centre of Excellence in Molecular Biology, University of the Punjab, Lahore, Pakistan*
[2] *Montclair State University, Montclair, New Jersey, NJ 07043, USA*

Abstract: Plant molecular farming (PMF) aims to develop plants that express and accumulate proteins of our interest in considerable quantities. Transgenic plants produce edible vaccines, antibodies, therapeutic proteins for human and animal health, and other recombinant proteins required for industrial purposes. Plant systems (PS) to produce pharmaceutical products are preferred over microbial and mammalian systems as they require less input to grow and produce higher biomass. Hence, a variety of proteins are synthesized by plants that are completely free from human pathogens and mammalian toxins. Additionally, they have immunity against infectious and other life-threatening diseases such as cancer. In this review, plant-inferred therapeutic and non-therapeutic protein items that are in the position of clinical progression or commercialization are summarized. Available plant production platforms are also compared along with associated biosafety and regulatory issues. Further, plant transformation techniques are also analyzed for the development of genetically modified organisms in vaccine production. The use of PMF on a commercial scale is still a long way to go before it is achievable. New methods and techniques are needed to be developed to solve the problems of low yield, scalability, stability, and efficacy of the recombinant proteins, as well as biosafety and regulatory issues. Hence, this strategy will be the ultimate proposed solution to protect humans and animals from health threats in the future.

Keywords: Genetic transformation, Plant molecular farming, Pharmaceutical products, Plant systems, Recombinant Proteins, Transient Expression.

* **Corresponding author Bushra Rashid:** Centre of Excellence in Molecular Biology, University of the Punjab Lahore, Pakistan; Tel: +92 42 35293141-6; Ext 142; E-mail: bushra.cemb@pu.edu.pk

Muhammad Sarwar Khan (Ed)

1. INTRODUCTION

The production of many useful products and recombinant proteins from genetically modified plants is known as plant molecular farming. The term has many other names such as pharming, plant-made pharmaceuticals, plant bio-pharming, bio-manufacturing, bio-pharmaceuticals, plant-derived products of interest, and plants with novel traits [1]. It is a unique application of genetic engineering as it involves plants to manufacture different valuable proteins at the industrial level. Recombinant DNA technology has made it possible to isolate a gene of interest from any organism and its transformation in the plant expression system to alter the specific trait. A worldwide increase in demand for therapeutic products leads the industries and governments towards an alternative approach of pharmaceutical production, plant systems (PS). This basic research field has global worth in the future [2].

The concept of using genetically modified plants is not recent; it dates back to 1986, when tobacco was genetically modified for the first time to produce a human growth hormone [3]. Human serum albumin is reported to be the second plant-derived product obtained from genetically modified tobacco and sunflower [4]. One big breakthrough in developing the field of PMF was brought about in 2012 when the first plant-derived recombinant protein "taliglucerase alfa", a recombinant human glucocerebrosidase was synthesized and developed by Protalix Bio-therapeutics. Later on, that was approved for clinical trials for the treatment of Gaucher's disease. Afterward, many other plant-derived products, such as antibodies, vaccines, enzymes, biocatalysts, biosensors, diagnostic reagents, growth factors, and cosmetic reagents, were commercialized [5].

Several living systems like yeast, bacterial, and animal cell cultures are being used for the production of recombinant proteins. Methods of extraction and purification, along with cold storage, short shelf-lives, and transportation, make these production systems expensive [6, 7]. Plants can be used as "bioreactors" and can replace fermenters, which will reduce the upstream facility and processing of plant tissues for the oral delivery of edible vaccines. This will reduce the downstream processing [8]. The production cost of biological molecules by using plants is less as compared to other systems [9, 10]. Moreover, they also minimize the health risks with additional benefits of high stability of recombinant proteins at large-scale production [11]. The advantage of the production of therapeutically active bio-molecules in edible crops is that they are taken orally as a regular diet, without any hesitation or change in daily habits. It also offers swift scale-up and suitable storage of unprocessed plant materials. The life-threatening diseases and infections in humans and animals due to viruses, bacteria, and other pathogens are now possible to treat with edible vaccines [12, 13]. The availability of cheaper, easy to consume, and easy to store plant products (seeds, fruits, leaves, *etc.*) that contained edible recombinant proteins, made it possible to prevent life-threatening diseases such as HIV, HBV, *etc.* [14, 15].

Presently, PMF is becoming a very profitable industry with the synthesis of several new recombinant proteins, and many biotechnological companies and governments of different countries are adopting it. Profit associated with PMF can be envisaged by comparing the development of Bt-cotton with that of Bt-corn. An international PMF society has also been formed, and it will become a foundation to support the production of recombinant proteins from PS [16]. In upcoming years, the main focus of scientists will be the development of pharmaceutical products from PS against complex diseases like metabolic disorders, infectious and neurodegenerative diseases and cancer, *etc.* [17].

The current study summarizes the applications and benefits of PMF associated with health issues in humans and animals. It also summarizes available PS, their benefits and limitations, and the solutions to solve these limitations in the future. It also covers health and environmental concerns associated with PMF and different approaches to minimize these limitations so that PMF products may not be rejected by Genetically Modified Organisms' legislation bodies and the general public. Furthermore, it also covers available methods of genetic transformation to develop transgenic plants that synthesize the required proteins on a small as well as on a commercial scale because benefits associated with PMF will only be realized when production is taken up at a commercial scale.

2. SIGNIFICANCE OF PLANT MOLECULAR FARMING (PMF) APPLICATIONS

Several reports have shown benefits associated with PMF in comparison with other production systems [18]. The pattern of protein synthesis in plants is not very much different from the mammalian pattern but plant systems are very cost-effective. Plants also have protein disulfide isomerases and certain chaperons that ensure proper folding and assembly of native as well as non-native proteins and this capability makes them superior over bacterial production platforms [19]. Types of post-transcriptional modifications specifically N- and O- type glycosylation offered by different plant species make PS the most suitable and cost-effective platforms for the synthesis of human-like proteins as well as bio-betters. Such modifications play their role in efficiency, storage as well as downstream processing [20]. Production cost is also reduced to a considerable extent because downstream processing is not required in some cases like edible vaccines and it is also free from human pathogens. Many protein products are successfully produced by PMF by using different expression systems and the applications are discussed as below.

2.1. Recombinant Antibodies

Monoclonal antibodies have high market value due to their unique properties and specificity approved for several targets and life-threatening diseases, most commonly Cancer [21]. The ability of plants to produce recombinant antibodies resulted in the development of many derivative products like single-chain variable fragments (scFv) and variable heavy chain antibody fragments (also known as nanobodies) [22, 23]. Now it is possible to produce recombinant proteins in the form of fusion proteins which results in improved functionality of proteins; for example, some reports have shown selective killing of cancer cells when the binding domain of a protein was combined with a visual marker or a toxin [24]. Mammalian cells are used to produce some small antibody fragments like scFv due to their ability to produce high-titers, but this platform is most expensive. Plants are economical and scalable systems for the production of such antibodies and they are preferred nowadays. Production of full-size IgG [25] was the first success of PS and after this, many other recombinant proteins, like full-size secretory IgA and IgM, were also synthesized successfully. Nowadays, the monoclonal antibodies and relevant proteins where soluble cytokine receptors are bonded with antibody's constant regions have interestingly led to the advancement in the treatment of chronic inflammatory diseases and cancer [21].

One of the unique characteristics of plants is their ability to synthesize a polyclonal mixture of antibodies under a phenomenon known as superinfection exclusion. Julve *et al.* [26] demonstrated this as leaf extract containing a polyclonal mixture of antibodies synthesized from different types of cells, each infected by a specific *Tobacco mosaic virus* vector. There is no plant-derived pharmaceutical antibody in the market. Some products are approved for marketing, such as a purification reagent for the hepatitis B vaccine and an antibody developed for the treatment of dental caries caused by *Streptococcus* mutants [27]. Some other examples of recombinant antibodies developed in plant expression systems are Anti-CD20 developed in transgenic Duckweed for therapy of Non-Hodgkin's lymphoma, rheumatoid arthritis; Anti-aCCR5 developed in tobacco against HIV by using transient expression system. Moreover, Anti-HIV gp120 was raised in transgenic maize and tobacco transient expression systems against HIV [28]. Several reports have shown the successful production of many non-pharmaceutical antibodies, which are being analyzed for their application in food processing, quality validation, and diagnostic purposes [22, 29].

Certain auto-immune diseases, such as Diabetes type 1, need regular treatment for which insulin is the ultimate choice, but that does not completely cure. Therefore, plant-derived medications are used to induce oral tolerance and targeted such types of diseases, like autoimmune diabetes type 1 [30, 31]. Using oral tolerance as prophylactic tolerance induction prevents the rejection responses that hamper the replacement therapies in consanguineous genetic disorders like hemophilia A and B [32 - 34] and Pompe disease [35]. Sometimes the antibodies are neutralized in a few patients against the non-self-antigenic therapeutic proteins during regular replacement therapy, which may be problematic [31]. Coagulation factors VIII, IX, and acid alpha-glucosidase (GAA) expressed in transplastomic leaves were tested to treat these diseased mouse models that suppressed out the surplus adaptive immune response and induced the oral tolerance in all the cases. It has also been observed that transgenic plants are developing allergen-specific immunotherapy by producing heterologous allergens to treat allergic asthma [36]. Other therapeutic and dietary proteins are reported. Glucocerebrosidase was developed in transgenic carrot cell suspension culture for the therapy of Gaucher disease. Gastric lipase was developed in transgenic maize for Cystic fibrosis and pancreatitis therapy. Lactoferrin in transgenic maize was developed against Gastrointestinal infections. Transgenic Safflower and *Arabidopsis thaliana* were produced harboring Insulin and Human Intrinsic factor to be used for the therapy of diabetes and Vitamin B12 deficiency, respectively [28]. A brief description of various diseases in humans and animals against which edible vaccines have been developed by introducing specific antigens in different host plant expression systems [37 - 46] is shown in Table **1**.

Table 1. Edible vaccines developed against different diseases in animals and plants.

Disease/Causal organism	Specific antigen	Host plant expression system	References
Cholera/*Vibrio cholera*	Cholera toxin B (CTB)	Potato/Rice	[37]
Diarrhoea/ *Enterotoxigenic E. coli*	Heat-labile enterotoxin B (LT-B)	Potato and tobacco	[38]
Malaria/*Plasmodium falciparum*	Merozoite surface protein (MSP)	Tomato	[39]
Hepatitis B/*HBV*	HBsAg	Potato	[40]
Procrine reproductive and respiratory syndrome/ *(PRRSV)*	Porcine reproductive and respiratory syndrome virus (PRRSV)	Maize	[41]
Foot and mouth disease virus/*FMDV*	Polyprotein P1	Rice	[42]
Tuberculosis/*Mycobacterium tuberculosis*	CFP10, ESAT6	Carrot	[43]

(Table 1) cont.....

Disease/Causal organism	Specific antigen	Host plant expression system	References
AIDS/HIV	p24, Nef	Tobacco	[44]
Anthrax/ ***Bacillus anthracis***	Protective antigen (PA)	Tobacco	[45]
Influenza/*Influenza virus*	M2e peptide	Tobacco	[46]

2.2. Edible Vaccines

Plant-derived vaccines have been developed against various infectious and metabolic diseases in humans and animals. In properly refined form, plant-derived recombinant proteins can be used as injectable vaccines, but oral administration of such proteins as edible vaccines also proves to be useful. They could be termed as edible "Bio-factories" as they have no or less toxic compounds as well as response to allergies [47]. Rarely, the edible vaccine is more shielded as they are cost-effective than with the economically available vaccines. Injections, needles, and pre-refrigeration are not the limitations for edible vaccines, and they are successfully tested in banana, potato, tomato, and carrot [48 - 50] and other crop plants as well as model plants such as Arabidopsis, tobacco, and alfalfa. The edible plant parts are the fruits (banana, tomato), seeds (corn, peanuts, rice, wheat, pea), leaves (spinach, lettuce, alfalfa), tubers (potato), roots (carrots) [13]. These crops are used to raise the vaccines against measles, cholera, foot and mouth disease, *etc* [51]. However, according to some reports, proteins may become inactive during cooking, such as in potatoes. Less preservation rate and long manufacturing cycle are some limitations in the common use of tomato and banana as a plant manufacturing system. There is no preservation issue in the use of carrots but they can only be used when they are fresh. Different plant species used to produce edible vaccines have been used as expression systems against different pathogens with proven benefits, and possible limitations are summarized [14, 15, 48 - 50, 52 - 58] in Table **2**.

Table 2. Edible vaccines produced in plants against various pathogens in humans and animals.

Host Plant Species	Resistance Against Pathogens	Specific Benefits of Crop	Possible Disadvantages	References
Potato	Narovirus and *Escherichia coli*	Easy transformation, propagation and storage of crop	Cooking may degrade the antigen	[49]
Rice	*Escherichia coli*	Used in baby food, staple food in many countries, high expression of antigens	Crop requires special conditions to grow	[14, 52]

(Table 2) cont.....

Host Plant Species	Resistance Against Pathogens	Specific Benefits of Crop	Possible Disadvantages	References
Pea	Rinderpest virus	High in protein content, easy storage	May degrade while cooking	[15]
Banana	HBV	Used raw, stable protein while cooking, inexpensive	Plants take 2-3 years to mature, fruit spoils shortly	[48]
Tomato	Norwalk virus; HBV	Used raw, easy and quickly cultivated	Fruit's shelf life is short	[53 - 55]
Lettuce	*Escherichia coli*	Easy and quickly consumed; eaten raw	Perishable, short shelf life	[56]
Alfalfa	*Echinococcus granulosus*	Mostly used in animal feed	Immunity is still not known to much extent	[57]
Carrots	*Helicobactor pylori, Escherichia coli,* HIV	Easy to consume as raw, healthy, delicious	Short seasonal crop in some countries	[50]
Corn/Maize	NDV (Newcastle disease virus)	Easy to make diet of chicks	Longer storage of crop may degrade the antigen	[58]

2.3. Biocatalysts

The production of enzymes in plants is of great concern to scientists because it is quite inexpensive and a bulk amount of product is produced within a limited time. A variety of enzymes as recombinant protein ranges from laundry detergents to textile dying to leather tanning and therapeutic proteins used to treat cancer [59]. The first-ever enzyme produced on an industrial scale by using a plant system was trypsin which was produced from the transgenic corn that comes under the brand name TrypZean™ [5]. The ability of trypsin to degrade proteins into peptides makes it a valuable reagent in food processing, pharmaceutical protein processing, and leather tanning. It is also used to digest proteins when a proteomic analysis is required. Biochemical properties like optimum pH, stability, K_m, and V_{max} of plant-derived trypsin are similar to the native enzyme. Although some international companies have successfully produced cost-effective trypsin by using mammalian production platforms plant-derived trypsin is of better quality, therefore it is valued for the production of cosmetics and tissue culture products [60]. Many reports have shown that some commercial companies are selling/using plant-based enzyme products for commercial use in Research reagents, cosmetics, food products, and the diagnostics such as Avidin produced in maize for diagnostic purposes by Sigma-Aldrich; Human epidermal growth factor produced in barley seeds for use in cosmetics by Sif Cosmetics [61]. By continuing the research in this area, soon it will be possible that synthetic or naturally occurring proteins would be produced easily, safely, and conveniently in plants.

2.4. Biopolymers

Starch and lignocellulose synthesized by plants are the most abundant polymers on the earth. Plants can be engineered to produce fibrous animal proteins like elastin, keratin, and collagen which will have proper functionality such as elasticity, toughness, strength, and biocompatibility [62]. Such valuable products can replace oil-based plastics if they are synthesized by plants and used to produce biopolymers [63]. The remarkable strength and toughness of dragline silk (produced by *Nephila clavipes*) make it a valuable product to be synthesized as a recombinant protein in available production systems. Bacterial and mammalian production platforms are failed to secrete this protein out of the cell due to its larger size. A report has shown the successful production of spider silk in the tobacco expression system [64]. Spider silk is considered as a prospective biomaterial for medical usage because of its immunogenicity and scientists are working to optimize the PS to produce recombinant silk in complete size and proper form [65]. Hence, spidroin-derived cytotoxicity indication was not observed in cytocompatibility studies which lead to the conclusion that these plant-derived synthetic biopolymers are the appropriate biomaterials [66].

Collagen is a class of structural proteins present in mammals. The structure of one collagen protein (type I collage) consists of three polypeptide chains that revolve around a common axis and form a heterotrimeric helix that interacts with each other and forms a 3D structure. Animal-derived collagens used in human medical practices have immunogenic and pathogenic risks. Recombinant DNA technology has revolutionized the industry for collagen production by using prokaryote and different eukaryote systems [67]. However, the recombinant plant collagen production system (Maize and tobacco) has the likely potential to surpass for its scalable production, cost-effectiveness, less toxicity, and less pathogenic risks. Moreover, it is a preferable option for its high-quality post-translational modification [68]. There is no risk of pathogens or cross-linking in plant-derived collagen and it can be used in adhesives, cosmetics, and food industries such as (gelatin).

2.5. Feed Additives

The addition of some chemicals such as feed fortifiers in the animals' feed increases the nutrient content of the feed whereas the addition of some feed enzymes makes it easy for the animals to digest the feed properly. Oral immunization of animals with feed pellets, seeds, or different plant parts as feed additives from transgenic plants provides immunity against different pathogens [12]. Production of feed additives (feed fortifier and feed enzymes) by the microbial or mammalian platform is an expensive process and the production trends have been shifted to express the feed additives directly in the feed crops [47]. Phytase is required to release bonded phosphorus from the phytate which also sequesters iron, calcium, and zinc present in other compounds. Phytase was produced in transgenic *Pichia pastoris* and then supplemented in animal feed but the recombinant phytase is very expensive. Direct expression of phytase in plants will be a cost-effective alternative to supplement the animal feed. Successful expression of *Aspergillus niger* phytase in maize is reported and the enzyme is functional and sufficient to release minerals from bonded forms and recombinant phytase survives in gastric digestion. Phytase expressing transgenic maize became the certified first transgenic crop of China [69]. Experiments on transgenic alfalfa containing Bovine Retrovirus (BRV) peptide eBRV4 provided the lactogenic immunity to cattle and the newborn which protected the newborn from the enteric pathogen [13]. Hence, animal feed containing plant recombinant proteins will be beneficial to induce immunogenicity in animals at a low cost and with less pathogenic risks.

2.6. Biofuel

The production of bioethanol is a valuable achievement of plant biotechnology. Plants having a higher content of sugars are preferred for bioethanol production and maize is one of the most commonly used plants for this purpose. Maize is primarily used as animal feed and only 20% is available for food and industrial applications. There are certain other plant products like cellulose and hemicellulose that are valuable alternatives for bioethanol production. A combination of many enzymes is required to degrade the complex structure of these compounds so that the simple sugars can be harvested. Zafar *et al.* [70] reported the synergistic saccharification of hemicellulases (endo-xylanase and β-xylosidase) evaluation for bioethanol production by using plant biomass. Endo-xylanase and β-xylosidase genes from *Bacillus licheniformis* were cloned and expressed in *E coli*. Production of xylose sugar by bioconversion of plant biomass verified the Saccharification potential of recombinant enzymes. Hence, recombinant hemicellulases are potential candidates for the conversion of complex agricultural residues into simple sugars for ultimate use in the biofuel industry. Replacement of fossil fuels and the use of bioethanol can significantly reduce the emission of toxic gases in the environment. Aftab *et al.* [71] reported the biofuel production by bioconversion of natural biomass by using the recombinant xylanase into simple sugars. Treatment of cellulosic and hemicellulosic biomass and other processes can reduce the production cost and enhance the yield of biofuels.

PMF offers a cost-effective system as compared to microbial production systems and it is possible to express polymer degrading enzymes in PS and they are then used in pure as well as crude form or they can be expressed directly in biofuel crops. A combination of three enzymes (β-D-glucosidase, Exo-1,4-β-glucanase, and endo-1,4-β-glucanase) is required to degrade cellulose. Some reports have shown successful development of maize lines for the expression of these enzymes by Agrivida (US) [72, 73]. Currently, Exo-1,4-β-glucanase is the only plant-derived hydrolase that is commercially available in the market and many others are under development. Plants are also genetically engineered to produce an improved quality enzyme including amylase, ligninase, cellulase, and hemicellulase so that they can also be used for the production of simple sugars, baking, and malting [59].

3. SUITABLE PLANT MOLECULAR FARMING PRODUCTION SYSTEM

One important factor for the success of PMF is the selection of a suitable host. For this purpose, certain economic factors like biomass production, maintenance costs, availability of labor, the requirement of land area, and cost of edibility along with the type of recombinant protein and methods of protein purification must be analyzed. The host must be susceptible to regeneration and transformation. Certain environmental factors including biosafety issues must also be considered. Following are the plant systems that are being used nowadays for recombinant protein production.

3.1. Food/Feed Crops

The use of leafy crops for the synthesis of recombinant proteins is reported to be beneficial due to their ability to produce higher biomass along with the high concentration of soluble proteins. Additionally, they require the minimum use of artificial fertilizers due to their ability to fix atmospheric nitrogen through a mutualistic relationship with rhizobacteria. The limitation of proteolytic activity by mature leaves can be minimized by the immediate processing of plant tissues after harvesting [74]. Several leafy crops including alfalfa, spinach, and lettuce have been successfully used for the production of recombinant proteins [56, 57]. Reports have shown successful production of HIV protein (C4(V3)6), pro-insulin, and F1-V fusion protein from lettuce that is being used as a vaccine against plague [16].

Safe storage and oral administration along with no need for fermenters and the cold chain for distribution make seed a promising platform for the production of recombinant proteins [75]. Different cereals like rice, maize, wheat, and barley along with different tissues of seed including endosperm and cotyledons can be used to produce and store recombinant proteins. The presence of chaperones and enzymes in seeds ensures the proper folding of proteins as reported for maize seeds [58]. Maize can easily be transformed and it also has greater biomass yield as compared to other food crops while rice is self-pollinating with fewer chances of involuntary gene flow that makes them a promising platform for recombinant protein production. Soybean has also been used to produce many antibodies and human growth hormones [76]. Successful production of a recombinant monoclonal antibody 2F5 for humans is reported which is used in seeds to neutralize HIV and recombinant coagulation factors IX. Seeds are harvested only once in a crop season and the low yield of recombinant proteins is the limitation in this production system.

3.2. Non-Food/Feed Crops

Certain environmental risks associated with food/feed crops including health issues for humans and animals can be minimized by using non-food/feed crops. Such plants are being used as a host for the expression of required protein. Tobacco has superiority over others because of the availability of well-established and optimized *in-vitro* protocols and the first edible vaccine against poultry was developed in tobacco cultures which have also been approved by USDA [28]. Another PMF platform based on the *Nicotiana* production system has been well established, it is also very promising to solve global health challenges along with limited manufacturing capacity [77]. Certain non-crop plants like model species *Arabidopsis* have also been exploited as a production platform due to its well-established genetic model and short regeneration and production time. Successful production of an immunogenic protein VP2 against viruses, human glucocerebrosidase, and recombinant virus-like particles for HIV-1/HBV has been reported. Animal or bacterial cell suspension cultures hold several advantages that include a high level of protein production which are then secreted into the liquid medium [78]. Researchers are working to combine the whole plant with mammalian or bacterial cell cultures that can easily be grown in bioreactors [79]. Along with the advantage of easy purification from a liquid medium, cell suspension systems also have some limitations that include the degradation of recombinant proteins by proteases present in the liquid medium.

Duckweeds are known as aquatic higher plants as their capability to synthesize complex protein at a higher level. They have fast clonal growth and harvesting is easy which makes them promising candidates for PMF. Duckweed expression technology known as LEX system-SM developed by Synthon is optimized to produce pharmaceutically active compounds including veterinary medicines, biosimilars, and monoclonal antibodies. A report has shown successful production of IFN-alpha2B (Locteron) against hepatitis C [80]. *Chlamydomonas reinhardtii* is one of the promising microalgae species for PMF because of the availability of well-established and optimized cloning and expression toolkit [81]. Successful production of immunotoxin protein, a large chain antibody, and human glutamic acid decarboxylase (hGAD65) against cancer, herpes simplex virus, and diabetes respectively has been reported [82].

4. LIMITATIONS AND OPTIMIZATIONS OF PMF PLATFORMS

Although the field of PMF is not very old, however, several obstacles have been worn down and some products have been approved for human therapy and many are in clinical trials [83]. But some hurdles including the problem of low yield is still needed to be addressed [84]. It is not possible to predict the yield of fully assembled and active protein products synthesized during any production system, because it is defined by many factors including properties of proteins, production environment, and type of host platform. Following section deals with the commonly used approaches that have shown promising results for the development of PMF.

4.1. Optimizing Transcript Expression

Efficiently designed suitable expression cassette that has a promoter, a cloning site, and 3` un-translated region plays a major role in proper transcription and translation of GOI in host plants. Gene promoter plays an important role because it binds to transcription factors and regulates gene expression [13]. Cauliflower mosaic virus (CaMV) promoter for dicots and Ubiquitin-I promoter for monocots are optimized for constant expression of the transcript [85]. Organ-specific promoters can also be used for the expression of a transgene in specific organs like tubers, seeds, or fruits. When vegetative organs are used as a site of storage and transgene expression is initiated by the tissue-specific promoter, then the accumulation of recombinant protein in tissues downregulates the plant development. The problem of lethality can be avoided by inducible promoters. In addition, the expression level of transgenes can be enhanced by the use of transcription factors [59].

4.2. Optimizing Protein Stability

Recombinant protein instability is considered to be a major limitation of PS and the ratio of the rate of proteolysis and biosynthesis define the overall yield, whereas proteolysis is the cell's natural and inevitable turnover process. A specific type of subcellular targeting is needed for optimizing the stability of the protein. The micro-environment of cell compartments may not, or poorly supports proteolysis due to different pH and salt levels. Therefore, such compartmentalization makes easy and effective processing of associated proteins [83]. It is also used to enhance the downstream processing by the addition of affinity tags and fusions. The stability of protein can be enhanced by targeting the subcellular compartments of secretory pathways like the endoplasmic reticulum for storage. Some proteins can be targeted to the chloroplast, whereas, others can also be directed to the surface of membranes and their cytosolic breakdown can be prevented [21]. Recombinant proteins can also be accumulated in vacuoles. When producing pharmaceuticals from plants, extracellular photolytic breakdown must also be considered. Extracellular recombinant proteins should be expressed in such host plants which are free of peptidases. It is useful to develop antibodies displaying their activity along with high resistance to peptidases [86]. Development of protease deficient cell lines by RNA interference (RNAi) approaches, gene silencing, and gene knockout are also helpful to minimize protein degradation [84, 87].

An approach based upon co-expression of recombinant protein along with protease inhibitors and optimization of physical and chemical conditions of PS also proves to be useful (Table **3**) [88 - 95]. Successful production of the Human Serum Albumin and α(1)-antitrypsin (AAT) (a human recombinant protein) under the high level of stability at alleviated pH has been reported [96]. Along with these, some additional strategies like strain optimization and selection of best performing transformants can also play a big role in yield enhancement. Optimization of production processes like media optimization, design, and parameters of processing also interfere with the final yield. Only 1-2°C change in the temperature range of the production platform can lower the final yield to a considerable extent, therefore an optimized and controlled environment of the production platform is very important [97]. Some reports have shown considerable enhancement like 10.6%, 36%, and 72% in yield of Human Serum Albumin, a murine antibody expressed in Rice, and some proteins are expressed in tobacco chloroplast [98, 99]. Statistical experimental designs are available to compare and analyze all these parameters so that a well-optimized system can be developed to enhance the yield to a considerable extent [100].

Table 3. Approaches to increase protein stability for higher total protein production.

Methods		Production Platform	Protein of Interest	Yield Enhancement	References
By Using Stabilizing Agents	BSA	*N. tabacum* NT-1 cells	hGM-CSF	2-fold increase	[88]
	HAS	*P. patens*	Human VEGF	3-fold increase	[89]
Co-expression of Proteins	Cathepsin D inhibitor (SICDI)	*S. tuberosum*	α 1-anti-chymotrypsin	2.5-fold yield increase	[90]
	SICDI or Tomato Cystatin SICYS9	*N. benthamiana*	C5-1 IgG monoclonal antibody	Up to 80% increase in light chain yield	[91]
	SICDI	*N. benthamiana*	C5-1 IgG monoclonal antibody	Up to 85% increase in heavy chain yield	[91]
	Oryzacystatin-1	*N. tabacum*	Glutathione reductase (GR)	Increase in GR level is not consistent	[92]
Gene Knockdown	Expression of antisense sequence	*N. tabacum* BY-2 cells	IgG1κ antibody 2F5	4-fold increase in heavy chain yield	[93]
	RNAi construct specific to CysP6	*N. tabacum*	Human IL-10	1.6-fold yield increase	[84]
Fusion Proteins	Fusion of Zera domain	*N. tabacum* NT-1 cells	Zera domain fused subunit vaccine F1-V	3-fold yield increase	[94]
	Fusion of SICYS8	*N. benthamiana*	SICYS8 fused α 1-ACT	25-fold yield increase	[95]

5. BIOSAFETY AND REGULATORY ISSUES

The synthesis of biopharmaceutical products in plants has provided valuable vaccines, antibodies, drugs, and other recombinant proteins. Technology has contributed to the betterment of mankind but at the same time, it has considerable biosafety issues. The development of GMOs especially transgenic plants is a controversial issue, and PMF has raised novel environmental concerns. Vertical transfer of genetic material from a transgenic plant to the same or related species may result in harmful effects on animals at various levels of a food chain [101]. As far as there are benefits of the transgenic crops similarly, plant-based pharmaceuticals may cause an adverse immune reaction in people that accidentally consume them [102]. The data indicates that the US has allowed the adoption of GM products in the market but the perception of most of the public in European and Asian countries is the hurdle for the adoption of GM or plant molecular farming products [103]. Therefore, there should be strong social communication in society for the benefits of GM adaptability. All plants/plant-based products must go through a complete risk assessment before they become available in the market [101]. Similarly, it is to specify a threshold limit for PMF products that will be harmless even if persisting in the field. Biosafety issues and possible solutions are needed to be considered for the sustainable development of PMF.

One of the serious biosafety concerns is the development of antibiotic resistance in micro-flora. Vaccination or more specifically using edible vaccines can play a crucial role to reduce antibiotic intake immensely in the period where antibiotic resistance is becoming a major challenge [85]. There are possibilities that the transgene in plant-based edible vaccines may cause allergies during post-translational modifications, and oral tolerance when co-administered with oral adjuvants to ordinarily activate the mucosal immune system. This may aggravate the oversensitive reactions to other proteins contained in the food [104]. Since most of the vectors used for plant genetic transformation are antibiotic-resistant, so the GMOs developed may evolve accordingly. Therefore insertion of antigen is suggested under the promoter that could spontaneously remove the antigenic gene at a desired point [13].

Genetic pollution is also a considerable issue for the use of plant-based approaches to synthesize the recombinant proteins when viral or bacterial vectors are used for the transformation of crops. Cross-contamination occurs through the pollination between the transgenic plants and non-transgenic plants or the soil after growing the non-transgenic plants on the same land where transgenic plants were grown that may lead to the horizontal and vertical gene flow [105]. An ecological imbalance may occur in the environment as the transgene trait may get passed onto the wild-type population either through cross-contamination of the human food chain and then transfer to the animals/wildlife. The transgene may also be transferred to the water sources through contact with animals or birds and hence the water bodies may get contaminated [106]. Hence stability of transgene and proper isolation of the transgenic plants to avoid cross-contamination and to keep the purity of the genera is important [107].

6. POTENTIAL SOLUTIONS FOR BIOSAFETY CONCERNS

PMF is an advanced technology to provide humans with better health and to solve health problems by the development of cost-effective vaccines, antibiotics, and other valuable recombinant proteins [102]. But there are some potential threats and concerns associated with this novel technology such as cross-contamination, low stability, and allergic responses that must be assessed for their possible solutions [108]. Although there it is very little or no proven documentation for any harm from the GMOs to animals or humans but the myths in society are discouraging people to adopt this. However, the following approaches are proved to be useful, but there are no specific recommendations developed for the selection of host plants for synthesis of recombinant proteins, and plants are still selected on case-to-case bases [48].

6.1. Use of Non-Food Crops and Non-Crop Plants

The best solution to avoid food contamination is to use the non-food crops or the model plant species for PMF which will minimize the accidental contacts and transgene spread. One of the best and successfully used non-food crops for the expression of transgenes of pharmaceutical products are tobacco [109]. Some scientists also recommend the use of non-crop plants like duckweeds, *Arabidopsis* and mosses, or even microalgae. Hence, there will be rare chances of food and feed chain contaminations and these systems also ensure a higher yield of recombinant proteins [18, 86 - 88]. Green Microalgae (*Chlamydomonas reinhardtii*) has some benefits over the development of conventional plant systems for recombinant products such as there are no chances of adjacent yield cross-contamination limitation because green algae can be grown with encased bioreactors [110]. Moreover, algal biomass is quick and easy to accumulate and all of it can be used to raise the vaccine production system [85]. However, the use of non-food crops also has some limitations that include inadequate knowledge about genetics and biology of plants along with limited experience of cultivation, low yield and expression, and unsuitable glycosylation of antigen protein, *etc.* A few of the algae-based raised vaccines are in clinical trials such as the Human Papillomavirus, Hepatitis B virus, and Foot and mouth disease virus [51]. A comparison of the food/feed crop plants, as well as non-food/non-feed crop plants used to produce the edible vaccines, has been summarized in Table **4**.

Table 4. Comparison of edible plant molecular farming of crop plants.

Production Platform	Advantages	Disadvantages	Examples
Feed/food crops	High biomass production. Minimum use of artificial fertilizers. Good knowledge of cultivation practices and genetics of plants is required Well established transformation methods.	Risk of food chain contamination. Risk of trans-gene transfer to other crops. Low yield of product of interest.	Maize Rice Potato Tomato

(Table 4) cont.....

Production Platform	Advantages	Disadvantages	Examples
Non-food/non-feed crops	Minimum chance of food chain contamination. Easy to grow in restricted/ containment environment. Easy to maintain quality of the products.	Toxic products of some plants make downstream processing difficult. Knowledge about cultivation practices and genetic is not required. Higher cost of maintenance for cell cultures.	Tobacco Duckweeds and mosses Arabidopsis Carrot

6.2. Use of Cell Cultures of Transgenic Plants

Cell culture of transgenic plants is a valuable method to avoid food chain contamination. Cell cultures expressing transgene are propagated in the form of cell suspensions in closed bioreactors. Environmental exposure of cell cultures and their products is highly restricted. This method also improves the yield of recombinant protein by making downstream processing easy and efficient [111]. Successful production of human glucocerebrosidase protein in carrot suspension culture for the treatment of Gaucher's disease was approved by FDA in 2012 and is now available in the market by Pfizer [12, 28]. The development of cell lines requires extensive knowledge about the genetics and biology of plant tissues in terms of yield, protein stability, and scalability, batch processing time and duration, and biological containment [83]. Therefore, cell cultures of only a few plants (rice, tobacco, and *Arabidopsis*) are developed and being used.

6.3. Use of Physical and Spatial Containments

Methods to restrict the exposure of GM plants to the environment are proved to be useful to minimize the contamination of food or feed chains. Banana is the ideal example in this case as it is grown throughout the year in the tropical and subtropical regions and the fruit is consumed raw. Banana plants are propagated vegetatively through suckers and therefore provide inbuilt biological containment that restricts the risk of environmental gene flow. However, this species has the limitation as the plants take a long time to be developed as transgenic containing recombinant proteins which reduce its feasibility for the oral delivery of biopharmaceuticals [112]. Green-house facilities, plastic tunnels, and growth chambers in laboratories are commonly used methods under physical containments. These methods are considered to be environment friendly and proved to be helpful to minimize environmental safety concerns. Close monitoring of the crops is necessary to avoid these ailments [83]. But they have not replaced open-field production systems which are still preferred, because large quantities of products are required and for those plants that are not able to grow in isolated systems. Labeling of edible pharmaceutical plants should be done with unique features to preserve their identity, expedite traceability and prevent the cross-contamination of food and feed supply. To prove this conception, transgenic tomatoes containing IgA antibodies against Rotavirus were crossed with another tomato transgenic line expressing *Antirrhinum majus* Rosea1 and Delila transcription factors in tomato fruit which led to the activation of anthocyanin biosynthesis and produced the purple fruits. The resultant crossbreed (purple transgenic tomato fruit) was expressing the newly recombinant antibodies which made them easy to classify [113].

Cultivation of PMF crops in a specified region that is away from an area where other crops are being grown is also proved to be useful to minimize the transgene spread through cross-pollination. Such approaches are useful alternatives against crop rotation between non-transgenic and GM crops. Some other methods developed to stop cross-pollination between different varieties can also be used to minimize the crossing between GM and other crops. The use of "trap" plants and minimum isolation distance to separate the pharmaceutical plants from other plants are reported to reduce the gene flow *via* pollen dispersal [106].

7. PLANT TRANSFORMATION TECHNIQUES

Many transformation systems are exploited for PMF to introduce the gene of interest in specific plant species which can be transformed stably. The trans-protein is stored in a specific tissue (seeds) or secreted into the hydroponic medium and can be recovered easily from the genetically modified plant. Sometimes, it is favorable to transform a plant transiently to check the suitable host and expression level of a protein. A brief description of the most commonly used plant transformation techniques is summarized in Table **5** [114 - 118].

Table 5. Summary of plant transformation techniques.

Transformation Method	Advantages	Disadvantages	References
Stable Nuclear Transformation	*Agrobacterium* mediated transformation is established and universal method. Transgene expression is 1-2% of total soluble proteins	Non-homologous recombination results in irregular gene expression.	[114]
Stable Plastid Transformation	Cross breeding is unlikely due to maternal inheritance. Transgene expression is 5-25% of total soluble proteins.	Homologous recombination makes vectors construction laborious.	[115]

(Table 5) cont.....

Transformation Method		Advantages	Disadvantages	References
Transient Expression Systems	Agroinfiltration method	After successful expression of GOI, same vector can be used for genome integration.	Limited to *Agrobacterium* host species only.	[116]
	Virus Infection Method	High level of recombinant protein expression due to high rate of viral propagation.	Limited to host range Manipulation of viral genome is restricted.	[117]
	Biolistic Gene Delivery	No limitation of host range or tissue.	Possible tissue damage. Highly expensive.	[118]

Stable Nuclear Transformation is the only example of natural inter-kingdom DNA transfer [117]. Mechanism of *Agrobacterium tumefaciens* transformation has been exploited to use it as a vehicle to transfer GOI into plant nuclear genome and this is considered to be the most effective tool for plant genetic transformation [119]. When a new gene is integrated into the genome, the genetic material of the host plant is altered and it becomes a GMO and starts to form a specific protein of interest. A successful transformation means a new stable trait is conferred to a plant that is not present in wild type [114].

Stable Plastid Transformation is considered to be another promising method for PMF in tobacco. The chloroplast genome is 120-220 kb and it contains 110-120 genes [117]. Integration of transgene by homologous recombination makes plastid transformation superior over nuclear transformation where GOI is integrated by non-homologous recombination. Tobacco is reported to be most commonly used to check the validity of the stable plastid transformation method [120, 121]. Natural bio-contaminants delivery of the transgene flow by the out-crossing makes it more profitable as compared to the stable nuclear transformation. In most of the species, plastids are of maternal origin, that's why pollens do not have any traces of transgene which lowers the public concerns about transgenic plants [122].

Transient Expression Systems are used to observe the expression of GOI in the biopharmaceutical plant as this is the most suitable and easy method to get desired protein through PMF [11]. These systems are most commonly used to get a higher level of transgene expression for a short time and these platforms also ensure that the next generations of host plants do not have any remains of GOI [123]. Transient expression of proteins in plant cells may be determined as;

1) ***Agroinfiltration***: A leaf surface is inoculated with *A. tumefaciens* harboring genetically engineered binary vectors for GOI expression by using a needle-less syringe or vacuum infiltration. Transgene begins to express itself shortly after the

T-DNA has been delivered to the cell as the integration of transgene to the cellular genome is not required [124]. Successful transcription and translation of transgene can be visualized by fluorescent microscopy and Western blot analysis within 1-2 days of inoculation.

2) Virus Infection Method: Different viruses of plants such as potato virus X (PVX) and tobacco mosaic virus (TMV) can be used as vectors for gene delivery to plant cells [116]. Immediate processing for degradation and stability of the recombinant protein in the cell are drawbacks of this system. The success of this system is the production of an idiotype vaccine that is effective against B-cell non-Hodgkin's lymphoma [125]. Limitation of host range is also associated with viral vector system. Moreover, viruses do not allow substantial manipulation of the genome concerning the size and place of GOI which might interfere with viral replication and genetic stability [124].

3) Biolistic Gene Delivery: Genetically engineered plant DNA is coated on micron-sized metal particles and these particles are then bombarded onto target plant cells. Plant cells survived after particle bombardment will express GOI within 24 to 48h. With the development of low-pressure gene guns, tissue damage has been reduced to a minimum extent [118].

8. PURIFICATION OF RECOMBINANT PROTEIN PRODUCTS

For the development of recombinant proteins used in the pharmaceutical industry, 80% of the total production cost required for purification is due to several chemicals and physical steps (downstream processing). Some plant-derived products like edible vaccines or many non-pharmaceutical products do not require any downstream processing [5, 16]. The major advantage of lower cost is that the less developed countries may also contribute to the production of plant-derived recombinant proteins [10]. To meet the leading need for greater containment of transgenic plants, different cost-effective plant species are being analyzed and industrial and pharmaceutical purposes are compared [59]. It is beneficial to use production systems containing watery tissues like tomatoes, instead of dry tissues like cereals. This will reduce the purification cost, because, it is easy to extract protein from watery tissues than from dry tissues. Tomatoes can be grown in a contained environment in green-house which will overcome the biosafety concerns. Steps in protein purification involve harvesting the plant material followed by extraction and purification of required protein. Chromatography-based methods are also used in some cases like purification of a vaccine designed to treat the human papillomavirus (HPV16) [126]. Although chromatography-based methods are not properly scalable and generally expensive up to 99% purity can be achieved with 0.1g/kg of protein yield. With the advancement of technology, some novel purification strategies like the use of fusion tags *e.g.* HIS_6 are also being used for protein purification [5]. Although, plants are free from mammalian viruses and bacterial endotoxins downstream processing is still necessary due to current good manufacturing practices (cGMP) compliance for pharmaceutical products [27].

9. FUTURE PROSPECTIVE

The introduction of PMF in the field of plant biotechnology has resulted in considerable improvement of many molecular and biochemical techniques being used to solve human health issues related to infectious, auto-immune, and life-threatening diseases such as cancer. These technological advances include protein targeting to a specific tissue, regulation of gene expression, transformation methods, and the use of different crops as production platforms. There is a need to engineer plants for proper post-transcriptional modifications so that allergic reactions and immunogenicity issues can be solved for plant-derived pharmaceutical products. Problems of low yield and protein instability are being analyzed to promote PMF as a future medication. Now it becomes possible to knockout any gene from a living cell with the help of engineered nucleases specifically the CRISPR/Cas9 system and this advancement will prove to be helpful to save the recombinant product from proteases [127]. Combining nano-technologies with the development of transgenic plants containing edible vaccines successfully coated before its expression will bring a revolutionary change in this area. PMF offers a cost-effective alternative for several recombinant products used in industrial processes. Plants offer a highly scalable and inexpensive platform for the production of industrial enzymes at minimum expense scale-up capacity as compared with fermenters [128].

Plant scientists are working to control the expression of recombinant proteins in the host plants so that a constant final yield can be assured in the final product. Many regulatory issues will automatically be solved if researchers become successful to control the expression of a transgene in any specific tissue of the host. Regulation of transgene expression will also solve the problem of gene dispersal through pollens or other unintentional exposure to the environment and there will be no need to keep the transgenic plants in isolation from other organisms. Still, a few issues need critical analysis like the justification of hereditary control of expression so that public acceptability issues can be solved. Proper R&D and clinical trials of the drugs/ vaccines/ therapeutic proteins will lead to safely solving the public health issues.

CONCLUSION

PMF is considered to be a favorable biotechnological method to synthesize edible recombinant proteins (antibodies/therapeutic proteins) against infectious or life-threatening diseases. Plants are valuable bio-systems for the production of pharmaceutical proteins for human health along with other important industrial products like biocatalysts, feed additives, and biopolymers. Many plant species are being tested for the production of recombinant products and non-feed/nonfood crops offer several advantages over other crops. The complexity of the plant genome makes it difficult to establish a single platform for all types of recombinant products as well as to predict the final yield of the product of interest. A plant cell has several proteases that degrade non-native products within no time. Certain genetic changes in transgene expression cassette along with some approaches to control proteases' activity are being tested to enhance the expression of recombinant products so that the final yield can be increased to a considerable extent. PMF has some biosafety, regulatory and R&D issues like genetic pollution and antibiotic resistance development. The use of non-feed/non food crops along with physical and spatial containments results in the safe development of transgenic plants. Advancement in molecular biology has resulted in the development of several well-established protocols to transfer GOI into any host plant. Purification of protein of interest from whole plant proteome is a difficult task but the use of affinity tags with the product along with expression in watery tissue makes it quite efficient and easy. Quick adaptability, the ability to fold the complex proteins precisely, protection from human pathogens make plants a preferable system for recombinant protein production. Proper R&D and clinical trials of the vaccine on the host will surpass the other production systems and will produce such products that are cheaper and easily available to everyone.

CONSENT FOR PUBLICATION

Not applicable.

CONFLICT OF INTEREST

The author declares no conflict of interest, financial or otherwise.

ACKNOWLEDGEMENTS

Declared none.

REFERENCES

[1]   Shinmyo A, Kato K. Molecular farming: production of drugs and vaccines in higher plants. J Antibiot (Tokyo) 2010; 63(8): 431-3.
[http://dx.doi.org/10.1038/ja.2010.63] [PMID: 20588301]

[2]   Drake PM, Thangaraj H. Molecular farming, patents and access to medicines. Expert Rev Vaccines 2010; 9(8): 811-9.
[http://dx.doi.org/10.1586/erv.10.72] [PMID: 20673006]

[3]   Barta A, Sommergruber K, Thompson D, Hartmuth K, Matzke MA, Matzke AJ. The expression of a nopaline synthase - human growth hormone chimaeric gene in transformed tobacco and sunflower callus tissue. Plant Mol Biol 1986; 6(5): 347-57.
[http://dx.doi.org/10.1007/BF00034942] [PMID: 24307385]

[4]   Sijmons PC, Dekker BM, Schrammeijer B, Verwoerd TC, van den Elzen PJ, Hoekema A. Production of correctly processed human serum albumin in transgenic plants. Biotechnology (N Y) 1990; 8(3): 217-21.
[PMID: 1366404]

[5]   Tschofen M, Knopp D, Hood E, Stöger E. Plant molecular farming: much more than medicines. Annu Rev Anal Chem (Palo Alto, Calif) 2016; 9(1): 271-94.
[http://dx.doi.org/10.1146/annurev-anchem-071015-041706] [PMID: 27049632]

[6]   Kwon KC, Verma D, Singh ND, Herzog R, Daniell H. Oral delivery of human biopharmaceuticals, autoantigens and vaccine antigens bioencapsulated in plant cells. Adv Drug Deliv Rev 2013; 65(6): 782-99.
[http://dx.doi.org/10.1016/j.addr.2012.10.005] [PMID: 23099275]

[7]   Daniell H, Singh ND, Mason H, Streatfield SJ. Plant-made vaccine antigens and biopharmaceuticals. Trends Plant Sci 2009; 14(12): 669-79.
[http://dx.doi.org/10.1016/j.tplants.2009.09.009] [PMID: 19836291]

[8]   Ganapathy M. Plants as bioreactors- A review. Adv Tech Biol Med 2016; 4: 161.

[9]   Merlin M, Gecchele E, Capaldi S, Pezzotti M, Avesani L. Comparative evaluation of recombinant protein production in different biofactories: the green perspective. BioMed Res Int 2014; 2014: 136419.
[http://dx.doi.org/10.1155/2014/136419] [PMID: 24745008]

[10]   Tsekoa TL, Singh AA, Buthelezi SG. Molecular farming for therapies and vaccines in Africa. Curr Opin Biotechnol 2020; 61: 89-95.
[http://dx.doi.org/10.1016/j.copbio.2019.11.005] [PMID: 31786432]

[11]   Rybicki EP. Plant-made vaccines for humans and animals. Plant Biotechnol J 2010; 8(5): 620-37.
[http://dx.doi.org/10.1111/j.1467-7652.2010.00507.x] [PMID: 20233333]

[12]   Shahid N, Daniell H. Plant-based oral vaccines against zoonotic and non-zoonotic diseases. Plant Biotechnol J 2016; 14(11): 2079-99.
[http://dx.doi.org/10.1111/pbi.12604] [PMID: 27442628]

[13]   Aryamvally A, Gunasekaran V, Narenthiran KR, Pasupathi R. New strategies toward edible vaccines: An overview. J Diet Suppl 2017; 14(1): 101-16.
[http://dx.doi.org/10.3109/19390211.2016.1168904] [PMID: 27065206]

[14]   Oszvald M, Kang TJ, Tomoskozi S, *et al.* Expression of a synthetic neutralizing epitope of porcine epidemic diarrhea virus fused with synthetic B subunit of Escherichia coli heat labile enterotoxin in rice endosperm. Mol Biotechnol 2007; 35(3): 215-23.
[http://dx.doi.org/10.1007/BF02686007] [PMID: 17652785]

[15]   Mikschofsky H, Broer I. Feasibility of Pisum sativum as an expression system for pharmaceuticals. Transgenic Res 2012; 21(4): 715-24.
[http://dx.doi.org/10.1007/s11248-011-9573-z] [PMID: 22057506]

[16]   Moustafa K, Makhzoum A, Trémouillaux-Guiller J. Molecular farming on rescue of pharma industry for next generations. Crit Rev Biotechnol 2016; 36(5): 840-50.
[http://dx.doi.org/10.3109/07388551.2015.1049934] [PMID: 26042351]

[17]   Kwon KC, Daniell H. Oral delivery of protein drugs bioencapsulated in plant cells. Mol Ther 2016; 24(8): 1342-50.
[http://dx.doi.org/10.1038/mt.2016.115] [PMID: 27378236]

[18]   Avesani L, Merlin M, Gecchele E, *et al.* Comparative analysis of different biofactories for the production of a major diabetes autoantigen. Transgenic Res 2014; 23(2): 281-91.
[http://dx.doi.org/10.1007/s11248-013-9749-9] [PMID: 24142387]

[19]   Baeshen NA, Baeshen MN, Sheikh A, *et al.* Cell factories for insulin production. Microb Cell Fact 2014; 13(1): 141.
[http://dx.doi.org/10.1186/s12934-014-0141-0] [PMID: 25270715]

[20]   Forthal DN, Gach JS, Landucci G, *et al.* Fc-glycosylation influences Fcγ receptor binding and cell-mediated anti-HIV activity of monoclonal antibody 2G12. J Immunol 2010; 185(11): 6876-82.
[http://dx.doi.org/10.4049/jimmunol.1002600] [PMID: 21041724]

[21]   Shepard HM, Phillips GL, D Thanos C, Feldmann M. Developments in therapy with monoclonal antibodies and related proteins. Clin Med (Lond) 2017; 17(3): 220-32.
[http://dx.doi.org/10.7861/clinmedicine.17-3-220] [PMID: 28572223]

[22]   De Meyer T, Muyldermans S, Depicker A. Nanobody-based products as research and diagnostic tools. Trends Biotechnol 2014; 32(5): 263-70.
[http://dx.doi.org/10.1016/j.tibtech.2014.03.001] [PMID: 24698358]

[23]   Geering B, Fussenegger M. Synthetic immunology: modulating the human immune system. Trends Biotechnol 2015; 33(2): 65-79.
[http://dx.doi.org/10.1016/j.tibtech.2014.10.006] [PMID: 25466879]

[24]   Justino CI, Duarte AC, Rocha-Santos TA. Analytical applications of affibodies. Trends Analyt Chem 2015; 65: 73-82.
[http://dx.doi.org/10.1016/j.trac.2014.10.014]

[25]   Hiatt A, Cafferkey R, Bowdish K. Production of antibodies in transgenic plants. Nature 1989; 342(6245): 76-8.
[http://dx.doi.org/10.1038/342076a0] [PMID: 2509938]

[26]   Julve JM, Gandía A, Fernández-Del-Carmen A, *et al.* A coat-independent superinfection exclusion rapidly imposed in *Nicotiana benthamiana* cells by tobacco mosaic virus is not prevented by depletion of the movement protein. Plant Mol Biol 2013; 81(6): 553-64.
[http://dx.doi.org/10.1007/s11103-013-0028-1] [PMID: 23417583]

[27]   Ma JK, Drossard J, Lewis D, *et al.* Regulatory approval and a first-in-human phase I clinical trial of a monoclonal antibody produced in transgenic tobacco plants. Plant Biotechnol J 2015; 13(8): 1106-20.
[http://dx.doi.org/10.1111/pbi.12416] [PMID: 26147010]

[28]   Yusibov V, Streatfield SJ, Kushnir N. Clinical development of plant-produced recombinant pharmaceuticals: vaccines, antibodies and beyond. Hum Vaccin 2011; 7(3): 313-21.
[http://dx.doi.org/10.4161/hv.7.3.14207] [PMID: 21346417]

[29]   Ritala A, Leelavathi S, Oksman-Caldentey KM, Reddy VS, Laukkanen ML. Recombinant barley-produced antibody for detection and immunoprecipitation of the major bovine milk allergen, β-lactoglobulin. Transgenic Res 2014; 23(3): 477-87.
[http://dx.doi.org/10.1007/s11248-014-9783-2] [PMID: 24497085]

[30]   Avesani L, Bortesi L, Santi L, Falorni A, Pezzotti M. Plant-made pharmaceuticals for the prevention and treatment of autoimmune diseases: where are we? Expert Rev Vaccines 2010; 9(8): 957-69.
[http://dx.doi.org/10.1586/erv.10.82] [PMID: 20673017]

[31]   Wang X, Sherman A, Liao G, *et al.* Mechanism of oral tolerance induction to therapeutic proteins. Adv Drug Deliv Rev 2013; 65(6): 759-73.
[http://dx.doi.org/10.1016/j.addr.2012.10.013] [PMID: 23123293]

[32]   Sherman A, Su J, Lin S, Wang X, Herzog RW, Daniell H. Suppression of inhibitor formation against FVIII in a murine model of hemophilia A by oral delivery of antigens bioencapsulated in plant cells. Blood 2014; 124(10): 1659-68.
[http://dx.doi.org/10.1182/blood-2013-10-528737] [PMID: 24825864]

[33]   Su J, Zhu L, Sherman A, *et al.* Low cost industrial production of coagulation factor IX bioencapsulated in lettuce cells for oral tolerance induction in hemophilia B. Biomaterials 2015; 70: 84-93.
[http://dx.doi.org/10.1016/j.biomaterials.2015.08.004] [PMID: 26302233]

[34]   Verma D, Moghimi B, LoDuca PA, *et al.* Oral delivery of bioencapsulated coagulation factor IX prevents inhibitor formation and fatal anaphylaxis in hemophilia B mice. Proc Natl Acad Sci USA 2010; 107(15): 7101-6.
[http://dx.doi.org/10.1073/pnas.0912181107] [PMID: 20351275]

[35]   Su J, Sherman A, Doerfler PA, Byrne BJ, Herzog RW, Daniell H. Oral delivery of Acid Alpha Glucosidase epitopes expressed in plant chloroplasts suppresses antibody formation in treatment of Pompe mice. Plant Biotechnol J 2015; 13(8): 1023-32.
[http://dx.doi.org/10.1111/pbi.12413] [PMID: 26053072]

[36]   Suzuki K, Kaminuma O, Yang L, *et al.* Prevention of allergic asthma by vaccination with transgenic rice seed expressing mite allergen: induction of allergen-specific oral tolerance without bystander suppression. Plant Biotechnol J 2011; 9(9): 982-90.
[http://dx.doi.org/10.1111/j.1467-7652.2011.00613.x] [PMID: 21447056]

[37]   Baldauf KJ, Royal JM, Hamorsky KT, Matoba N. Cholera toxin B: one subunit with many pharmaceutical applications. Toxins (Basel) 2015; 7(3): 974-96.
[http://dx.doi.org/10.3390/toxins7030974] [PMID: 25802972]

[38]   Zhou Z, Gong S, Li XM, *et al.* Expression of Helicobacter pylori urease B on the surface of Bacillus subtilis spores. J Med Microbiol 2015; 64(Pt 1): 104-10.
[http://dx.doi.org/10.1099/jmm.0.076430-0] [PMID: 25355934]

[39]   Chen Q, Lai H. Gene delivery into plant cells for recombinant protein production. BioMed Res Int 2015; 2015: 932161.
[http://dx.doi.org/10.1155/2015/932161] [PMID: 26075275]

[40]   Rukavtsova EB, Rudenko NV, Puchko EN, Zakharchenko NS, Buryanov YI. Study of the immunogenicity of hepatitis B surface antigen synthesized in transgenic potato plants with increased biosafety. J Biotechnol 2015; 203: 84-8.
[http://dx.doi.org/10.1016/j.jbiotec.2015.03.019] [PMID: 25840367]

[41]   Hu J, Ni Y, Dryman BA, Meng XJ, Zhang C. Immunogenicity study of plant-made oral subunit vaccine against porcine reproductive and respiratory syndrome virus (PRRSV). Vaccine 2012; 30(12): 2068-74.
[http://dx.doi.org/10.1016/j.vaccine.2012.01.059] [PMID: 22300722]

[42]   Wang Y, Shen Q, Jiang Y, *et al.* Immunogenicity of foot-and-mouth disease virus structural polyprotein P1 expressed in transgenic rice. J Virol Methods 2012; 181(1): 12-7.
[http://dx.doi.org/10.1016/j.jviromet.2012.01.004] [PMID: 22274594]

[43]   Uvarova EA, Belavin PA, Permyakova NV, *et al.* Oral Immunogenicity of plant-made Mycobacterium tuberculosis ESAT6 and CFP10. BioMed Res Int 2013; 2013: 316304.
[http://dx.doi.org/10.1155/2013/316304] [PMID: 24455687]

[44]   Gonzalez-Rabade N, McGowan EG, Zhou F, *et al.* Immunogenicity of chloroplast-derived HIV-1 p24 and a p24-Nef fusion protein following subcutaneous and oral administration in mice. Plant Biotechnol J 2011; 9(6): 629-38.
[http://dx.doi.org/10.1111/j.1467-7652.2011.00609.x] [PMID: 21443546]

[45]   Gorantala J, Grover S, Rahi A, *et al.* Generation of protective immune response against anthrax by oral

immunization with protective antigen plant-based vaccine. J Biotechnol 2014; 176: 1-10.
[http://dx.doi.org/10.1016/j.jbiotec.2014.01.033] [PMID: 24548460]

[46]   Mardanova ES, Kotlyarov RY, Kuprianov VV, *et al.* Rapid high-yield expression of a candidate influenza vaccine based on the ectodomain of M2 protein linked to flagellin in plants using viral vectors. BMC Biotechnol 2015; 15: 42.
[http://dx.doi.org/10.1186/s12896-015-0164-6] [PMID: 26022390]

[47]   Shakoor S, Rao AQ, Shahid N, *et al.* Role of oral vaccines as an edible tool to prevent infectious diseases. Acta Virol 2019; 63(3): 245-52.
[http://dx.doi.org/10.4149/av_2019_301] [PMID: 31507189]

[48]   Guan ZJ, Guo B, Huo YL, Guan ZP, Dai JK, Wei YH. Recent advances and safety issues of transgenic plant-derived vaccines. Appl Microbiol Biotechnol 2013; 97(7): 2817-40.
[http://dx.doi.org/10.1007/s00253-012-4566-2] [PMID: 23447052]

[49]   Rigano MM, De Guzman G, Walmsley AM, Frusciante L, Barone A. Production of pharmaceutical proteins in solanaceae food crops. Int J Mol Sci 2013; 14(2): 2753-73.
[http://dx.doi.org/10.3390/ijms14022753] [PMID: 23434646]

[50]   Zhang H, Liu M, Li Y, *et al.* Oral immunogenicity and protective efficacy in mice of a carrot-derived vaccine candidate expressing UreB subunit against Helicobacter pylori. Protein Expr Purif 2010; 69(2): 127-31.
[http://dx.doi.org/10.1016/j.pep.2009.07.016] [PMID: 19651219]

[51]   Gunasekaran B, Gothandam KM. A review on edible vaccines and their prospects. Braz J Med Biol Res 2020; 53(2): e8749.
[http://dx.doi.org/10.1590/1414-431x20198749] [PMID: 31994600]

[52]   Qian B, Shen H, Liang W, *et al.* Immunogenicity of recombinant hepatitis B virus surface antigen fused with preS1 epitopes expressed in rice seeds. Transgenic Res 2008; 17(4): 621-31.
[http://dx.doi.org/10.1007/s11248-007-9135-6] [PMID: 17882531]

[53]   Zhang X, Buehner NA, Hutson AM, Estes MK, Mason HS. Tomato is a highly effective vehicle for expression and oral immunization with Norwalk virus capsid protein. Plant Biotechnol J 2006; 4(4): 419-32.
[http://dx.doi.org/10.1111/j.1467-7652.2006.00191.x] [PMID: 17177807]

[54]   Lou XM, Yao QH, Zhang Z, Peng RH, Xiong AS, Wang HK. Expression of the human hepatitis B virus large surface antigen gene in transgenic tomato plants. Clin Vaccine Immunol 2007; 14(4): 464-9.
[http://dx.doi.org/10.1128/CVI.00321-06] [PMID: 17314228]

[55]   Srinivas L, Kumar GS, Ganapathi TR, Revathi CJ, Bapat VA. Transient and stable expression of hepatitis B surface antigen in tomato (*Lycopersicon esculentum* L.). Plant Biotechnol Rep 2008; 2(1): 1-6.
[http://dx.doi.org/10.1007/s11816-008-0041-z]

[56]   Kim TG, Kim MY, Kim BG, *et al.* Synthesis and assembly of Escherichia coli heat-labile enterotoxin B subunit in transgenic lettuce (*Lactuca sativa*). Protein Expr Purif 2007; 51(1): 22-7.
[http://dx.doi.org/10.1016/j.pep.2006.05.024] [PMID: 16919472]

[57]   Yan-Ju Y, Wen-Gui L. Immunoprotection of transgenic alfalfa (Medicago sativa) containing Eg95-EgA31 fusion gene of Echinococcus granulosus against Eg protoscoleces. Redai Yixue Zazhi 2010; 10(3): 235-7.

[58]   Shahid N, Samiullah TR, Shakoor S, *et al.* Early stage development of a newcastle disease vaccine candidate in corn. Front Vet Sci 2020; 7: 499.
[http://dx.doi.org/10.3389/fvets.2020.00499] [PMID: 33062645]

[59]   Puetz J, Wurm FM. Recombinant proteins for industrial versus pharmaceutical purposes: a review of process and pricing. Processes (Basel) 2019; 7(8): 476.

[http://dx.doi.org/10.3390/pr7080476]

[60]   Krishnan A, Woodard SL. TrypZean™: an animal-free alternative to bovine trypsin.Commercial plant-produced recombinant protein products. Springer-Verlag Berlin Heidelberg 2014; pp. 43-66.
[http://dx.doi.org/10.1007/978-3-662-43836-7_4]

[61]   Schillberg S, Raven N, Spiegel H, Rasche S, Buntru M. Critical analysis of the commercial potential of plants for the production of recombinant proteins. Front Plant Sci 2019; 10: 720.
[http://dx.doi.org/10.3389/fpls.2019.00720] [PMID: 31244868]

[62]   Davison-Kotler E, Marshall WS, García-Gareta E. Sources of collagen for biomaterials in skin wound healing. Bioengineering (Basel) 2019; 6(3): 56.
[http://dx.doi.org/10.3390/bioengineering6030056] [PMID: 31261996]

[63]   Mohammadinejad R, Shavandi A, Raie DS, *et al.* Plant molecular farming: production of metallic nanoparticles and therapeutic proteins using green factories. Green Chem 2019; 21: 1845-65.
[http://dx.doi.org/10.1039/C9GC00335E]

[64]   Hauptmann V, Menzel M, Weichert N, Reimers K, Spohn U, Conrad U. In planta production of ELPylated spidroin-based proteins results in non-cytotoxic biopolymers. BMC Biotechnol 2015; 15(1): 9.
[http://dx.doi.org/10.1186/s12896-015-0123-2] [PMID: 25888206]

[65]   Weichert N, Hauptmann V, Menzel M, *et al.* Transglutamination allows production and characterization of native-sized ELPylated spider silk proteins from transgenic plants. Plant Biotechnol J 2014; 12(2): 265-75.
[http://dx.doi.org/10.1111/pbi.12135] [PMID: 24237483]

[66]   Hauptmann V, Weichert N, Rakhimova M, Conrad U. Spider silks from plants - a challenge to create native-sized spidroins. Biotechnol J 2013; 8(10): 1183-92.
[http://dx.doi.org/10.1002/biot.201300204] [PMID: 24092675]

[67]   Xu X, Gan Q, Clough RC, *et al.* Hydroxylation of recombinant human collagen type I alpha 1 in transgenic maize co-expressed with a recombinant human prolyl 4-hydroxylase. BMC Biotechnol 2011; 11(1): 69.
[http://dx.doi.org/10.1186/1472-6750-11-69] [PMID: 21702901]

[68]   Yan J, Hu K, Xiao Y, *et al.* Preparation of recombinant human-like collagen/fibroin scaffold and its promoting effect on vascular cells biocompatibility. J Bioact Compat Polym 2018; 33(4): 416-25.
[http://dx.doi.org/10.1177/0883911518769680]

[69]   Xu X, Zhang Y, Meng Q, *et al.* Overexpression of a fungal β-mannanase from Bispora sp. MEY-1 in maize seeds and enzyme characterization. PLoS One 2013; 8(2): e56146.
[http://dx.doi.org/10.1371/journal.pone.0056146] [PMID: 23409143]

[70]   Zafar A, Aftab MN, Saleem MA. Pilot scale production of recombinant hemicellulases and their saccharification potential. Prep Biochem Biotechnol 2020; 50(10): 1063-75.
[http://dx.doi.org/10.1080/10826068.2020.1783679] [PMID: 32594842]

[71]   Aftab MN, Zafar A, Iqbal I, Kaleem A, Zia KM, Awan AR. Optimization of saccharification potential of recombinant xylanase from Bacillus licheniformis. Bioengineered 2018; 9(1): 159-65.
[http://dx.doi.org/10.1080/21655979.2017.1373918] [PMID: 28886289]

[72]   Li Q, Song J, Peng S, *et al.* Plant biotechnology for lignocellulosic biofuel production. Plant Biotechnol J 2014; 12(9): 1174-92.
[http://dx.doi.org/10.1111/pbi.12273] [PMID: 25330253]

[73]   Zhang D, VanFossen AL, Pagano RM, *et al.* Consolidated pretreatment and hydrolysis of plant biomass expressing cell wall degrading enzymes. BioEnergy Res 2011; 4(4): 276-86.
[http://dx.doi.org/10.1007/s12155-011-9138-2]

[74]   Pillay P, Schlüter U, van Wyk S, Kunert KJ, Vorster BJ. Proteolysis of recombinant proteins in bioengineered plant cells. Bioengineered 2014; 5(1): 15-20.

[http://dx.doi.org/10.4161/bioe.25158] [PMID: 23778319]

[75]   Sabalza M, Madeira L, van Dolleweerd C, Ma JK, Capell T, Christou P. Functional characterization of the recombinant HIV-neutralizing monoclonal antibody 2F5 produced in maize seeds. Plant Mol Biol 2012; 80(4-5): 477-88.
[http://dx.doi.org/10.1007/s11103-012-9962-6] [PMID: 22965278]

[76]   Obembe OO, Popoola JO, Leelavathi S, Reddy SV. Advances in plant molecular farming. Biotechnol Adv 2011; 29(2): 210-22.
[http://dx.doi.org/10.1016/j.biotechadv.2010.11.004] [PMID: 21115109]

[77]   Whaley KJ, Hiatt A, Zeitlin L. Emerging antibody products and Nicotiana manufacturing. Hum Vaccin 2011; 7(3): 349-56.
[http://dx.doi.org/10.4161/hv.7.3.14266] [PMID: 21358287]

[78]   Schillberg S, Raven N, Fischer R, Twyman RM, Schiermeyer A. Molecular farming of pharmaceutical proteins using plant suspension cell and tissue cultures. Curr Pharm Des 2013; 19(31): 5531-42.
[http://dx.doi.org/10.2174/13816128113199310008] [PMID: 23394569]

[79]   Xu J, Kieliszewski MJ. A novel plant cell bioproduction platform for high-yield secretion of recombinant proteins. Methods Mol Biol 2012; 824: 483-500.
[http://dx.doi.org/10.1007/978-1-61779-433-9_26] [PMID: 22160916]

[80]   Paul M, Ma JKC. Plant-made pharmaceuticals: leading products and production platforms. Biotechnol Appl Biochem 2011; 58(1): 58-67.
[http://dx.doi.org/10.1002/bab.6] [PMID: 21446960]

[81]   Ma S, Jevnikar A, Hüner N. Microalgae as bioreactors for production of pharmaceutical proteins. 2011.
[http://dx.doi.org/10.1016/B978-0-08-088504-9.00406-2]

[82]   Tran M, Van C, Barrera DJ, *et al.* Production of unique immunotoxin cancer therapeutics in algal chloroplasts. Proc Natl Acad Sci USA 2013; 110(1): E15-22.
[http://dx.doi.org/10.1073/pnas.1214638110] [PMID: 23236148]

[83]   Merlin M, Pezzotti M, Avesani L. Edible plants for oral delivery of biopharmaceuticals. Br J Clin Pharmacol 2017; 83(1): 71-81.
[http://dx.doi.org/10.1111/bcp.12949] [PMID: 27037892]

[84]   Mandal MK, Ahvari H, Schillberg S, Schiermeyer A. Tackling unwanted proteolysis in plant production hosts used for molecular farming. Front Plant Sci 2016; 7: 267.
[http://dx.doi.org/10.3389/fpls.2016.00267] [PMID: 27014293]

[85]   Kurup VM, Thomas J. Edible vaccines: Promises and challenges. Mol Biotechnol 2020; 62(2): 79-90.
[http://dx.doi.org/10.1007/s12033-019-00222-1] [PMID: 31758488]

[86]   Schiermeyer A. Optimizing product quality in molecular farming. Curr Opin Biotechnol 2020; 61: 15-20.
[http://dx.doi.org/10.1016/j.copbio.2019.08.012] [PMID: 31593785]

[87]   Tremblay R, Diao H, Hüner N, Jevnikar AM, Ma S. The development of a high-yield recombinant protein bioreactor through RNAi induced knockdown of ATP/ADP transporter in *Solanum tuberosum*. J Biotechnol 2011; 156(1): 59-66.
[http://dx.doi.org/10.1016/j.jbiotec.2011.08.005] [PMID: 21864587]

[88]   James EA, Wang C, Wang Z, *et al.* Production and characterization of biologically active human GM-CSF secreted by genetically modified plant cells. Protein Expr Purif 2000; 19(1): 131-8.
[http://dx.doi.org/10.1006/prep.2000.1232] [PMID: 10833400]

[89]   Baur A, Reski R, Gorr G. Enhanced recovery of a secreted recombinant human growth factor using stabilizing additives and by co-expression of human serum albumin in the moss Physcomitrella patens. Plant Biotechnol J 2005; 3(3): 331-40.
[http://dx.doi.org/10.1111/j.1467-7652.2005.00127.x] [PMID: 17129315]

[90]   Goulet C, Benchabane M, Anguenot R, Brunelle F, Khalf M, Michaud D. A companion protease inhibitor for the protection of cytosol-targeted recombinant proteins in plants. Plant Biotechnol J 2010; 8(2): 142-54.
[http://dx.doi.org/10.1111/j.1467-7652.2009.00470.x] [PMID: 20051033]

[91]   Goulet C, Khalf M, Sainsbury F, D'Aoust MA, Michaud D. A protease activity-depleted environment for heterologous proteins migrating towards the leaf cell apoplast. Plant Biotechnol J 2012; 10(1): 83-94.
[http://dx.doi.org/10.1111/j.1467-7652.2011.00643.x] [PMID: 21895943]

[92]   Pillay P, Kibido T, du Plessis M, *et al.* Use of transgenic oryzacystatin-I-expressing plants enhances recombinant protein production. Appl Biochem Biotechnol 2012; 168(6): 1608-20.
[http://dx.doi.org/10.1007/s12010-012-9882-6] [PMID: 22965305]

[93]   Mandal MK, Fischer R, Schillberg S, Schiermeyer A. Inhibition of protease activity by antisense RNA improves recombinant protein production in *Nicotiana tabacum* cv. Bright Yellow 2 (BY-2) suspension cells. Biotechnol J 2014; 9(8): 1065-73.
[http://dx.doi.org/10.1002/biot.201300424] [PMID: 24828029]

[94]   Alvarez ML, Topal E, Martin F, Cardineau GA. Higher accumulation of F1-V fusion recombinant protein in plants after induction of protein body formation. Plant Mol Biol 2010; 72(1-2): 75-89.
[http://dx.doi.org/10.1007/s11103-009-9552-4] [PMID: 19789982]

[95]   Sainsbury F, Varennes-Jutras P, Goulet MC, D'Aoust MA, Michaud D. Tomato cystatin SlCYS8 as a stabilizing fusion partner for human serpin expression in plants. Plant Biotechnol J 2013; 11(9): 1058-68.
[http://dx.doi.org/10.1111/pbi.12098] [PMID: 23911079]

[96]   Sun QY, Ding LW, Lomonossoff GP, *et al.* Improved expression and purification of recombinant human serum albumin from transgenic tobacco suspension culture. J Biotechnol 2011; 155(2): 164-72.
[http://dx.doi.org/10.1016/j.jbiotec.2011.06.033] [PMID: 21762733]

[97]   Buyel JF. Plant molecular farming- Integration and exploitation of side streams to achieve sustainable biomanufacturing. Front Plant Sci 2019; 9: 1893.
[http://dx.doi.org/10.3389/fpls.2018.01893] [PMID: 30713542]

[98]   He Y, Ning T, Xie T, *et al.* Large-scale production of functional human serum albumin from transgenic rice seeds. Proc Natl Acad Sci USA 2011; 108(47): 19078-83.
[http://dx.doi.org/10.1073/pnas.1109736108] [PMID: 22042856]

[99]   Oey M, Lohse M, Kreikemeyer B, Bock R. Exhaustion of the chloroplast protein synthesis capacity by massive expression of a highly stable protein antibiotic. Plant J 2009; 57(3): 436-45.
[http://dx.doi.org/10.1111/j.1365-313X.2008.03702.x] [PMID: 18939966]

[100]   Fischer R, Vasilev N, Twyman RM, Schillberg S. High-value products from plants: the challenges of process optimization. Curr Opin Biotechnol 2015; 32: 156-62.
[http://dx.doi.org/10.1016/j.copbio.2014.12.018] [PMID: 25562816]

[101]   Mishra M, Kumari S. Biosafety issues related to genetically engineered crops. MOJ Res Rev 2018; 1(6): 272-6.
[http://dx.doi.org/10.15406/mojcrr.2018.01.00045]

[102]   Smyth SJ. The human health benefits from GM crops. Plant Biotechnol J 2020; 18(4): 887-8.
[http://dx.doi.org/10.1111/pbi.13261] [PMID: 31544299]

[103]   Menary J, Hobbs M, Mesquita de Albuquerque S, *et al.* Shotguns vs Lasers: Identifying barriers and facilitators to scaling-up plant molecular farming for high-value health products. PLoS One 2020; 15(3): e0229952.
[http://dx.doi.org/10.1371/journal.pone.0229952] [PMID: 32196508]

[104]   Swapna LA. Edible vaccines: A new approach for immunization in plant biotechnology. Sch Acad J Pharm 2013; 2: 227-32.

[105] Zapanta PE, Ghorab S. Age of bioterrorism: are you prepared? Review of bioweapons and their clinical presentation for otolaryngologists. Otolaryngol Head Neck Surg 2014; 151(2): 208-14.
[http://dx.doi.org/10.1177/0194599814531907] [PMID: 24757076]

[106] Hirlekar R, Bhairy SR. Edible vaccines: An advancement in oral immunization. Asian J Pharm Clin Res 2017; 16: 20.

[107] Maxwell S. Analysis of laws governing combination products, transgenic food, pharmaceutical products and their applicability to edible vaccines. BYU Prelaw Rev 2014; 28(1): 65-82.

[108] Zhu Q, Berzofsky JA. Oral vaccines: directed safe passage to the front line of defense. Gut Microbes 2013; 4(3): 246-52.
[http://dx.doi.org/10.4161/gmic.24197] [PMID: 23493163]

[109] Moon KB, Park JS, Park YI, *et al.* Development of systems for the production of plant-derived biopharmaceuticals. Plants 2019; 9(1): 30.
[http://dx.doi.org/10.3390/plants9010030] [PMID: 31878277]

[110] Specht EA, Mayfield SP. Algae-based oral recombinant vaccines. Front Microbiol 2014; 5: 60.
[http://dx.doi.org/10.3389/fmicb.2014.00060] [PMID: 24596570]

[111] Shanmugaraj B, I Bulaon CJ, Phoolcharoen W. Plant molecular farming: a viable platform for recombinant biopharmaceutical production. Plants 2020; 9(7): 842.
[http://dx.doi.org/10.3390/plants9070842] [PMID: 32635427]

[112] Chan HT, Chia MY, Pang VF, Jeng CR, Do YY, Huang PL. Oral immunogenicity of porcine reproductive and respiratory syndrome virus antigen expressed in transgenic banana. Plant Biotechnol J 2013; 11(3): 315-24.
[http://dx.doi.org/10.1111/pbi.12015] [PMID: 23116484]

[113] Juárez P, Presa S, Espí J, *et al.* Neutralizing antibodies against rotavirus produced in transgenically labelled purple tomatoes. Plant Biotechnol J 2012; 10(3): 341-52.
[http://dx.doi.org/10.1111/j.1467-7652.2011.00666.x] [PMID: 22070155]

[114] Fischer U, Kuhlmann M, Pecinka A, Schmidt R, Mette MF. Local DNA features affect RNA-directed transcriptional gene silencing and DNA methylation. Plant J 2008; 53(1): 1-10.
[http://dx.doi.org/10.1111/j.1365-313X.2007.03311.x] [PMID: 17971044]

[115] Staub JM, Garcia B, Graves J, *et al.* High-yield production of a human therapeutic protein in tobacco chloroplasts. Nat Biotechnol 2000; 18(3): 333-8.
[http://dx.doi.org/10.1038/73796] [PMID: 10700152]

[116] Porta C, Lomonossoff GP. Viruses as vectors for the expression of foreign sequences in plants. Biotechnol Genet Eng Rev 2002; 19(1): 245-91.
[http://dx.doi.org/10.1080/02648725.2002.10648031] [PMID: 12520880]

[117] Meyers B, Zaltsman A, Lacroix B, Kozlovsky SV, Krichevsky A. Nuclear and plastid genetic engineering of plants: comparison of opportunities and challenges. Biotechnol Adv 2010; 28(6): 747-56.
[http://dx.doi.org/10.1016/j.biotechadv.2010.05.022] [PMID: 20685387]

[118] Ueki S, Lacroix B, Krichevsky A, Lazarowitz SG, Citovsky V. Functional transient genetic transformation of Arabidopsis leaves by biolistic bombardment. Nat Protoc 2009; 4(1): 71-7.
[http://dx.doi.org/10.1038/nprot.2008.217] [PMID: 19131958]

[119] Karmakar S, Molla KA, Gayen D, *et al.* Development of a rapid and highly efficient *Agrobacterium*-mediated transformation system for pigeon pea [*Cajanus cajan* (L.) Millsp]. GM Crops Food 2019; 10(2): 115-38. [Cajanus cajan (L.) Millsp].
[http://dx.doi.org/10.1080/21645698.2019.1625653] [PMID: 31187675]

[120] Svab Z, Maliga P. High-frequency plastid transformation in tobacco by selection for a chimeric aadA gene. Proc Natl Acad Sci USA 1993; 90(3): 913-7.

[http://dx.doi.org/10.1073/pnas.90.3.913] [PMID: 8381537]

[121] Daniell H, Khan MS, Allison L. Milestones in chloroplast genetic engineering: an environmentally friendly era in biotechnology. Trends Plant Sci 2002; 7(2): 84-91.
[http://dx.doi.org/10.1016/S1360-1385(01)02193-8] [PMID: 11832280]

[122] Cardi T, Lenzi P, Maliga P. Chloroplasts as expression platforms for plant-produced vaccines. Expert Rev Vaccines 2010; 9(8): 893-911.
[http://dx.doi.org/10.1586/erv.10.78] [PMID: 20673012]

[123] Vézina LP, Faye L, Lerouge P, *et al.* Transient co-expression for fast and high-yield production of antibodies with human-like N-glycans in plants. Plant Biotechnol J 2009; 7(5): 442-55.
[http://dx.doi.org/10.1111/j.1467-7652.2009.00414.x] [PMID: 19422604]

[124] Rybicki EP. Plant molecular farming of virus-like nanoparticles as vaccines and reagents. Wiley Interdiscip Rev Nanomed Nanobiotechnol 2020; 12(2): e1587.
[http://dx.doi.org/10.1002/wnan.1587] [PMID: 31486296]

[125] McCormick AA, Reddy S, Reinl SJ, *et al.* Plant-produced idiotype vaccines for the treatment of non-Hodgkin's lymphoma: safety and immunogenicity in a phase I clinical study. Proc Natl Acad Sci USA 2008; 105(29): 10131-6.
[http://dx.doi.org/10.1073/pnas.0803636105] [PMID: 18645180]

[126] Buyel JF, Bautista JA, Fischer R, Yusibov VM. Extraction, purification and characterization of the plant-produced HPV16 subunit vaccine candidate E7 GGG. J Chromatogr B Analyt Technol Biomed Life Sci 2012; 880(1): 19-26.
[http://dx.doi.org/10.1016/j.jchromb.2011.11.010] [PMID: 22134037]

[127] Belhaj K, Chaparro-Garcia A, Kamoun S, Nekrasov V. Plant genome editing made easy: targeted mutagenesis in model and crop plants using the CRISPR/Cas system. Plant Methods 2013; 9(1): 39.
[http://dx.doi.org/10.1186/1746-4811-9-39] [PMID: 24112467]

[128] Howard JA, Nikolov Z, Hood E. Enzyme production systems for biomass conversion.Plant biomass conversion. 1st ed. London: John Wiley & Sons 2011; pp. 227-53.
[http://dx.doi.org/10.1002/9780470959138.ch10]

CHAPTER 11

Proteins and Peptides as Biomarkers for Diagnosis of Cardiovascular Diseases

Sehar Aslam[1], Samman Munir[1], Muhammad Shareef Masoud[1], Usman Ali Ashfaq[1], Nazia Nahid[1], Mohsin Khurshid[2] and Muhammad Qasim[1,*]

[1] Department of Bioinformatics and Biotechnology, Government College University, Faisalabad, Pakistan

[2] Department of Microbiology, Government College University, Faisalabad, Pakistan

Abstract: With the increase in the prevalence of cardiovascular diseases internationally, particularly cardiac failure (CF) and atherosclerosis, the investigation for new biological markers remains one of the main priorities. In contrast to complicated diagnostic methods that might not be appropriate to be employed on a larger population, biological markers are effective for the screening of the population. Owing to their non-invasive detection with typically high accuracy and sensitivity, circulating biomarkers have become increasingly significant for routine medical practice. Cardiac troponins and natriuretic peptides (NPs), specifically brain NP (BNP), mid-regional pro arterial NP, and N-terminal (NT) pro BNP, are validated blood biological markers in the diagnosis of CF and prediction of CF-associated outcomes. Inflammatory proteins like C-reactive protein can also have increased importance in anti-inflammatory treatment guidance. Moreover, next-generation biological markers like galectin-3, growth/differentiation factor 15, diverse miRNAs, and soluble suppression of tumorigenicity-2 might have additional value in the analysis of ventricular remodeling and differentiation of CF subtypes. In this chapter, we will first discuss the biological markers as per the major categories of cardiovascular disease, *i.e.*, myocardial stress, inflammation, plaque instability, myocardial injury, systemic stress, calcium homeostasis, and platelet activation. Lastly, we will describe the multi-marker methods, including various combinations of novel and established biomarkers that may improve the risk prediction of CF at the population level.

Keywords: Acute cardiac infractions, Cardiovascular diseases (CVD), Cardiac failure (CF), Congenital heart diseases (CHD), Natriuretic peptides (NPs).

1. INTRODUCTION

Cardiovascular disease (CVD) is the leading cause of death internationally, in

---

* **Corresponding author Muhammad Qasim:** Department of Bioinformatics and Biotechnology, Government College University, Faisalabad, Pakistan; E-mail: qasemawan@gmail.com both developing as well as industrial countries, and involves arteriosclerotic (for example, peripheral arterial disease (PAD), cerebrovascular disease, coronary heart disease (CHD)), and nonarterioosclerotic disease (venous thromboembolism, congenital heart defect (CHD) and valvular disease). The incidence of CVD rises with age, and numerous risk factors, like obesity, alcohol abuse, physical inactivity, and tobacco use, are involved in the progression and development of CVD [1]. Moreover, an additional increase by almost 50% is expected till 2030, thus rising social and economic challenges with cost explosion in the coming decades [2]. The lifetime risk for cardiac failure (CF) is still 20-45 percent and highly age-dependent [3]. The rates of CF hospitalization declined significantly over the years; however, the one-year mortality rate due to CF has not yet substantially improved [4].

One key issue in the development of preventive approaches is that modern risk assessment tools have reduced predictive capacity and undergo miscalibration when employed on various population groups. Biomarkers have been increasingly attracting the scientist, as they enable the non-invasive identification of high-risk patients. The development of unique and highly prognostic CVD biomarkers possesses the potential of improving risk stratification and facilitate the targeted prevention approaches in the pre-clinical phase when treatment would most probably be effective. The use of markers in a clinical setting is also recommended in modern CF guidelines, which indicates their growing importance in the field of CF. Furthermore, circulating markers can provide valuable information regarding causal pathways involved in disease and possess the potential of pathway-targeted treatments and personalized treatment approaches.

This chapter aims to describe both the novel as well as established biological markers in CVD risk assessment. Biological markers will be characterized based on various pathophysiologic processes they indicated and placed within the framework of unmet needs in the area of CVD risk estimation and prevention.

2. BIOMARKERS FOR CVD

A biological marker can perform various functions when utilized within a clinical setting. It is described as a "characteristic that is objectively measured and evaluated as an indicator of normal biological processes, pathogenic processes, or pharmacologic responses to a therapeutic intervention" and performs a variety of functions according to different phases in the evolution of disease (Fig. **1**) [5]. Therefore, biomarkers are considered as an indicator of a disease state, disease rate, or disease trait [6]. A biological marker for cardiovascular disease is not restricted to a particular molecule, for example, RNA, metabolite, or protein, measured within a biospecimen like tissue or bodily fluids (cerebrospinal fluids, urine, plasma), but also a determination of various physical parameters, for example blood pressure, echocardiogram or electrocardiogram.

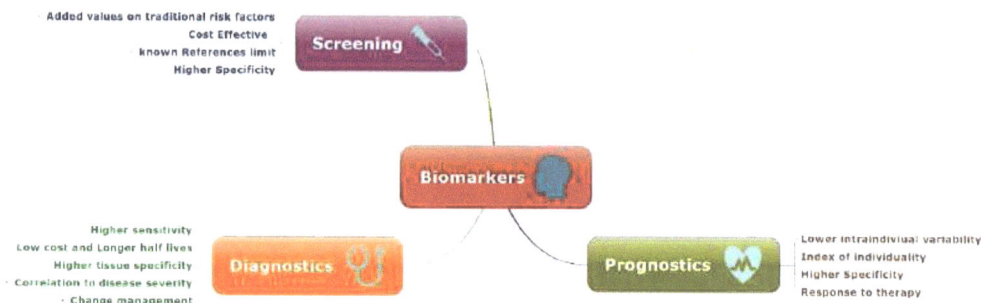

Fig (1). Characteristics and use of Biomarkers.

Preferably, a CVD biomarker should improve the capability of the clinician to treat the affected patient in the best possible way. Though numerous studies have suggested many protein markers for cardiovascular diseases, only a few of them have been employed successfully within cardiology practice – most of them are used for diagnostic purpose (Table 1).

Table 1. Commonly used diagnostic biomarkers for CVDs.

Biomarkers	Pathology	References
Mammary derived growth inhibitor	Critical illness	[114]
	Myocardial infarction	[115]
Cardiac troponins	Coronary heart disease	[116]
	Myocardial infarction	[29]
Copeptin	Critical illness	[117]
	Myocardial infarction	[118]
Atrial natriuretic peptide	Ischemic stroke	[119]
Brain natriuretic peptide	Heart failure	[120]
Myosin-binding protein-C	Myocardial infarction	[121]

Biomarkers for cardiovascular disease are usually categorized according to various pathological processes they indicate. As many biological markers reflect numerous pathophysiological processes involving extra-cardiac pathology (for example, kidney failure), this method can result in oversimplification. Yet, to be in accordance with the current research, the cardiac biological markers in this chapter are categorized based on the major classes of CVDs, *i.e.*, myocardial stress, myocardial injury, inflammation, plaque instability, calcium homeostasis, platelet activation, systemic stress.

2.1. Myocardial Stress

2.1.1. Atrial Natriuretic Peptides

Natriuretic peptides (NPs) comprise A-type/atrial NP (ANP), C-type NP (CNP), and brain NP (BNP). ANP was initially described in 1981 when a diuretic and natriuretic effect was observed *via* injection of the atrial extracts within rats [7]. ANP is produced as prepro-ANP, which is then converted to proANP and finally stored within the granules. In response to an atrial wall stretching, the proANP is secreted and transformed into the inactive N terminal-proANP and an active alpha-ANP by the corin (serine protease) (Fig. **2**) [8]. Owing to the in-vitro instability and shorter half-life of ANP, the measurement of the midregional portion of pro-ANP can function as a powerful surrogate for atrial-NP activity [9]. ANP levels are strongly connected to the diagnosis of heart failure in people presenting with acute dyspnea, with the diagnostic properties similar to BNP [10]. For people with chronic heart failure, the concentrations of ANP can also provide additional prognostic data to NT-proBNP concentrations [11]. In patients affected with acute myocardial infarction (AMI), stable coronary artery disease (CAD), and chest pain, concentrations of ANP serve as a strong predictor of adverse effects and mortality [12, 13].

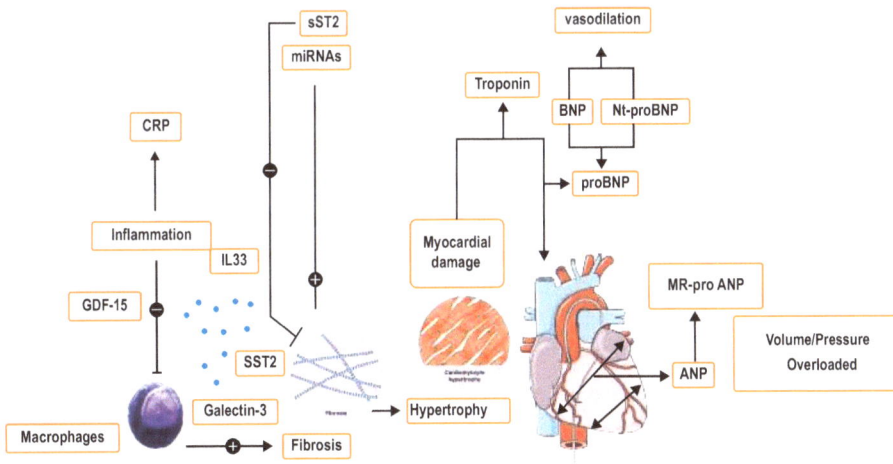

Fig. (2). Biomarkers involved in mechanism of cardiac injury and heart failure.

2.1.2. B-type Natriuretic Peptides

With regard to both cardiovascular and clinical cardiology research, NT-proBNP and BNP are one of the most widely used NPs. Upon hemodynamic stress (mechanical stress/or myocardial ischemia), a precursor BNP is produced within the ventricular cardiomyocytes which is then processed into signal peptide and proBNP. The proBNP is later cleaved into an inactive N terminal-proBNP and active C-terminal BNP [14]. Recently, it was found that the proBNP1-108 peptide is glycosylated, affecting the concentrations of N terminal-proBNP and processing of the short peptide fragments as evaluated by the commercially available N terminal-proBNP test (Fig. 2) [15]. procBNP promotes arterial vasodilation and diuresis/natriuresis and also opposes the action of the adrenergic system and renin-angiotensin system [16]. N terminal-proBNP might be better than BNP regarding risk estimation of future clinical outcomes, possibly due to its long in-vitro as well as in-vivo half-life [17]. Numerous investigations have evaluated the role of NPs in directing medical treatment of cardiac failure, with mixed outcomes [18].

In addition to the well-known role of diagnosing chronic and acute heart failure, concentrations of N terminal-proBNP are strongly connected to the risk of CVD as well as death, both within the overall population, in AMI patients, and stable CAD patients [19, 20]. N-terminal-proBNP measurement also seems important in the evaluation of prognosis in and risk of atrial fibrillation (AF). Several modifiable as well as non-modifiable determinants or risk factors are found to be connected with the NPs concentrations. In healthy people, women usually display higher NPs concentrations, probably due to the differences in hormones among males and females [21]. The concentrations of NPs rise with the age, declining kidney function, and occurrence of AF.

2.1.3. Copeptin

Copeptin (glycopeptide), a pro-hormone to antidiuretic hormone or arginine vasopressin (AVP), and a nonspecific endogenous depression marker [22]. Similar to heart-type fatty acid-binding protein (h-FABP), it increases rapidly after the myocardial injury (MI), with a subsequent quick decrease. In clinical settings, it is the only biomarker along with cardiac troponins (cTn) which is presently advised for routine medical practice in diagnosis of heart attack [23]. Copeptin also provides enhanced diagnostic accuracy for heart attack when used along with cTnI and cTnT measured with traditional assays, however, the increased diagnostic importance is less obvious when coupled with high sensitivity (hs) cTn assays [24]. For prognosis, the concentration of copeptin is associated with negative outcomes in people infected with AMI, sepsis, and heart failure [25, 26]. Copeptin concentrations have been robustly associated with death in people affected with less severe communicable diseases [27].

2.2. Myocardial Injury

2.2.1. Cardiac Troponin

Troponin/troponin complex is present in the striated muscle and composed of three subunits: troponin T, troponin I, and troponin C (Fig. **2**). These isoforms are present in both skeletal and cardiac muscle, however, the cTnI and cTnT isoforms are only found in the myocardium [28]. For the diagnosis of AMI, a measurement of cTn above the 99th percentage value of a certain test with a considerable change within the concentrations of cTn among two measurements is compulsory [29]. Apart from significant change and high cTn concentrations, the patients need to exhibit one or several of these features: pathological changes within electrocardiogram, angiographic evidence of coronary occlusion, clinical symptoms in accordance with myocardial ischemia, and echocardiographic (echo) indices of myocardial necrosis.

With the rising sensitivity of cTn tests, numerous methods for rapid diagnosis of acute coronary syndrome (ACS) have been reported [30]. Chest pain (discomfort) is a common symptom among patients visiting the emergency department, and advanced diagnostic approaches possess the capability to shorten the diagnosis and treatment time, reducing the stay of patients in the emergency room and subsequently decrease overall expenditures. The major advantage of this enhanced sensitivity is immediate and quicker rule out, because the time of detecting cTn after AMI is gradually getting lower [31].

High sensitivity cTn assays have enabled the measurement of cTnTI and cTnT concentrations below the value of diagnostic cut-offs for heart attack. The majority of the next-generation assays can detect the circulating cTn within a considerable amount of the suspected healthy people. The low-level concentrations of cTn are associated with several traditional risk factors of CVD, including blood pressure, kidney failure, diabetes mellitus, and age [32, 33]. cTnT and cTnI are also found to be connected with several risk factors of cardiovascular death and heart failure development in CAD patients, while the connection with atherosclerotic vascular events seems to be a little weaker [34, 35]. In presumed stable CAD patients, the cTnT concentrations reflect reversible ischemic heart disease and left ventricular (LV) mass [36]. Moreover, in the overall population, the concentrations of cTn were independently and strongly associated with adverse outcomes [37].

Even though not yet prepared for medical practice, latest studies have proposed that measurements of cTn with the hs assays can be utilized to observe the outcome of statin treatment. In 2016, a study determined the potential effects of the statin treatment in CVD risk and the subsequent cTn concentrations [38]. Men with elevated levels of bad cholesterol were randomly assigned to placebo or statin treatment for five years. Baseline concentrations of cTnI estimated the adverse cardiac effects and were significantly reduced *via* statin treatment with a corresponding decrease in risk of cardiovascular independently of changes in cholesterol level. These results strongly suggest that repeated measurements of cTn can mirror the outcome of the treatment in CVD risk management.

2.2.2. Cardiac Myosin-binding Protein C

Myosin-binding protein C (MyBP-C) was first found in 1973 and consist of 3 isoforms: cardiac (cMyC), slow skeletal, and fast skeletal [39]. Several research groups have reported the role of cMyC as an ideal biomarker of myocardial infarction necrosis [40]. Similar to cTn, high sensitivity assays have been designed for cMyC, enabling the quantification above the detection limit in patients who are not known to have CVD [41]. Owing to the rapid rise in plasma levels of cMyC after acute myocardial infarction, recent studies have started to concentrate on cMyC for quick diagnosis of AMI [42]. Recently, a comparative study was performed between cTn and cMyC with the use of high-sensitivity assays for the diagnosis of acute myocardial infarction [43]. Similar diagnostic characteristics were observed for cardiac troponin and cMyC, however, cMyC exhibited excellent diagnostic properties in patients presented with chest pain. To predict the risk in the people affected with left ventricular hypertrophy resulting from aortic stenosis (AS), cMyC concentrations are strongly associated with the inferior prognosis [44]. The prognostic potentials of cMyC have appeared similar in participants which were enrolled from the overall population [45]. Nevertheless, no commercially available cMyC assay exists yet, largely preventing implementation into modern clinical practice.

2.3. Inflammation

2.3.1. Interleukin-6

Interleukin 6 (IL-6) is mainly produced by T-cells and macrophages in reaction to any trauma and infections with subsequent release of C Reactive Protein (CRP) [46]. Moreover, the studies have exhibited the links between IL-6 and cardiovascular diseases as well as an adverse prognosis among the persons suffering from ACS [47]. Further, the evidence has also suggested the association between CAD and IL-6 levels (Fig. **2**) [48]. Assisting this model, immunomodulatory treatment for inhibiting an IL-6 inductor, interleukin 1b that significantly lowered the chances of recurrent AMI in a patient with a medical record of AMI [49].

2.3.2. C-reactive Protein

C-reactive protein (CRP) belongs to the pentraxin family and mainly synthesized in the liver in reaction to any systemic inflammation, triggering the complementary system and afterward promoting/stimulating the phagocytosis of bacteria and cellular debris mediated by macrophages [50].

Concerning cardiovascular disease, the concentrations of CRP are associated with peripheral artery disease, AMI, and ischemic stroke as well as cardiac infarction [51]. Increased CRP level has been consequently linked to worse outcomes [52]. Remarkably, in healthy people CRP is the best predictor of unfavorable outcomes but not in the persons infected with HF [53]. Moreover, serial measurement of CRP exhibited well associated with various outcomes related to HF [54]. In general, a higher concentration of CRP reflects inflammatory conditions associated with adverse outcomes of heart failure, however, CRP has not yet proved to be a useful marker in the management of clinical HF (Fig. **2**) [50].

2.3.3. Galectin-3

Galectin-3 belongs to β-galactoside-binding lectin members of the galectin family and plays an important role in cell-to-cell interaction, macrophage phagocytosis, apoptosis, and angiogenesis [55]. Concerning CVD, galectin-3 has an association with the inflammatory changes and early fibrotic observed in heart failure and both conditions of acute and chronic decompensated heart failure (DHF) [56, 57]. Increased expression of galectin-3 is observed in mature macrophages, which stimulates collagen deposition, the proliferation of heart fibroblasts, and ventricular dysfunction [57]. Evidence-based on experiment has exhibited that galectin-3 directly stimulates pathological heart remodeling and is thus taken as culprit protein in the proliferation of heart fibrosis in CF [58]. pharmaceutical inhibition of galectin-3 reduced LV dysfunction, heart fibrosis, and consecutive CF [59]. In patients infected with chronic heart failure, an increased level of galectin-3 indicated more dangerous CF (Fig. **2**) [60]. Overall, even if galectin-3 is potentially useful in CF subtypes, but still its use in case of clinical CF risk management and prediction is unsatisfactory [50].

2.3.4. Growth Differentiation Factor 15 (GDF-15)

GDF-15 also named as macrophage inhibitory cytokine-1 (MIC-1) of serum, is a member of the superfamily of TGF-β (transforming growth factor-beta) [61]. GDF-15 is a biomarker of inflammation and cell injury. In the physiological condition, the elevated level of GDF-15 is exhibited in the placenta. A rise in concentrations of blood is associated with various pathological states, including chronic CF and atherosclerosis (Fig. **2**) [62]. GDF-15 was revealed to independently predict mortality or first CF rehospitalization in both subtypes of CF even after modifying for significant clinical predictors consist of Nt-proBNP and hs-TnT [63]. In healthy people with no medical history of heart failure, GDF-15 along with echocardiographic measurements and common risk factors exhibited considerable improvement in the prediction of risk [64]. In summary, GDF-15 revealed a powerful predictor of death but with no benefit in the diagnosis of CF or disease-specific consequences [50].

2.3.5. Suppressor of Tumorigenicity 2 (ST2)

ST2 belongs to the interleukin (IL)-1 superfamily of cytokine receptor and found in both soluble and membrane-bounded forms. Both types serve as receptors for IL-33 *i.e.* the inflammatory cytokine. IL-33 exhibited protective cardiac effects that include reduced cardiac hypertrophy, decreased atherosclerotic burden [65]. The sST2 *i.e.* soluble ST2 inhibits IL-33 binding to the membrane-bound form of ST2 causing tissue fibrosis, accelerated disease progression, and reduced cardiovascular function (Fig. **2**) [66]. In mice, the ablation of IL-33 led to aggravated LVH, increased heart chamber dilation, exacerbated fibrosis, and reduced cardiac function [67]. However, in patients with Cardiac failure (CF) levels of sST2 are higher as compared to persons without CF but this marker is not helpful to diagnose CF [68]. In patients of dyspnoeic, concentrations of sST2 were a strong predictor of death. Inline, elevated concentrations of sST2 in acute CF patients specifically predicted 12 months all-cause deaths [69]. After cardiac infarction, increased values of ST2 were linked with a high risk of CF and death [70]. While the normal range of sST2 levels, did not reveal significant prognostic evidence in HFrEF, but an elevated level of sST2 within a year was strongly linked with elevated risk factors of poor consequences [71]. In general, although in the diagnosis of HF, sST2 has no significant clinical evidence, it looks to be a useful biomarker of heart remodeling [50].

2.4. Plaque Instability

2.4.1. Matrix Metalloproteinase-9 (MMP-9)

MMP-9 belongs to the family of zinc-dependent endopeptidase, these all are involved in the degradation of the extracellular matrix such as the basement and interstitial membrane collagen. Regarding cardiac diseases, elevated MMP-9 concentrations have been predicted in vulnerable arteriosclerosis [72]. With regard to enzymatic actions of MMP-9, particularly on basement membrane collagen its activity is linked with a high risk of plaque rupturing and resulting in ischemia [73]. In support of this prediction, MMP-9 concentration has been associated with ischemic stroke and CAD [74]. Though, the most relevant clinical feature of MMP-9 measurement in cardiac disease looks to be relevant with prognosis following CAD incident [75]. More particularly, MMP-9 concentration seems to be associated with the progression of CF and in the proliferation of post congestive heart failure [76]. This hypothesis is further supported by experimentation on mice with MMP-9 deficiency and in rabbit by inhibiting MMP-9 with lowered ventricular dilation [77].

2.4.2. Lipoprotein-associated Phospholipase A2 (Lp-PLA2)

Lp-PLA2 (Lipoprotein-associated phospholipase A2) a hormone secreted by inflammatory cells associated with the metabolism of phospholipids of the arterial wall. The activity of Lp-PLA2 produces proatherogenic metabolites and is strongly associated with inflammation of vulnerable plaque related to the development of arteriosclerosis disease [78]. Lp-PLA2 concentrations are linked with the risk of cardiovascular events, but not associated with the concentration of LDL cholesterol [79]. For arteriosclerosis disease, concentrations of Lp-PLA2 are linked with ischemic stroke incident [80]. Despite significant associations with stable as well as unstable arteriosclerosis disease, findings from a clinical trial with inhibitors of Lp-PLA2 have been unsatisfactory [81]. In patients with both ACS and stable CAD, therapy with direct inhibitors of Lp-PLA2 did not affect the non-fatal and fatal cardiac event [82].

2.5. Calcium Homeostasis

2.5.1. Secretoneurin

Secretoneurin, a tiny peptide derivative of SgII, which becomes highly conserved by evolution and is regarded as the functional peptide of granin protein [83]. The secretoneurin, and granin proteins, specifically, seem to be linked with the handling of cardiomyocyte Ca2þ. *In vitro* experimentation has exhibited secretoneurin to be a significant endogenous regulator of calmodulin-dependent kinase II(CaMKII)\Ca2þ, a multifunctional enzyme that is fundamental in the regulation of myocardial hypertrophy, apoptosis, cardiac ECC (excitation-contraction coupling), and inflammation [84, 85]. CaMKII is directly inhibited by Secretoneurin with a further reduction of CaMKII-dependent phosphorylation of the Ca2þ release channel in cardiac myocytes, ryanodine receptor 2 [84]. Gene expression of secretoneurin is 10-fold upregulated in the cardiac muscles of animals with CF [86]. In patients with myocardial infraction such as acute CF, acute respiratory failure related to cardiac, sepsis, and after CF, concentrations of secretoneurin are increased and associated independently with death risk [87, 88]. The positive effects of secretoneurin observed on the handling of myocardial Ca2þ suggested that the upregulation and association with poor consequences in patients with a cardiac infarction are due to a mechanism of compensation. Nevertheless, the stimulus that upregulates secretoneurin in vivo, or whether it is associated with ventricular arrhythmias risk and cardiac Ca2þ-handling, remains confusing [89].

2.6. Platelet Activation

2.6.1. P-selectin

P-selectin is a member of the selectin family, is the largest molecule of cell adhesion, and is activated in response to different inflammatory mediators expressed on endothelial cells and platelets [90]. The association between glycoprotein ligand-1 of P-selectin and P-selectin starts the recruitment of leukocyte, its attachment, extravasation, and rolling at inflammation sites. Expression of P-selectin is elevated in arteriosclerotic endothelial cells and P-selectin knocked-down mice revealed evidence of reduced arteriosclerotic burden, both assisting its role in the progression of arteriosclerosis [91]. In both patients infected with peripheral vascular disease and after an acute cardiac infarction, increased concentrations of P-selectin have been associated with worse consequences. Moreover, P-selectin concentrations are linked with AMI risk, cardiovascular death, and stroke in healthy women [79].

2.6.2. CD40 Ligand

CD40 is present on APC (antigen-presenting cells) and is stimulated by T-cell CD40 ligand. This association enhances the activation of ECs (endothelial cells), tissue factor production, and release of various inflammatory cytokines [92]. Furthermore, CD40L in a soluble form (sCD40L) is present in platelets and activated form of T-cells and is involved in activation pathways of B-cell. With respect to cardiac disease, evidence suggested an association between the signaling pathway of CD40L and CD40, proliferation, and progression of arteriosclerosis. Further promoting this theory, both CD40L deficiency and CD40L antibodies have been associated with the inflammatory burden and decreased arteriosclerosis [93]. Concentrations of sCD40L are linked with ACS prognosis and the positive effects of antagonist glycoprotein IIb/IIIa and aspirin treatment in atherothrombosis may be explored by sCD40L inhibition [94, 95].

2.7. Systemic Stress

2.7.1. Granin Proteins

It belongs to a family of proteins that play significant roles associated secretion of growth factors, peptides, neurotransmitters, and hormones [96]. The characterization of granin proteins is based on bulky size (50 kDa) and the high extent of amino acids acidic in nature [97]. In the protein family, chromogranin CgA is the basic member and along with secretogranin II and CgB, regarded as classical members from the eight total members of granin protein [96]. Moreover, studies have reported that blood tests of CgA, as well as catecholamines, exhibited activity of sympathetic nerve and that levels of CgA and catecholamine are upregulated in CF and ACS. Reciprocal relation between the variability of heart rate and CgA levels in CF patients has been also documented [98]. The concentration of CgA is associated with poor outcomes of acute CF and ACS but not chronic CF [98, 99]. CgA is more stable as compare to catecholamine for the repetition of both thawing, freezing, and incubating for a prolonged period at room temperature, and therefore CgA has been considered to have an elevated signal to noise ratio [100].

2.7.2. Catecholamines

Catecholamines have been broadly investigated in CF patients as a prognostic marker. Cohn *et.al*, (1984) was the first to report that urinary norepinephrine and elevated plasma level are independent predictors of death in chronic CF patients [101]. Later, a large population of CF patients was examined to check the prognostic capability of norepinephrine. Norepinephrine concentrations were considerably associated with disease and death [102]. Furthermore, change in norepinephrine with time-correlated with changes in risk [103]. In analytical processes, the use of catecholamines at the clinical level has high limitations due to the instability of substances requiring complicated and rapid analysis [104].

3. EXPERIMENTAL TECHNIQUES DESIGN FOR SELECTION OF PROTEIN BIOMARKERS OF CARDIOVASCULAR DISEASE

With accuracy towards pharmaceutical proposals are trying for extensive supply of calculable placeholders having unique features, a request for surveys to identify recent competitor biological marker that will presume to enhance. Flourishing shields concerning protein competitor of biomarker need information about:

i. With virtuous covering protein existing in specimen; along with
ii. In an adequate number about point to discuss suitable strength for revelation.

In contrast to DNA microarrays along with RNA-Seq technique (RNA sequencing) is a solitary proteome test which often could not concurrently acquire full coverage along with degree of efficiency and constraining exchanges in observational formulates. The pros and cons of several technical forums should consequently be calculated comprehensively in virtue of the range and astuteness of evaluation is needed, and uncompromising preceding analysis of chemical substance. There are three standards formulates that deliver quantitative data and qualitative data about proteins are:

i. Universal discovery
ii. Specific target and
iii. Target detection.

Each of these have diverse advantages and they are superior to various experimental objectives and everyone is executed in adequate technical forums.

4. FROM DISCOVERY TO CLINICAL VALUE: STATUS AND PERSPECTIVE

When a biological marker has been developed it needs to be validated and confirmed in hundreds or thousands of samples before being utilized for clinical trials. Furthermore, a biomarker should be reproducible, specific as well as sensitive. Regardless of the exponential increase in the amount of biomarker-based publications over the past several years, only a few useful protein biomarkers have successfully been approved for utilization in research or clinical practice [105]. The deficiencies in the discovery process of proteomic biomarkers are due to the absence of suitable platforms (analytical) for the validation of candidate biomarkers in a large set of sample/cohort, lack of properly organized pipeline for biomarker development, restricted access to samples that matched well to cardiovascular risk factors, age, gender and medication, and absence of empirical approach towards biomarker investigation [106, 107].

One more reason that relates to the deficiency of clinically useful biomarkers is that the sampling of plasma from controls and cases are mostly not done in optimal conditions when considering the clinical point of view. It is quite clear that factors such as time of the day, comorbidities, fasting, and various others affect the concentrations of protein, therefore, it is important that samples obtained for discovery researches, aimed at searching clinically relevant protein biomarkers, must be sampled carefully. For instance, if the plan is to design a biomarker that can help the physician to differentiate between arterial occlusion vs. intracranial hemorrhage in patients exhibiting stroke symptoms, it is crucial to look for markers in those samples that were taken at the exact point of time when this research question was relevant, not taken after hours/days. One problem is that such ideal samples of plasma are not often available. Furthermore, in such researches, it is essential to have the correct information regarding the current criterion for a clinical condition for which the new biological marker is designed.

Not every obstacle toward the development of novel protein biomarkers are, nevertheless, associated with the challenges in proteomics and technology, it must also be noted that the clinical community needs to define that which biomarkers can be of real use in clinical settings or research studies. It needs to be understood that there is no need to find and look for accurate estimates for a particular risk in a given condition if it is not useful for clinicians in suggesting a better treatment of a patient. Aspects regarding this matter should be the part of discovery studies design, *i.e.* the specimens that are intended to be used for research must reflect the condition in which the biomarkers will be utilized. This is significant for all types of biological markers, *i.e.* prognosis, companion diagnostics, prediction, and diagnosis.

5. MULTI-MARKER APPROACHES

There is yet no clinically available multiple biomarker panel that can be able to strongly rule-in CF, even though the combined use of novel and established biomarkers has shown to enhance the accuracy for CF diagnosis [108]. In CF patients, a multiple biomarker approach containing 5 biomarkers (including hs-TnT, GDF-15, and Nt-proBNP) exhibited better results for prediction of CF-related events compared to Nt-proBNP alone [109]. Moreover, a biomarker panel consisting of TnI, BNP, renal function, and hs-CRP markers considerably improved the prediction of undesirable effects in ambulatory patients with chronic CF [110].

In a consortium, thirty biomarkers were assessed in FINRISK 1997 study, and the score of a biomarker out of CRP, TnI, and CRP was confirmed in the 'Prospective Epidemiological Study of Myocardial Infarction' trial. The score might greatly enhance the ten-year risk estimation of cardiac events [111]. In another study, people with a multiple marker score consisting of GDF-15, hs-TnI, and sST2 in the upper quartile, exhibited a six-fold greater incident of CF risk and three-fold greater mortality risk [112]. In accordance, a combination of 6 biomarkers *i.e.* serpin E1, BNP, aldosterone/renin ratio, urinary albumin/creatinine ratio, homocysteine, and CRP was closely related to CF risk, while BNP appeared as a major biological marker in the prediction of now-onset CF risk with additional predictive efficacy over conventional risk factors [113].

6. LIMITATIONS OF PROTEINS/PEPTIDES AS BIOMARKERS FOR CV DISEASES

BiomarCare (Biomarker for Cardiovascular Risk Assessment across Europe) is a federation which is funded by Europeon union; comprising of 30 distinct biomarkers for which an experimental trial survey was performed in 1997 on a large group of Finnish population. Furthermore a biological marker is effaced of N-terminal pro hormone B-type natriuretic peptide (Nt-proBNP), troponin I (TnI) and C-reactive protein (CRP) were endorsed within Belfast PRIME (Prospective Epidemiological Study of Myocardial Infarction) study. In a time period of decade there is a significant development in the prediction of risk factors for CVD (cardiovascular disease) [115]. Cardiovascular study was held in Framingham, where people are composed of multiple biological marker score having growth differentiation factor-15 (GDF-15), soluble ST-2 (sST-2) and high sensitivity troponin I (hs-TnI); is leading quadrant, having 6 times increased the risk of cardiac failure along with 3 times increased the risk of death [116]. Six various types of cardiac biomarkers such as C-reactive protein (CRP), PAI-1 (plasminogen activator inhibitor-1), oxidative stress (homocysteine), UACR (urinary albumin- to-creatinine ratio), ARR (aldosterone-to-renin ratio) and B-type natriuretic peptide *(BNP)* were substantially relevant to the risk of heart failure (HF), despite the fact that B-type natriuretic peptide *(BNP)* emerge as the essential biomarker [117].

Appertaining to initial method (age, biomarkers, and clinical history) discussed earlier, having various riveting information has been made public elaborating multiple biological markers evaluating concerning unpropitious consequences in diseased patients having fibrillating trials. Data is used from Eliquis (Apixaban) for the diminution in heart attack along with various other thromboembolic complications within auricular fibrillation in ARISTOTLE trial, commencement of darapladib therapy trials lead to the stabilization of cholesterol on artery walls along with a stable and regular anti-coagulation therapy, novel ABC auricular fibrillation biomarker was initially suggested by Hijazi *et al* Within 3 different case reports, researchers scrutinize various aspects that are affiliated with anticoagulant therapy during auricular fibrillation are ischemic attack, death and at risk of serious hemorrhage [119]. Risk paragons comprising of historical data along with age and gender of critical ischemic attack conjugation with proportions concerning Hb (hemoglobin), cardiac *troponin T (cTnT),* N-terminal pro-B-type natri- uretic peptide (NT-proBNP) and growth differentiation factor-15 (GDF-15) surpass the already fixed risk score from all types of ischemic attacks (CHA_2DS_2-VASC) as well as critical bleeding in people affected with auricular fibrillation upon blood thinner therapies [120]. In spite of critical hemorrhages and ischemic attacks, the principal reason of mortality in people affected with auricular fibrillation causes heart failure (HF) along with abrupt cardiac arrest. Precautionary measures over anticoagulant medication, suchlike statin drugs along with ACE (angiotensin-converting enzymes), ARBs (angiotensin-II receptor blockers) might consequently be suitable within auricular fibrillation sufferers along with increase in death rate, that could also be measured by means of related algorithmic rule [121]. Altogether, extensive assessing risk grade through artificial bee colony (ABC) algorithm be able to provide predictive aptitudes were vulnerable to customary risk predicting exemplars, specifically in cases where evaluation of multifarious detrimental consequences are merited.

CONCLUSION

Currently, numerous cardiovascular biomarkers are available that have clinical applications as predictive or diagnostic biomarkers. However, a number of them need to be intensively tested to evaluate their clinical usefulness across a variable range of people affected by arteriosclerosis or other heart diseases. The majority of the biological markers described in this chapter are not myocardium specific and their concentrations are affected by other medical conditions. Therefore, a multi-marker approach could potentially improve this absence of specificity.

Moreover, biomarkers assessing prognostic outcomes must report calibration, reclassification, and discrimination in patients to validate their added usefulness over conventional and other routinely used biomarkers. The utilization of biological markers as surrogate markers for prognostics and predictive values within clinical studies will probably dictate the future of cardiovascular therapy, and be able to open new avenues for the evaluation of biomarkers as potential targets for the development and delivery of drugs.

CONSENT FOR PUBLICATION

Not applicable.

CONFLICTS OF INTEREST

The author declares no conflict of interest, financial or otherwise.

ACKNOWLEDGEMENTS

Declared none.

REFERENCES

[1]   Savji N, Rockman CB, Skolnick AH, *et al.* Association between advanced age and vascular disease in different arterial territories: a population database of over 3.6 million subjects. J Am Coll Cardiol 2013; 61(16): 1736-43.
        [http://dx.doi.org/10.1016/j.jacc.2013.01.054] [PMID: 23500290]

[2]   Heidenreich PA, Albert NM, Allen LA, *et al.* Forecasting the impact of heart failure in the United States: a policy statement from the American Heart Association. Circ Heart Fail 2013; 6(3): 606-19.
        [http://dx.doi.org/10.1161/HHF.0b013e318291329a] [PMID: 23616602]

[3]   Huffman MD, Berry JD, Ning H, *et al.* Lifetime risk for heart failure among white and black Americans: cardiovascular lifetime risk pooling project. J Am Coll Cardiol 2013; 61(14): 1510-7.
        [http://dx.doi.org/10.1016/j.jacc.2013.01.022] [PMID: 23500287]

[4]   Chen J, Normand SL, Wang Y, Krumholz HM. National and regional trends in heart failure hospitalization and mortality rates for Medicare beneficiaries, 1998-2008. JAMA 2011; 306(15): 1669-78.
        [http://dx.doi.org/10.1001/jama.2011.1474] [PMID: 22009099]

[5]   Biomarkers and surrogate endpoints: preferred definitions and conceptual framework. Clin Pharmacol Ther 2001; 69(3): 89-95.
        [http://dx.doi.org/10.1067/mcp.2001.113989] [PMID: 11240971]

[6]   Fox N, Growdon JH. Biomarkers and Surrogates. NeuroRx 2004; 1: 181.
        [http://dx.doi.org/10.1602/neurorx.1.2.181]

[7]   de Bold AJ, Borenstein HB, Veress AT, Sonnenberg H. A rapid and potent natriuretic response to intravenous injection of atrial myocardial extract in rats. Reprinted from Life Sci 2001.

[8]   Yan W, Wu F, Morser J, Wu Q. Corin, a transmembrane cardiac serine protease, acts as a pro-atrial natriuretic peptide-converting enzyme. Proc Natl Acad Sci USA 2000; 97(15): 8525-9.
        [http://dx.doi.org/10.1073/pnas.150149097] [PMID: 10880574]

[9]   von Haehling S, Jankowska EA, Morgenthaler NG, *et al.* Comparison of midregional pro-atrial natriuretic peptide with N-terminal pro-B-type natriuretic peptide in predicting survival in patients with chronic heart failure. J Am Coll Cardiol 2007; 50(20): 1973-80.
        [http://dx.doi.org/10.1016/j.jacc.2007.08.012] [PMID: 17996563]

[10]   Maisel A, Mueller C, Nowak R, *et al.* Mid-region pro-hormone markers for diagnosis and prognosis in acute dyspnea: results from the BACH (Biomarkers in Acute Heart Failure) trial. J Am Coll Cardiol 2010; 55(19): 2062-76.
        [http://dx.doi.org/10.1016/j.jacc.2010.02.025] [PMID: 20447528]

[11]   Masson S, Latini R, Carbonieri E, *et al.* The predictive value of stable precursor fragments of vasoactive peptides in patients with chronic heart failure: data from the GISSI-heart failure (GISSI-HF) trial. Eur J Heart Fail 2010; 12(4): 338-47.
        [http://dx.doi.org/10.1093/eurjhf/hfp206] [PMID: 20097683]

[12]   Tzikas S, Keller T, Ojeda FM, *et al.* MR-proANP and MR-proADM for risk stratification of patients with acute chest pain. Heart 2013; 99(6): 388-95.
        [http://dx.doi.org/10.1136/heartjnl-2012-302956] [PMID: 23213173]

[13]　Sabatine MS, Morrow DA, de Lemos JA, *et al.* Evaluation of multiple biomarkers of cardiovascular stress for risk prediction and guiding medical therapy in patients with stable coronary disease. Circulation 2012; 125(2): 233-40.
[http://dx.doi.org/10.1161/CIRCULATIONAHA.111.063842] [PMID: 22179538]

[14]　de Lemos JA, McGuire DK, Drazner MH. B-type natriuretic peptide in cardiovascular disease. Lancet 2003; 362(9380): 316-22.
[http://dx.doi.org/10.1016/S0140-6736(03)13976-1] [PMID: 12892964]

[15]　Røsjø H, Dahl MB, Jørgensen M, *et al.* Influence of glycosylation on diagnostic and prognostic accuracy of N-terminal pro-B-type natriuretic peptide in acute dyspnea: data from the Akershus Cardiac Examination 2 Study. Clin Chem 2015; 61(8): 1087-97.
[http://dx.doi.org/10.1373/clinchem.2015.239673] [PMID: 26056354]

[16]　Braunwald E. Biomarkers in heart failure. N Engl J Med 2008; 358(20): 2148-59.
[http://dx.doi.org/10.1056/NEJMra0800239] [PMID: 18480207]

[17]　Masson S, Latini R, Anand IS, *et al.* Direct comparison of B-type natriuretic peptide (BNP) and amino-terminal proBNP in a large population of patients with chronic and symptomatic heart failure: the Valsartan Heart Failure (Val-HeFT) data. Clin Chem 2006; 52(8): 1528-38.
[http://dx.doi.org/10.1373/clinchem.2006.069575] [PMID: 16777915]

[18]　Persson H, Erntell H, Eriksson B, Johansson G, Swedberg K, Dahlström U. Improved pharmacological therapy of chronic heart failure in primary care: a randomized Study of NT-proBNP Guided Management of Heart Failure--SIGNAL-HF (Swedish Intervention study--Guidelines and NT-proBNP AnaLysis in Heart Failure). Eur J Heart Fail 2010; 12(12): 1300-8.
[http://dx.doi.org/10.1093/eurjhf/hfq169] [PMID: 20876734]

[19]　Omland T, Persson A, Ng L, *et al.* N-terminal pro-B-type natriuretic peptide and long-term mortality in acute coronary syndromes. Circulation 2002; 106(23): 2913-8.
[http://dx.doi.org/10.1161/01.CIR.0000041661.63285.AE] [PMID: 12460871]

[20]　Patton KK, Ellinor PT, Heckbert SR, *et al.* N-terminal pro-B-type natriuretic peptide is a major predictor of the development of atrial fibrillation: the Cardiovascular Health Study. Circulation 2009; 120(18): 1768-74.
[http://dx.doi.org/10.1161/CIRCULATIONAHA.109.873265] [PMID: 19841297]

[21]　Gupta DK, de Lemos JA, Ayers CR, Berry JD, Wang TJ. Racial Differences in Natriuretic Peptide Levels: The Dallas Heart Study. JACC Heart Fail 2015; 3(7): 513-9.
[http://dx.doi.org/10.1016/j.jchf.2015.02.008] [PMID: 26071618]

[22]　Christ-Crain M, Fenske W. Copeptin in the diagnosis of vasopressin-dependent disorders of fluid homeostasis. Nat Rev Endocrinol 2016; 12(3): 168-76.
[http://dx.doi.org/10.1038/nrendo.2015.224] [PMID: 26794439]

[23]　Roffi M, Patrono C, Collet JP, *et al.* 2015 ESC Guidelines for the management of acute coronary syndromes in patients presenting without persistent ST-segment elevation: Task Force for the Management of Acute Coronary Syndromes in Patients Presenting without Persistent ST-Segment Elevation of the European Society of Cardiology (ESC). Eur Heart J 2016; 37(3): 267-315.
[http://dx.doi.org/10.1093/eurheartj/ehv320] [PMID: 26320110]

[24]　Maisel A, Mueller C, Neath SX, *et al.* Copeptin helps in the early detection of patients with acute myocardial infarction: primary results of the CHOPIN trial (Copeptin Helps in the early detection Of Patients with acute myocardial INfarction). J Am Coll Cardiol 2013; 62(2): 150-60.
[http://dx.doi.org/10.1016/j.jacc.2013.04.011] [PMID: 23643595]

[25]　Voors AA, von Haehling S, Anker SD, *et al.* C-terminal provasopressin (copeptin) is a strong prognostic marker in patients with heart failure after an acute myocardial infarction: results from the OPTIMAAL study. Eur Heart J 2009; 30(10): 1187-94.
[http://dx.doi.org/10.1093/eurheartj/ehp098] [PMID: 19346228]

[26]   Khan SQ, Dhillon OS, O'Brien RJ, *et al.* C-terminal provasopressin (copeptin) as a novel and prognostic marker in acute myocardial infarction: Leicester Acute Myocardial Infarction Peptide (LAMP) study. Circulation 2007; 115(16): 2103-10.
[http://dx.doi.org/10.1161/CIRCULATIONAHA.106.685503] [PMID: 17420344]

[27]   Müller B, Morgenthaler N, Stolz D, *et al.* Circulating levels of copeptin, a novel biomarker, in lower respiratory tract infections. Eur J Clin Invest 2007; 37(2): 145-52.
[http://dx.doi.org/10.1111/j.1365-2362.2007.01762.x] [PMID: 17217381]

[28]   Mohammed AA, Januzzi JL Jr. Clinical applications of highly sensitive troponin assays. Cardiol Rev 2010; 18(1): 12-9.
[http://dx.doi.org/10.1097/CRD.0b013e3181c42f96] [PMID: 20010334]

[29]   Thygesen K, Alpert JS, Jaffe AS, *et al.* Fourth Universal Definition of Myocardial Infarction (2018). Circulation 2018; 138(20): e618-51.
[http://dx.doi.org/10.1161/CIR.0000000000000617] [PMID: 30571511]

[30]   Cervellin G, Mattiuzzi C, Bovo C, Lippi G. Diagnostic algorithms for acute coronary syndrome-is one better than another? Ann Transl Med 2016; 4(10): 193.
[http://dx.doi.org/10.21037/atm.2016.05.16] [PMID: 27294089]

[31]   Morrow DA. Clinician's Guide to Early Rule-Out Strategies With High-Sensitivity Cardiac Troponin. Circulation 2017; 135(17): 1612-6.
[http://dx.doi.org/10.1161/CIRCULATIONAHA.117.026717] [PMID: 28438803]

[32]   Gore MO, Seliger SL, Defilippi CR, *et al.* Age- and sex-dependent upper reference limits for the high-sensitivity cardiac troponin T assay. J Am Coll Cardiol 2014; 63(14): 1441-8.
[http://dx.doi.org/10.1016/j.jacc.2013.12.032] [PMID: 24530665]

[33]   Saunders JT, Nambi V, de Lemos JA, *et al.* Cardiac troponin T measured by a highly sensitive assay predicts coronary heart disease, heart failure, and mortality in the Atherosclerosis Risk in Communities Study. Circulation 2011; 123(13): 1367-76.
[http://dx.doi.org/10.1161/CIRCULATIONAHA.110.005264] [PMID: 21422391]

[34]   Vavik V, Pedersen EKR, Svingen GFT, *et al.* Usefulness of higher levels of cardiac troponin t in patients with stable angina pectoris to predict risk of acute myocardial infarction. Am j cardiol 2018; 122(7): 1142-7.
[http://dx.doi.org/10.1016/j.amjcard.2018.06.027] [PMID: 30146101]

[35]   McQueen MJ, Kavsak PA, Xu L, Shestakovska O, Yusuf S. Predicting myocardial infarction and other serious cardiac outcomes using high-sensitivity cardiac troponin T in a high-risk stable population. Clin Biochem 2013; 46(1-2): 5-9.
[http://dx.doi.org/10.1016/j.clinbiochem.2012.10.003] [PMID: 23063983]

[36]   Myhre PL, Omland T, Sarvari SI, *et al.* Cardiac troponin t concentrations, reversible myocardial ischemia, and indices of left ventricular remodeling in patients with suspected stable angina pectoris: a doppler-cip substudy. Clin Chem 2018; 64(9): 1370-9.
[http://dx.doi.org/10.1373/clinchem.2018.288894] [PMID: 29959147]

[37]   Willeit P, Welsh P, Evans JDW, *et al.* High-sensitivity cardiac troponin concentration and risk of first-ever cardiovascular outcomes in 154,052 participants. J Am Coll Cardiol 2017; 70(5): 558-68.
[http://dx.doi.org/10.1016/j.jacc.2017.05.062] [PMID: 28750699]

[38]   Ford I, Shah AS, Zhang R, *et al.* High-sensitivity cardiac troponin, statin therapy, and risk of coronary heart disease. J am coll cardiol 2016; 68(25): 2719-28.
[http://dx.doi.org/10.1016/j.jacc.2016.10.020] [PMID: 28007133]

[39]   Offer G, Moos C, Starr R. A new protein of the thick filaments of vertebrate skeletal myofibrils. Extractions, purification and characterization. J Mol Biol 1973; 74(4): 653-76.
[http://dx.doi.org/10.1016/0022-2836(73)90055-7] [PMID: 4269687]

[40]   Baker JO, Tyther R, Liebetrau C, *et al.* Cardiac myosin-binding protein C: a potential early biomarker

of myocardial injury. Basic Res Cardiol 2015; 110(3): 23.
[http://dx.doi.org/10.1007/s00395-015-0478-5] [PMID: 25837837]

[41]   Marjot J, Liebetrau C, Goodson RJ, *et al.* The development and application of a high-sensitivity immunoassay for cardiac myosin-binding protein C. Transl Res 2016; 170: 17-25.e5.
[http://dx.doi.org/10.1016/j.trsl.2015.11.008] [PMID: 26713894]

[42]   Kaier TE, Anand A, Shah AS, Mills NL, Marber M. Temporal relationship between cardiac myosin-binding protein c and cardiac troponin I in type 1 myocardial infarction. Clin chem 2016; 62(8): 1153-5.
[http://dx.doi.org/10.1373/clinchem.2016.257188] [PMID: 27324735]

[43]   Kaier TE, Twerenbold R, Puelacher C, *et al.* Direct comparison of cardiac myosin-binding protein c with cardiac troponins for the early diagnosis of acute myocardial infarction. Circulation 2017; 136(16): 1495-508.
[http://dx.doi.org/10.1161/CIRCULATIONAHA.117.028084] [PMID: 28972002]

[44]   Anand A, Chin C, Shah ASV, *et al.* Cardiac myosin-binding protein C is a novel marker of myocardial injury and fibrosis in aortic stenosis. Heart 2018; 104(13): 1101-8.
[http://dx.doi.org/10.1136/heartjnl-2017-312257] [PMID: 29196542]

[45]   Tong CW, Dusio GF, Govindan S, *et al.* Usefulness of released cardiac myosin binding protein-c as a predictor of cardiovascular events. Am j cardiol 2017; 120(9): 1501-7.
[http://dx.doi.org/10.1016/j.amjcard.2017.07.042] [PMID: 28847594]

[46]   Kaptoge S, Seshasai SR, Gao P, *et al.* Inflammatory cytokines and risk of coronary heart disease: new prospective study and updated meta-analysis. Eur Heart J 2014; 35(9): 578-89.
[http://dx.doi.org/10.1093/eurheartj/eht367] [PMID: 24026779]

[47]   Zamani P, Schwartz GG, Olsson AG, *et al.* Inflammatory biomarkers, death, and recurrent nonfatal coronary events after an acute coronary syndrome in the MIRACL study. J Am Heart Assoc 2013; 2(1): e003103.
[http://dx.doi.org/10.1161/JAHA.112.003103] [PMID: 23525424]

[48]   Swerdlow DI, Holmes MV, Kuchenbaecker KB, *et al.* The interleukin-6 receptor as a target for prevention of coronary heart disease: a mendelian randomisation analysis. Lancet 2012; 379(9822): 1214-24.
[http://dx.doi.org/10.1016/S0140-6736(12)60110-X] [PMID: 22421340]

[49]   Ridker PM, Everett BM, Thuren T, *et al.* Antiinflammatory Therapy with Canakinumab for Atherosclerotic Disease. N Engl J Med 2017; 377(12): 1119-31.
[http://dx.doi.org/10.1056/NEJMoa1707914] [PMID: 28845751]

[50]   Magnussen C, Blankenberg S. Biomarkers for heart failure: small molecules with high clinical relevance. J Intern Med 2018; 283(6): 530-43.
[http://dx.doi.org/10.1111/joim.12756] [PMID: 29682806]

[51]   Ridker PM, Hennekens CH, Buring JE, Rifai N. C-reactive protein and other markers of inflammation in the prediction of cardiovascular disease in women. N Engl J Med 2000; 342(12): 836-43.
[http://dx.doi.org/10.1056/NEJM200003233421202] [PMID: 10733371]

[52]   Minami Y, Kajimoto K, Sato N, Hagiwara N, Takano T. C-reactive protein level on admission and time to and cause of death in patients hospitalized for acute heart failure. Eur Heart J Qual Care Clin Outcomes 2017; 3(2): 148-56.
[PMID: 28927169]

[53]   Lourenço P, Paulo Araújo J, Paulo C, *et al.* Higher C-reactive protein predicts worse prognosis in acute heart failure only in noninfected patients. Clin Cardiol 2010; 33(11): 708-14.
[http://dx.doi.org/10.1002/clc.20812] [PMID: 21089116]

[54]   van Boven N, Akkerhuis KM, Anroedh SS, *et al.* In search of an efficient strategy to monitor disease status of chronic heart failure outpatients: added value of blood biomarkers to clinical assessment.

Neth Heart J 2017; 25(11): 634-42.
[http://dx.doi.org/10.1007/s12471-017-1040-x] [PMID: 28983818]

[55]   Funasaka T, Raz A, Nangia-Makker P. Galectin-3 in angiogenesis and metastasis. Glycobiology 2014; 24(10): 886-91.
[http://dx.doi.org/10.1093/glycob/cwu086] [PMID: 25138305]

[56]   Yancy CW, Jessup M, Bozkurt B, *et al.* 2013 ACCF/AHA guideline for the management of heart failure: a report of the American College of Cardiology Foundation/American Heart Association Task Force on practice guidelines. Circulation 2013; 128(16): e240-327.
[http://dx.doi.org/10.1161/CIR.0b013e31829e8776] [PMID: 23741058]

[57]   de Boer RA, Lok DJ, Jaarsma T, *et al.* Predictive value of plasma galectin-3 levels in heart failure with reduced and preserved ejection fraction. Ann Med 2011; 43(1): 60-8.
[http://dx.doi.org/10.3109/07853890.2010.538080] [PMID: 21189092]

[58]   Gehlken C, Suthahar N, Meijers WC, de Boer RA. Galectin-3 in Heart Failure: An Update of the Last 3 Years. Heart Fail Clin 2018; 14(1): 75-92.
[http://dx.doi.org/10.1016/j.hfc.2017.08.009] [PMID: 29153203]

[59]   Yu L, Ruifrok WP, Meissner M, *et al.* Genetic and pharmacological inhibition of galectin-3 prevents cardiac remodeling by interfering with myocardial fibrogenesis. Circ Heart Fail 2013; 6(1): 107-17.
[http://dx.doi.org/10.1161/CIRCHEARTFAILURE.112.971168] [PMID: 23230309]

[60]   Anand IS, Rector TS, Kuskowski M, Adourian A, Muntendam P, Cohn JN. Baseline and serial measurements of galectin-3 in patients with heart failure: relationship to prognosis and effect of treatment with valsartan in the Val-HeFT. Eur J Heart Fail 2013; 15(5): 511-8.
[http://dx.doi.org/10.1093/eurjhf/hfs205] [PMID: 23291728]

[61]   Corre J, Hébraud B, Bourin P. Concise review: growth differentiation factor 15 in pathology: a clinical role? Stem Cells Transl Med 2013; 2(12): 946-52.
[http://dx.doi.org/10.5966/sctm.2013-0055] [PMID: 24191265]

[62]   Wischhusen J, Melero I, Fridman WH. Growth/Differentiation Factor-15 (GDF-15): From Biomarker to Novel Targetable Immune Checkpoint. Front Immunol 2020; 11: 951-1.
[http://dx.doi.org/10.3389/fimmu.2020.00951] [PMID: 32508832]

[63]   Chan MM, Santhanakrishnan R, Chong JP, *et al.* Growth differentiation factor 15 in heart failure with preserved vs. reduced ejection fraction. Eur J Heart Fail 2016; 18(1): 81-8.
[http://dx.doi.org/10.1002/ejhf.431] [PMID: 26497848]

[64]   Pareek M, Bhatt DL, Vaduganathan M, *et al.* Single and multiple cardiovascular biomarkers in subjects without a previous cardiovascular event. Eur J Prev Cardiol 2017; 24(15): 1648-59.
[http://dx.doi.org/10.1177/2047487317717065] [PMID: 28644092]

[65]   Miller AM, Xu D, Asquith DL, *et al.* IL-33 reduces the development of atherosclerosis. J Exp Med 2008; 205(2): 339-46.
[http://dx.doi.org/10.1084/jem.20071868] [PMID: 18268038]

[66]   Kakkar R, Lee RT. The IL-33/ST2 pathway: therapeutic target and novel biomarker. Nat Rev Drug Discov 2008; 7(10): 827-40.
[http://dx.doi.org/10.1038/nrd2660] [PMID: 18827826]

[67]   Veeraveedu PT, Sanada S, Okuda K, *et al.* Ablation of IL-33 gene exacerbate myocardial remodeling in mice with heart failure induced by mechanical stress. Biochem Pharmacol 2017; 138: 73-80.
[http://dx.doi.org/10.1016/j.bcp.2017.04.022] [PMID: 28450225]

[68]   Januzzi JL Jr, Peacock WF, Maisel AS, *et al.* Measurement of the interleukin family member ST2 in patients with acute dyspnea: results from the PRIDE (Pro-Brain Natriuretic Peptide Investigation of Dyspnea in the Emergency Department) study. J Am Coll Cardiol 2007; 50(7): 607-13.
[http://dx.doi.org/10.1016/j.jacc.2007.05.014] [PMID: 17692745]

[69]   Mueller T, Dieplinger B, Gegenhuber A, Poelz W, Pacher R, Haltmayer M. Increased plasma

concentrations of soluble ST2 are predictive for 1-year mortality in patients with acute destabilized heart failure. Clin Chem 2008; 54(4): 752-6.
[http://dx.doi.org/10.1373/clinchem.2007.096560] [PMID: 18375488]

[70]   Jenkins WS, Roger VL, Jaffe AS, *et al.* Prognostic value of soluble st2 after myocardial infarction: a community perspective. Am j med 2017; 130(9): 1112.e9-1112.e15.
[http://dx.doi.org/10.1016/j.amjmed.2017.02.034] [PMID: 28344136]

[71]   Anand IS, Rector TS, Kuskowski M, Snider J, Cohn JN. Prognostic value of soluble ST2 in the Valsartan Heart Failure Trial. Circ Heart Fail 2014; 7(3): 418-26.
[http://dx.doi.org/10.1161/CIRCHEARTFAILURE.113.001036] [PMID: 24622243]

[72]   Sluijter JP, Pulskens WP, Schoneveld AH, *et al.* Matrix metalloproteinase 2 is associated with stable and matrix metalloproteinases 8 and 9 with vulnerable carotid atherosclerotic lesions: a study in human endarterectomy specimen pointing to a role for different extracellular matrix metalloproteinase inducer glycosylation forms. Stroke 2006; 37(1): 235-9.
[http://dx.doi.org/10.1161/01.STR.0000196986.50059.e0] [PMID: 16339461]

[73]   Yabluchanskiy A, Ma Y, Iyer RP, Hall ME, Lindsey ML. Matrix metalloproteinase-9: Many shades of function in cardiovascular disease. Physiology (Bethesda) 2013; 28(6): 391-403.
[http://dx.doi.org/10.1152/physiol.00029.2013] [PMID: 24186934]

[74]   Jefferis BJ, Whincup P, Welsh P, *et al.* Prospective study of matrix metalloproteinase-9 and risk of myocardial infarction and stroke in older men and women. Atherosclerosis 2010; 208(2): 557-63.
[http://dx.doi.org/10.1016/j.atherosclerosis.2009.08.018] [PMID: 19748093]

[75]   Ramos-Fernandez M, Bellolio MF, Stead LG. Matrix metalloproteinase-9 as a marker for acute ischemic stroke: a systematic review. J Stroke Cerebrovasc Dis 2011; 20(1): 47-54.
[http://dx.doi.org/10.1016/j.jstrokecerebrovasdis.2009.10.008] [PMID: 21044610]

[76]   Morishita T, Uzui H, Mitsuke Y, *et al.* Association between matrix metalloproteinase-9 and worsening heart failure events in patients with chronic heart failure. ESC Heart Fail 2017; 4(3): 321-30.
[http://dx.doi.org/10.1002/ehf2.12137] [PMID: 28772055]

[77]   Hayashidani S, Tsutsui H, Shiomi T, *et al.* Fluvastatin, a 3-hydroxy-3-methylglutaryl coenzyme a reductase inhibitor, attenuates left ventricular remodeling and failure after experimental myocardial infarction. Circulation 2002; 105(7): 868-73.
[http://dx.doi.org/10.1161/hc0702.104164] [PMID: 11854129]

[78]   Caslake MJ, Packard CJ, Suckling KE, Holmes SD, Chamberlain P, Macphee CH. Lipoprotein-associated phospholipase A(2), platelet-activating factor acetylhydrolase: a potential new risk factor for coronary artery disease. Atherosclerosis 2000; 150(2): 413-9.
[http://dx.doi.org/10.1016/S0021-9150(99)00406-2] [PMID: 10856534]

[79]   Ridker PM, Buring JE, Rifai N. Soluble P-selectin and the risk of future cardiovascular events. Circulation 2001; 103(4): 491-5.
[http://dx.doi.org/10.1161/01.CIR.103.4.491] [PMID: 11157711]

[80]   Thompson A, Gao P, Orfei L, *et al.* Lipoprotein-associated phospholipase A(2) and risk of coronary disease, stroke, and mortality: collaborative analysis of 32 prospective studies. Lancet 2010; 375(9725): 1536-44.
[http://dx.doi.org/10.1016/S0140-6736(10)60319-4] [PMID: 20435228]

[81]   White HD, Held C, Stewart R, *et al.* Darapladib for preventing ischemic events in stable coronary heart disease. N Engl J Med 2014; 370(18): 1702-11.
[http://dx.doi.org/10.1056/NEJMoa1315878] [PMID: 24678955]

[82]   O'Donoghue ML, Braunwald E, White HD, *et al.* Effect of darapladib on major coronary events after an acute coronary syndrome: the SOLID-TIMI 52 randomized clinical trial. JAMA 2014; 312(10): 1006-15.
[http://dx.doi.org/10.1001/jama.2014.11061] [PMID: 25173516]

[83]   Fischer-Colbrie R, Kirchmair R, Kähler CM, Wiedermann CJ, Saria A. Secretoneurin: a new player in angiogenesis and chemotaxis linking nerves, blood vessels and the immune system. Curr Protein Pept Sci 2005; 6(4): 373-85.
       [http://dx.doi.org/10.2174/1389203054546334] [PMID: 16101435]

[84]   Ottesen AH, Louch WE, Carlson CR, *et al.* Secretoneurin is a novel prognostic cardiovascular biomarker associated with cardiomyocyte calcium handling. J Am Coll Cardiol 2015; 65(4): 339-51.
       [http://dx.doi.org/10.1016/j.jacc.2014.10.065] [PMID: 25634832]

[85]   Anderson ME, Brown JH, Bers DM. CaMKII in myocardial hypertrophy and heart failure. J Mol Cell Cardiol 2011; 51(4): 468-73.
       [http://dx.doi.org/10.1016/j.yjmcc.2011.01.012] [PMID: 21276796]

[86]   Røsjø H, Stridsberg M, Florholmen G, *et al.* Secretogranin II; a protein increased in the myocardium and circulation in heart failure with cardioprotective properties. PLoS One 2012; 7(5): e37401-1.
       [http://dx.doi.org/10.1371/journal.pone.0037401] [PMID: 22655045]

[87]   Røsjø H, Masson S, Caironi P, *et al.* Prognostic value of secretoneurin in patients with severe sepsis and septic shock: data from the albumin italian outcome sepsis study. Crit care med 2018; 46(5): e404-10.
       [http://dx.doi.org/10.1097/CCM.0000000000003050] [PMID: 29481425]

[88]   Ottesen AH, Carlson CR, Eken OS, *et al.* Secretoneurin is an endogenous calcium/calmodulin-dependent protein kinase II inhibitor that attenuates Ca^{2+}-dependent arrhythmia. Circ arrhythm electrophysiol 2019; 12(4): e007045.
       [http://dx.doi.org/10.1161/CIRCEP.118.007045] [PMID: 30943765]

[89]   Lyngbakken MN, Myhre PL, Røsjø H, Omland T. Novel biomarkers of cardiovascular disease: Applications in clinical practice. Crit Rev Clin Lab Sci 2019; 56(1): 33-60.
       [http://dx.doi.org/10.1080/10408363.2018.1525335] [PMID: 30457415]

[90]   Romano SJ. Selectin antagonists : therapeutic potential in asthma and COPD. Treat Respir Med 2005; 4(2): 85-94.
       [http://dx.doi.org/10.2165/00151829-200504020-00002] [PMID: 15813660]

[91]   Dong ZM, Brown AA, Wagner DD. Prominent role of P-selectin in the development of advanced atherosclerosis in ApoE-deficient mice. Circulation 2000; 101(19): 2290-5.
       [http://dx.doi.org/10.1161/01.CIR.101.19.2290] [PMID: 10811597]

[92]   Schönbeck U, Libby P. The CD40/CD154 receptor/ligand dyad. Cell Mol Life Sci 2001; 58(1): 4-43.
       [http://dx.doi.org/10.1007/PL00000776] [PMID: 11229815]

[93]   Lutgens E, Gorelik L, Daemen MJ, *et al.* Requirement for CD154 in the progression of atherosclerosis. Nat Med 1999; 5(11): 1313-6.
       [http://dx.doi.org/10.1038/15271] [PMID: 10546000]

[94]   Varo N, de Lemos JA, Libby P, *et al.* Soluble CD40L: risk prediction after acute coronary syndromes. Circulation 2003; 108(9): 1049-52.
       [http://dx.doi.org/10.1161/01.CIR.0000088521.04017.13] [PMID: 12912804]

[95]   Nannizzi-Alaimo L, Alves VL, Phillips DR. Inhibitory effects of glycoprotein IIb/IIIa antagonists and aspirin on the release of soluble CD40 ligand during platelet stimulation. Circulation 2003; 107(8): 1123-8.
       [http://dx.doi.org/10.1161/01.CIR.0000053559.46158.AD] [PMID: 12615789]

[96]   Bartolomucci A, Possenti R, Mahata SK, Fischer-Colbrie R, Loh YP, Salton SRJ. The extended granin family: structure, function, and biomedical implications. Endocr Rev 2011; 32(6): 755-97.
       [http://dx.doi.org/10.1210/er.2010-0027] [PMID: 21862681]

[97]   Helle KB. The granin family of uniquely acidic proteins of the diffuse neuroendocrine system: comparative and functional aspects. Biol Rev Camb Philos Soc 2004; 79(4): 769-94.
       [http://dx.doi.org/10.1017/S146479310400644X] [PMID: 15682870]

[98]   Røsjø H, Masson S, Latini R, *et al.* Prognostic value of chromogranin A in chronic heart failure: data from the GISSI-Heart Failure trial. Eur J Heart Fail 2010; 12(6): 549-56.
       [http://dx.doi.org/10.1093/eurjhf/hfq055] [PMID: 20388648]

[99]   Ottesen AH, Carlson CR, Louch WE, *et al.* Glycosylated chromogranin a in heart failure: implications for processing and cardiomyocyte calcium homeostasis. Circ heart fail 2017; 10(2): 10.
       [http://dx.doi.org/10.1161/CIRCHEARTFAILURE.116.003675] [PMID: 28209766]

[100] Stridsberg M, Eriksson B, Oberg K, Janson ET. A comparison between three commercial kits for chromogranin A measurements. J Endocrinol 2003; 177(2): 337-41.
       [http://dx.doi.org/10.1677/joe.0.1770337] [PMID: 12740022]

[101] Cohn JN, Rector TS. Prognosis of congestive heart failure and predictors of mortality. Am J Cardiol 1988; 62(2): 25A-30A.
       [http://dx.doi.org/10.1016/S0002-9149(88)80081-X] [PMID: 3389302]

[102] Latini R, Masson S, Anand I, *et al.* The comparative prognostic value of plasma neurohormones at baseline in patients with heart failure enrolled in Val-HeFT. Eur Heart J 2004; 25(4): 292-9.
       [http://dx.doi.org/10.1016/j.ehj.2003.10.030] [PMID: 14984917]

[103] Anand IS, Fisher LD, Chiang YT, *et al.* Changes in brain natriuretic peptide and norepinephrine over time and mortality and morbidity in the Valsartan Heart Failure Trial (Val-HeFT). Circulation 2003; 107(9): 1278-83.
       [http://dx.doi.org/10.1161/01.CIR.0000054164.99881.00] [PMID: 12628948]

[104] Sievert LL, Brown DE. Biological measures of human experience across the lifespan: making visible the invisible. Springer 2016.
       [http://dx.doi.org/10.1007/978-3-319-44103-0]

[105] Poste G. Bring on the biomarkers. Nature 2011; 469(7329): 156-7.
       [http://dx.doi.org/10.1038/469156a] [PMID: 21228852]

[106] Anderson NL. The clinical plasma proteome: a survey of clinical assays for proteins in plasma and serum. Clin Chem 2010; 56(2): 177-85.
       [http://dx.doi.org/10.1373/clinchem.2009.126706] [PMID: 19884488]

[107] Rifai N, Gillette MA, Carr SA. Protein biomarker discovery and validation: the long and uncertain path to clinical utility. Nat Biotechnol 2006; 24(8): 971-83.
       [http://dx.doi.org/10.1038/nbt1235] [PMID: 16900146]

[108] Sherwi N, Pellicori P, Joseph AC, Buga L. Old and newer biomarkers in heart failure: from pathophysiology to clinical significance. J Cardiovasc Med (Hagerstown) 2013; 14(10): 690-7.
       [http://dx.doi.org/10.2459/JCM.0b013e328361d1ef] [PMID: 23846675]

[109] Jungbauer CG, Riedlinger J, Block D, *et al.* Panel of emerging cardiac biomarkers contributes for prognosis rather than diagnosis in chronic heart failure. Biomarkers Med 2014; 8(6): 777-89.
       [http://dx.doi.org/10.2217/bmm.14.31] [PMID: 25224934]

[110] Ky B, French B, Levy WC, *et al.* Multiple biomarkers for risk prediction in chronic heart failure. Circ Heart Fail 2012; 5(2): 183-90.
       [http://dx.doi.org/10.1161/CIRCHEARTFAILURE.111.965020] [PMID: 22361079]

[111] Blankenberg S, Zeller T, Saarela O, *et al.* Contribution of 30 biomarkers to 10-year cardiovascular risk estimation in 2 population cohorts: the MONICA, risk, genetics, archiving, and monograph (MORGAM) biomarker project. Circulation 2010; 121(22): 2388-97.
       [http://dx.doi.org/10.1161/CIRCULATIONAHA.109.901413] [PMID: 20497981]

[112] Wang TJ, Wollert KC, Larson MG, *et al.* Prognostic utility of novel biomarkers of cardiovascular stress: the Framingham Heart Study. Circulation 2012; 126(13): 1596-604.
       [http://dx.doi.org/10.1161/CIRCULATIONAHA.112.129437] [PMID: 22907935]

[113] Velagaleti RS, Gona P, Larson MG, *et al.* Multimarker approach for the prediction of heart failure

incidence in the community. Circulation 2010; 122(17): 1700-6.
[http://dx.doi.org/10.1161/CIRCULATIONAHA.109.929661] [PMID: 20937976]

[114] Vasile VC, Chai HS, Abdeldayem D, Afessa B, Jaffe AS. Elevated cardiac troponin T levels in critically ill patients with sepsis. Am J Med 2013; 126(12): 1114-21.
[http://dx.doi.org/10.1016/j.amjmed.2013.06.029] [PMID: 24083646]

[115] McCann CJ, Glover BM, Menown IB, *et al.* Novel biomarkers in early diagnosis of acute myocardial infarction compared with cardiac troponin T. Eur Heart J 2008; 29(23): 2843-50.
[http://dx.doi.org/10.1093/eurheartj/ehn363] [PMID: 18682444]

[116] Samman Tahhan A, Sandesara P, Hayek SS, *et al.* High-sensitivity troponin i levels and coronary artery disease severity, progression, and long-term outcomes. J am heart assoc 2018; 7(5): 7.
[http://dx.doi.org/10.1161/JAHA.117.007914] [PMID: 29467150]

[117] Travaglino F, De Berardinis B, Magrini L, *et al.* Utility of Procalcitonin (PCT) and Mid regional pro-Adrenomedullin (MR-proADM) in risk stratification of critically ill febrile patients in Emergency Department (ED). A comparison with APACHE II score. BMC Infect Dis 2012; 12: 184.
[http://dx.doi.org/10.1186/1471-2334-12-184] [PMID: 22874067]

[118] Reichlin T, Hochholzer W, Stelzig C, *et al.* Incremental value of copeptin for rapid rule out of acute myocardial infarction. J Am Coll Cardiol 2009; 54(1): 60-8.
[http://dx.doi.org/10.1016/j.jacc.2009.01.076] [PMID: 19555842]

[119] Shah RV, Truong QA, Gaggin HK, Pfannkuche J, Hartmann O, Januzzi JL Jr. Mid-regional pro-atrial natriuretic peptide and pro-adrenomedullin testing for the diagnostic and prognostic evaluation of patients with acute dyspnoea. Eur Heart J 2012; 33(17): 2197-205.
[http://dx.doi.org/10.1093/eurheartj/ehs136] [PMID: 22645194]

[120] Maisel AS, Krishnaswamy P, Nowak RM, *et al.* Rapid measurement of B-type natriuretic peptide in the emergency diagnosis of heart failure. N Engl J Med 2002; 347(3): 161-7.
[http://dx.doi.org/10.1056/NEJMoa020233] [PMID: 12124404]

[121] Kaier TE, Alaour B, Marber M. Cardiac myosin-binding protein c-from bench to improved diagnosis of acute myocardial infarction. Cardiovasc drugs ther 2019; 33(2): 221-30.
[http://dx.doi.org/10.1007/s10557-018-6845-3] [PMID: 30617437]

SECTION III: Molecular Pharming for Animals

CHAPTER 12

Veterinary Nutraceutics, Pharmaceutics and Vaccine

Amjad Islam Aqib[1,*]**, Muhammad Shoaib**[2]**, Muhammad Aamir Naseer**[3]**, Saad Ahmad**[4]**, Mubashrah Mahmood**[5]**, Faisal Siddique**[6]**, Tean Zaheer**[7]**, Aisha Mahmood**[8] **and Iqra Muzammil**[9]

[1] *Department of Medicine, Faculty of Veterinary Science, Cholistan University of Veterinary and Animal Sciences, Bahawalpur, 63100, Pakistan*

[2] *Institute of Microbiology, Faculty of Veterinary Science, University of Agriculture, Faisalabad, 38000, Pakistan*

[3] *Department of Clinical Medicine and Surgery, Faculty of Veterinary Science, University of Agriculture, Faisalabad, 38000, Pakistan*

[4] *Lanzhou Institute of Husbandry and Pharmaceutical Sciences, Lanzhou, China*

[5] *Department of Theriogenology, Faculty of Veterinary Science, University of Agriculture, Faisalabad, 38000, Pakistan*

[6] *Department of Microbiology, Faculty of Veterinary Science, Cholistan University of Veterinary and Animal Sciences, Bahawalpur, 63100, Pakistan*

[7] *Department of Parasitology, Faculty of Veterinary Science, University of Agriculture, Faisalabad, 38000, Pakistan*

[8] *Department of Physiology, Faculty Of Veterinary And Animal Sciences, The Islamia University of Bahawalpur, Bahawalpur, Pakistan*

[9] *Department of Veterinary Medicine, Faculty of Veterinary Science, University of Veterinary and Animal Sciences, Lahore, 54000, Pakistan*

Abstract: Animals have been utilized extensively as part and parcel of pharmaceutical and vaccine development. Many studies from animal models for human diseases have re-affirmed inventions in the field of medicine. Animal nutraceuticals of biological origin have marked exceptional promise by enhancing the production and performance of commercial animals. Transgenic animals have helped transform laboratory-scale developments into clinical applications. The nutraceutical potential of animal products is a fascinating area of research with considerable anti-microbial, anti-cancer, anti-inflammatory, anti-diabetic and neuroprotective functions. Vaccines in veterinary sciences have been revolutionized based on the efficacy demonstrated by animal models. Vaccines are being routinely used against bacteria, viruses and some parasites at commercial levels. Third-generation vaccines that were thought to be very expensive

---

[*] **Corresponding author Amjad Islam Aqib:** Department of Medicine, Faculty of Veterinary Science, Cholistan University of Veterinary and Animal Sciences, Bahawalpur, 63100, Pakistan; Tel: +9237474583; E-mail: amjadislamaqib@cuvas.edu.pk

in the last century are now being commercially produced and marketed worldwide for animal health. Most recently, many avenues have opened that encourage the use of biologically derived pharmaceuticals and vaccine products. This chapter deals with a very comprehensive contrast of history and recent trends in veterinary pharmaceuticals and vaccines. It concludes that more research focus is required to come up with more efficient treatment and prophylactic approaches amidst mutating pathogens of concern.

Keywords: Animal peptides and proteins, Nutraceutics, Passive immunization, Pharmaceutics, Vaccines.

1. INTRODUCTION

Over the 300 years of the history of biomedical and pharmaceutical work, animals are of prime importance in regard to testing models as well as the source of pharmaceutical products [1]. It is an essential part of toxicological studies to utilize animal paradigms to explore the pharmaceutical pros and cons of the underlined products [2]. Animal carcass and its by-products have been used as a potent medicinally important biochemical for a long time. These biochemicals are highly labile to proteolytic enzymes, microbial as well as mal-handling degradation, which renders their quality and quantity. Market analyses showed that animal health pharmaceutics has emerged as a progressive and proactive industry in world-leading forums [3]. Animal pharmaceutics has a collection of nutrition, reproduction, and production-related entities; entitled in Chinese, herbal and modern medications [4]. After the down shock from Transmissible Bovine Spongiform Encephalopathy (TBSE), quality maintenance has become an optimistic point of concern in the animal pharmaceuticals and veterinary products market. Provision and critical analysis of the anatomical, biological, and geographical justification of animal products and by-products are of prime importance in the drug development industry. Reports from the Swiss Market showed 438 out of 655 pharmaceutical products of animal origin undergone the strict screening of TBSE [5]. Another study revealed that 530 out of 535 laboratory synthesized pharmaceutical products in the market are positive to at least one carcinogenic testing assay. Moreover, about half of the pharmaceutical ingredients used in Germany are reported to be of environmental relevance and are potent contaminants. This indicates the safety ranges of artificial products and empowers the concept of utilizing natural resources in this regard [6].

Pharmaceutical proteins of animal origin have been used since the 1920s, when insulin was extracted from the pig pancreas for human use. Later in the 1980s, this job was done using biotechnologically prepared recombinant microbes. In 2006, the pharmacologically active protein "antithrombin III" was approved as a medicinal product by European Medicinal Evaluation agencies. Other pharmaceutically important proteins include human growth factors, hormones, coagulation factors, monoclonal antibodies, interferons, enzymes, and collagen [7].

2. VETERINARY ORIGIN NUTRACEUTICS

Milk peptides are proven excellent dietary supplements as well as potent pharmaceutics. According to numerous studies, milk contains several biologically active peptides that can enhance antibacterial activity. These peptides have a variety of physiological functions, including metabolic, immunomodulatory, antibacterial, thrombolytic, and prebiotic/probiotic functions [8]. Alpha1-casei--derived peptides are the first antibacterial peptides to be effective against a wide variety of gram-positive bacteria, including *Staphylococcus aureus*. The N-terminal α1-casein peptide can be used *in vitro* to counteract lactobacilli and gram-positive bacteria. It also provides effective protection against *Streptococcus pyogenes*, *Staphylococcus aureus*, and *Listeria monocytogenes* [9]. Therefore, according to the literature, these biologically active peptides change their biological function due to their ability to bind specific receptors to target cells and induce various biological reactions in the host organism, as shown in Table **1**. Bovine casein can produce peptides such as SKVLPVPQK (β-CN; f168-176), YQKFPQY (αs2-CN; f89-95), LPYPYY (κ-CN; f56-61), LPQNIPPL (β-CN; f70). I will. -77), FLPYPYY (κ-CN; f55-61), and two novel angiotensin-converting enzyme (ACE) inhibitor peptides [10]. Compared to milk casein, milk casein contains more α-S2 and β casein than κ180 casein. Alpha S2-casein consists of two main components and several secondary components with varying degrees of post-translational phosphorylation. Kappa-casein can inhibit toxin invasion and pathogenic adhesion to the cell wall and protect cells from infections mediated by *Porphyromonas gingivalis*, *Salmonella*, *Rhodococcus*, *Streptococcus mutans*. It is hydrolyzed by mineral binding peptide 1 and binds zinc and iron ions to form a soluble dialyzable complex. This product is best at removing 2,2-diphenyl-1-pyridhydrazino (DPPH), which strengthens the immune system [11].

Table 1. Therapeutic usage of milk constituents [11 - 19].

Sr. No.	Milk Derived Bio-Active Constituents	Action Spectrum
1.	Opioid peptides	Agonist of opioid receptors
2.	Angiotensin inhibitory peptides	Antihypertensive effect
3.	Antimicrobial peptides	Bactericidal
4.	Immunogenic casein peptides	Lymphocytic and macrophage proliferation
5.	Antithrombotic peptides	Inhibit fibrin binding and platelets aggregation

(Table 1) cont.....

Sr. No.	Milk Derived Bio-Active Constituents	Action Spectrum
6.	Casein phospho-peptides	Mineral carriers, especially calcium
7.	7-h hydrolysate casein	Immune booster
8.	TR – 35	Antitumor
9.	Small hydrolyzed peptides	Anti-biofilm
10.	Peptidoglycan recognition protein`15	Antibacterial
11.	Lactoferrin	Anti-microbial, Antiviral
12.	Lacto-peroxidase	Antibacterial
13.	Lysozymes	Anti-microbial
14.	Lactophorin	Immune booster

Biological activities of camel milk proteins as anti-diabetic, antitumor, antioxidant, and heat tolerant potency have also been scientifically approved [20]. Ration energy and protein supply to animals in lactation can be improved by the inclusion of rumen-protected fat and protein. Heat stress in lactating cows has been dealt with successfully by feeding rumen by-pass fat. A study showed that all the heat stress indicators and serum changes are noted to be significantly lower in animals, supplemented with rumen by-pass fat [21]. Inclusion of rumen-protected fat in the ration of lactating cattle showed changes in serum amino acid profile without any significant effect on the reproductive behavior of the animal [22]. Rumen-protected lysine supplementation in postpartum animals resulted in higher intake, higher milk yield, and reduced risk of ketosis [23]. The inclusion of rumen-protected protein in lamb feed helps in immune-boosting and lowers the gastro-intestinal worm load [24]. Supplementation of rumen-protected betadine enhanced the fat content in carcass meat by regulating the mTOR signaling pathway [25]. Carcass quality was enhanced in culled ewes, supplemented with rumen-protected fat [26]. The addition of rumen-protected glucose during the transient period of lactating cows lowers the ketosis risk and improves body energetics [27]. L-carnitine was successfully utilized in rumen by-pass to improve milk quality [28]. The inclusion of rumen-protected lysine and histidine in the feed of high-yielding cattle showed lower body protein mobilization, improved milk quality, and better body metabolism [29]. Supplementation of rumen-protected betadine in lambs improves the antioxidant and immune status of the host [30]. The nutraceuticals of veterinary origin can improve the production and performance parameters in farmed animals. The cost-effective production, metabolism dynamics and consumer awareness are some areas to be addressed in nutraceuticals of veterinary origin.

3. VETERINARY PHARMACEUTICS

3.1. Role of Transgenic Animals in Pharmaceutical Industry

Transgenic animals are animals that undergo genetic transformation by splicing foreign genes (animal or human) and inserting them into chromosomes. With the successful integration of genes, animals can produce the proteins that express in milk, urine, blood, semen, or eggs, as well as excrete organs that are resistant to transplant rejection. For the expression of a protein of interest, the recombinant DNA is designed to express a functional protein when introduced into the nucleus of a transgenic animal of interest. Regardless of the type of animal, microorganism, or even plant obtained, the genes encoded in the resulting DNA construct can produce the same type of protein. These well-known recombinant proteins have therapeutic potential in treating, for example, cystic fibrosis, hemophilia, osteoporosis, arthritis and parasites, or infectious diseases such as malaria, AIDS, Hepatitis B. Transgenic animals can also produce monoclonal antibodies that specifically target disease proteins for vaccine development to meet global demand. In this era of next-generation sequencing, the rapidly growing knowledge of DNA and its function has dramatically changed the speed at which the necessary animal models are created [31]. In particular, the use of new and powerful technologies such as the CRISPR system (CRISPR = Clustered regularly inter-spaced short palindromic repeats) and CRISPR / Cas9 allows the creation of transgenic animals quickly and easily [32]. Transgenic animals can be obtained in a variety of ways, including microinjection [33], embryonic stem cell transplantation [34], and retrovirus-mediated gene transplantation [35]. Transgenic animals of various types include disease models, heterologous transplants, drug delivery, food sources, and animals designed as models for scientific research. Various disease models have been genetically engineered to mimic the symptoms of human disease. Examples of such models include Oncomouse (mouse model for cancer research), AIDS mice, Alzheimer's mice, and HLA-A2.1 / DTR mice used for research purposes, which are not usually represented by mouse antigens. Transgenic animal models used in scientific research are usually created by embedding transgenes in their DNA in order to study the effect of gene overexpression on the physiological processes of animals. The gene under study can be removed to determine its effect on normal human metabolism. Well-known examples of research models are ANDi transgenic monkeys, young supermice with smart mice, and flu-resistant mice [36].

Transgenic animals can better serve as an important model for transforming fundamental scientific advances into potential clinical applications. In addition, the use of genetically engineered animals as bioreactors ranges from various forms of protein production (such as milk, blood, urine, and other tissues) to tissue and organ modifications for transplantation in the pharmaceutical industry. The use of gene recombination technology may be important, but it is not yet fully understood. Research on transgenic animals is widespread in the field of human medicine. Various therapeutic proteins or peptides used to treat human diseases require effective modifications specific to animal cells and are usually produced in mammalian cell bioreactors. Starting a new cell culture manufacturing facility to produce a single therapeutic protein or peptide could cost more than $ 600 million, and most patients may not be able to afford the drug. The therapeutic protein industry is finding it difficult to keep up with the rapid growth in drug discovery and development, resulting in a surge in unmet patient needs and drug costs. Since the production of recombinant proteins in animal secrets (milk, blood, eggs, *etc.* from genetically modified animals) is much cheaper than the production of therapeutic proteins, genetically modified animals will be able to produce these protein / peptide drugs in the future. Anti-thrombin III (ATryn, GTC Bio-therapeutics, Framingham, Massachusetts), the first intracellular therapeutic protein, was extracted from the milk of transgenic goats in 2006 and is the European Commission for the treatment of patients with hereditary anti-thrombin deficiency. Approved Biopharmaceutical serum products are also available from genetically engineered animals such as antibodies that can be used to treat infections, cancer, organ transplant rejection, and autoimmune diseases (such as rheumatoid arthritis) [37]. Human blood is currently donated as the main production system for this blood product, but due to disease problems (HIV / AIDS / Hepatitis B / Hepatitis C, *etc.*) Lack of qualified donors and regulatory issues. Genetically modified animals such as cattle carrying the human antibody gene provide stable polyclonal antibodies for the treatment of various infectious diseases and other diseases. Transgenic mouse models also provide a lot of information about human diseases, and these models are of great importance in biological and biomedical research. Genetically modified sheep have been used to produce protease inhibitor 1 (PI-1) protein, which is released into serum and binds to elastase secreted by neutrophils in response to specific spores. A large amount of elastase released can damage the elastin of the alveolar wall and cause severe emphysema. There are two ways to support people with abnormally regulated production of α1-antitrypsin (emphysema or cystic fibrosis). Gene therapy using the functional α1-antitrypsin gene or the administration of high doses of α1-PI aerosol. Gene therapy is still controversial. Therefore, the only way to produce large amounts of α1-PI is to produce transgenic animals carrying the α1-P--producing gene. Pharmaceutical Protein Ltd. Originally from Midlogian, Scotland, I tried to make this enzyme from goat's milk. The use of sheep has potential benefits as sheep are mammals and can produce the same α1PI as humans. Sheep are cheaper and mature faster than cows. Enzymes are produced only in milk and are easy to harvest, which keeps sheep healthy for a long time. In addition, because these sheep are easy to raise, they can produce a large number of enzymes, and enzymes refined from milk are inexpensive. Once a purified enzyme is produced, it must be clinically tested and approved by a regulatory body before it can be released to the market [31].

Several biomedical research models have been genetically engineered, including pet species specifically designed to treat various human diseases (such as Alzheimer's and conjunctivitis disease) and the potential for heterologous cell, tissue, and organ transplantation. Genetically modified animals are also useful for research on animal diseases such as mad cow disease (bovine encephalopathy) and breast infections (mammoth). Currently, scientists have developed genetically engineered animals for agriculture, including animals with increased productivity and disease resistance, but Aqua Bounty (Boston, Massachusetts) is the only company who produces growth enhancer salmon for commercialization of agriculture applications.. The economic benefit from the production of genetically modified / transgenic animals for human medicine is higher than for agriculture. Concerns about costs and timelines associated with regulatory processes and consumer acceptability issues have hindered the commercialization of genetically modified animals for agricultural use. In addition, potential investors are hesitant because the general acceptance of genetic engineering for agricultural applications is usually lower than for medical applications (such as recombinant insulin) [38].

3.2. Hormones and Enzymes of Animal Origin Use in Pharmaceutical Industry

Animal tissue and organs are known as rich sources of biochemical, some of which are of medicinal importance. This is particularly true when proper care is taken in handling, collecting, preserving and storing organs and tissues which are the raw materials for further processing. The biochemical contained within them are very labile. They are susceptible to the action and proteolytic enzymes and to microbial degradation often associated with the improper handling of these organs and tissues prior to chilling or freezing. It is therefore important to maintain the content of the labile biochemical within these by-products if one hopes to derive from them their full monetary value. Throughout history, the inter-dependency between man arid animals has been profound for food. In addition to our dependence on animals for nutritional needs, our dependence on animals for maintenance of our health is greater today than ever before. Animal tissue extracts have been used to meet various needs of medicine over the past century. Drugs of animal origin designed to meet the present unfilled needs of medicine are under development in both academic and industrial laboratories at this time. One new hormone derived from animal tissue, expected to be of considerable therapeutic importance has just recently been introduced into medical practice and will be discussed later. Today, many of our major drugs are derived directly from animal tissue, primarily the tissue of cattle and hogs. In general, these tissues provide drugs that are agents in the regulation of functions of the body. They are the catalysts of the biochemical reactions of our cells and are such agents as enzymes, vitamins or hormones [39].

One common general metabolic deficiency occurs is hypothyroidism. Assessment of blood chemistry indicates a lowered triiodothyronine binding capacity and a low serum thyroxine. This is one of the medical conditions that is almost completely correctable by medication. The medication in this case is thyroid tablets which are usually prepared from pork thyroid tissue. Beef thyroid can also be used for this purpose [40]. Another gland of interest is the pituitary gland. This gland yields a number of regulatory hormones that are used in human medicine. It is estimated that in the United States, the pituitaries from 60,000,000 hogs are processed annually for this hormone and that 16,000,000 human doses of ACTH are administered per year is administered by injection and, unlike the steroids, in the U.S., all of this hormone is still produced from animal tissue. This hormone, which is peptide in structure, is used to treat inflammatory diseases in a manner similar to the synthetic steroids. The hormone has been found especially useful in rheumatic disorders such as arthritis, collagen diseases such as lupus erythematosus, dermatologic diseases and allergic states such' as bronchial asthma [41].

Insulin is a protein hormone concerned with the regulation of the rate of carbohydrate metabolism. The lives of diabetics depend on insulin. The hormone is produced in the pancreas as the inactive form known as proinsulin. This is converted into the active insulin and then secreted into general circulation in the body. It is estimated that we produce between 1-3 mg of insulin/day. This quantity is probably of the same order of magnitude that an animal such as a cow or a pig produces. In the case of a cow of 500 kg body weight, the weight of the pancreas is about 0.2 kg or 0.04% of the body weight. The insulin content is about 20 mg, or 0.01% of the weight of the pancreas [42]. A bone disease involving abnormal calcium metabolism called "osteitis deformans (Paget's disease)". The disease was found to be progressive and in the advanced stages, produced severe bone pain, lead to spontaneous fractures of weakened bone and occasionally to high output cardiac failure due to increased vascularity of the diseased bone. A hormone named as calcitonin was found to be effective for treatment of this disease and could be extracted from mammlian thyroid tissue especially pork thyroid tissue. The hormone was found to be a peptide containing thirty-two amino acids [43].

Unlike insulin, heparin is a non-protein pharmaceutical. It is a very acidic, highly sulfated anionic polysaccharide, also known as glycosaminoglycan. . It is a potent anticoagulant. Heparin is widely distributed in mammalian tissues and fluids, the largest amounts being in lung, spleen, liver, muscle and intestine. Beef lung was used earlier as a source of heparin [44]. It is now customary to isolate heparin from hog intestinal mucosa [45]. A hormone known as Corifollitropin alfa belong to pituitary hormones is obtained from Chinese Hamster Ovary cells. It was recently introduced into an IVF (In Vitro Fertilization) treatment to stimulate the patient's ovaries. It contains the same α-subunit as FSH and is linked to the terminal peptides of human FSH-β and carboxyl hybrid subunits and subunit of chorionic gonadotropin-β.Follicular stimulating hormone alpha is the same hormone as follicular stimulating hormone (FSH), produced by the pituitary gland that helps ovarian development [46]. Another hormone known asFollitropin alfa also belong to pituitary hormones and is obtained from Chinese Hamster Ovary cells.Follitropin alfa is the same hormone as follicular stimulating hormone (FSH), produced by the pituitary gland which helps ovarian development [47].

Besides insulin and kallikren, the pancreas also produces, stores and secretes many more enzymes and hormones. One of these enzymes is called kallikrein, a Greek term for the pancreas. Kallikrein is a carbohydrate-containing protein of about 30,000 molecular weight. It is a very specific proteolytic enzyme. The only known substrate, present in plasma, is called kininogen. Thus the enzyme is also known as kininogenase. A nine-amino acid peptide, bradykinin, is generated by another well-known pancreatic enzyme, trypsin that can be obtained from pig pancreas .These kinins are the vasoactive substances, capable of dilating smooth-muscle tissues, which lead to a reduction in blood pressure. Kallikrein, like insulin, is present in the pancreas of hog and cow in small quantities, about 100 mg/kg tissue. It can be isolated by a variety of methods. In Japan, the enzyme powder is formulated with other ingredients in tablet form containing the equivalent of 10-50 μg of pure kallikrein/ tablet [48].

The enzymes such as Agalsidase, Imiglucerase, and Laronidase are obtained from Chinese Hamster Ovary cells and used as enzyme replacement therapy. Enzymatic Replacement Therapy Agalsidase alfa (Replagal®) has the same amino acid sequence as natural alpha-galactosidase and 0.2 mg / kg alfa agarcidase is used every two weeks for long-term treatment of patients with confirmed Fabry disease [49]. Imiglucerase, a recombinant glucocerebrosidase, can be used for enzyme replacement therapy (ERT) in patients with a non-neurological disease called Gocher's disease. ERT can reduce liver constipation and increase the number of red blood cells and platelets. In Gocher's disease, the patient lacks the normal functions of macrophages and monocytes, β-glucosidase or glucosylceramidase, which are necessary for the destruction of old red blood cells

[50]. Other enzymes such as pancrelipase, lipase, protease, and amylase are obtained from porcine tissues and used as digestive supplement as well as cholelitholytics [51].

The therapeutic products I have mentioned were developed over the past several decades. I have mentioned only a few. The list of drugs from animal tissue includes such other important drugs as, bile salts from ox gall [52], liver extracts from beef liver [53], the spreading enzyme hyaluronidase from bull testes [54], posterior pituitary hormone from beef pituitaries and glucagon from beef pancreas [55]. All have been introduced in this century and each has a special place in medicine.

3.3. Milk Proteins of Pharmaceutical Importance

Milk is the opaque lacteal secretion, other than colostrum, that is acquired by the complete milking of the mammary gland of adult female mammals and is used for the nourishment of their young ones. Milk is a liquid that is enriched with almost all nutritional components, including carbohydrates, proteins, lipids, minerals, and vitamins. Among the various constituents of milk, proteins are of great significance. According to [56], an important source of protein in human diets is milk that accounts for the provision of 32g/liter. This protein fraction of milk mainly accounts for bioactive molecules that play a defensive role for human health. According to a report by FAO, cattle account for 81% of the total milk produced in the world, buffaloes for 15%, goats for 2%, sheep for 1%, and camel for 0.5% while the rest is produced by equines and yaks [56]. have broadly classified this milk protein into three categories namely whey proteins, caseins, and milk fat globule membrane (MFGM) origin proteins. Caseins account for about 78% of total milk proteins and include α_{s1}-casein, α_{s2}-casein, β-casein, and κ-casein. Whey proteins account for 18% of total milk proteins and include α-lactalbumin,β-lactoglobulin, enzymes, immunoglobulins,glycomacropeptides, growth factors, lactoferrin, and serum albumin. Milk fat globule membrane (MFGM) origin proteins account for <4% [57].

3.3.1. Bovine Milk Proteins

Protein content for cattle milk is reported to be 3.2–3.8%. Most of the proteins of milk origin have nutraceutical importance. Milk peptides have pharmacological properties like antihypertensive, anti-inflammatory, antilipemic, antioxidant, antithrombotic, antineoplastic, antimutagenic, antibacterial, cytomodulatory, immunomodulatory, and antithrombotic actions. Milk origin proteins are regarded as supplementary nutraceuticals. They also offer promising avenues for prevention, control, and treat various disorders and diseases with the regular intake. Scientists are trying to develop edible films based on bovine milk immunoglobulins to get benefits from its immune effect. α-lactorphin, β-lactorphin, β-lactosin B and Casein-derived lactotripeptides impart antihypertensive properties to the bovine milk. Lactoferrin, lactoperoxidase, milk growth factor, and immunoglobulin G are involved in the immunomodulatory action of milk proteins. β-lactoglobulin, lactoferrin, and α-lactalbumin have antineoplastic action. Glyco macro peptides account for the antiviral activity of milk. Lactoperoxidase, lactoferrins, lysozyme, lactophorins, κ-casein, α_{S1}-casein and α_{S2}-casein impart antimicrobial properties to milk [58].

3.3.2. Camel Milk Proteins

Camel milk differs from other dairy milk both in concentration and its protein content (3.0 – 3.9%). It has low sugar, cholesterol and lactose content as compared to bovine milk, but is rich in lactoferrin, protein and insulin [59]. It also differs in terms of its nutraceutical importance. Camel milk contains a high concentration of immunoglobulins and insulin. Thus, it has good antibacterial properties. Camel milk lacks β-lactoglobulin and has a high content of α-lactalbumin just like human milk, this bestows anti-allergic properties to the camel milk. Camel milk has renowned antioxidant, anti-inflammatory, hepato-protective, anti-diabetic, ACE inhibitory and anti-autism properties [60]. Camel milk lactoferrin is unique and is involved in a variety of cellular physiological (proliferation and homeostasis) and immune functions (antioxidant, antibacterial, antifungal, antiviral, antineoplastic and immunomodulatory). Camel milk has therapeutic effects for many diseases on account of its lactoproteins, such as diabetes mellitus, diarrhea, asthma, anemia, autism spectral disorders, coronary heart diseases, jaundice, piles, and tuberculosis. Camel milk insulin makes it a potential candidate for its therapeutic use in diabetic patients. Its anti-autistic effect also makes it a suitable choice for children with atrial septic defects (ASD) [59].

The fight against human cancer remains a serious problem in modern medicine. Natural products have been recognized as one of the sources of many therapeutic agents. Camel milk is one of these natural foods and is rich in molecules that are safe for humans and have profound anti-cancer effects. *In vitro* studies of the anticancer effect of these products are primarily associated with inhibition of cancer and mutagenesis, induction of proliferation and apoptosis. These data suggest that camel milk and its exosomes may have anti-cancer effects, inducing apoptosis and suppressing oxidative stress, inflammation, angiogenesis, and metastasis in the tumor microenvironment. Therefore, camel milk and its exosomes can be used as anti-cancer agents for cancer treatment [61, 62].

3.3.3. Sheep / Goat Milk Proteins

Sheep milk is regarded as a nutritional powerhouse. It has a protein content of around 5.6 - 6.7% while goat milk has protein around 2.9–3.7%. The bioactive peptides that originate from sheep milk impart antihypertensive, antimicrobial, antioxidant, antithrombotic, and immunomodulatory properties. Goat milk is enriched with β- casein while ewe milk is rich in αs1- casein. Sheep milk has established its space in the cosmetic industry in the domain of anti-aging products and cosmetic soaps for treating chronic conditions such as skin eczema and psoriasis [63]. Caprine milk has the potential as an efficient probiotic's carrier. Caprine milk has a specific "goaty" odor and unpleasant taste. This limits the use of caprine milk to some extent, but this issue can be overcome with fortification. α-s1 casein content of caprine milk is almost nil and overall casein content of caprine milk is also low. While α lactalbumin and β Lactoglobulin, the whey proteins, are high in concentrations compared to bovine milk [64]. Both caprine and ovine milk have the potential to be used as a substitute for bovine milk in lactose-intolerant persons with cow milk allergy [63]. Its use for people suffering from diseases like eczema, asthma, acidity, colitis, digestive disorders, peptic ulcers, constipation, migraine, insomnia, and neurotic indigestion has been well documented. Caprine milk proteins have potential antioxidant, anti-aging, and therapeutic properties [65].

4. VETERINARY PEPTIDE AND PROTEIN BASED VETERINARY VACCINES

Vaccines are effective public health interventions that are used to provide a product that can be used to protect animals and humans against a given particular disease. Veterinary vaccines have made their impact on animal production, health, and welfare, and these vaccines are also found to be beneficial for human health [66]. Vaccines play a vital role in the control and prevention of infectious diseases in animals and humans. Veterinary vaccines in particular have contributed to the control and eradication of various diseases of pet animals, poultry, and livestock, such as rinderpest, foot and mouth disease, canine distemper, parvovirus, Newcastle disease and classical swine fever, *etc*. Moreover, veterinary vaccines play important role in the control of various zoonotic infectious diseases, such as rabies and brucellosis. The first vaccine was developed by Edward Jenner in 1796 for cowpox which was administered to humans and provided immunity against smallpox. Louis Pasteur after one hundred years adapted the term immunogenic for the protective inoculum, use to provoke an immune response against infect-

ious diseases [67]. Vaccination is found to be more cost-effective than that treating animals in veterinary practice.

Peptide vaccines represent an attractive alternative strategy that uses short peptide fragments to direct the induction of highly targeted immune responses and avoid allergens and / or reactive sequences. Only peptide vaccines containing epitopes that can elicit a positive and ideal immune response mediated by T cells and B cells. The "peptides" used in these vaccines are 20-30 amino acid sequences that are synthesized to form immunogenic peptide molecules, which are specific epitopes of an antigen. On the other hand, it is believed that these peptides are sufficient to activate the corresponding cellular and humoral responses. This is because of a larger protein epitope as an antigenic determinant [68]. Few examples of peptide/protein nature vaccines used in veterinary are listed below in Table **2**.

Table 2. Proteins/Peptide subunits used in veterinary for vaccine production.

Species	Disease	Protein Subunit	References
Cattle	Histophilosis / Bovine respiratory disease	Outer membrane protein (OMP)	[69]
	Bovine viral diarrhea (BVD)	Surface E2 glycoprotein	
	Rabies	Virus G protein	[70]
	FMD	VP1	[69]
	Black leg	Flagellin protein	[71]
	Bovine Rota corona virus	VP8-S2 protein	[72]
	Hemorrhagic septicemia	Fimbrial protein	[73]
	Bovine ephemeral fever	G protein	[74]
	Bovine brucellosis	OMP16, OMP19, liposomized protein L7/L12, OMP25, p39 (a putative periplasmic binding protein), and AsnC	[75]
Goats	Goat pox	VP3 protein	[76]
Camel	MERS-CoV	Spike S1 protein	[77]
Equine	Glanders	outer membrane proteins	[78]
	Strangles	M protein	[79]
Poultry	Influenza A virus	Hemagglutinin (HA) +nucleoprotein (NP)	[69]
Dog	Canine parvovirus	VP2	

(Table 2) cont.....

Species	Disease	Protein Subunit	References
Pigs	Classical swine fever virus	E_2 fusion protein	[80]

4.1. Role of Animals in Passive Immunization

Passive immunization is the instant yet short lived administration of antibodies, demonstrating no role in immunological memory, in contrast with active immunization [81]. The concept of passive immunity holds significance against many pathogens of concern. It could be utilized to curtail the non-judicious use of anti-microbials and the AMR [82]. These immunoglobulins may be used to specifically target the pathogens of concern during production phases of commercial animals. The production of immunoglobulins at cost-efficient scale is still a major concern.The concept of passive immunization itself is more than a century old. Passive immunization is a natural process shown to prevent pathogens in neonates, by transfer of immunoglobulins to the offspring via placenta/yolk sac or milk [83]. In some animal species, there's lack of trans-placental immunoglobulin transfer. The lactation stage of these animals has higher concentrations of circulating immunoglobulins for a rapid immunological protection [84].

The lactogenic immunity profiles of camel and bovine milk are high. Immunoglobulins from camel milk have been anticipated as potential candidate milk replacers in allergic human infants [85]. Their purified forms could be used in immunocompromised patients, for preventing attack by opportunistic pathogens. Milk of animals vaccinated with human antigens can prevent future infections by the consumption of immunoglobulin rich constituents. Similarly, the equine derived anti-toxins preventing diphtheria are being used in humans [86]. These preparations have been found to successfully prevent the infection, in conjunction with anti-microbials. Anti-botulinum toxoid is another development in the animal-origin passive immunization for humans.

Murine models have been extensively utilized to develop anti-cancerous, anti-HIV and anti-inflammatory monoclonal antibodies, approved for human use. Hamsters, rabbits and llama have been deployed to develop antibodies against auto-immune disorders and cancers in humans [87]. Animals have intrinsic ability of producing very specific antibodies, exhibiting lower immunoreactivity when humanized [88]. However, the high degree of batch to batch variability and animal welfare concerns have led to the development of alternative methods of clonal antibodies. The European Union has emphasized on developing non-animal-derived antibodies for immunization [89].

Livestock origin immunization by passive route has shown promise against infectious and allergic and metabolic diseases of humans. However, there's an emergent concern of shortage in supply of animal-derived anti-toxins for human use as point of care therapies [86]. This trend is also indicative of the costs incurred in the production of anti-toxins. The consumption of raw immunogenic milk in humans has led to concerns of gastrointestinal pathogens. Safe utilization of lactogenic immunoglobulins of animal origin is an important area of prospective research [89]. Furthermore, it is necessary to carry more clinical research on the safety evaluation of lactogenic immunoglobulins proposed for human use. The EU has discouraged the use of animal derived antibodies owing to the 3Rs of animal welfare. It is therefore anticipated that the animal-driven immunological products would be replaced by non-animal, alternative products.

4.2. Animal Derived Antibodies

Animal-derived monoclonal and polyclonal antibodies are in use against bacterial and viral diseases as anti-infective antibodies [90]. Utilizing biotechnological tools, vesicular antibodies have been prepared for immunotherapy against a wide range of diseases such as viral and bacterial infections, cancer, and immune system disorders [91]. Highly efficient immuno-diagnostic tools (biomarkers, immunoassays), based on the recombinant antigen or antibody detection principals, have been manufactured and utilized with accuracy and reliability [91].

Hyper-immune serum against *Mycobacterium tuberculosis* showed strong curative as well as protective potential in model animals. Moreover, relapse of disease experimentally induced reinfection was also prohibited [92]. Mammaglobin is considered a reliable marker in the case of human breast cancer patients and bitches. Hyperimmune serum against this protein is raised in model animals and used in confirmatory diagnostic ELISA of breast cancer identification with 90% sensitivity and 95% specificity [93]. Hyperimmune serum against hydatidosis, raised in model animals, has shown promising results in curing and immune-boosting aspects [94]. Horse-originated hyperimmune serum against bothropic venom has been successfully isolated and implemented in the patients. Treatment of amyotrophic lateral sclerosis using hyperimmune goat serum showed promising results by shortening the disease course and declining the symptoms shown [95]. The hyper-immune serum has been successfully used as an alternative to antibiotics to treat endometritis. Hyperimmune rabbit serum has been proved potent for the inhibition of hepatitis C virus by recognition of hypervariable-I region of virus. A patient with myaesthenia gravis has been successfully treated using hyper-immune goat serum. Krabbes disease patients had also shown promising results when treated with caprine-originated serum

[96]. Endometritis in mare has been treated with model animal raised serum with success [97]. Rabbit-raised hyper-immune serum has been used to treat Bovine viral diarrhea [98]. Equine-originated hyper-immune serum and purified antibodies have been used to combat *Crotalusatrox* venom [18]. Research is going on to utilize hyper-immune sera, raised in model animals, against emerging diseases including Ebola [99] and pandemic COVID-19 [100].

Medical prophylaxis against tetanus includes immunization with formaldehyde inactivated tetanus neurotoxin (TeNT). It takes two injections every 3-4 weeks to develop effective immunity. There is no special medicine for the tetanus. TeNT antibodies can prevent free TeNT from entering neurons through serum, but toxins absorbed by neurotoxin cells cannot be produced with antitoxin antibodies [101].

4.3. Animal Derived Nanobodies

In addition to classic IgG, researchers have also found that some animals, such as camels and sharks, can produce their own antibodies that naturally lack a light chain, so-called heavy chain-only antibodies (HCAb). The antigen binding site of the HCAb is composed of two different variable domains called heavy chain variable domains (VHH) of heavy chain antibodies only. The difference between VHH and traditional IgG and various IgG derivatives lies in the unique properties of their molecular size (2.5 nm in diameter and 4 nm in height). Therefore it is also called nanobodies (Nb). Since its inception, the production of nanobodies has become increasingly standardized and includes subsequent steps such as immunization of camels, collection of lymphocytes and phage display. It is noteworthy that the high-throughput combination of DNA sequencing and mass spectrometry technologies has resulted in the formation of high-affinity nanobodies. Nanobody (Nbs) is the smallest fully functional antigen-binding fragment. Due to its small size (15 kDa) and single domain, Nbs is an ideal device for basic research, biosensors and therapeutic applications [102]. The antigen-binding fragment of this monomer has high thermal stability and configuration stability, high solubility in water produced directly by microorganisms, and excellent resistance to acidic and alkaline pH values. It has good affinity and specificity, easy to manufacture and low immunogenicity. In addition, Nbs has excellent distribution in the body, tissue permeability and faster blood cleansing. These favorable properties make Nb an attractive and valuable tool for a variety of applications. In general, Nb can be isolated from the Nb immune library. The VHHs gene can be reverse transcribed from peripheral blood lymphocyte (PBL) mRNA in animals immunized with the antigen of interest. These superior properties greatly expand the use of nanobodies in complex operations such as drug delivery, biomarker detection, molecular imaging and disease management [103].

4.3.1. Role of Nanobodies in Drug Delivery and Biomarker Detection

To improve the therapeutic effect of diseases such as cancer, it is necessary to administer targeted drugs [104]. The nanometer antibody can be used as an excellent alternative to mAbs as a recognition molecule for targeted drug delivery applications. For example, nanoscale albumin-bound paclitaxel has shown excellent therapeutic effects in breast cancer [105]. In addition [106], have developed a drug delivery system based on functionalized albumin nanoparticles nanobodies to deliver a variety of kinase inhibitors to epidermal growth factors receptor (EGFR). The receptor (EGFR) over expresses tumor cells in which EGFR-specific nano-bodies bind to nanoparticles of the active substance albumin through chemical binding. Likewise, nanoparticles (NPs) of polylactic acid (lactic acid-glycolic acid-hydroxymethylglycolic acid) (PLGHMGA) are specific for the selective delivery of human epidermal growth factor receptor 2 (HER2) in saponin-mediated endocytosis.

In combination with a branched dendrimer, nanobodies can also be effectively used to deliver DNA for gene transfection. Jamnani *et al.* created an improved polyamide amide dendrimer (PAMAM) linked to an anti-HER2 nanobody to achieve targeted DNA delivery [107]. Liposome-based nanocarriers are often used as delivery matrices because they resemble the morphology of cell membranes. One of the first attempts was to link nanobodies. For example, EGFR antagonists are on the surface of liposomes (do not load drug). After receptor-mediated internalization of this binding, suppression of EGFR was observed due to partial degradation of the cell membrane, which significantly inhibited the growth of tumor cells. As the smallest but most stable recognition molecule, nanobodies have the potential to recognize biomarkers in the early detection of disease. Many fast and powerful detection platforms have been developed from a combination of probes to amplify nanobodies and molecular signals [108].

Using screen-printed disposable electrodes (SPE), developed an immunoassay based on nanoantibody with a metal chelate His tag and combined it with Co^{2+}, nitrile triacetic acid transition metal ion magnetic beads (MB) formed to capture recognition molecules [109]. Considerable efforts have been made to create nanobodies in combination with semiconductor quantum dots (QDs) to create a sensitive detection platform [110]. Nanobody can also be used in Förster Resonance Energy Transfer (FRET) immunoassay, since the distance between the formed donor-acceptor complexes are in the FRET range and conventional antibodies are not suitable.

4.3.2. Nanobody as an Indicator in Diagnostics and Molecular Imaging

Anti-tumor antibody-based radiolabeled probes have been used for non-invasive imaging in a clinical setting. In contrast, nanobodies are ideal molecular indicators for clinical imaging, as they quickly accumulate on the target, are evenly distributed, and quickly removed. Nanobodies are usually combined with chelating agents to achieve metal binding capacity. In preliminary studies of nano-bodies labeled with radioactive nuclei, an attempt was made to label nano-bodies against EGFR using 99mTc-tricarbonyl intermediates for non-invasive single photon emission computed tomography (SPECT) expression of EGFR [111]. Similarly, 99mTc-labeled antiprostatic specific antibodies have reported nanoantibodies to membrane antigen (PSMA) for targeting tumors *in vivo* [112]. Such imaging studies have been successfully performed using a heterologous graft mouse model in which cancerous human cells are transplanted directly into immunodeficient mice [113]. Notably, nanobodies showed accurate contrast images of the tumor in preclinical studies and were not significantly toxic. In addition to animal imaging, the first open-label Phase I study was conducted in 20 patients with HER2-expressing breast cancer to assess probe safety, biodistribution, and potential tumor targeting, of which 68Ga is the HER2 nanobody label used for Positron Emission Tomography / Computed Tomography (PET / CT) [114]

In addition to radioactive nuclei, the use of nanobody-based markers for optical imaging is also widespread in tumor diagnosis and image-guided surgery. The use of carbonate anhydrase IX (CAIX) as a target for near infrared fluorophore-labeled nanobodies for optical imaging can aid in breast cancer imaging is strictly defined as a single domain, which allows nanobodies to be predominantly intracellularly expressed to explore inaccessible proteins. Nanobodies are used not only for diagnostic imaging, but also successfully fuse with fluorescent proteins to form chromosomes that can track intracellular compartments of living cells [103].

4.3.3. The Role of Nanobodies in the Treatment of Diseases

Another important biomedical application of nanobodies is the treatment of neurodegenerative diseases or diseases associated with improper protein folding. Nanobody has been studied for the treatment of dialysis-induced amyloidosis, Parkinson's disease and cancer treatment. However, many pharmaceutical companies are in various stages of clinical trials using nanobodies to create treatments for various diseases such as rheumatoid arthritis, osteoporosis and bone metastases [103].

CONCLUSION

Magnanimous research and developments have been carried out in the field of veterinary pharmacology and vaccinology. Practical, economical and commercially applicable products have been developed owing to the research based on animal models of diseases. There's need to further amplify these approaches and trial the efficacy against human diseases of concern. Similar products showing large promise on animal models could be developed and optimized for use in Humans. The potential of biologically, especially botanically derived therapeutics and vaccines needs more industry focus for funding the ongoing research developments. Nanotechnology, like wise is a modern way forward to the pathogens of animal and human concern.

CONSENT FOR PUBLICATION

Not applicable.

CONFLICT OF INTEREST

The author declares no conflict of interest, financial or otherwise.

ACKNOWLEDGEMENTS

Declared none.

REFERENCES

[1]   Maheshwari R, *et al.* Guiding Principles for Human and Animal Research During Pharmaceutical Product Development 2018.
        [http://dx.doi.org/10.1016/B978-0-12-814421-3.00018-X]

[2]   Mangipudy R, Burkhardt J, Kadambi VJ. Use of animals for toxicology testing is necessary to ensure patient safety in pharmaceutical development. Regul Toxicol Pharmacol 2014; 70(2): 439-41.
        [http://dx.doi.org/10.1016/j.yrtph.2014.07.014] [PMID: 25058855]

[3]   Carnevale RA, Shryock TR. Animal health pharmaceutical industry. Prev Vet Med 2006; 73(2-3): 217-20.
        [http://dx.doi.org/10.1016/j.prevetmed.2005.09.009] [PMID: 16266763]

[4]   Clement M, *et al.* Veterinary pharmaceuticals and antimicrobial resistance in developing countries.Veterinary Medicine and Pharmaceuticals. IntechOpen 2019.

[5]   Berger CN, Le Donne P, Windemann H. Use of substances of animal origin in pharmaceutics and compliance with the TSE-risk guideline -- a market survey. Biologicals 2005; 33(1): 1-7.
        [http://dx.doi.org/10.1016/j.biologicals.2004.10.002] [PMID: 15713551]

[6]   Oluwaseun AC, *et al.* Characterization and optimization of a rhamnolipid from Pseudomonas aeruginosa C1501 with novel biosurfactant activities. Sustain Chem Pharm 2017; 6: 26-36.
        [http://dx.doi.org/10.1016/j.scp.2017.07.001]

[7]   Houdebine L-M. Production of pharmaceutical proteins by transgenic animals. Comp Immunol Microbiol Infect Dis 2009; 32(2): 107-21.

[http://dx.doi.org/10.1016/j.cimid.2007.11.005] [PMID: 18243312]

[8]   Korhonen H, Pihlanto A. Bioactive peptides: production and functionality. Int Dairy J 2006; 16(9): 945-60.
        [http://dx.doi.org/10.1016/j.idairyj.2005.10.012]

[9]   Anusha R, Bindhu O. Bioactive peptides from milk. MILK PROTEINS 2016; p. 101.

[10]   Jiang J, Chen S, Ren F, Luo Z, Zeng SS. Yak milk casein as a functional ingredient: preparation and identification of angiotensin-I-converting enzyme inhibitory peptides. J Dairy Res 2007; 74(1): 18-25.
        [http://dx.doi.org/10.1017/S0022029906002056] [PMID: 16987434]

[11]   Kulyar MF-A, *et al.* Bioactive potential of yak's milk and its products; pathophysiological and molecular role as an immune booster in antibiotic resistance. Food Biosci 2020; 100838.

[12]   Berlutti F, Pantanella F, Natalizi T, *et al.* Antiviral properties of lactoferrin--a natural immunity molecule. Molecules 2011; 16(8): 6992-7018.
        [http://dx.doi.org/10.3390/molecules16086992] [PMID: 21847071]

[13]   Gizachew A, *et al.* Review on medicinal and nutritional values of camel milk. Nat Sci 2014; 12(12): 35-41.

[14]   El-Hatmi H, *et al.* Characterisation of whey proteins of camel (*Camelus dromedarius*) milk and colostrum. Small Rumin Res 2007; 70(2-3): 267-71.
        [http://dx.doi.org/10.1016/j.smallrumres.2006.04.001]

[15]   Konuspayeva G, Faye B, Loiseau G. The composition of camel milk: a meta-analysis of the literature data. J Food Compos Anal 2009; 22(2): 95-101.
        [http://dx.doi.org/10.1016/j.jfca.2008.09.008]

[16]   Mati A, *et al.* Dromedary camel milk proteins, a source of peptides having biological activities–A review. Int Dairy J 2017; 73: 25-37.
        [http://dx.doi.org/10.1016/j.idairyj.2016.12.001]

[17]   Meisel H. Biochemical properties of bioactive peptides derived from milk proteins: potential nutraceuticals for food and pharmaceutical applications. Livest Prod Sci 1997; 50(1-2): 125-38.
        [http://dx.doi.org/10.1016/S0301-6226(97)00083-3]

[18]   Park H, *et al.* Hyperimmune horse serum and immunopurified antibodies impact on human fibroblast cytokine response to Crotalus atrox venom. Toxicon 2020; 182: S11-2.
        [http://dx.doi.org/10.1016/j.toxicon.2020.04.033]

[19]   Yan S, *et al.* Bifidobacterium longum subsp. longum YS108R fermented milk alleviates DSS induced colitis via anti-inflammation, mucosal barrier maintenance and gut microbiota modulation. J Funct Foods 2020; 73: 104153.
        [http://dx.doi.org/10.1016/j.jff.2020.104153]

[20]   Ayyash M, Abu-Jdayil B, Itsaranuwat P, *et al.* Exopolysaccharide produced by the potential probiotic Lactococcus garvieae C47: Structural characteristics, rheological properties, bioactivities and impact on fermented camel milk. Food Chem 2020; 333: 127418.
        [http://dx.doi.org/10.1016/j.foodchem.2020.127418] [PMID: 32653680]

[21]   Guo W, *et al.* Rumen-bypassed tributyrin alleviates heat stress by reducing the inflammatory responses of immune cells. Poult Sci 2020.
        [PMID: 33357699]

[22]   Long N, *et al.* Reproductive performance and serum fatty acid profiles of underdeveloped beef heifers supplemented with saturated or unsaturated rumen bypass fat compared to an isocaloric control. Prof Anim Sci 2014; 30(5): 502-9.
        [http://dx.doi.org/10.15232/pas.2014-01311]

[23]   Girma DD, Ma L, Wang F, *et al.* Effects of close-up dietary energy level and supplementing rumen-protected lysine on energy metabolites and milk production in transition cows. J Dairy Sci 2019;

102(8): 7059-72.
[http://dx.doi.org/10.3168/jds.2018-15962] [PMID: 31178198]

[24] Crawford CD, *et al.* Effects of supplementation containing rumen by-pass protein on parasitism in grazing lambs. Small Rumin Res 2020; 106161.
[http://dx.doi.org/10.1016/j.smallrumres.2020.106161]

[25] Dong L, Jin Y, Cui H, *et al.* Effects of diet supplementation with rumen-protected betaine on carcass characteristics and fat deposition in growing lambs. Meat Sci 2020; 166: 108154.
[http://dx.doi.org/10.1016/j.meatsci.2020.108154] [PMID: 32330830]

[26] Bruijnis M, *et al.* Dairy farmers' attitudes and intentions towards improving dairy cow foot health. Livest Sci 2013; 155(1): 103-13.
[http://dx.doi.org/10.1016/j.livsci.2013.04.005]

[27] McCarthy CS, Dooley BC, Branstad EH, *et al.* Energetic metabolism, milk production, and inflammatory response of transition dairy cows fed rumen-protected glucose. J Dairy Sci 2020; 103(8): 7451-61.
[http://dx.doi.org/10.3168/jds.2020-18151] [PMID: 32448574]

[28] Cao Q-R, Lee ES, Choi YJ, Cho CS, Lee BJ. Rumen bypass and biodistribution of l-carnitine from dual-layered coated pellets in cows, in vitro and in vivo. Int J Pharm 2008; 359(1-2): 87-93.
[http://dx.doi.org/10.1016/j.ijpharm.2008.03.017] [PMID: 18448287]

[29] Morris DL, Kononoff PJ. Effects of rumen-protected lysine and histidine on milk production and energy and nitrogen utilization in diets containing hydrolyzed feather meal fed to lactating Jersey cows. J Dairy Sci 2020; 103(8): 7110-23.
[http://dx.doi.org/10.3168/jds.2020-18368] [PMID: 32505393]

[30] Yu L, *et al.* Effects of dietary rumen-protected betaine supplementation on the antioxidant status of lambs. Livest Sci 2020; 104026.
[http://dx.doi.org/10.1016/j.livsci.2020.104026]

[31] Nishu N, *et al.* Transgenic animals in research and industry.Animal Biotechnology. 2nd ed. Boston: Academic Press 2020; pp. 463-80.
[http://dx.doi.org/10.1016/B978-0-12-811710-1.00021-5]

[32] Bonafont J, Mencía Á, García M, *et al.* Clinically relevant correction of recessive dystrophic epidermolysis bullosa by dual sgRNA CRISPR/Cas9-mediated gene editing. Mol Ther 2019; 27(5): 986-98.
[http://dx.doi.org/10.1016/j.ymthe.2019.03.007] [PMID: 30930113]

[33] DeLay BD, Krneta-Stankic V, Miller RK. Technique to target microinjection to the developing Xenopus kidney. JoVE (Journal of Visualized Experiments), 2016; e53799.
[http://dx.doi.org/10.3791/53799]

[34] Bai M, Wu Y, Li J. Generation and application of mammalian haploid embryonic stem cells. J Intern Med 2016; 280(3): 236-45.
[http://dx.doi.org/10.1111/joim.12503] [PMID: 27138065]

[35] Liao J, Wei Q, Fan J, *et al.* Characterization of retroviral infectivity and superinfection resistance during retrovirus-mediated transduction of mammalian cells. Gene Ther 2017; 24(6): 333-41.
[http://dx.doi.org/10.1038/gt.2017.24] [PMID: 28387759]

[36] Sagar D, Masih S, Schell T, *et al.* In vivo immunogenicity of Tax(11-19) epitope in HLA-A2/DTR transgenic mice: implication for dendritic cell-based anti-HTLV-1 vaccine. Vaccine 2014; 32(26): 3274-84.
[http://dx.doi.org/10.1016/j.vaccine.2014.03.087] [PMID: 24739247]

[37] Patel TB, *et al.* Transgenic avian-derived recombinant human interferon-alpha2b (AVI-005) in healthy subjects: an open-label, single-dose, controlled study. 2007.

[38] Richt JA, Kasinathan P, Hamir AN, *et al.* Production of cattle lacking prion protein. Nat Biotechnol

2007; 25(1): 132-8.
[http://dx.doi.org/10.1038/nbt1271] [PMID: 17195841]

[39]   Khouw BT, Rubin LJ, Berry B. Meat animal by-products of pharmaceutical and food interest Proceedings-Annual Reciprocal Meat Conference of the American Meat Science Association (USA).

[40]   Hennessey JV. Historical and current perspective in the use of thyroid extracts for the treatment of hypothyroidism. Endocr Pract 2015; 21(10): 1161-70.
[http://dx.doi.org/10.4158/EP14477.RA] [PMID: 26121440]

[41]   Lowry P. 60 YEARS OF POMC: Purification and biological characterisation of melanotrophins and corticotrophins. J Mol Endocrinol 2016; 56(4): T1-T12.
[http://dx.doi.org/10.1530/JME-15-0260] [PMID: 26643914]

[42]   Baragob AEA, *et al.* Determination of the potency of extracted, purified and formulated insulin from the pancreatic organs of the sudanese beef cattle. Pharmacol pharm 2013; 4: 467.
[http://dx.doi.org/10.4236/pp.2013.46067]

[43]   Foster GV, Baghdiantz A, Kumar MA, Slack E, Soliman HA, Macintyre I. Thyroid origin of calcitonin. Nature 1964; 202: 1303-5.
[http://dx.doi.org/10.1038/2021303a0] [PMID: 14210962]

[44]   Sarwar MI, Hussain MS, Leghari AR. Heparin can be isolated and purified from bovine intestine by different techniques. Int J Pharm Sci Invent 2013; 2: 21-5.

[45]   Lee DY, Lee SY, Kang HJ, Park Y, Hur SJ. Development of effective heparin extraction method from pig by-products and analysis of their bioavailability. J Anim Sci Technol 2020; 62(6): 933-47.
[http://dx.doi.org/10.5187/jast.2020.62.6.933] [PMID: 33987573]

[46]   Loutradis D, *et al.* Corifollitropin alfa, a long-acting follicle-stimulating hormone agonist for the treatment of infertility. Current opinion in investigational drugs (London, England: 2000), 2009; 10: 372-80.

[47]   Humaidan P, Chin W, Rogoff D, *et al.* Efficacy and safety of follitropin alfa/lutropin alfa in ART: a randomized controlled trial in poor ovarian responders. Hum Reprod 2017; 32(3): 544-55.
[PMID: 28137754]

[48]   FRIrz , H , *et al.* Specific Isolation and Modification Methods for Proteinase, in Proceedings of the International Research Conference on Proteinase Inhibitors, Munich, November 4--6, 1970. 28.

[49]   Keating GM. Agalsidase alfa: a review of its use in the management of Fabry disease. BioDrugs 2012; 26(5): 335-54.
[http://dx.doi.org/10.1007/BF03261891] [PMID: 22946754]

[50]   Flaherty K, D . Immunology for pharmacy. Elsevier 2012.

[51]   Szwiec K, *et al.* Novel potential of pancreatic-like enzymes of microbial origin in exocrine pancreatic insufficiency-study on a pig model. Journal of Pre-Clinical and Clinical Research, 2015; 9
[http://dx.doi.org/10.5604/18982395.1157568]

[52]   Hu P-L, Yuan YH, Yue TL, Guo CF. Bile acid patterns in commercially available oxgall powders used for the evaluation of the bile tolerance ability of potential probiotics. PLoS One 2018; 13(3): e0192964.
[http://dx.doi.org/10.1371/journal.pone.0192964] [PMID: 29494656]

[53]   Ercan P, El SN. Changes in content of coenzyme Q10 in beef muscle, beef liver and beef heart with cooking and in vitro digestion. J Food Compos Anal 2011; 24: 1136-40.
[http://dx.doi.org/10.1016/j.jfca.2011.05.002]

[54]   Kaya MO, Arslan O, Guler OO. A new affinity method for purification of bovine testicular hyaluronidase enzyme and an investigation of the effects of some compounds on this enzyme. J Enzyme Inhib Med Chem 2015; 30(4): 524-7.
[http://dx.doi.org/10.3109/14756366.2014.949253] [PMID: 25373501]

[55]　Adeva-Andany MM, Funcasta-Calderón R, Fernández-Fernández C, Castro-Quintela E, Carneiro-Freire N. Metabolic effects of glucagon in humans. J Clin Transl Endocrinol 2018; 15: 45-53.
[http://dx.doi.org/10.1016/j.jcte.2018.12.005] [PMID: 30619718]

[56]　Vargas-Bello-Pérez E, Márquez-Hernández RI, Hernández-Castellano LE. Bioactive peptides from milk: animal determinants and their implications in human health. J Dairy Res 2019; 86(2): 136-44.
[http://dx.doi.org/10.1017/S0022029919000384] [PMID: 31156082]

[57]　Murgiano L, Timperio AM, Zolla L, Bongiorni S, Valentini A, Pariset L. Comparison of milk fat globule membrane (MFGM) proteins of Chianina and Holstein cattle breed milk samples through proteomics methods. Nutrients 2009; 1(2): 302-15.
[http://dx.doi.org/10.3390/nu1020302] [PMID: 22253986]

[58]　Aqib , A , et al. Reconnoitering Milk Constituents of Different Species, Probing and Soliciting Factors to Its Soundness, in Milk Production, Processing and Marketing. IntechOpen. 2019.
[http://dx.doi.org/10.5772/intechopen.82852]

[59]　Aqib , A.I , et al. Camel milk insuline: Pathophysiological and molecular repository. Trends Food Sci Technol 2019; 88: 497-504.
[http://dx.doi.org/10.1016/j.tifs.2019.04.009]

[60]　Izadi A, Khedmat L, Mojtahedi SY. Nutritional and therapeutic perspectives of camel milk and its protein hydrolysates: A review on versatile biofunctional properties. J Funct Foods 2019; 60: 103441.
[http://dx.doi.org/10.1016/j.jff.2019.103441]

[61]　Alebie G, Yohannes S, Worku A. Therapeutic applications of camel's milk and urine against cancer: current development efforts and future perspectives. J Cancer Sci Ther 2017; 9: 468-78.
[http://dx.doi.org/10.4172/1948-5956.1000461]

[62]　Badawy AA, El-Magd MA, AlSadrah SA. Therapeutic effect of camel milk and its exosomes on MCF7 cells *in vitro* and *in vivo*. Integr Cancer Ther 2018; 17(4): 1235-46.
[http://dx.doi.org/10.1177/1534735418786000] [PMID: 29986606]

[63]　Mohapatra A, Shinde AK, Singh R. Sheep milk: A pertinent functional food. Small Rumin Res 2019; 181: 6-11.
[http://dx.doi.org/10.1016/j.smallrumres.2019.10.002]

[64]　Ranadheera CS, Evans CA, Baines SK, *et al.* Probiotics in goat milk products: Delivery capacity and ability to improve sensory attributes. Compr Rev Food Sci Food Saf 2019; 18(4): 867-82.
[http://dx.doi.org/10.1111/1541-4337.12447] [PMID: 33337004]

[65]　Meena S, *et al.* Effect of goat and camel milk *vis a vis* cow milk on cholesterol homeostasis in hypercholesterolemic rats. Small Rumin Res 2019; 171: 8-12.
[http://dx.doi.org/10.1016/j.smallrumres.2018.12.002]

[66]　Balke I, Zeltins A. Recent advances in the use of plant virus-like particles as vaccines. Viruses 2020; 12(3): 270.
[http://dx.doi.org/10.3390/v12030270] [PMID: 32121192]

[67]　Podda A, Del Giudice G. MF59-adjuvanted vaccines: increased immunogenicity with an optimal safety profile. Expert Rev Vaccines 2003; 2(2): 197-203.
[http://dx.doi.org/10.1586/14760584.2.2.197] [PMID: 12899571]

[68]　Li W, Joshi MD, Singhania S, Ramsey KH, Murthy AK. Peptide vaccine: progress and challenges. Vaccines (Basel) 2014; 2(3): 515-36.
[http://dx.doi.org/10.3390/vaccines2030515] [PMID: 26344743]

[69]　Francis MJ. Recent advances in vaccine technologies. Vet Clin North Am Small Anim Pract 2018; 48(2): 231-41.
[http://dx.doi.org/10.1016/j.cvsm.2017.10.002] [PMID: 29217317]

[70]　Wunner WH, Dietzschold B, Curtis PJ, Wiktor TJ. Rabies subunit vaccines. J Gen Virol 1983; 64(Pt

8): 1649-56.
[http://dx.doi.org/10.1099/0022-1317-64-8-1649] [PMID: 6348210]

[71]   Ziech RE, *et al.* Blackleg in cattle: current understanding and future research needs. Cienc Rural 2018; 48(5)
[http://dx.doi.org/10.1590/0103-8478cr20170939]

[72]   Nasiri K, Nassiri M, Tahmoorespur M, Haghparast A, Zibaee S. Design and construction of chimeric VP8-S2 antigen for bovine rotavirus and bovine coronavirus. Adv Pharm Bull 2016; 6(1): 91-8.
[http://dx.doi.org/10.15171/apb.2016.014] [PMID: 27123423]

[73]   Mohd Yasin IS, Mohd Yusoff S, Mohd ZS, Abd Wahid Mohd E. Efficacy of an inactivated recombinant vaccine encoding a fimbrial protein of Pasteurella multocida B:2 against hemorrhagic septicemia in goats. Trop Anim Health Prod 2011; 43(1): 179-87.
[http://dx.doi.org/10.1007/s11250-010-9672-5] [PMID: 20697957]

[74]   Walker P. Bovine ephemeral fever in Australia and the world. The World of Rhabdoviruses. Springer 2005; pp. 57-80.
[http://dx.doi.org/10.1007/3-540-27485-5_4]

[75]   Gheibi A, Khanahmad H, Kashfi K, Sarmadi M, Khorramizadeh MR. Development of new generation of vaccines for *Brucella abortus.* Heliyon 2018; 4(12): e01079.
[http://dx.doi.org/10.1016/j.heliyon.2018.e01079] [PMID: 30603712]

[76]   Kitching RP. Vaccines for lumpy skin disease, sheep pox and goat pox. Dev Biol (Basel) 2003; 114: 161-7.
[PMID: 14677686]

[77]   Zhou Y, Jiang S, Du L. Prospects for a MERS-CoV spike vaccine. Expert Rev Vaccines 2018; 17(8): 677-86.
[http://dx.doi.org/10.1080/14760584.2018.1506702] [PMID: 30058403]

[78]   Johnson MM, Ainslie KM. Vaccines for the prevention of melioidosis and glanders. Curr Trop Med Rep 2017; 4(3): 136-45.
[http://dx.doi.org/10.1007/s40475-017-0121-7] [PMID: 29242769]

[79]   Robinson C, Frykberg L, Flock M, Guss B, Waller AS, Flock JI. Strangvac: A recombinant fusion protein vaccine that protects against strangles, caused by Streptococcus equi. Vaccine 2018; 36(11): 1484-90.
[http://dx.doi.org/10.1016/j.vaccine.2018.01.030] [PMID: 29398274]

[80]   Park Y, An DJ, Choe S, *et al.* Development of recombinant protein-based vaccine against classical swine fever virus in pigs using transgenic Nicotiana benthamiana. Front Plant Sci 2019; 10: 624.
[http://dx.doi.org/10.3389/fpls.2019.00624] [PMID: 31156681]

[81]   Slifka MK, Amanna IJ. Passive immunization. Plotkin's Vaccines 2018; p. 84.

[82]   Hedegaard CJ, Heegaard PM. Passive immunisation, an old idea revisited: Basic principles and application to modern animal production systems. Vet Immunol Immunopathol 2016; 174: 50-63.
[http://dx.doi.org/10.1016/j.vetimm.2016.04.007] [PMID: 27185263]

[83]   Palmeira P, *et al.* IgG placental transfer in healthy and pathological pregnancies. Clinical and Developmental Immunology, 2012.
[http://dx.doi.org/10.1155/2012/985646]

[84]   Cervenak J, Kacskovics I. The neonatal Fc receptor plays a crucial role in the metabolism of IgG in livestock animals. Vet Immunol Immunopathol 2009; 128(1-3): 171-7.
[http://dx.doi.org/10.1016/j.vetimm.2008.10.300] [PMID: 19027179]

[85]   Maryniak NZ, Hansen EB, Ballegaard AR, Sancho AI, Bøgh KL. Comparison of the Allergenicity and Immunogenicity of Camel and Cow's Milk-A Study in Brown Norway Rats. Nutrients 2018; 10(12): 1903.
[http://dx.doi.org/10.3390/nu10121903] [PMID: 30518040]

[86]   Smith HL, Saia G, Lobikin M, Tiwari T, Cheng SC, Molrine DC. Characterization of serum anti-diphtheria antibody activity following administration of equine anti-toxin for suspected diphtheria. Hum Vaccin Immunother 2017; 13(11): 2738-41.
[http://dx.doi.org/10.1080/21645515.2017.1362516] [PMID: 28933665]

[87]   Kurosawa N, Yoshioka M, Fujimoto R, Yamagishi F, Isobe M. Rapid production of antigen-specific monoclonal antibodies from a variety of animals. BMC Biol 2012; 10(1): 80.
[http://dx.doi.org/10.1186/1741-7007-10-80] [PMID: 23017270]

[88]   Zaroff S, Tan G. Hybridoma technology: the preferred method for monoclonal antibody generation for in vivo applications. Future Science 2019.

[89]   Barroso J, Halder M, Whelan M. EURL ECVAM recommendation on non-animal-derived antibodies. EU Science Hub-European Commission 2020.

[90]   Berry JD, Gaudet RG. Antibodies in infectious diseases: polyclonals, monoclonals and niche biotechnology. N Biotechnol 2011; 28(5): 489-501.
[http://dx.doi.org/10.1016/j.nbt.2011.03.018] [PMID: 21473942]

[91]   Ren E, *et al.* Vesicular antibodies for immunotherapy: The blooming intersection of nanotechnology and biotechnology. Nano Today 2020; 34: 100896.
[http://dx.doi.org/10.1016/j.nantod.2020.100896]

[92]   Guirado E, Amat I, Gil O, *et al.* Passive serum therapy with polyclonal antibodies against Mycobacterium tuberculosis protects against post-chemotherapy relapse of tuberculosis infection in SCID mice. Microbes Infect 2006; 8(5): 1252-9.
[http://dx.doi.org/10.1016/j.micinf.2005.12.004] [PMID: 16702016]

[93]   Pandey M, Kumar BV, Verma R. Mammaglobin as a diagnostic serum marker of complex canine mammary carcinomas. Res Vet Sci 2015; 103: 187-92.
[http://dx.doi.org/10.1016/j.rvsc.2015.10.010] [PMID: 26679816]

[94]   Bazl R. Production of hyperimmune serum against hydatidosis in order to evaluate hydatid cyst fluid immunogenicity. Clin Biochem 2011; 13(44): S189.

[95]   Mackenzie R, Kiernan M, McKenzie D, Youl BD. Hyperimmune goat serum for amyotrophic lateral sclerosis. J Clin Neurosci 2006; 13(10): 1033-6.
[http://dx.doi.org/10.1016/j.jocn.2006.03.009] [PMID: 16996272]

[96]   Youl B, Crum J. Clinical response in Krabbe's disease case treated with hyperimmune goat serum product Aimspro. 2005.

[97]   Watson ED, Stokes CR. Use of hyperimmune serum in treatment of endometritis in mares. Theriogenology 1988; 30(5): 893-9.
[http://dx.doi.org/10.1016/S0093-691X(88)80051-7] [PMID: 16726531]

[98]   Chen KS, Johnson DW. Neutralization kinetics of bovine viral diarrhea virus by hyperimmune serum: one or multi-hit mechanism. Comp Immunol Microbiol Infect Dis 1986; 9(1): 37-45.
[http://dx.doi.org/10.1016/0147-9571(86)90073-1] [PMID: 3021386]

[99]   Almansa R, Eiros JM, Fedson D, Bermejo-Martin JF. Hyperimmune serum from healthy vaccinated individuals for Ebola virus disease? Lancet Glob Health 2014; 2(12): e686.
[http://dx.doi.org/10.1016/S2214-109X(14)70341-9] [PMID: 25433619]

[100]   Annamaria P, Eugenia Q, Paolo S. Anti-SARS-CoV-2 hyperimmune plasma workflow. Transfus Apheresis Sci 2020; 59(5): 102850.
[http://dx.doi.org/10.1016/j.transci.2020.102850] [PMID: 32540345]

[101]   Popoff MR. Tetanus in animals. J Vet Diagn Invest 2020; 32(2): 184-91.
[http://dx.doi.org/10.1177/1040638720906814] [PMID: 32070229]

[102]   Yan J, Wang P, Zhu M, *et al.* Characterization and applications of Nanobodies against human procalcitonin selected from a novel naïve Nanobody phage display library. J Nanobiotechnology 2015;

13(1): 33.
[http://dx.doi.org/10.1186/s12951-015-0091-7] [PMID: 25944262]

[103] Yu X, Xu Q, Wu Y, *et al.* Nanobodies derived from Camelids represent versatile biomolecules for biomedical applications. Biomater Sci 2020; 8(13): 3559-73.
[http://dx.doi.org/10.1039/D0BM00574F] [PMID: 32490444]

[104] Nicolas J, Mura S, Brambilla D, Mackiewicz N, Couvreur P. Design, functionalization strategies and biomedical applications of targeted biodegradable/biocompatible polymer-based nanocarriers for drug delivery. Chem Soc Rev 2013; 42(3): 1147-235.
[http://dx.doi.org/10.1039/C2CS35265F] [PMID: 23238558]

[105] Petrelli F, Borgonovo K, Barni S. Targeted delivery for breast cancer therapy: the history of nanoparticle-albumin-bound paclitaxel. Expert Opin Pharmacother 2010; 11(8): 1413-32.
[http://dx.doi.org/10.1517/14656561003796562] [PMID: 20446855]

[106] Martínez-Jothar L, Beztsinna N, van Nostrum CF, Hennink WE, Oliveira S. Selective cytotoxicity to HER2 positive breast cancer cells by saporin-loaded nanobody-targeted polymeric nanoparticles in combination with photochemical internalization. Mol Pharm 2019; 16(4): 1633-47.
[http://dx.doi.org/10.1021/acs.molpharmaceut.8b01318] [PMID: 30817164]

[107] Jafari Iri Sofla F, *et al.* Specific gene delivery mediated by poly (ethylene glycol)-grafted polyamidoamine dendrimer modified with a novel HER2-targeting nanobody. J Bioact Compat Polym 2015; 30(2): 129-44.
[http://dx.doi.org/10.1177/0883911515569005]

[108] Wang H, Li G, Zhang Y, *et al.* Nanobody-based electrochemical immunoassay for ultrasensitive determination of apolipoprotein-A1 using silver nanoparticles loaded nanohydroxyapatite as label. Anal Chem 2015; 87(22): 11209-14.
[http://dx.doi.org/10.1021/acs.analchem.5b04063] [PMID: 26522241]

[109] Campuzano S, Salema V, Moreno-Guzmán M, *et al.* Disposable amperometric magnetoimmunosensors using nanobodies as biorecognition element. Determination of fibrinogen in plasma. Biosens Bioelectron 2014; 52: 255-60.
[http://dx.doi.org/10.1016/j.bios.2013.08.055] [PMID: 24060974]

[110] Cai E, Ge P, Lee SH, *et al.* Stable small quantum dots for synaptic receptor tracking on live neurons. Angew Chem Int Ed Engl 2014; 53(46): 12484-8.
[PMID: 25255882]

[111] Huang L, Gainkam LO, Caveliers V, *et al.* SPECT imaging with 99mTc-labeled EGFR-specific nanobody for in vivo monitoring of EGFR expression. Mol Imaging Biol 2008; 10(3): 167-75.
[http://dx.doi.org/10.1007/s11307-008-0133-8] [PMID: 18297364]

[112] Evazalipour M, D'Huyvetter M, Tehrani BS, *et al.* Generation and characterization of nanobodies targeting PSMA for molecular imaging of prostate cancer. Contrast Media Mol Imaging 2014; 9(3): 211-20.
[http://dx.doi.org/10.1002/cmmi.1558] [PMID: 24700748]

[113] De Vos J, Devoogdt N, Lahoutte T, Muyldermans S. Camelid single-domain antibody-fragment engineering for (pre)clinical in vivo molecular imaging applications: adjusting the bullet to its target. Expert Opin Biol Ther 2013; 13(8): 1149-60.
[http://dx.doi.org/10.1517/14712598.2013.800478] [PMID: 23675652]

[114] Keyaerts M, Xavier C, Heemskerk J, *et al.* Phase I study of 68Ga-HER2-nanobody for PET/CT assessment of HER2 expression in breast carcinoma. J Nucl Med 2016; 57(1): 27-33.
[http://dx.doi.org/10.2967/jnumed.115.162024] [PMID: 26449837]

Frontiers in Protein and Peptide Sciences, 2021, *Vol. 2*, 349-383

CHAPTER 13

Plant Molecular Pharming For Livestock And Poultry

Rimsha Riaz[1], **Saher Qadeer**[1], **Faiz Ahmad Joyia**[1], **Ghulam Mustafa**[1] and **Muhammad Sarwar Khan**[1,*]

[1] *Center of Agricultural Biochemistry and Biotechnology (CABB), University of Agriculture, Faisalabad – 38040, Pakistan*

Abstract: Plants are exploited as bioreactors for the cost-effective production of pharmaceuticals, predominantly for the expression and accumulation of antigenic proteins, to be used as vaccines for livestock and poultry. Due to the high body mass of large animals and large population of poultry and other birds, a larger quantity of vaccines is needed continuously. It increases the production costs of vaccines for these animals. Under high biomass production ability, plants represent promising biofactory with added advantages of pathogen-free production of desired proteins in bulk quantities. Hence, plant-based transient, as well as stable expression systems, have been exceedingly applied to express immunogenic proteins. We have been using various plants like soybean and *Trifolium* to produce edible vaccines for poultry and livestock, respectively. Here we have reviewed various types of vaccines with a special focus on their plant-based expression with examples of Infectious Bursal Disease (IBD), New Castle Disease (ND), and Foot and Mouth Disease (FMD).

Keywords: Biopharming, Molecular pharming, IBD, ND, FMD, Vaccine, Edible vaccine, DNA vaccine, Subunit vaccine.

1. INTRODUCTION

Plants have been used for medicinal purposes since ancient times. More than 120 plant-derived drugs/pharmaceuticals have been approved for commercial utilization. Plant-derived pharmaceuticals include pain killers, wound healers, diagnostic reagents, antibiotics, therapeutic proteins, antimalarial and anticancer drugs [1]. These exogenous proteins are mostly produced as monoclonal antibodies, drugs, and vaccines not only to treat diseases but also to provide

---

* **Corresponding author Muhammad Sarwar Khan:** Center of Agricultural Biochemistry and Biotechnology (CABB), University of Agriculture, Faisalabad – 38040, Pakistan; E-mail: sarwarkhan_40@hotmail.com

Muhammad Sarwar Khan (Ed)

additive nutrients. Plant molecular farming gained more importance after the production of insulin using a bacterial expression system in 1982 [2]. Since then, numerous medicinal products have been produced by using different expression systems, *i.e.*, mammalian cell lines, insect cell lines, and yeast cell lines. Compared to the conventionally produced pharmaceuticals, recombinant therapeutic proteins can be complex with an intricate mode of action. Owing to the involvement of complex biochemical pathways, their chemical synthesis is quite difficult and has numerous limitations so they must be produced in a living system to exploit the host cellular machinery for protein synthesis [3]. Each of the production systems has certain advantages and pitfalls regarding production cost, risks of contamination, post-translational modifications, downstream processing cost, regulatory and approval issues, *etc.*

Most of the commercial pharmaceuticals are produced through mammalian cell lines, yet the plant expression system has its significance [4]. Pioneering research on plant expression systems (plant-based pharmaceuticals) has proved its competitiveness for commercial applications. Compared to other production platforms, plants have several distinctive attributes to produce a wide variety of valuable pharmaceuticals. Various industrial enzymes, vaccine candidates, and monoclonal antibodies have been successfully expressed in plants because they are easy to scale up and do not require capital investment for infrastructure. They can be cultivated in a greenhouse or open fields, and as a result, the cost of production is very low. They are also free from human and animal pathogens; therefore, they are biosafe with limited chances of contamination [5]. Furthermore, plants can produce recombinant proteins with proper post-translational modifications. Hence, effective recombinant therapeutic proteins can be produced at the commercial level in a short period [6 - 8].

Elelyso® was the first recombinant therapeutic protein that was commercially produced in plants in 2012 and was approved for enzyme replacement therapy. Numerous plant species have been explored to produce various proteins of pharmaceutical importance. The most notable are: production of diagnostic reagent (avidin) in maize, commodity chemicals in rice, veterinary medicine in strawberry, and nutraceuticals in barley [9]. A large number of whole plant-produced pharmaceuticals are in pipeline for commercialization, *i.e.*, influenza vaccine is in 3rd phase of the trial. A large number of proteins including pharmaceuticals have been attempted to be produced through plant expression systems. This chapter highlights the use of plants as a valuable source for the production of biopharmaceuticals with special emphasis on the proteins of worth for livestock and poultry. Further, comparisons among different expression systems and transmission of diseases from animals to humans have also been discussed.

2. GLOBAL IMPORTANCE OF LIVESTOCK AND POULTRY SECTORS

Animal-derived food demand has drastically increased owing to the rapidly increasing population, increased income, and urbanization. The livestock species play a critical role not only to fulfill protein needs but also in the well-being of poor farmers, all over the world [10]. Besides meat and milk, several other products (animal fat, skin, hides, and horns) are used for domestic consumption as well as for the manufacture of various industrial products. Total world meat production has increased to 342.42 million tonnes where major contributors are poultry, pork, cow, buffalo, sheep, goat, camel, horse, and ducks. In 2018, approximately 302 million cattle, 479 million goats, 574 million sheep, 456 million turkeys, 1.5 billion pigs, and 69 billion chickens were killed or slaughtered to obtain meat. Likewise, milk production has increased to 800 million tonnes, but still, there is a continuous increase in the demand for animal protein and milk [11].

Compared with other sources of animal proteins, poultry meat has shown the fastest trend of growth during the last decades. During the last 50 years, the growth rate of poultry meat was 5%, that of beef was 1.5%, for pork it was 3.1%, and for small ruminants, it was 1.7% [12]. This growth trend is even higher in developing countries, particularly South East and East Asia (7.4%). The top-most producer of poultry meat is United States (20 million tons), followed by China (18 million tons), European Union (EU), and Brazil (13 million tons per annum). More than 23 billion poultry birds are present on the planet whereas efforts are in progress for its improvement through breeding and advanced molecular research. Poultry meat and eggs are the most common animal food, consumed all over the world. It is categorized as the most efficient sub-sector regarding protein provision and use of natural resources. Increased per capita consumption is a clear-cut indication of the importance of poultry. Egg consumption has increased from 4.55 kg to 8.92 kg whereas meat consumption has increased from 2.88 kg to 14.13 kg. Though the poultry sector has contributed a lot to ensure food security by providing proteins, energy, and essential nutrients to humans yet further efforts are direly needed to fulfill the ever-increasing demand for eggs and meat [13]. Further, it has widespread implications in the manufacture of non-food products which need to be explored and worked out. Thus, producing more from less without affecting the environment is the key to meet the sharply increasing demands of animal protein and its by-products.

3. ANIMALS TO HUMAN SPREAD OF DISEASES

Vertebrates are one of the major contributors to the transmission of infectious pathogens in humans. Pets (domestic and wild animals) are involved in these transmissions. Diseases of animal origin account for millions of deaths and about a billion cases of illness each year, thus posing serious threats to human life. These include bacterial pathogens (involved in Lyme disease, anthrax, brucellosis, salmonellosis, plague, and tuberculosis), viral pathogens (involved in rabies, AIDS, avian influenza, ebola), parasitic pathogens (involved in malaria, trematodes, trichinosis, giardiasis, echinococcosis, toxoplasmosis), fungal pathogens (causing ringworm), rickettsial pathogens (causing Q-fever), chlamydial pathogens (causing psittacosis), mycoplasma (causing mycoplasma pneumonia), different protozoa and non-viral pathogenic agents which are involved in the transmission of mad cow disease and spongiform encephalopathies [14]. Hence, most of the devastating pandemics of history have an animal origin, including the Black Death pandemic in Europe during 1347, yellow fever in South America during the 16th century, global flu during 1918, severe acute respiratory syndrome during 2003, and modern-age pandemics such as AIDS, HIV and COVID [15].

More than 61% of these pathogens are zoonotic and include both Gram-positive and Gram-negative bacteria which can be transmitted directly or indirectly to humans. Direct transmission of pathogens may be through air or droplets or fomites. Avian influenza is one such disease that is transmitted through fomites or droplets. Rabies is one of the deadliest diseases that can be transmitted through bites. Whenever a rabid animal (bat, raccoon, dog, monkey, fox, or skunk) bites, the virus enters the human body through saliva. The indirect transmission includes transmission through vectors. The most common vectors involved in indirect disease transmission are ticks and mosquitoes [16]. Domestic animals are also of critical importance in the transmission of these diseases. They act as amplifiers of the pathogen transmitted from wild animals to domestic animals thus, are major contributors to the transmission of diseases through inhalation, ingestion, direct contact, biting, and conjunctiva [17].

4. EMERGENCE AND RE-EMERGENCE OF NEW DISEASES

Emerging infectious diseases (EIDs) are newly emerged diseases that are detected in any population for the first time. Two third of the EIDs are contributed by wild animals and in most cases, they originate from wild animals. For any of the newly emerging diseases, epidemiological factors and the source of the outbreak need to be explored and identified [18]. Hence, the development of cost-effective and realistic tools for the prevention and prediction of EID is a major scientific challenge [19]. Variations in human civilization and so-called developmental activities are thought to be the core source for the emergence of new diseases. A rapid increase in urbanization, deforestation, agricultural intensification, and habitat fragmentation have led to the loss of natural habitats and the environment. This has altered human-wildlife interactions thus forcing numerous species to live in more proximity to humans. If interacting species is the host of some new pathogen, it results in the new disease infection and transmission [20]. Ease of traveling, more exposure to animal products or animals, farming of wild animals, hunting and transportation of wildlife, and consumption of wild meat are other factors responsible for the re-emergence of new diseases [21].

Re-emergence infectious diseases are a major global concern for the last two decades. The outbreak of infectious pathogens affects economical activities posing a huge financial burden on the infected communities. Trade activities are also affected adversely owing to the fear of the spread of diseases among the nations. Most of the newly emerging diseases of the modern age are viruses from animal reservoirs and 75% of the emerging human diseases originate from wild animals or domestic animals. This demands close interaction and collaboration between human health organizations and animal health organizations for the culmination of the emergence of new diseases [22]. Examples of newly emerging diseases are feline cowpox, avian influenza, rotavirus infection, bovine spongiform encephalopathy, ebola, norovirus infection, West Nile fever, hantavirus infection, MRSA infection, canine leptospirosis, severe fever with thrombocytopenia syndrome, cat scratch disease, severe acute respiratory syndrome and COVID-19 [23, 24]. In addition to the emergence of new pathogens, some of the diseases are re-emerging in certain regions of The World. These include Japanese encephalitis, rabies, tuberculosis, *Schistosoma japonica* infection, and brucellosis.

5. LIMITATIONS OF CURRENTLY USED VACCINES FOR LIVESTOCK AND POULTRY

Vaccines are considered the most valuable means of avoiding infectious diseases. To date, conventional vaccines have been used in most cases. Though, numerous disadvantages of conventional vaccines are evident like the deficiency to differentiate the infected animals from the vaccinated ones. Although many efficacious conventional veterinary vaccines are available but cannot be administered because of the fear that the vaccinated animals will be found positive during disease surveillance based on serological testing. In this way, that country will lose the disease-free status which will harm its trade. A characteristic instance is FMD in cattle. Though numerous countries have conventional (inactivated) vaccines for FMD and those vaccines have been found quite effective in disease control [25], however, such vaccines are not being consumed in FMD-free countries, as it would harm the disease-free status of those countries and hence their international trade will be affected severely. The development of marker vaccines or DIVA strategy may help in this regard.

Another major issue is the limited efficacy of commonly used vaccines. Many microorganisms especially viruses have a very fast mutation rate due to which new strains arise continuously. This limited efficacy poses the biggest challenge for a scientist to develop preventive and therapeutic interventions against disease outbreaks [26]. The development of universal vaccines against such problematic infections may help in disease control.

Another major problem includes the constraints and restrictions in antigenic protein production. In this context, it is very astonishing that work on antigenic protein production in plants is continued for the last 26 years but only two plant-based recombinant proteins had been able to clear the regulatory procedures and got licensed: the first one was a monoclonal antibody against HBsAg while the second one was a vaccine against New Castle Disease (ND). Another idea was to develop vaccines in the edible parts of plants. That idea gathered much publicity. Hence, passionate work on the expression of various antigenic proteins in plants started. Yet, there were many slips between the cup and lips [27]. There was plenty of information about the success of expressing desired antigens in plants yet, most of the cases were unsuccessful in achieving the expression levels feasible for cots effective production and commercialization [28] Additionally, the expression of vaccines in important food plants budded another fear of contaminating the food chain with these unusual proteins. Biosafety regulations have also discouraged the progress in plant-based production of recombinant vaccines [29]. It can be said that with improved control over the gene expression system of plants, deeper knowledge of immunity, and basic biology, the development of perfect plant-made edible vaccines is not too far [30].

6. TYPES OF VACCINES FOR ANIMALS AND POULTRY

Vaccines comprise the major part of the biopharmaceuticals approved for livestock and poultry because vaccination programs of massive scale are inevitable to avert quick and unchecked breakouts of many contagious diseases in highly dense populations of poultry and livestock maintained in contemporary intensive agriculture systems. For instance, only during the year 2021, the total broiler meat production in the USA was expected to reach around 20.4 million metric tons and in China, total production was expected to reach around 15 million metric tons (https://www.statista.com/statistics/237597/leading-10-

countries-worldwide-in-poultry-meat-production-in-2007/). Similarly, the number of cattle around the globe is expected to reach 1000.97 million during the year 2021 (https://www.statista.com/statistics/263979/global-cattle-population-since-1990/). Hence, the threat of a disease outbreak can be rapid as well as severe. It is reported that 50 million birds died in avian influenza (HPAI) outbreak that proved to be severely pathogenic. Hence, it was declared the largest animal-health emergency in the USA [31]. Moreover, an estimated amount of $610M was expended by the US government in emergency plans to restrain the disease from spread to other areas [32]. Similarly, during 2001, FMD outbreaks in the United Kingdom and during 2010-11 in South Korea & Japan brought caused the death of a huge livestock population. The efforts of disease control in South Korea resulted in mass-scale culling of around 20% of the country's livestock herds. In Japan, the spread of FMD into high-density livestock areas led to complications in finding appropriate burial sites, which caused interruptions in disease eradication exercises and compelled the Japanese government to allow a vaccinate-to-kill method [33, 34]. Keeping in view the above incidences, it is concluded that the continuous and uninterrupted production, supply, and availability of livestock and poultry vaccine is of utmost importance to develop sustainable food production and malnutrition management [35]. From a scientific viewpoint, animal vaccines are far more diverse than their counterparts for humans.

Following are various types of vaccines most commonly used for various diseases of livestock and poultry.

6.1. Conventional Live and Inactivated Vaccines:

Inspired from the vaccine against smallpox in humans (first vaccines discovered), live vaccines for veterinary were used and gave mild infections while using live microbes isolated from nontarget hosts or attenuated by series of cultures in diverse cell lines or embryos found in eggs. Another method to obtain attenuated viral strains was to induce random mutations in infectious agents and selecting microbes with reduced virulence. Because live organisms are still able to infect the target tissues, such vaccines can replicate themselves and trigger both cellular as well as humoral immunity and usually do not need any adjuvant for effective uptake. Moreover, the live microbes may also exhibit user-friendly administration. Nonetheless, they may have some risks including (i) residual virulence and reversion of attenuated microbes to pathogenic ones and (ii) vaccines as a potential source of infection in the environment [36].

Despite these shortcomings, live viral vaccines have significantly provided effective control of diseases and their elimination. For example, it is generally accepted that the de facto global elimination of the rinderpest virus largely depended on the administration of the famous "Plowright" vaccine [37, 38]. An example of this was a novel vaccine (*Enterisol Ileitis*) against porcine proliferative enteropathy, a causal organism for which is *Lawsonia intracellularis* [39]. It was a Kabete O strain-derived attenuated vaccine [40]. It has been found that only one point mutation in the gene coding for polymerase may give rise to numerous stable attenuations. Still, the high spontaneous mutation rate in the genome of RNA viruses increased the risk of reversal of vaccine to a pathogen.

Contrary to the live viral vaccines, the entire inactivated or killed viral vaccines normally show better stability and very meager chances of reversal to virulent strain. However, such vaccines have a lesser ability to trigger an immune response by infecting cells and activating cytotoxic T cells. Hence, such vaccines are found less effective in disease control [41]

6.2. Subunit Vaccines

Identification of viral protein with antigenic potential facilitates the extraction and production of recombinant subunit vaccines for pathogen-free delivery of non-replicating viral vaccines. The major drawback of isolated antigens is their weak antigenic ability to protect against diseases hence; subunit vaccines normally need strong adjuvant making them more effective and less competitive. Despite these shortcomings, there are various cases of subunit vaccines considerably effective against diseases. Numerous killed vaccines comprising of crude extracts of a complete organism or well-defined antigenic parts have been used for mild control. Generally, such vaccines are less effective than live organisms however can be used to slow down the disease spread up to a certain level. There is an example of a vaccine produce by PalA protein (peptidoglycan-associated lipoprotein) and the most "immunodominant" outer membrane protein of *A. pleuropneumoniae*. Certain safety measures are necessary while using such vaccines as antibodies induced in response to this PalA antigenic protein intensified the after-effects of a challenge infection, and vaccination with PalA, when combined with RTX toxins, was able to counteract the protecting effect of anti-ApxI and anti-ApxII antibodies [42].

6.3. Genetically Modified Organisms

With the advancement in the subject of molecular biology, more and more information is available regarding genome sequences, gene functions, and ways of genetic modification in microbial genomes. Hence by using this knowledge, scientists have been able to make desired modifications, including deletions and insertions into the genomes of pathogenic organisms making them suitable to be used as vaccines. For example, the DIVA strategy has been applied to regulate Aujeszky's disease in swine by using vaccine produced after deletion of glycoprotein I and/or glycoprotein X genes [43, 44]. In another example, the deletion of thymidine kinase from the BHV-1 vaccine has resulted in inactivation and later reactivation after treatment with dexamethasone [45]. Furthermore, the removal of numerous genes has been postulated to improve animal safety [46]. The production of chimera viruses is the most interesting development in genetically engineered viral vaccines. By this strategy, we can combine various portions of different infective viral genomes resulting in chimera viruses. For example, a chimera vaccine (PCV1-2) was produced by ligating the immunogenic capsid gene of PCV2 into the backbone of PCV1 a nonpathogenic virus. The resultant chimeric virus is known to induce an immune response against wild-type PCV2 challenge in pigs [47]. Another relatively more sophisticated approach has been developed against avian influenza virus with the name Poulvac FluFend vaccines. In this vaccine, the gene for hemagglutinin was deleted from an H5N1 virus, then it was deactivated by deleting the polybasic amino acid sequences, and ligated to the NA gene from an H2N3 virus in an H1N1 "backbone" virus. In another example, the structural genes have been swapped between yellow fever YF-17D and West Nile virus (WNV). In this chimera vaccine, the genes of PreM and E proteins of WNV were expressed, whereas the nucleocapsid (C) protein, nonstructural proteins, and untranslated regions including termini which control virus replication were taken from the original yellow fever 17D virus [48]. This was found so effective that only one vaccine dose was able to stimulate both the cell-mediated as well as humoral responses. As a result, the horses remained protected against West Nile virus WNV for up to one year. Moreover, there were no clinical symptoms in horses and there was no spread to sentinel horses. This chimera vaccine was officialized in the USA in 2006 against the West Nile virus (WNV) [49]. In another example, a chimeric vaccine composed of inactivated H5N3-expressing virus administered in a water-in-oil emulsion was reported to protect chickens and ducks against H5N1 which is known as the highly pathogenic strain.

6.4. DNA-Based Vaccines

Vaccinating poultry and livestock with nucleotide sequences encoding protective viral proteins offers several benefits, as it not only rules out the biosafety apprehensions associated with live vaccines but also encourages the trigging of cytotoxic T cells after intracellular expression of the antigens. Moreover, the stability of DNA vaccines is very high as they do not need a cold chain for short-term storage and transport. However, DNA vaccination efforts of large animals have not been found as efficient as primarily found in mice. Although some research groups have demonstrated substantial progress in eliciting immune responses using advanced techniques such as the precise direction of the vaccine antigen to antigen-presenting cells [50], priming-boosting with stimulating CpG oligo-deoxynucleotides [51], and *in vivo* electroporation of DNA [52].

A DNA vaccine has been developed against viremia disease in horses caused by the West Nile virus under a license from the USDA. West Nile disease is caused by a flavivirus belonging to the Japanese encephalitis virus complex. It is enzootic in various parts of the African and Asian continents. However, it was detected in the USA in 1999, for the first time, during an outbreak involving birds, horses, and humans in New York. Later, it reached speedily to other states [53]. The DNA plasmid used as a vaccine coded for the outer coat proteins of WNV and was delivered with a patented adjuvant [54, 55]. In this case, the viral protein was found particularly effective as it can produce highly immunogenic virus-like particles naturally [56]. It can be concluded that broad spectrum implementation of DNA vaccines will likely entail additional developments and optimization for each host-pathogen combination.

7. PLANT-BASED EXPRESSION PLATFORM FOR VACCINE PRODUCTION

Plant-based express systems offer numerous distinctive attributes to produce valuable pharmaceuticals. Since the late 1980s, biotechnologists have been trying to use plants as bioreactors for the economical production of recombinant proteins [57]. Both pharmaceutical and non-pharmaceutical proteins have been manufactured in plant-based expression systems. For example, growth factors, various industrial enzymes, monoclonal antibodies, diagnostic agents, vaccine candidates, *etc.* During recent decades, the industry has rapidly flourished and the pitfalls associated with the high level of protein expression and the purification of the desired product during plant molecular pharming have now been managed to a large extent [58]. The advantages, challenges, and common applications of different expression systems are summarized in Table **1**.

Table 1. Merits and demerits of different expression systems used in molecular farming.

Expression System	Examples	Reproduction Time (Hours)	Applications	Advantages	Challenges	References
Bacterial	• *Escherichia coli* • Enterobacter (*Pseudomonas* spp.) • Firmicutes (*Bacillus* spp.)	0.3-0.5	• Structural and functional assays • Protein expression and interaction • Antigen and antibody production	• Rapid expression with low cost • Ease to manipulate and scale-up • High expression rate • Simple cultural conditions and practices	• No post-translational modifications (PTMs) • Improper protein folding • Difficulties in expressing mammalian proteins	[2, 59 - 61]
Yeast	• Unicellular Methylotrophic (*Pichia pastoris, Hansenula polymorpha*) • Unicellular Non-methylotrophic (*Saccharomyces cerevisiae, Kluyveromyces lactis*)	1.3-2.0	• Functional and structural analyses • Protein interaction • Antibody production	• Eukaryotic protein expression • Rapid expression and less expensive • PTMs to maintain biological activity • Well for intracellular and secreted proteins • Simple media requirements • Ease and rapid scale-up • Optimized growth conditions and procedures	• Growth requires optimization • Requires secretion of signal peptides • Different N-linked glycan structures of protein • For high yield, fermentation is required • Rigidity in disruption of thick cell wall • Limited glycosylation • Protein hyperglycosylation may occur	[59, 62, 63]

(Table 1) cont.....

Expression System	Examples	Reproduction Time (Hours)	Applications	Advantages	Challenges	References
Insect	• *Danau plexippis* • Baculovirus-infected insect cells • *Xenopus* oocytes	16-72	• Expression of intracellular proteins and protein complexes • Functional and structural analyses • Virus production	• Good for secreted, intracellular, and membrane proteins • Protein folding and PTMs are possible	• Expensive • Difficulties in scaling-up • More demanding cultural requirements • Time-consuming • Un-wanted PTMs • Immunogenic sugars during non-human glycosylation	[61, 64]
Mammalian	• Chinese Hamster Ovary (CHO), • Human Embryonic Kidney Cells (HEK293)	14-36	• Expression of intracellular proteins and protein complexes • Functional and structural analysis • Virus production • Transient and stable expression	• PTMs and protein folding is possible • Used well for membrane and secreted proteins • High and rapid expression profile • Presence of regulatory system • Highest protein processing	• Expensive • Time-consuming expression system • Difficulties in scaling-up • Low yield of intracellular proteins • More demanding and expensive cultural requirements • Human pathogen contamination • Unstable cell lines	[6, 65, 66]
Plants	• Tobacco • Tomato • Potato • Soybean • Rice • Lettuce • Carrot	24-48	• Biopolymer production • Industrial enzymes and protein production • Therapeutic proteins	• Optimized growth conditions • Economical • Free of human, and bacterial pathogens • Occurrence of PTMs • Maximum scale-up possibility • Less cost on growth	• Lack of regulatory approval • Immunogenic sugars during non-human glycosylation • Limited protein glycosylation	[6, 67]

The plant-based expression platform is an attractive alternative to produce vaccines due to the following reasons: plants are easy to scale up and economical because they do not need capital investment for infrastructure and fermenters. They can be grown in fields or in contained facilities which are not much expensive. The production cost from the plant system is about 10-50 times less than microbial systems, *i.e.*, bacterial or yeast expression systems, while it is found negligible as compared to mammalian cell lines [29, 30]. Plants are also free from pathogen-causing diseases in animals and humans, so the plant-based vaccine will be biosafe and have fewer chances of contamination [5, 68]. Furthermore, the mass production of plant-made vaccines is possible at the commercial level in a short period [6 - 8]. Moreover, in plants, any plant part, like stem, root, leaf, fruit, and seeds can be utilized to produce medicinal drugs. The application of a plant-derived pharmaceutical production was initially conceptualized to produce plant-based edible prophylactic and therapeutic medicines against various human, veterinary, and poultry diseases [28] because there would be no need for cold-chain and injections during vaccine logistics and vaccine delivery/administration, respectively. In the case of plant-based edible vaccines, the plants harboring vaccine antigens could be consumed either directly or after slight processing. Orally administered vaccines help in eliciting both mucosal and serum immune responses by stimulating IgG production that leads to cell-mediated immunity. To retain the biological activity of therapeutic proteins, the proper folding and structure of a therapeutic protein are mandatory and luckily plants can maintain the pharmacokinetics of the drug by post-translational modifications (PTMs) such as glycosylation and disulfide bond formation [69, 70]. In a plant-based expression system, the vaccine antigen is encapsulated in plant cells to protect it from degradation by gut microbes when it is exposed to the harsh environment of the animal digestive tract. The antigen is slowly released into the body, ensuring effective immunity [71, 72]. *Nicotiana tabacum* has been mostly used as a model plant for the expression of therapeutic proteins [73]. Cereals (*Zea mays*, *Oryza sativa*), tubers (*Solanum tuberosum*, *Daucus carota*), fruits (*Solanum lycopersicum*, *Musa velutina*), leaves (*Lactuca sativa*), and legumes (*Glycine max*) have also been frequently used in producing pharmaceuticals [74]. For edible vaccine production, the crop plants must be easily accessible, palatable, and digestible. Cereals and fodder crops are ideal for producing veterinary vaccines because they can be easily administered in their feed without any processing.

An efficient biotechnological approach is a prerequisite for optimum production and purification of the vaccine candidate in its native conformation. The vaccine antigen can be integrated into the plant genome either through stable or transient expression. Stable nuclear integration of antigen is a frequently used approach that can be achieved by *Agrobacterium tumefaciens* (vector-mediated) and through biolistic transformation (physically). The schematic illustration of plant-based expression systems is shown in Fig. (**1**).

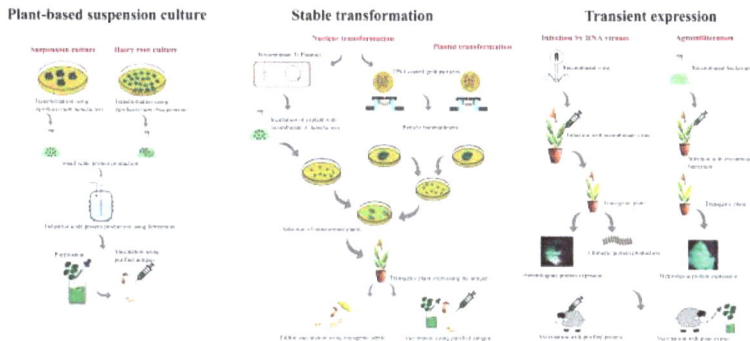

Fig. (1). The schematic illustration of plant-based expression systems.

During *Agrobacterium*-mediated gene delivery, the plant tissue is incubated with bacterium inoculum having a transgene cloned into the T-DNA borders of a binary vector. Bacterial "vir" genes then help the transgene to stably integrate into the nuclear genome and result in the development of stable transgenic lines [75]. This method is most suitable for the expression of the transgene in dicotyledons as the bacterium easily and efficiently transfects them. Some monocots have also been successfully transformed *via Agrobacterium*-mediated transformation like *Oryza sativa* [76]. The members of the Solanaceae family (potato, tomato, tobacco) have been frequently transformed by *Agrobacterium* for stable transgene integration [77]. Large DNA fragments can be inserted into the nuclear genome either at the same loci or different loci by non-homologous recombination with great efficiency [78]. This approach is very extensive, time-consuming and its expression is also not sufficient for commercial vaccine production. But in the case of edible vaccines, it will be cost-effective due to lack of protein purification or minimal downstream processing [79, 80]. Further, the exogenous protein is modified post-translationally during its secretion from the nucleus to several organelles (Endoplasmic reticulum and Golgi apparatus). Hence, it is suitable for accomplishing the continuing manufacture of glycoproteins such as antibodies. The vaccine antigen can be physically bombarded into the nuclear genome using a gene gun. Both monocot and dicot plants can be transformed with DNA-coated gold particles. The major limitation of nuclear expression is the low level of foreign protein due to RNAi, but it can be addressed by engineering strong promoters, enhancer sequences at 5`, use of 3` UTRs, subcellular localization signals, endoplasmic reticulum (ER) retention signal "KDEL" at the C-terminus, and codon-optimized transgenes according to target plant species [81].

Chloroplast transformation has the potential to address the concerns about nuclear transformation. Firstly, the chloroplasts of both monocots and dicots can be stably transformed using microprojectile bombardment. The integration of candidate vaccine antigens into the chloroplast genome is site-specific due to homologous recombination [82, 83]. During chloroplast transformation, transgene DNA-coated gold particles are physically bombarded at high pressure against target plant tissue for the successful incorporation of exogenous DNA into the chloroplast genome. A high level of antigen expression is promising due to multiple copies of plastids in each cell and many chloroplasts in each plastid cell. No gene silencing due to RNAi, no epigenetic effect, and biocontainment of transgene due to maternal inheritance of chloroplast are some of its advantages that hinder high transgene expression during stable nuclear transformation [84 - 86]. Moreover, the chloroplast genome is a potent target for integrating multiple antigenic genes in an operon under a single promoter [87]. Some important pharmaceuticals that have been produced through tobacco chloroplast transformation include human growth hormone, serum albumin, and tetanus-toxin fragment. Besides all these benefits, post-translational modifications are not possible in chloroplast because it lacks ER chaperons and glycosylation does not occur. Thus, human glycoproteins cannot be produced through chloroplast transformation [88].

For rapid and mass production of therapeutic proteins, the transient expression system is the best option because a large amount of foreign protein can be obtained within a few days after infection. It can be mediated either through Agroinfiltration or using viral vectors. But, like nuclear transformation, the transient expression has almost the same limitations, for example, it is time-consuming, inadequate protein expression, inconsistency, *etc*. Nevertheless, it is a highly appropriate platform to produce monoclonal antibodies and vaccine antigens in an emergency against infectious diseases or to analyze the rapid response of vaccines. For example, during the Ebola outbreak in 2014, Mapp Biopharmaceutical Inc., USA developed the ZMapp drug in the tobacco expression system. It was an emergency antibody cocktail produced using three chimeric monoclonal antibodies [89]. The single-stranded RNA viruses are feasible vectors for the transient expression of recombinant proteins. The most used viral vectors during transient expression are Cauliflower mosaic virus (CMV), Tobacco *etc*hes virus (TEV), and Tobacco mosaic virus (TMV). During transient expression, the transgene is introduced in viral capsids and allowed to infect the susceptible plant hosts. Gene of interest and genomes of the virus are not integrated into the plant genome rather they are expressed by infected generation only [90, 91]. Contrarily, the expression of foreign proteins in the leaves is carried out by agroinfiltration in which the suspension having a binary vector is injected into the leaf interstitials [92]. This technique allows the characterization of protein expression and is inherently scalable and flexible. Heterologous proteins can be obtained using plant suspensions. The cell suspensions prepared from calli are used for large-scale protein production using fermenters under sterile conditions. In 2012, the first FDA-approved plant-based pharmaceutical "Elelyso" was produced suspension cultures of carrot whereas the first USDA-approved poultry vaccine was produced using tobacco suspension cultures [80, 93 - 95]. The hairy roots are also used to produce recombinant pharmaceuticals as it is very economical, and the proteins can be continually recovered from the culture medium. Further, there is no need for cell lysis during protein extraction [96].

8. IMPORTANT LIVESTOCK AND POULTRY DISEASES AND THEIR PLANT-BASED VACCINES

The agricultural development of any country is heavily reliant on its livestock and poultry industries. Livestock plays a substantial role in the sustainability of the agriculture sector and is a crucial part of poverty reduction initiatives while poultry farming is a sub-sector of animal husbandry and has become one of agriculture's most essential parts. They give nutrition in the form of meat, milk, milk products, and eggs, as well as providing manure for crop farming. The production of both sub-sectors is heavily affected due to their infectious diseases. The major deadly diseases of poultry are Infectious Bursal Disease (IBD), Avian Reo Virus Infections (ARV), Newcastle Disease (ND), Infectious Bronchitis, Avian Influenza (AI), whereas livestock threatening diseases include Foot-an--mouth disease virus (FMDV), Bovine Tuberculosis, hemorrhagic septicemia (HS), Peste des Petits Ruminants (PPR), *etc.* For sustainable agricultural development, the treatment of such diseases is compulsory. Currently administered vaccines are unable to overcome disease outbreaks. With the introduction of transgenic technology, the development of plant-based vaccines has become a new possibility for overcoming the obstacles of conventional vaccines and they will confer better immunity against infection. The current status of the development of plant-based vaccines against a few economically important poultry and livestock diseases and their application as diagnostic tools to treat the diseases are discussed below.

8.1. Infectious Bursal Disease (IBD)

IBD is a poultry disease that is commonly known as Gumboro. It is a viral disease that affects three to six-week-old chickens. The diseased chickens face severe damage to the major lymphoid organ- the bursa of Fabricius (BF) that leads to immunosuppression due to weakening of B lymphocytes and thus the infected ones become more prone to secondary infections [97]. Its etiological agent is the Infectious Bursal Disease Virus (IBDV) that belongs to the genus *Avibirnavirus* and the family *Birnaviridae* [98]. It has double-stranded bi-segmented RNA and out of its five viral proteins, VP2 is a highly antigenic protein and has been widely characterized during subunit vaccine production. The virus is transmitted among the flocks through the fecal-oral route [99]. The virus is further divided into two serotypes and three pathotypes based on different hosts and virulence, respectively. Chickens are the hosts of serotype I while turkeys are mostly affected by serotype II [99]. There are three pathotypes of serotype I: classical virulent (cv), antigenic variant, and very virulent (vv) IBDVs [100].

Oral vaccination is the most effective way to combat infectious bursal disease because it has various improvements over traditional vaccines. That is why multiple examples of producing VP2 protein in various plant species (both stably and transiently) have been documented in the last 15 years. The first-ever attempt to develop a plant-based vaccine was the expression of VP2 protein in *Arabidopsis thaliana* using the *Agrobacterium tumefaciens*-mediated transformation approach [101]. Immunization with this plant-based edible vaccine to one and three-week-old chickens followed by a booster at four weeks of age has revealed that the oral vaccine can elicit a wide range of immune responses with an 80% protection level that was comparable to the commercially available vaccines. Simultaneously, the specific pathogen-free (SPF) chickens were immunized subcutaneously with a commercial IBD vaccine in the first week and orally with a plant-based vaccine in the third week, resultantly 90% protection was elicited [102]. Then rice seeds expressed with the same VP2 protein were administered orally to SPF chickens that provoked a significant number of neutralizing antibodies. On viral exposure, these chickens were protected from the highly virulent strains of IBD [103]. Thus, it has been concluded that these biomolecules are extremely stable for a long time without any need for cold storage [104]. Furthermore, oral immunization has the potential to trigger both systemic and mucosal immune responses, making it an excellent line of defense against viral infections. Considering all these findings, it has been shown that the plant-made antigenic protein can stimulate better IBDV-specific immunity in chickens, and it will not deteriorate even when exposed to harsh conditions of the avian digestive system. This oral immunization is gained without any adjuvant. However, the low expression level of antigenic protein is a major hurdle in using this approach, thus necessitating the development of a protocol to quantify the antigen so that a particular dose for oral administration is standardized for each animal [105].

During stable transformation, the concern about the low expression of the antigenic protein can be resolved by expressing the antigenic proteins in the chloroplast. Due to the aforementioned advantages of transplastomic technology, tobacco chloroplast(s) were targeted with *VP2* through microprojectile bombardment. The expression and selection cassettes are flanked by *trnI/trnA* chloroplast target sequences that are transcriptionally active regions and will be able to induce high antigenic expression in chloroplasts. During the experiment, tobacco was used as a model plant to investigate the expression and stability of the antigenic protein in successive generations. Now experiments are being performed to transform soybean [*Glycine max* (L.), Merrill] with *VP2* through biolistic transformation. It is an ideal crop plant for integrating the vaccine antigen of poultry disease because it can be processed for oral administration during edible vaccine production [106]. Modern techniques based on transient expression systems have been established to increase heterologous protein expression in plants and to get over the constraints imposed by stable expression. It will further help in improving yields and substantially lower the production time. In *Nicotiana benthamiana*, up to 1% of a total soluble protein antigen (VP2) was obtained using the agroinfiltration method [107]. In chickens, a three-dose oral vaccination with a double amount of plant extract (12g protein) was found enough to generate a neutralizing antibody response. Recently, transgenic tobacco was developed from Cauliflower mosaic virus-mediated transgenesis for the production of VP2 antigen [108]. In this experiment, about 260 µg of VP2 protein per gram of fresh leaf was obtained and for immunogenicity analysis, seven injections of 50 µg/dose were injected intramuscularly along with an adjuvant. It was found that the plant-based antigen is as efficacious as commercial vaccines, in inducing immunity in mice. In another experiment, the crude extract of *Nicotiana tabacum* harboring 30 µg of VP2 protein was fed to breeder hens that were already hyperimmunized with a commercially available vaccine. The boosted immune response was observed without using any adjuvant. Finally, these experiments showed that the transient expression system provides a larger amount of antigenic protein (VP2) as compared to stable expression. It also does not matter that either the crude extract or the purified protein is being used for the immunization. In addition, even when the adjuvant is not included in the formulation, the plant extracts are equally efficacious when delivered intramuscularly. There are several polysaccharides found in the plant extracts that induce the maturation of dendritic cells, thus affecting/enhancing the immunological response [109 - 111].

Subunit vaccines are less effective in provoking life-long immunity rather than inactivated and live attenuated vaccines because they need frequent boosters to maintain the immune response. Chen *et al.* defined the procedure for producing an IBDV VP2 epitope-based vaccine in *Chenopodium quinoa* using a plant viral epitope (Bamboo mosaic virus) presentation method for the first time [112]. The susceptible chickens were then injected intramuscularly with 600 µg of chimeric viral particles coupled with an adjuvant. The protective immune response was observed against the very virulent strain of IBDV that demonstrated the effectiveness of this technique.

8.2. Newcastle Disease (ND)

Newcastle disease (ND) is another viral disease of poultry and causes huge economic losses to the industry. Its causative agent (Newcastle Disease Virus) belongs to the family *Paramyxoviridae* and genus *Avulavirus*. The virus has a negative-sense single-stranded RNA genome. The virus infects the host cells through its surface glycoproteins: Fusion protein (F) and haemagglutinin-neuraminidase (HN). Both these proteins are highly antigenic and play an important role during immunization [113]. The isolates are divided into three virulent groups: lentogenic (less virulent), mesogenic (moderately virulent), and velogenic (highly virulent). Many inactivated and live-attenuated vaccines are commercially available to fight against disease outbreaks. But they failed to confer complete protection due to some limitations. In the past few decades, plants have been utilized to produce antigenic proteins. Transgenic *Nicotiana tabacum* plants were developed through *Agrobacterium tumefaciens for the* introduction of HN genes. The plant extract was orally administered to six-wee--old chickens for immunization that provided significant protection against viral exposure [114]. As the surface proteins are thought to be antigenic so both HN and F proteins were produced through the transgenic potato. Upon oral immunization, it successfully provoked both mucosal and systemic immune responses. Dow AgroSciences LLC developed the first plant cell suspension-derived ND vaccine in 2006 by integrating HN protein in tobacco. The crude extract was injected into chickens to stimulate the protective immune response against the viral challenge. Although it was approved as an ND vaccine by the existing USDA regulatory system due to conferring 90% protection upon viral challenge, Dow AgroSciences decided not to commercialize it [115]. NDV was selected as a proof of concept study, and a single antigenic protein was targeted, a well-defined antibody titer was observed upon hemagglutination inhibition assays, showing that antigenic proteins were biologically active in plant cells [116]. Other stably transformed plant-based NDV vaccines are in challenge trials. Maize was stably transformed by Guerrero-Andrade *et al.* [117] to express fusion protein (F) in its kernels and fed to chickens to evaluate the immunogenicity. Forty-five days post-vaccinated chickens were compared with the control treatments (intranasal and oral delivery of commercial vaccine and feed non-transformed maize as negative control). A significant amount of neutralizing antibody titer was observed in the chicken fed with transformed maize kernels. All the chickens to which transformed maize was orally given survived the NDV challenge, while those treated as negative control were killed. In other studies, F and HN glycoproteins were expressed together in *Solanum tuberosum* [118, 119], while F and HN proteins were also separately expressed in rice [120] and tobacco [114], respectively, without viral challenge studies.

In another study, it was tried to enhance the yield of protein by fusing HN glycoprotein with the KDEL sequence to accumulate the protein in the ER and prevent its degradation in cellular secretory pathways. The results demonstrated that the expression level is greatly affected by genetic constructs, irrespective of stable or transient plant transformation [121]. Thus, it could be utilized as an effective immunogenic construct in vaccine development.

8.3. Foot-and-Mouth Disease (FMD)

FMD is a highly fatal and economically devastating livestock disease that mainly affects cloven-hoofed animals. It comes under the list-A of OIE classification of diseases and is characterized by high fever, and lesions in the mouth, resultantly the infected animals being unable to eat and move [122, 123]. It severely affects the production of the diseased animals even after the recovery that is why a high rate of morbidity has been observed. The mortality rate is greater among calves due to myocardiopathy but low in adult animals [124]. Its causative agent-FMDV is a single-stranded RNA with a positive polarity that belongs to the genus *Aphthovirus* and family *Picornaviridae*. The wide host range, high mutation rate during viral replication, and its multi-routed transmission (by air, water, contaminated feed and consumables, direct exposure with the diseased animal) are the significant attributes of its epidemiology that neutralize the effects of conventional vaccines [125]. Its genome encodes four capsid proteins (VP1-4), of which VP1 is highly immunogenic and comprises multiple antigenic determinants. VP1 protein has a conserved "RGDLXXL" motif within a hypervariable GH loop that mediates infection by interacting with the specific receptors of host cells [126]. FMDV has been classified into seven serotypes (O, A, Asia-1, C, SAT1-3) and more than 60 subtypes. Serotype O, A, and Asia-1 are the most prevalent serotypes worldwide [127]. Effective and multivalent vaccination is the only solution to combat disease outbreaks because commercially available vaccines have failed to completely eradicate the disease, and cannot control the virus shedding.

The utilization of plants as green bioreactors to produce recombinant proteins is a favorable approach for the mass production of anti-FMDV vaccines and immunizing animals against the disease [128]. Thus, many experiments have been conducted in the past decade to develop a plant-based FMDV vaccine. As mentioned earlier, VP1 contains multiple neutralizing epitopes, so this protein must be engineered in experiments to manufacture immunogenic and cost-effective FMDV vaccines. Aiming that, both complete and partial *VP1* has been utilized in plant expression systems to develop an edible vaccine. Table **2** summarizes the transgenic plant-based vaccines against FMD.

Table 2. Plant-based recombinant vaccines against FMD.

Serotype	Target Protein	Expression System	Transformation Strategy	Efficacy	Expression Level	References
O1 Campos (O1C)	Complete VP1	*Nicotiana tabacum*	Transient expression using TMV vector	Significant immunization upon intraperitoneal administration into mice	50-150 mg/gram of freshly harvested leaves	[129]
O1 Campos (O1C)	Complete VP1	*Medicago sativa*	*Agrobacterium*-mediated nuclear transformation	Strong antibody response in intraperitoneally immunized mice	Not quantified but expression seemed too low	[130]
O1Campos (O1C)	Residue135-160 of VP1	*Arabidopsis thaliana*	Stable nuclear transformation via *Agrobacterium tumefaciens*	Strong anti-FMDV antibody response upon intraperitoneal inoculation of mice with plant extracts	High level of GUS expression	[131]
O1Campos (O1C)	P1 polyprotein +3C^{pro}	*Medicago sativa*	*Agrobacterium*-mediated nuclear transformation	Strong FMDV-specific antibody response in mice administered with 15–20 mg emulsified plant tissue	Low expression	[132]
O	VP1 (residue 140-160)	*Nicotiana tabacum*	*Agrobacterium* mediated genetic transformation	Detected significant antibodies in mice	100 μg of TSPs	[133]
O	CompleteVP1	*Nicotiana tabacum*	Chloroplast transformation	Not performed	2-3% of TSPs	[134]
O1Campos (O1C)	VP1+CTB	*Solanum tuberosum*	*Agrobacterium* mediated stable expression	Not performed	0.1-0.13% of TSPs	[135]

(Table 2) cont.....

Serotype	Target Protein	Expression System	Transformation Strategy	Efficacy	Expression Level	References
O	Complete VP1	*Stylosanthes guianensis*	*Agrobacterium* mediated nuclear transformation	Virus-specific immune response by oral administration to mice	0.1-0.5% of TSPs	[136]
O strain China/1/99	P1+2A+3C^{pro}	*Solanum lycopersicum*	*Agrobacterium* mediated nuclear transformation	Anti-FMDV serum antibodies detected in guinea pigs	Not quantified	[137]
O1Campos (O1C)	VP1(residue 135-160)	*Nicotiana tabacum*	Chloroplast transformation	Anti-FMDV serum antibodies observed in mice immunized intraperitoneally	51% of TSPs	[138]
O/ China/99	Complete VP1	*Arabidopsis thaliana*	Transient expression via floral dip method	Not performed	Low expression in every plant tissue but not quantified	[139]
O/ES/2001	P1 polyprotein	*Oryza sativa*	*Agrobacterium*-mediated nuclear transformation	Specific neutralizing antibodies detected after intraperitoneal immunization of mice	0.6-1.3 µg/mg of TSPs	[140]
O	VP1(129-169aa)	*Nicotiana tabacum*	Agroinfiltration	Specific antigen/antibody reaction by dot blot assay	-	[141]
O	VP1(129-169aa)	*Spinacia oleracea*	Agroinfiltration	Specific antigen/antibody reaction by dot blot assay	-	[142]
O	VP1(129-169aa)	*Medicago sativa*	Agroinfiltration	Specific antigen/antibody reaction by dot blot assay	-	[143]

(Table 2) cont.....

Serotype	Target Protein	Expression System	Transformation Strategy	Efficacy	Expression Level	References
O	VP1(129-169aa)	*Nicotiana tabacum*	*Agrobacterium* mediated genetic transformation	Anti FMDV antibodies detected in rabbits upon intraperitoneal immunization	0.65-0.72% of TSPs	[144]

However, in all the experimentations, the expression level of the transgene was too low to use as an edible plant-based vaccine. The requirement for several boosters during immunization and problems in detecting the foreign protein in plant extracts during blot analysis confirms the relatively poor expression level. Though it is a promising alternative to produce valuable antigenic proteins to treat infectious diseases, it is a matter of great concern in cases where the crude extract has to be used for immunization without any downstream processing. Thus, the best approach is that the vaccine candidate should be genetically manipulated for a detectable high yield [145].

Considering all the concerns, we have (MS Khan, personal communication) engineered the empty capsid (P12A) along with 3C^{pro} of FMDV serotype O and Asia-1 in such a way that they will surely result in high expression levels in desired crop plants. The viral intrinsic protease is used in the study for the efficient processing of empty capsid (P1) while the whole empty capsid is used to induce a wide range of immune responses against the disease. Further, the Kozak sequence, ER retention signal (KDEL), and Cholera Toxin B subunit (CTB) as a mucosal adjuvant are engineered to enhance protein expression. The immunogenic profiling of these antigenic proteins is also performed *in silico* to get confirmation of antigenic epitopes present in these capsid proteins [146, 147]. In Asian countries, both FMDV serotype O and Asia-1 are responsible for causing destructive outbreaks, thus the antigenic proteins against both the serotypes are expressed *via* nuclear transformation in *Nicotiana tabacum*. The route of administration and antigen of interest play a decisive role in defining the host plant for efficient vaccine production. For veterinary diseases, oral immunization is the best option as it shields the mucosal lining of the digestive tract, thus preventing viral invasions. Therefore, the candidate vaccine antigens against both the serotypes are also expressed in *Trifolium alexandrinum* (Berseem) to develop a plant-based edible vaccine (Unpublished Data). The experiment is in the pipeline, and it will hopefully help in eliciting a better immune response against the disease outbreak because *in silico* analyses have proved the presence of multiple antigenic determinants in the selected transgene. Moreover, Multiepitope Subunit Vaccine (MESV) constructs against both the serotypes have also been designed [146, 147] that could be expressed in *Trifolium alexandrinum* shortly.

9. A CURRENT MARKET SCENARIO OF PLANT-BASED THERAPEUTICS FOR LIVESTOCK AND POULTRY DISEASES

Globally, the market for veterinary vaccines is in its early stages of development but a considerable gap is found between the production of plant-based human and veterinary vaccines. As per estimation, it will expand at a rate of 5.5 to 8.1% CAGR (Compound Annual Growth Rate) and will achieve 14,000 million US$ by 2026. There are only 47 veterinary biopharmaceuticals approved by the EU (European Union) and USDA for commercialization, while 211 analogous products for humans are in the market. Plant-based therapeutics can be categorized into two sections based on their use: for medical use and veterinary use. Several studies have successfully immunized the model animals against pathogen challenges, for example, FMDV, canine parvovirus, mink enteritis virus, chicken infectious bronchitis virus, rabbit hemorrhagic disease virus, and murine hepatitis virus. In 2006, the first plant-based poultry vaccine against ND was approved for release by USDA regulatory authority but its manufacturer "Dow AgroSciences LLC" refused to commercialize it. A lot of clinical trials are in the pipeline against various veterinary diseases. Later in 2012, a live vector-based vaccine was developed against NDV and Avian Influenza Virus (AIV) and approved by the USA, USDA, and Avimex. It is named "NewH5". Attenuated NDV vaccine was expressed with LaSota strain antigens of avian influenza virus (HA, H5) and successfully immunized the chickens against AIV and NDV [148]. ProdiGene, Inc. first time exhibited that plant-based edible vaccines can confer protective immunity in livestock against virus exposure. The clinical trials of an oral vaccine against the gastroenteritis virus are under trial. Few licensed veterinary vaccines are described below in Table **3**.

Table 3. Commercially available veterinary vaccines.

Pathogen	Brand name	Distributor	Target Animal	Reference(s)
Classical swine fever virus	Porcilis Pesti	Intervet	Pigs	[149]
Classical swine fever virus	Bayovac CSF E2	Bbayer Leverkusen	Pigs	[150]
Equine influenza virus	PROTEQ-FLU (EU), Recombitek (US)	Merial	Horses	[151]
NDV and AIV	Vectormune FP-ND	Intervet	Poultry	[152]
Rabies virus	Purevax Feline Rabies	Merial	Cats	[153]

(Table 3) cont.....

Pathogen	Brand name	Distributor	Target Animal	Reference(s)
Rabies virus	PUREVAX Feline Rabies	Merial	Cats	[154]
Caine parvovirus1	RECOMBITEK Canine Parvo	Merial	Dogs	[155]

CONSENT FOR PUBLICATION

Not applicable.

CONFLICT OF INTEREST

The author declares no conflict of interest, financial or otherwise.

ACKNOWLEDGEMENTS

Declared none.

REFERENCES

[1]   White RJ, Razgour O. Emerging zoonotic diseases originating in mammals: a systematic review of effects of anthropogenic land-use change. Mammal Rev 2020; 50(4): 336-52.
[http://dx.doi.org/10.1111/mam.12201] [PMID: 32836691]

[2]   Kamionka M. Engineering of therapeutic proteins production in Escherichia coli. Curr Pharm Biotechnol 2011; 12(2): 268-74.
[http://dx.doi.org/10.2174/138920111794295693] [PMID: 21050165]

[3]   Ramalingam S, Iyappan G, Priya SH, Kadirvelu K. Rapid production of therapeutic proteins using plant system. Def Life Sci J 2017; 2(2): 95.
[http://dx.doi.org/10.14429/dlsj.2.11372]

[4]   Lalonde M-E, Durocher Y. Therapeutic glycoprotein production in mammalian cells. J Biotechnol 2017; 251: 128-40.
[http://dx.doi.org/10.1016/j.jbiotec.2017.04.028] [PMID: 28465209]

[5]   Moustafa K, Makhzoum A, Trémouillaux-Guiller J. Molecular farming on rescue of pharma industry for next generations. Crit Rev Biotechnol 2016; 36(5): 840-50.
[http://dx.doi.org/10.3109/07388551.2015.1049934] [PMID: 26042351]

[6]   Burnett MJ, Burnett AC. Therapeutic recombinant protein production in plants: Challenges and opportunities. Plants, People. Planet 2020; 2(2): 121-32.

[7]   Shanmugaraj B, Ramalingam S. Plant expression platform for the production of recombinant pharmaceutical proteins. Austin J Biotechnol Bioeng 2014; 1(4)

[8]   Sainsbury F, Lomonossoff GP. Transient expressions of synthetic biology in plants. Curr Opin Plant Biol 2014; 19: 1-7.
[http://dx.doi.org/10.1016/j.pbi.2014.02.003] [PMID: 24631883]

[9]   Sack M, Hofbauer A, Fischer R, Stoger E. The increasing value of plant-made proteins. Curr Opin Biotechnol 2015; 32: 163-70.
[http://dx.doi.org/10.1016/j.copbio.2014.12.008] [PMID: 25578557]

[10]   Herrero M, Grace D, Njuki J, *et al.* The roles of livestock in developing countries. Animal 2013;7:3–18. 11. Bettencourt EMV, Tilman M, Narciso V, Carvalho MLdS, Henriques PDdS. The livestock roles in the wellbeing of rural communities of Timor-Leste. Rev Econ Sociol Rural 2015; 53: 63-80.

[12] Henchion M, Hayes M, Mullen AM, Fenelon M, Tiwari B. Future protein supply and demand: strategies and factors influencing a sustainable equilibrium. Foods 2017; 6(7): 53.
[http://dx.doi.org/10.3390/foods6070053] [PMID: 28726744]

[13] Mottet A, Tempio G. Global poultry production: current state and future outlook and challenges. Worlds Poult Sci J 2017; 73(2): 245-56.
[http://dx.doi.org/10.1017/S0043933917000071]

[14] Chomel B. Zoonoses. Encyclopedia of Microbiology. 2009; p. 820.

[15] Wang C, Wang Z, Wang G, Lau JY, Zhang K, Li W. COVID-19 in early 2021: current status and looking forward. Signal Transduct Target Ther 2021; 6(1): 114.
[http://dx.doi.org/10.1038/s41392-021-00527-1] [PMID: 33686059]

[16] Huang YS, Higgs S, Vanlandingham DL. Arbovirus-mosquito vector-host interactions and the impact on transmission and disease pathogenesis of arboviruses. Front Microbiol 2019; 10: 22.
[http://dx.doi.org/10.3389/fmicb.2019.00022] [PMID: 30728812]

[17] Klous G, Huss A, Heederik DJJ, Coutinho RA. Human-livestock contacts and their relationship to transmission of zoonotic pathogens, a systematic review of literature. One Health 2016; 2: 65-76.
[http://dx.doi.org/10.1016/j.onehlt.2016.03.001] [PMID: 28616478]

[18] DiEuliis D, Johnson KR, Morse SS, Schindel DE. Opinion: Specimen collections should have a much bigger role in infectious disease research and response. Proc Natl Acad Sci USA 2016; 113(1): 4-7.
[http://dx.doi.org/10.1073/pnas.1522680112] [PMID: 26733667]

[19] Lendak D, Preveden T, Kovaeevic N, Tomic S, Ruzic M, Fabri M. Novel infectious diseases in Europe. Med Pregl 2017; 70(11-12): 385-90.
[http://dx.doi.org/10.2298/MPNS1712385L]

[20] Jones BA, Grace D, Kock R, *et al.* Zoonosis emergence linked to agricultural intensification and environmental change. Proc Natl Acad Sci USA 2013; 110(21): 8399-404.
[http://dx.doi.org/10.1073/pnas.1208059110] [PMID: 23671097]

[21] Daszak P, Cunningham AA, Hyatt AD. Emerging infectious diseases of wildlife--threats to biodiversity and human health. Science 2000; 287(5452): 443-9.
[http://dx.doi.org/10.1126/science.287.5452.443] [PMID: 10642539]

[22] Bass D, Stentiford GD, Wang H-C, Koskella B, Tyler CR. The pathobiome in animal and plant diseases. Trends Ecol Evol 2019; 34(11): 996-1008.
[http://dx.doi.org/10.1016/j.tree.2019.07.012] [PMID: 31522755]

[23] Kruse H, kirkemo A-M, Handeland K. Wildlife as source of zoonotic infections. Emerg Infect Dis 2004; 10(12): 2067-72.
[http://dx.doi.org/10.3201/eid1012.040707] [PMID: 15663840]

[24] Rahman MT, Sobur MA, Islam MS, *et al.* Zoonotic diseases: etiology, impact, and control. Microorganisms 2020; 8(9): 1405.
[http://dx.doi.org/10.3390/microorganisms8091405] [PMID: 32932606]

[25] Doel TR. FMD vaccines. Virus Res 2003; 91(1): 81-99.
[http://dx.doi.org/10.1016/S0168-1702(02)00261-7] [PMID: 12527439]

[26] Sautto GA, Kirchenbaum GA, Ross TM. Towards a universal influenza vaccine: different approaches for one goal. Virol J 2018; 15(1): 17.
[http://dx.doi.org/10.1186/s12985-017-0918-y] [PMID: 29370862]

[27] Rybicki EP. Plant-produced vaccines: promise and reality. Drug Disy Today 2009; 14(1-2): 16-24.
[http://dx.doi.org/10.1016/j.drudis.2008.10.002] [PMID: 18983932]

[28] Mason HS, Warzecha H, Mor T, Arntzen CJ. Edible plant vaccines: applications for prophylactic and therapeutic molecular medicine. Trends Mol Med 2002; 8(7): 324-9.
[http://dx.doi.org/10.1016/S1471-4914(02)02360-2] [PMID: 12114111]

[29]　Mett V, Farrance CE, Green BJ, Yusibov V. Plants as biofactories. Biologicals 2008; 36(6): 354-8.
[http://dx.doi.org/10.1016/j.biologicals.2008.09.001] [PMID: 18938088]

[30]　Giddings G. Transgenic plants as protein factories. Curr Opin Biotechnol 2001; 12(5): 450-4.
[http://dx.doi.org/10.1016/S0958-1669(00)00244-5] [PMID: 11604319]

[31]　Spackman E, Pantin-Jackwood MJ, Kapczynski DR, Swayne DE, Suarez DL. H5N2 Highly Pathogenic Avian Influenza Viruses from the US 2014-2015 outbreak have an unusually long pre-clinical period in turkeys. BMC Vet Res 2016; 12(1): 260.
[http://dx.doi.org/10.1186/s12917-016-0890-6] [PMID: 27876034]

[32]　McKenna M. Bird flu cost the US $3.3 billion and worse could be coming. Natl Geogr Mag 2015; 15: •••.https://www.nationalgeographic.com/science/article/bird-flu-2

[33]　Flory G, Peer R. Mesophillic static pile composting of animal carcasses. Biocycle 2017; 58: 65-8.

[34]　Kim S, Kwon H, Park S, Jeon H, Park J-k, Park J. Pilot-scale bio-augmented aerobic composting of excavated foot-and-mouth disease carcasses. Sustainability 2017; 9(3): 445.
[http://dx.doi.org/10.3390/su9030445]

[35]　Kapczynski DR, King DJ. Protection of chickens against overt clinical disease and determination of viral shedding following vaccination with commercially available Newcastle disease virus vaccines upon challenge with highly virulent virus from the California 2002 exotic Newcastle disease outbreak. Vaccine 2005; 23(26): 3424-33.
[http://dx.doi.org/10.1016/j.vaccine.2005.01.140] [PMID: 15837366]

[36]　Meeusen EN, Walker J, Peters A, Pastoret P-P, Jungersen G. Current status of veterinary vaccines. Clin Microbiol Rev 2007; 20(3): 489-510.
[http://dx.doi.org/10.1128/CMR.00005-07] [PMID: 17630337]

[37]　Barrett T, Forsyth MA, Inui K, *et al.* Rediscovery of the second African lineage of rinderpest virus: its epidemiological significance. Vet Rec 1998; 142(24): 669-71.
[http://dx.doi.org/10.1136/vr.142.24.669] [PMID: 9670447]

[38]　Taylor WP. Rinderpest and peste des petits ruminants: virus plagues of large and small ruminants. Elsevier 2005.

[39]　McOrist S, Jasni S, Mackie RA, MacIntyre N, Neef N, Lawson GH. Reproduction of porcine proliferative enteropathy with pure cultures of ileal symbiont intracellularis. Infect Immun 1993; 61(10): 4286-92.
[http://dx.doi.org/10.1128/iai.61.10.4286-4292.1993] [PMID: 8406817]

[40]　Kalunda M. The production and use of rinderpest cell culture vaccine. Dev Biol Stand 1976; 36: 305-6.
[PMID: 1030429]

[41]　Hardham J, Reed M, Wong J, *et al.* Evaluation of a monovalent companion animal periodontal disease vaccine in an experimental mouse periodontitis model. Vaccine 2005; 23(24): 3148-56.
[http://dx.doi.org/10.1016/j.vaccine.2004.12.026] [PMID: 15837214]

[42]　van den Bosch H, Frey J. Interference of outer membrane protein PalA with protective immunity against *Actinobacillus pleuropneumoniae* infections in vaccinated pigs. Vaccine 2003; 21(25-26): 3601-7.
[http://dx.doi.org/10.1016/S0264-410X(03)00410-9] [PMID: 12922088]

[43]　Lehman JR, Weigel RM, Siegel AM, Herr LG, Taft AC, Hall WF. Progress after one year of a pseudorabies eradication program for large swine herds. J Am Vet Med Assoc 1993; 203(1): 118-21.
[PMID: 8407443]

[44]　Maes RK, Sussman MD, Vilnis A, Thacker BJ. Recent developments in latency and recombination of Aujeszky's disease (pseudorabies) virus. Vet Microbiol 1997; 55(1-4): 13-27.
[http://dx.doi.org/10.1016/S0378-1135(96)01305-3] [PMID: 9220593]

[45]   Whetstone CA, Miller JM, Seal BS, Bello LJ, Lawrence WC. Latency and reactivation of a thymidine kinase-negative bovine herpesvirus 1 deletion mutant. Arch Virol 1992; 122(1-2): 207-14.
[http://dx.doi.org/10.1007/BF01321129] [PMID: 1309641]

[46]   Belknap EB, Walters LM, Kelling C, *et al.* Immunogenicity and protective efficacy of a gE, gG and US2 gene-deleted bovine herpesvirus-1 (BHV-1) vaccine. Vaccine 1999; 17(18): 2297-305.
[http://dx.doi.org/10.1016/S0264-410X(98)00466-6] [PMID: 10403598]

[47]   Fenaux M, Opriessnig T, Halbur PG, Elvinger F, Meng XJ. A chimeric porcine circovirus (PCV) with the immunogenic capsid gene of the pathogenic PCV type 2 (PCV2) cloned into the genomic backbone of the nonpathogenic PCV1 induces protective immunity against PCV2 infection in pigs. J Virol 2004; 78(12): 6297-303.
[http://dx.doi.org/10.1128/JVI.78.12.6297-6303.2004] [PMID: 15163723]

[48]   Monath TP, Arroyo J, Miller C, Guirakhoo F. West Nile virus vaccine. Curr Drug Targets Infect Disord 2001; 1(1): 37-50.
[http://dx.doi.org/10.2174/1568005013343254] [PMID: 12455232]

[49]   Monath TP, Liu J, Kanesa-Thasan N, *et al.* A live, attenuated recombinant West Nile virus vaccine. Proc Natl Acad Sci USA 2006; 103(17): 6694-9.
[http://dx.doi.org/10.1073/pnas.0601932103] [PMID: 16617103]

[50]   Kennedy NJ, Spithill TW, Tennent J, Wood PR, Piedrafita D. DNA vaccines in sheep: CTLA-4 mediated targeting and CpG motifs enhance immunogenicity in a DNA prime/protein boost strategy. Vaccine 2006; 24(7): 970-9.
[http://dx.doi.org/10.1016/j.vaccine.2005.08.076] [PMID: 16242220]

[51]   Liang R, van den Hurk JV, Babiuk LA, van Drunen Littel-van den Hurk S. Priming with DNA encoding E2 and boosting with E2 protein formulated with CpG oligodeoxynucleotides induces strong immune responses and protection from Bovine viral diarrhea virus in cattle. J Gen Virol 2006; 87(Pt 10): 2971-82.
[http://dx.doi.org/10.1099/vir.0.81737-0] [PMID: 16963756]

[52]   Scheerlinck JP, Karlis J, Tjelle TE, Presidente PJ, Mathiesen I, Newton SE. In vivo electroporation improves immune responses to DNA vaccination in sheep. Vaccine 2004; 22(13-14): 1820-5.
[http://dx.doi.org/10.1016/j.vaccine.2003.09.053] [PMID: 15068866]

[53]   Gerhardt R. West Nile virus in the United States (1999-2005). J Am Anim Hosp Assoc 2006; 42(3): 170-7.
[http://dx.doi.org/10.5326/0420170] [PMID: 16611928]

[54]   Lorenzen N, LaPatra SE. DNA vaccines for aquacultured fish. Rev Sci Tech 2005; 24(1): 201-13.
[http://dx.doi.org/10.20506/rst.24.1.1565] [PMID: 16110889]

[55]   Powell K. DNA vaccines--back in the saddle again? Nat Biotechnol 2004; 22(7): 799-801.
[http://dx.doi.org/10.1038/nbt0704-799] [PMID: 15229530]

[56]   Seregin A, Nistler R, Borisevich V, *et al.* Immunogenicity of West Nile virus infectious DNA and its noninfectious derivatives. Virology 2006; 356(1-2): 115-25.
[http://dx.doi.org/10.1016/j.virol.2006.07.038] [PMID: 16935318]

[57]   Sala F, Manuela Rigano M, Barbante A, Basso B, Walmsley AM, Castiglione S. Vaccine antigen production in transgenic plants: strategies, gene constructs and perspectives. Vaccine 2003; 21(7-8): 803-8.
[http://dx.doi.org/10.1016/S0264-410X(02)00603-5] [PMID: 12531364]

[58]   Shanmugaraj B, I Bulaon CJ, Phoolcharoen W. I Bulaon CJ, Phoolcharoen W. Plant molecular farming: a viable platform for recombinant biopharmaceutical production. Plants 2020; 9(7): 842.
[http://dx.doi.org/10.3390/plants9070842] [PMID: 32635427]

[59]   Chen R. Bacterial expression systems for recombinant protein production: E. coli and beyond. Biotechnol Adv 2012; 30(5): 1102-7.

[http://dx.doi.org/10.1016/j.biotechadv.2011.09.013] [PMID: 21968145]

[60]   Lueking A, Holz C, Gotthold C, Lehrach H, Cahill D. A system for dual protein expression in *Pichia pastoris* and *Escherichia coli.* Protein Expr Purif 2000; 20(3): 372-8.
[http://dx.doi.org/10.1006/prep.2000.1317] [PMID: 11087676]

[61]   Tripathi NK, Shrivastava A. Recent developments in bioprocessing of recombinant proteins: expression hosts and process development. Front Bioeng Biotechnol 2019; 7: 420.
[http://dx.doi.org/10.3389/fbioe.2019.00420] [PMID: 31921823]

[62]   Baghban R, Farajnia S, Rajabibazl M, *et al.* Yeast expression systems: overview and recent advances. Mol Biotechnol 2019; 61(5): 365-84.
[http://dx.doi.org/10.1007/s12033-019-00164-8] [PMID: 30805909]

[63]   Çelik E, Çalık P. Production of recombinant proteins by yeast cells. Biotechnol Adv 2012; 30(5): 1108-18.
[http://dx.doi.org/10.1016/j.biotechadv.2011.09.011] [PMID: 21964262]

[64]   Hitchman RB, Possee RD, King LA. Baculovirus expression systems for recombinant protein production in insect cells. Recent Pat Biotechnol 2009; 3(1): 46-54.
[http://dx.doi.org/10.2174/187220809787172669] [PMID: 19149722]

[65]   Aricescu AR, Lu W, Jones EY. A time- and cost-efficient system for high-level protein production in mammalian cells. Acta Crystallogr D Biol Crystallogr 2006; 62(Pt 10): 1243-50.
[http://dx.doi.org/10.1107/S0907444906029799] [PMID: 17001101]

[66]   Barnes LM, Bentley CM, Dickson AJ. Stability of protein production from recombinant mammalian cells. Biotechnol Bioeng 2003; 81(6): 631-9.
[http://dx.doi.org/10.1002/bit.10517] [PMID: 12529877]

[67]   Daniell H, Kulis M, Herzog RW. Plant cell-made protein antigens for induction of Oral tolerance. Biotechnol Adv 2019; 37(7): 107413.
[http://dx.doi.org/10.1016/j.biotechadv.2019.06.012] [PMID: 31251968]

[68]   Schillberg S, Twyman RM, Fischer R. Opportunities for recombinant antigen and antibody expression in transgenic plants--technology assessment. Vaccine 2005; 23(15): 1764-9.
[http://dx.doi.org/10.1016/j.vaccine.2004.11.002] [PMID: 15734038]

[69]   Chen Q, Davis KR. The potential of plants as a system for the development and production of human biologics. F1000 Res 2016; 5: 5.
[http://dx.doi.org/10.12688/f1000research.8010.1] [PMID: 27274814]

[70]   Streatfield SJ, Howard JA. Plant-based vaccines. Int J Parasitol 2003; 33(5-6): 479-93.
[http://dx.doi.org/10.1016/S0020-7519(03)00052-3] [PMID: 12782049]

[71]   Kwon K-C, Verma D, Singh ND, Herzog R, Daniell H. Oral delivery of human biopharmaceuticals, autoantigens and vaccine antigens bioencapsulated in plant cells. Adv Drug Deliv Rev 2013; 65(6): 782-99.
[http://dx.doi.org/10.1016/j.addr.2012.10.005] [PMID: 23099275]

[72]   Lakshmi PS, Verma D, Yang X, Lloyd B, Daniell H. Low cost tuberculosis vaccine antigens in capsules: expression in chloroplasts, bio-encapsulation, stability and functional evaluation *in vitro.* PLoS One 2013; 8(1): e54708.
[http://dx.doi.org/10.1371/journal.pone.0054708] [PMID: 23355891]

[73]   Dhama K, Wani MY, Deb R, *et al.* Plant based oral vaccines for human and animal pathogens-a new era of prophylaxis: current and future perspectives. J Exp Biol Agric Sci 2013; 1(1): 1-12.
[http://dx.doi.org/10.18006/2018.6(1).1.31]

[74]   Aswathi PB, Bhanja SK, Yadav AS, *et al.* Plant based edible vaccines against poultry diseases?: a review. Adv Anim Vet Sci 2014; 2: 305-11.
[http://dx.doi.org/10.14737/journal.aavs/2014/2.5.305.311]

[75] Jacob SS, Cherian S, Sumithra TG, Raina OK, Sankar M. Edible vaccines against veterinary parasitic diseases--current status and future prospects. Vaccine 2013; 31(15): 1879-85.
[http://dx.doi.org/10.1016/j.vaccine.2013.02.022] [PMID: 23485715]

[76] Chen Q. Expression and purification of pharmaceutical proteins in plants. Biol Eng Trans 2008; 1(4): 291-321.
[http://dx.doi.org/10.13031/2013.26854]

[77] Floss DM, Sack M, Stadlmann J, *et al.* Biochemical and functional characterization of anti-HIV antibody-ELP fusion proteins from transgenic plants. Plant Biotechnol J 2008; 6(4): 379-91.
[http://dx.doi.org/10.1111/j.1467-7652.2008.00326.x] [PMID: 18312505]

[78] Gelvin SB. Agrobacterium-mediated plant transformation: the biology behind the "gene-jockeying" tool. Microbiol Mol Biol Rev 2003; 67(1): 16-37.
[http://dx.doi.org/10.1128/MMBR.67.1.16-37.2003] [PMID: 12626681]

[79] Twyman RM, Stoger E, Schillberg S, Christou P, Fischer R. Molecular farming in plants: host systems and expression technology. Trends Biotechnol 2003; 21(12): 570-8.
[http://dx.doi.org/10.1016/j.tibtech.2003.10.002] [PMID: 14624867]

[80] Yusibov V, Rabindran S. Recent progress in the development of plant derived vaccines. Expert Rev Vaccines 2008; 7(8): 1173-83.
[http://dx.doi.org/10.1586/14760584.7.8.1173] [PMID: 18844592]

[81] Abiri R, Valdiani A, Maziah M, *et al.* A critical review of the concept of transgenic plants: insights into pharmaceutical biotechnology and molecular farming. Curr Issues Mol Biol 2016; 18: 21-42.
[PMID: 25944541]

[82] Daniell H, Khan MS, Allison L. Milestones in chloroplast genetic engineering: an environmentally friendly era in biotechnology. Trends Plant Sci 2002; 7(2): 84-91.
[http://dx.doi.org/10.1016/S1360-1385(01)02193-8] [PMID: 11832280]

[83] Daniell H, Muthukumar B, Lee SB. Marker free transgenic plants: engineering the chloroplast genome without the use of antibiotic selection. Curr Genet 2001; 39(2): 109-16.
[http://dx.doi.org/10.1007/s002940100185] [PMID: 11405095]

[84] Bock R. Plastid biotechnology: prospects for herbicide and insect resistance, metabolic engineering and molecular farming. Curr Opin Biotechnol 2007; 18(2): 100-6.
[http://dx.doi.org/10.1016/j.copbio.2006.12.001] [PMID: 17169550]

[85] Maliga P. Engineering the plastid genome of higher plants. Curr Opin Plant Biol 2002; 5(2): 164-72.
[http://dx.doi.org/10.1016/S1369-5266(02)00248-0] [PMID: 11856614]

[86] Zhang B, Shanmugaraj B, Daniell H. Expression and functional evaluation of biopharmaceuticals made in plant chloroplasts. Curr Opin Chem Biol 2017; 38: 17-23.
[http://dx.doi.org/10.1016/j.cbpa.2017.02.007] [PMID: 28229907]

[87] Bock R. Genetic engineering of the chloroplast: novel tools and new applications. Curr Opin Biotechnol 2014; 26: 7-13.
[http://dx.doi.org/10.1016/j.copbio.2013.06.004] [PMID: 24679252]

[88] Ma JK, Drake PM, Christou P. The production of recombinant pharmaceutical proteins in plants. Nat Rev Genet 2003; 4(10): 794-805.
[http://dx.doi.org/10.1038/nrg1177] [PMID: 14526375]

[89] Qiu X, Wong G, Audet J, *et al.* Reversion of advanced Ebola virus disease in nonhuman primates with ZMapp. Nature 2014; 514(7520): 47-53.
[http://dx.doi.org/10.1038/nature13777] [PMID: 25171469]

[90] Walmsley AM, Arntzen CJ. Plants for delivery of edible vaccines. Curr Opin Biotechnol 2000; 11(2): 126-9.
[http://dx.doi.org/10.1016/S0958-1669(00)00070-7] [PMID: 10753769]

[91]   Yusibov V, Modelska A, Steplewski K, *et al.* Antigens produced in plants by infection with chimeric plant viruses immunize against rabies virus and HIV-1. Proc Natl Acad Sci USA 1997; 94(11): 5784-8.
[http://dx.doi.org/10.1073/pnas.94.11.5784] [PMID: 9159151]

[92]   Kapila J, De Rycke R, Van Montagu M, Angenon G. An Agrobacterium-mediated transient gene expression system for intact leaves. Plant Sci 1997; 122(1): 101-8.
[http://dx.doi.org/10.1016/S0168-9452(96)04541-4]

[93]   Santos RB, Abranches R, Fischer R, Sack M, Holland T. Putting the spotlight back on plant suspension cultures. Front Plant Sci 2016; 7: 297.
[http://dx.doi.org/10.3389/fpls.2016.00297] [PMID: 27014320]

[94]   Tekoah Y, Shulman A, Kizhner T, *et al.* Large-scale production of pharmaceutical proteins in plant cell culture-the Protalix experience. Plant Biotechnol J 2015; 13(8): 1199-208.
[http://dx.doi.org/10.1111/pbi.12428] [PMID: 26102075]

[95]   Xu J, Zhang N. On the way to commercializing plant cell culture platform for biopharmaceuticals: present status and prospect. Pharm Bioprocess 2014; 2(6): 499-518.
[http://dx.doi.org/10.4155/pbp.14.32] [PMID: 25621170]

[96]   Gurusamy PD, Schäfer H, Ramamoorthy S, Wink M. Biologically active recombinant human erythropoietin expressed in hairy root cultures and regenerated plantlets of *Nicotiana tabacum* L. PLoS One 2017; 12(8): e0182367.
[http://dx.doi.org/10.1371/journal.pone.0182367] [PMID: 28800637]

[97]   Müller H, Mundt E, Eterradossi N, Islam MR. Current status of vaccines against infectious bursal disease. Avian Pathol 2012; 41(2): 133-9.
[http://dx.doi.org/10.1080/03079457.2012.661403] [PMID: 22515532]

[98]   Murphy FA, Gibbs EPJ, Horzinek MC, Studdert MJ. Veterinary virology. Elsevier 1999.

[99]   Alkie TN, Rautenschlein S. Infectious bursal disease virus in poultry: current status and future prospects. Vet Med (Auckl) 2016; 7: 9-18.
[PMID: 30050833]

[100]   Ture O, Saif YM, Jackwood DJ. Restriction fragment length polymorphism analysis of highly virulent strains of infectious bursal disease viruses from Holland, Turkey, and Taiwan. Avian Dis 1998; 42(3): 470-9.
[http://dx.doi.org/10.2307/1592673] [PMID: 9777147]

[101]   Wu H, Singh NK, Locy RD, Scissum-Gunn K, Giambrone JJ. Expression of immunogenic VP2 protein of infectious bursal disease virus in Arabidopsis thaliana. Biotechnol Lett 2004; 26(10): 787-92.
[http://dx.doi.org/10.1023/B:BILE.0000025878.30350.d5] [PMID: 15269548]

[102]   Wu H, Singh NK, Locy RD, Scissum-Gunn K, Giambrone JJ. Immunization of chickens with VP2 protein of infectious bursal disease virus expressed in *Arabidopsis thaliana.* Avian Dis 2004; 48(3): 663-8.
[http://dx.doi.org/10.1637/7074] [PMID: 15529992]

[103]   Wu J, Yu L, Li L, Hu J, Zhou J, Zhou X. Oral immunization with transgenic rice seeds expressing VP2 protein of infectious bursal disease virus induces protective immune responses in chickens. Plant Biotechnol J 2007; 5(5): 570-8.
[http://dx.doi.org/10.1111/j.1467-7652.2007.00270.x] [PMID: 17561926]

[104]   Fischer R, Stoger E, Schillberg S, Christou P, Twyman RM. Plant-based production of biopharmaceuticals. Curr Opin Plant Biol 2004; 7(2): 152-8.
[http://dx.doi.org/10.1016/j.pbi.2004.01.007] [PMID: 15003215]

[105]   Rage E, Marusic C, Lico C, Baschieri S, Donini M. Current state-of-the-art in the use of plants for the production of recombinant vaccines against infectious bursal disease virus. Appl Microbiol Biotechnol

2020; 104(6): 2287-96.
[http://dx.doi.org/10.1007/s00253-020-10397-2] [PMID: 31980920]

[106]   Fayyaz M. Engineering chloroplasts to accumulate antigenic protein(s) to immunize poultry birds against infectious bursal disease. Faisalabad: University of Agriculture 2018.

[107]   Gómez E, Lucero MS, Chimeno Zoth S, Carballeda JM, Gravisaco MJ, Berinstein A. Transient expression of VP2 in *Nicotiana benthamiana* and its use as a plant-based vaccine against infectious bursal disease virus. Vaccine 2013; 31(23): 2623-7.
[http://dx.doi.org/10.1016/j.vaccine.2013.03.064] [PMID: 23583894]

[108]   Taghavian O, Schillberg HDS. Expression and characterization of infectious bursal disease virus protein for poultry vaccine development and application in nanotechnology: Hochschulbibliothek der Rheinisch-Westfälischen Technischen Hochschule Aachen; 2013.

[109]   Di Bonito P, Grasso F, Mangino G, *et al.* Immunomodulatory activity of a plant extract containing human papillomavirus 16-E7 protein in human monocyte-derived dendritic cells. Int J Immunopathol Pharmacol 2009; 22(4): 967-78.
[http://dx.doi.org/10.1177/039463200902200412] [PMID: 20074460]

[110]   Feng H, McDonough SP, Fan J, *et al.* Phosphorylated radix cyathulae officinalis polysaccharides act as adjuvant via promoting dendritic cell maturation. Molecules 2017; 22(1): 106.
[http://dx.doi.org/10.3390/molecules22010106] [PMID: 28075416]

[111]   Yang Y, Wei K, Yang S, *et al.* Co-adjuvant effects of plant polysaccharide and propolis on chickens inoculated with Bordetella avium inactivated vaccine. Avian Pathol 2015; 44(4): 248-53.
[http://dx.doi.org/10.1080/03079457.2015.1040372] [PMID: 25989924]

[112]   Chen T-H, Chen T-H, Hu C-C, *et al.* Induction of protective immunity in chickens immunized with plant-made chimeric Bamboo mosaic virus particles expressing very virulent Infectious bursal disease virus antigen. Virus Res 2012; 166(1-2): 109-15.
[http://dx.doi.org/10.1016/j.virusres.2012.02.021] [PMID: 22406128]

[113]   Stone-Hulslander J, Morrison TG. Detection of an interaction between the HN and F proteins in Newcastle disease virus-infected cells. J Virol 1997; 71(9): 6287-95.
[http://dx.doi.org/10.1128/jvi.71.9.6287-6295.1997] [PMID: 9261345]

[114]   Hahn B-S, Jeon I-S, Jung Y-J, *et al.* Expression of hemagglutinin-neuraminidase protein of Newcastle disease virus in transgenic tobacco. Plant Biotechnol Rep 2007; 1(2): 85-92.
[http://dx.doi.org/10.1007/s11816-007-0012-9]

[115]   Fox JL. Turning plants into protein factories. Nat Biotechnol 2006; 24(10): 1191-3.
[http://dx.doi.org/10.1038/nbt1006-1191] [PMID: 17033647]

[116]   Mihaliak CA, Webb S, Miller T, Fanton M, Kirk D, Cardineau G, Eds. Development of plant cell produced vaccines for animal health applications. Proceedings of the 108th annual meeting of the United States Animal Health Association.

[117]   Guerrero-Andrade O, Loza-Rubio E, Olivera-Flores T, Fehérvári-Bone T, Gómez-Lim MA. Expression of the Newcastle disease virus fusion protein in transgenic maize and immunological studies. Transgenic Res 2006; 15(4): 455-63.
[http://dx.doi.org/10.1007/s11248-006-0017-0] [PMID: 16906446]

[118]   Berinstein A, Vazquez-Rovere C, Asurmendi S, *et al.* Mucosal and systemic immunization elicited by Newcastle disease virus (NDV) transgenic plants as antigens. Vaccine 2005; 23(48-49): 5583-9.
[http://dx.doi.org/10.1016/j.vaccine.2005.06.033] [PMID: 16099555]

[119]   Gómez E, Chimeno Zoth S, Carrillo E, Estela Roux M, Berinstein A. Mucosal immunity induced by orally administered transgenic plants. Immunobiology 2008; 213(8): 671-5.
[http://dx.doi.org/10.1016/j.imbio.2008.02.002] [PMID: 18950595]

[120]   Yang ZQ, Liu QQ, Pan ZM, Yu HX, Jiao XA. Expression of the fusion glycoprotein of Newcastle disease virus in transgenic rice and its immunogenicity in mice. Vaccine 2007; 25(4): 591-8.

[http://dx.doi.org/10.1016/j.vaccine.2006.08.016] [PMID: 17049688]

[121] Gómez E, Zoth SC, Asurmendi S, Vázquez Rovere C, Berinstein A. Expression of hemagglutinin-neuraminidase glycoprotein of newcastle disease Virus in agroinfiltrated Nicotiana benthamiana plants. J Biotechnol 2009; 144(4): 337-40.
[http://dx.doi.org/10.1016/j.jbiotec.2009.09.015] [PMID: 19799942]

[122] Donaldson A. Clinical signs of foot-and-mouth disease Foot and Mouth Disease. CRC Press 2019; pp. 93-102.
[http://dx.doi.org/10.1201/9780429125614-5]

[123] Jamal SM, Belsham GJ. Foot-and-mouth disease: past, present and future. Vet Res 2013; 44(1): 116.
[http://dx.doi.org/10.1186/1297-9716-44-116] [PMID: 24308718]

[124] Woodbury EL. A review of the possible mechanisms for the persistence of foot-and-mouth disease virus. Epidemiol Infect 1995; 114(1): 1-13.
[http://dx.doi.org/10.1017/S0950268800051864] [PMID: 7867727]

[125] Biswal JK, Sanyal A, Rodriguez LL, Subramaniam S, Arzt J, Sharma GK, *et al.* Foot-and-mouth disease: Global status and Indian perspective. Indian J Anim Sci 2012; 82(2): 109-31.

[126] Mason PW, Grubman MJ, Baxt B. Molecular basis of pathogenesis of FMDV. Virus Res 2003; 91(1): 9-32.
[http://dx.doi.org/10.1016/S0168-1702(02)00257-5] [PMID: 12527435]

[127] Knowles NJ, Samuel AR. Molecular epidemiology of foot-and-mouth disease virus. Virus Res 2003; 91(1): 65-80.
[http://dx.doi.org/10.1016/S0168-1702(02)00260-5] [PMID: 12527438]

[128] Ghaffar Shahriari A, Bagheri A, Bassami MR, Malekzadeh-Shafaroudi S, Afsharifar A, Niazi A. Expression of Hemagglutinin-Neuraminidase and fusion epitopes of Newcastle Disease Virus in transgenic tobacco. Electron J Biotechnol 2016; 19(4): 38-43.
[http://dx.doi.org/10.1016/j.ejbt.2016.05.003]

[129] Wigdorovitz A, Pérez Filgueira DM, Robertson N, *et al.* Protection of mice against challenge with foot and mouth disease virus (FMDV) by immunization with foliar extracts from plants infected with recombinant tobacco mosaic virus expressing the FMDV structural protein VP1. Virology 1999; 264(1): 85-91.
[http://dx.doi.org/10.1006/viro.1999.9923] [PMID: 10544132]

[130] Wigdorovitz A, Carrillo C, Dus Santos MJ, *et al.* Induction of a protective antibody response to foot and mouth disease virus in mice following oral or parenteral immunization with alfalfa transgenic plants expressing the viral structural protein VP1. Virology 1999; 255(2): 347-53.
[http://dx.doi.org/10.1006/viro.1998.9590] [PMID: 10069960]

[131] Dus Santos MJ, Wigdorovitz A, Trono K, *et al.* A novel methodology to develop a foot and mouth disease virus (FMDV) peptide-based vaccine in transgenic plants. Vaccine 2002; 20(7-8): 1141-7.
[http://dx.doi.org/10.1016/S0264-410X(01)00434-0] [PMID: 11803075]

[132] Dus Santos MJ, Carrillo C, Ardila F, *et al.* Development of transgenic alfalfa plants containing the foot and mouth disease virus structural polyprotein gene P1 and its utilization as an experimental immunogen. Vaccine 2005; 23(15): 1838-43.
[http://dx.doi.org/10.1016/j.vaccine.2004.11.014] [PMID: 15734052]

[133] Huang Y, Liang W, Wang Y, *et al.* Immunogenicity of the epitope of the foot-and-mouth disease virus fused with a hepatitis B core protein as expressed in transgenic tobacco. Viral Immunol 2005; 18(4): 668-77.
[http://dx.doi.org/10.1089/vim.2005.18.668] [PMID: 16359233]

[134] Li Y, Sun M, Liu J, Yang Z, Zhang Z, Shen G. High expression of foot-and-mouth disease virus structural protein VP1 in tobacco chloroplasts. Plant Cell Rep 2006; 25(4): 329-33.
[http://dx.doi.org/10.1007/s00299-005-0074-5] [PMID: 16320056]

[135] He D-M, Qian K-X, Shen G-F, *et al.* Stable expression of foot-and-mouth disease virus protein VP1 fused with cholera toxin B subunit in the potato (*Solanum tuberosum*). Colloids Surf B Biointerfaces 2007; 55(2): 159-63.
[http://dx.doi.org/10.1016/j.colsurfb.2006.11.043] [PMID: 17208421]

[136] Wang DM, Zhu JB, Peng M, Zhou P. Induction of a protective antibody response to FMDV in mice following oral immunization with transgenic Stylosanthes spp. as a feedstuff additive. Transgenic Res 2008; 17(6): 1163-70.
[http://dx.doi.org/10.1007/s11248-008-9188-1] [PMID: 18651235]

[137] Pan L, Zhang Y, Wang Y, *et al.* Foliar extracts from transgenic tomato plants expressing the structural polyprotein, P1-2A, and protease, 3C, from foot-and-mouth disease virus elicit a protective response in guinea pigs. Vet Immunol Immunopathol 2008; 121(1-2): 83-90.
[http://dx.doi.org/10.1016/j.vetimm.2007.08.010] [PMID: 18006078]

[138] Lentz EM, Segretin ME, Morgenfeld MM, *et al.* High expression level of a foot and mouth disease virus epitope in tobacco transplastomic plants. Planta 2010; 231(2): 387-95.
[http://dx.doi.org/10.1007/s00425-009-1058-4] [PMID: 20041332]

[139] Pan L, Zhang Y, Wang Y, Lv J, Zhou P, Zhang Z, *et al.* Expression and detection of the FMDV VP1 transgene and expressed structural protein in Arabidopsis thaliana. Turk J Vet Anim Sci 2011; 35(4): 235-42.

[140] Wang Y, Shen Q, Jiang Y, *et al.* Immunogenicity of foot-and-mouth disease virus structural polyprotein P1 expressed in transgenic rice. J Virol Methods 2012; 181(1): 12-7.
[http://dx.doi.org/10.1016/j.jviromet.2012.01.004] [PMID: 22274594]

[141] Habibi PM, Malekzadeh SS, Marashi H, Moshtaghi N, Nassiri M, Zibaee S. Transient expression of foot and mouth disease virus (FMDV) coat protein in Tobacco (*Nicotiana tabacum*) via Agroinfiltration. Iran J Biotechnol 2014; 12(3): 28-34.
[http://dx.doi.org/10.15171/ijb.1015]

[142] Habibi M, Malekzadeh-Shafaroudi S, Marashi H, Moshtaghi N, Nasiri M, Zibaee S. The transient expression of the coat protein of Foot and Mouth Disease Virus (FMDV) in spinach (*Spinacia oleracea*) using Agroinfiltration. J Plant Mol Breed 2014; 2(2): 18-27.

[143] Habibi-Pirkoohi M, Malekzadeh-Shafaroudi S, Marashi H, Zibaee S, Mohkami A. NejatizaXeh S. Transient expression of coat protein of Foot and Mouth Disease Virus (FMDV) in Alfalfa (Medicago sativa) by Agroinfiltration. J Cell Mol Res 2016; 8(2): 83-9.

[144] Habibi M, Malekzadeh S, Marashi H, Mohkami A. Expression of an epitope-based recombinant vaccine against Foot and Mouth Disease (FMDV) in tobacco plant (*Nicotiana tabacum*). J Plant Mol Breed 2019; 7(1): 1-9.

[145] Awale M, Mody S, Dudhatra G, *et al.* Transgenic plant vaccine: a breakthrough in immunopharmacotherapeutics. J Vaccines Vaccin 2012; 3(147): 1-7.

[146] Qadeer S, Khan MS, Joyia FA, Zia MA. Immunogenic profiling and designing of a novel vaccine from capsid proteins of FMDV serotype Asia-1 through reverse vaccinology. Infec, Genet Evo 2021; p. 104925.

[147] Riaz R, Khan MS, Joyia FA, Zia MA. Designing of Multiepitope-based Subunit Vaccine (MESV) against prevalent serotype of Foot and Mouth Disease Virus (FMDV) using Immunoinformatics Approach Pak J Vet. Inpress 2021.

[148] Ryan MP, Walsh G. Veterinary-based biopharmaceuticals. Trends Biotechnol 2012; 30(12): 615-20.
[http://dx.doi.org/10.1016/j.tibtech.2012.08.005] [PMID: 22995556]

[149] van Aarle P. Suitability of an E2 subunit vaccine of classical swine fever in combination with the E(rns)-marker-test for eradication through vaccination. Dev Biol (Basel) 2003; 114: 193-200.
[PMID: 14677689]

[150] Moormann RJ, Bouma A, Kramps JA, Terpstra C, De Smit HJ. Development of a classical swine fever subunit marker vaccine and companion diagnostic test. Vet Microbiol 2000; 73(2-3): 209-19.
[http://dx.doi.org/10.1016/S0378-1135(00)00146-2] [PMID: 10785329]

[151] Minke JM, Audonnet J-C, Fischer L. Equine viral vaccines: the past, present and future. Vet Res 2004; 35(4): 425-43.
[http://dx.doi.org/10.1051/vetres:2004019] [PMID: 15236675]

[152] Park M-S, Steel J, García-Sastre A, Swayne D, Palese P. Engineered viral vaccine constructs with dual specificity: avian influenza and Newcastle disease. Proc Natl Acad Sci USA 2006; 103(21): 8203-8.
[http://dx.doi.org/10.1073/pnas.0602566103] [PMID: 16717196]

[153] Rupprecht CE, Hanlon CA, Slate D. Oral vaccination of wildlife against rabies: opportunities and challenges in prevention and control. Dev Biol (Basel) 2004; 119: 173-84.
[PMID: 15742629]

[154] Slate D, Rupprecht CE, Rooney JA, Donovan D, Lein DH, Chipman RB. Status of oral rabies vaccination in wild carnivores in the United States. Virus Res 2005; 111(1): 68-76.
[http://dx.doi.org/10.1016/j.virusres.2005.03.012] [PMID: 15896404]

[155] Pardo M, Mackowlak M. Efficacy of a new canine origin, modified live virus vaccine against canine coronavirus. 1999.

SUBJECT INDEX

Muhammad Sarwar Khan (Ed.)
All rights reserved-© 2021 Bentham Science Publishers

www.ingramcontent.com/pod-product-compliance
Lightning Source LLC
Chambersburg PA
CBHW050759220326
41598CB00006B/69